Lecture Notes in Civil Engineering 510

Lecture Notes in Civil Engineering (LNCE) publishes the latest developments in Civil Engineering—quickly, informally and in top quality. Though original research reported in proceedings and post-proceedings represents the core of LNCE, edited volumes of exceptionally high quality and interest may also be considered for publication. Volumes published in LNCE embrace all aspects and subfields of, as well as new challenges in, Civil Engineering. Topics in the series include:

- Construction and Structural Mechanics
- Building Materials
- Concrete, Steel and Timber Structures
- Geotechnical Engineering
- Earthquake Engineering
- Coastal Engineering
- Ocean and Offshore Engineering; Ships and Floating Structures
- Hydraulics, Hydrology and Water Resources Engineering
- Environmental Engineering and Sustainability
- Structural Health and Monitoring
- Surveying and Geographical Information Systems
- Indoor Environments
- Transportation and Traffic
- Risk Analysis
- Safety and Security

To submit a proposal or request further information, please contact the appropriate Springer Editor:

- Pierpaolo Riva at pierpaolo.riva@springer.com (Europe and Americas);
- Swati Meherishi at swati.meherishi@springer.com (Asia—except China, Australia, and New Zealand);
- Wayne Hu at wayne.hu@springer.com (China).

All books in the series now indexed by Scopus and EI Compendex database!

Ioan Tuns · Radu Muntean · Dorin Radu ·
Christiana Cazacu · Teofil Gălățanu
Editors

Proceedings of CIBv 2023

Civil Engineering and Buildings Services

Springer

Editors
Ioan Tuns
Faculty of Civil Engineering
Transilvania University of Braşov
Braşov, Romania

Radu Muntean
Faculty of Civil Engineering
Transilvania University of Braşov
Braşov, Romania

Dorin Radu
Faculty of Civil Engineering
Transilvania University of Braşov
Braşov, Romania

Christiana Cazacu
Faculty of Civil Engineering
Transilvania University of Braşov
Braşov, Romania

Teofil Gălăţanu
Faculty of Civil Engineering
Transilvania University of Braşov
Braşov, Romania

ISSN 2366-2557 ISSN 2366-2565 (electronic)
Lecture Notes in Civil Engineering
ISBN 978-3-031-60764-6 ISBN 978-3-031-60765-3 (eBook)
https://doi.org/10.1007/978-3-031-60765-3

This Springer imprint is published by the registered company Springer Nature Switzerland AG
The registered company address is: Gewerbestrasse 11, 6330 Cham, Switzerland

Contents

Frost Resistance of Recycled Concrete: A Review

Yichao Zhang[1(✉)], Liguang Zhang[1], Chongchi Hou[1], Na Luo[1], and Yanhui Wang[2]

[1] Shenyang Jianzhu University, Shenyang, China
lunwenww@sina.com

[2] China Construction Third Engineering Bureau Group (Zhejiang) Co., Ltd., Hangzhou, China

Abstract. This paper presents a comprehensive analysis of the current state of research on the frost resistance of recycled aggregate concrete at home and abroad and finds that micro-cracks in recycled aggregates and the fragile interfacial transition zone of concrete have a significant impact on the frost resistance of recycled aggregate concrete. As the replacement rate of recycled coarse aggregate rises, recycled aggregate concrete becomes less resistant to frost. It is consequently advised that the replacement rate of recycled coarse aggregate not surpass 50%. At this rate, the frost resistance of recycled aggregate concrete is most similar to that of regular concrete. In addition, the addition of silica fume, ferronickel slag, glazed hollow bead, polypropylene fibres and air-entraining water-reducing agents can effectively improve the frost resistance of recycled aggregate concrete. In the future, more systematic and in-depth research on the freeze-thaw damage mechanism of recycled aggregate concrete is needed, and a more complete theoretical model should be established.

Keywords: recycled concrete · frost resistance · replacement rate · damage mechanism

1 Introduction

With the advancement of urban construction, both the demand for construction materials and the generation of construction waste have increased dramatically. Among them, concrete is one of the most widely used building materials, and waste concrete also accounts for a large proportion of construction waste. In 2022, China produced up to 2.35 billion tons of waste concrete. Nowadays, piling and landfilling are the primary methods used to dispose of leftover concrete, which seriously pollutes the environment. If waste concrete is crushed, cleaned, graded, and reused to make recycled concrete, at least 800 million tons of natural sand and gravel can be saved annually, reducing natural stone mining by one-third and releasing 2.33×10^8 square meters of landfill. Obviously, recycled concrete can effectively solve the problem of construction solid waste resource utilization and bring significant social benefits. However, as compared to regular concrete, recycled concrete has worse mechanical and durability qualities. Therefore, it is necessary to further study its mechanical properties and damage mechanisms for better application in practical engineering.

I. Tuns et al. (Eds.): CIBv 2023, LNCE 510, pp. 1–11, 2024.
https://doi.org/10.1007/978-3-031-60765-3_1

Frost resistance is one of the important indices to evaluate the durability and performance of concrete. In cold regions, concrete buildings are prone to performance deterioration and damage under the action of freeze-thaw cycles, which is a major problem in their use. Therefore, the study of the frost resistance of recycled concrete is of great significance for its promotion and application in cold regions. This paper combines the current research status of frost resistance of recycled concrete at home and abroad to discuss the damage mechanism and macro-micro-morphological changes of recycled concrete under the freeze-thaw cycle and analyze the deterioration process of its performance. At the same time, considering the influence of recycled aggregate content and freeze-thaw cycles on performance changes, relevant research suggestions are proposed to provide reference for the study of the frost resistance of recycled concrete.

2 Domestic Research Status

Niu et al. [1] conducted a study on the effect of recycled coarse aggregate and recycled fine aggregate on the compressive strength of concrete cubes at different replacement rates after 100 freeze-thaw cycles by performing freeze-thaw cycle tests of nine types of recycled aggregate concrete by the fast-freeze method and cube compressive strength tests. The findings demonstrate that, in comparison to regular concrete, recycled aggregate concrete has less freeze-thaw resistance; furthermore, this loss grows as the rate at which recycled aggregate is replaced rises. The rate of loss of compressive bearing capacity is comparable to regular concrete when the replacement rate of recycled coarse aggregate is 50% and the replacement rate of recycled fine aggregate is 0. Bai et al. [2] investigated the effect of the number of freeze-thaw cycles on the full stress-strain curve of recycled concrete by conducting uniaxial stress tests on cubic specimens of recycled aggregate concrete subjected to different numbers of freeze-thaw cycles (0, 25, 50, 75, 100, 125, 150). According to test data, recycled concrete's peak stress and elastic modulus gradually decline as the number of freeze-thaw cycles increases, peak strain gradually rises, and the stress-strain curve tends to be flat. Zou et al. [3] conducted a study on the effects of recycled coarse aggregate admixture and the number of freeze-thaw cycles on the mechanical properties of recycled aggregate concrete, such as mass loss rate, relative dynamic elastic modulus loss rate, peak stress, peak strain, elastic modulus, and transverse deformation coefficient, through freeze-thaw cycle tests and full curve tests of stress-strain after freeze-thaw cycles. The test results show that the mass loss rate and relative dynamic elastic modulus loss rate of recycled aggregate concrete increase and the peak stress and elastic modulus decrease with the increase of recycled coarse aggregate admixture and freeze-thaw times. The peak strain decreases with the increase of freeze-thaw cycles when the recycled coarse aggregate content is low and increases with the increase in freeze-thaw cycles when the recycled coarse aggregate content is high and the freeze-thaw cycles are more than 100. Wang et al. [4] conducted a study on the variation pattern of compressive strength, mass, dynamic elastic modulus, and ultrasonic loss rate of concrete cubes by freeze-thaw cycle tests on recycled aggregate concrete and ordinary concrete of different strengths. The findings indicate that the frost resistance of recycled concrete is similar to that of regular concrete of the same strength grade before 50 freeze-thaw cycles, and that it is slightly less than that of regular concrete after 50 freeze-thaw cycles. With the increase in the number of freeze-thaw cycles, the deficiency of the frost resistance of recycled concrete tends to be significant. The strength grade has a greater influence on the freezing resistance of the recycled concrete specimens, and the lower the strength grade, the more serious the freeze-thaw damage of the recycled concrete. Dai et al. [5] conducted a study on the mechanical properties of recycled aggregate concrete after freeze-thaw cycles by means of splitting tensile tests and compressive tests on concrete with different recycled coarse aggregate replacement rates after freeze-thaw cycles. The findings indicated that as the number of freeze-thaw cycles increased, recycled aggregate concrete's mass loss rate and elastic modulus loss rate progressively increased. Following 50 cycles of freeze-thaw, the recycled aggregate concrete's splitting tensile strength and compressive strength were found to be lower than those without freeze-thaw. Additionally, the trend of decreasing and then increasing with an increase in the recycled coarse aggregate replacement rate was observed. The lowest levels of both splitting tensile and compressive strengths were observed when the recycled coarse aggregate replacement rate was 50%.

Cao et al. [6] studied the apparent characteristics, mass loss rate, relative dynamic elastic modulus change, and compressive strength loss rate of cubic and prismatic specimens with different recycled coarse and recycled fine aggregate substitution rates after different number of freeze-thaw cycles, and investigated the changes of "mass loss rate of specimens - number of freeze-thaw cycles" and "relative dynamic elastic modulus change of specimens - number of freeze-thaw cycles". The results of the study showed that the specimens with 50% replacement of recycled coarse aggregate and ordinary sand as fine aggregate had the closest apparent characteristics, mass loss rate, and relative dynamic elastic modulus change to ordinary concrete after 100 freeze-thaw cycles. Chen et al. [7] conducted a freeze-thaw cycle test on concrete specimens with 50% replacement of recycled coarse aggregate mixed with polypropylene fiber and air -entraining water-reducing agent to investigate the effect of polypropylene fiber and air-entraining water-reducing agent on the frost durability of recycled aggregate concrete at different admixtures and the change law of internal damage variables of recycled concrete after freeze-thaw. The results show that the addition of polypropylene fiber and air-entraining water-reducing agent can effectively inhibit the freeze-thaw damage deterioration of recycled concrete, and the effect of air-entraining water-reducing agent is more obvious. The addition of these two substances can make the design strength grade of C30 recycled aggregate concrete achieve a freeze-resistant grade of F200 or higher. Wang et al. [8] established a freeze-thaw damage model with compressive strength loss rate, mass loss rate, and dynamic elastic modulus loss rate as damage variables by studying the influence law of mechanical properties and frost resistance of ordinary concrete and recycled aggregate concrete with 100% coarse aggregate replacement under different numbers of freeze-thaw cycles. According to the results, there was an inverse relationship between the number of freeze-thaw cycles and the compressive and splitting tensile strengths of recycled aggregate concrete. For every 50 times increase in freeze-thaw cycles, there was an average decrease of 33.1% in compressive strength and 33.0% in splitting tensile strength. In the early freeze-thaw period, the difference between the mechanical properties of ordinary concrete and recycled aggregate concrete is not significant, but with the increase in freeze-thaw times, both properties deteriorate, and the degree of deterioration of recycled concrete is greater than that of ordinary concrete. The freeze-thaw damage model based on the loss rate of dynamic elastic modulus has a higher fitting accuracy, and the establishment of the freeze-thaw damage model can directly and clearly reflect the changes in the macroscopic mechanical properties of recycled aggregate concrete. Fan [9] carried out the determination of freeze-thaw cycle parameters and stress-strain full curves of prismatic specimens with different numbers of cycles and different recycled aggregate admixtures by electron microscopy scanning analysis of 13 specimens and freeze-thaw cycle tests of 42 prismatic specimens. The study's findings indicate that the recycled coarse aggregate has a loose structure, numerous microcracks, and few cementite pores, all of which are detrimental to the recycled aggregate concrete's ability to withstand frost. The strength of recycled concrete after 28 days is approximately 25% less than that of ordinary concrete, and there is no correlation between the quantity of recycled coarse aggregate and the strength. Its stress-strain peak stress is about 10%–20% lower than that of ordinary concrete, and the peak stress decreases with the increase of recycled coarse aggregate. Shang [10] carried out a freeze-thaw cycle test, constitutive relationship test, and microstructure analysis of recycled aggregate concrete with air entraining aggregate and recycled aggregate concrete without air-entraining aggregate and analyzed the influence of aggregate type, air content, and aggregate treatment method of recycled aggregate concrete on strength, peak stress, peak strain, elastic modulus, and transverse deformation coefficient, as well as the interface transition zone of different types of specimens. The structure at the interface of aggregate and cement stone inside the specimen is relatively flimsy, and there are microcracks. The microcracks at the interface of the specimen have a tendency to expand after freezing and thawing, and the microcracks in the cement stone increase. The factors affecting the frost resistance and macro mechanical properties of concrete are mainly related to the transition zone at the interface between the aggregate and cement stone. Qi [11] conducted a study on the change law of stress-strain curve of freeze-thaw damaged recycled aggregate concrete by conducted a maximum number of 300 freeze-thaw cycles on 72 prism specimens of 100 × 100 × 400 mm, monotonic load tests and repeated load tests on recycled aggregate concrete with different recycled aggregate replacement rates. The test results show that with the increase in the number of freeze-thaw cycles, the mass loss rate of concrete in each group is less than 2%. The relative dynamic elastic modulus of ordinary concrete freeze-thaw cycles 300 times decreased to 95%, and the freeze-thaw damage was small; the freeze-thaw damage occurred 300 times for the recycled aggregate concrete with 50% recycled coarse aggregate substitution; the freeze-thaw damage occurred 150 times for the recycled aggregate concrete with 100% recycled coarse aggregate substitution, and the freeze-thaw deterioration was serious. As the replacement rate of recycled aggregates increases, the frost resistance of concrete decreases. The freeze-thaw-damaged recycled aggregate concrete repetitive load stress-strain full curve envelope is very close to the monotonic load stress-strain curve, and the equations applicable to the full stress-strain curve of recycled aggregate concrete can both describe the stress-strain curve envelope of recycled aggregate concrete under repetitive load more accurately.

3 Abroad Research Status

Wang et al. [12] conducted a study on the effect of recycled coarse aggregate and glazed hollow beads on the damage mode, ultimate bond strength, peak slip, and bond slip curves of concrete by performing freeze-thaw cycle tests on four types of concrete (plain concrete, recycled concrete, glazed hollow beads insulated concrete, and recycled aggregate insulated concrete) and a central pullout test to determine the adhesion properties of basalt fiber-reinforced polymer reinforcement to four types of concrete. The test results demonstrate that recycled coarse aggregate negatively impacts recycled concrete's mechanical qualities, resistance to freeze-thaw, and bonding. The reasonable addition of glazed hollow beads can make up for the adverse effects of recycled coarse aggregate and significantly improve the freeze-thaw resistance of concrete. Under the action of the freeze-thaw cycle, the glazed hollow beads can also improve the bonding performance between basalt fiber-reinforced polymer reinforcement and recycled concrete. Xiao et al. [13] conducted a study on the effects of recycled coarse aggregate replacement rate and number of freeze-thaw cycles on mass loss, compressive strength, relative dynamic elastic modulus, and capillary water absorption of recycled concrete by determining the capillary water absorption and uniaxial compressive stress-strain curves of recycled coarse aggregate replacement rates of 0%, 30%, and 50% of recycled concrete and the development of a damage constitutive model of recycled aggregate concrete based on the dual factors of replacement rate and number of freeze-thaw cycles. The test results showed that the mass, compressive strength, and relative dynamic elastic modulus of recycled concrete with different replacement rates tended to decrease with the increase of the number of freeze-thaw cycles, and the water absorption rate first increased and then tended to stabilize. After freeze-thaw, the peak stresses of the recycled concrete with different replacement rates were all reduced to different degrees, and the peak strains were slightly increased. Hao et al. [14] conducted a study on the durability performance of recycled aggregate thermal insulation concrete by testing the combined effect of freeze-thaw cycles and bending loads at different bending load ratios for recycled coarse aggregates with volume replacement rates of 0, 30, 50, 70, and 100% and the addition of glazed hollow bead particles with a volume percentage of 130%. The test results show that the bending load stress has a significant negative effect on the freeze-thaw resistance of recycled insulated concrete. When the bending load ratio was 45%, the recycled aggregate thermal insulation concrete fractured before the number of load cycles reached 30. The bending load stress accelerates the reduction in relative dynamic elastic modulus but has little effect on the mass loss rate. The durability performance of recycled aggregate thermal insulation concrete under combined action decreases with the increase in recycled coarse aggregate replacement rate and increases with the increase in glazed hollow bead particles admixture. Saberian et al. [15] conducted a study on the effect of the freeze-thaw process on the course performance of recycled coarse aggregates containing rubber powder for pavement base and subbase by conducting comparative tests on rubber-modified recycled concrete aggregates for pavement base/subbase at room temperature and rubber recycled concrete aggregates for pavement base/subbase after freeze-thaw. The test results showed that the resilient modulus of rubber-modified recycled concrete after freeze-thaw cycles was higher than that of the control specimens (prepared and tested at a constant room temperature of 25 °C). The unconfined compressive strength (UCS) of the F-T specimens (one day of freezing followed by one day of thawing) was slightly lower than that of the control specimens. The UCS values of the T-F samples (thawed for one day and subsequently frozen for one day) were found to be significantly higher than those of the control samples. Bao et al. [16] conducted freeze-thaw cycle tests, water absorption tests, SEM, MIP, and laser scanning microscopy (LSM) tests on recycled concrete with nickel-iron slag as fine aggregate by performing freeze-thaw cycle tests, water absorption tests, SEM, MIP, and laser scanning microscopy (LSM) tests on recycled concrete with different nickel-iron slag substitution rates (0%, 20%, 40%, 50%) and recycled coarse aggregate substitution rates (0%, 50%, 100%). The test results show that the frost resistance of concrete after freeze-thaw damage decreases with the increase of recycled coarse aggregate admixture, and increases and then decreases with the increase of ferronickel slag admixture. The microscopic test results showed that the addition of recycled aggregates formed multiple weak interfaces with loose structure inside the concrete. The recycled aggregate concrete with a 40% ferronickel slag admixture showed better frost resistance. Zou et al. [17] conducted a study on the modification mechanism of two modification methods, surface modification of recycled aggregates and monolithic modification of cement matrix, by measuring the compressive strength and freeze-thaw durability of recycled aggregate pervious concrete obtained by different silane modification methods and by microscopic analysis. The test results show that the surface modification of recycled aggregate concrete makes the surface of recycled aggregate form a hydrophobic organic film to prevent the accumulation of water on the surface of recycled aggregate, and the interface transition zone is more fully hydrated to form a dense structure. The bulk modification of the cement matrix leads to the presence of a hydrophobic silicone film on the surface of the cement particles, which inhibits cement hydration and increases the porosity of the cement matrix. Therefore, the surface modification of recycled aggregate is more effective in improving the compressive strength and freeze-thaw durability of recycled aggregate pervious concrete than the bulk modification of the cement matrix. Deng et al. [18] conducted a study on the microstructural damage process of air-entraining recycled coarse aggregate concrete and non-air-entraining recycled coarse aggregate concrete under freeze-thaw cycling conditions by conducting flexural strength tests and NMR tests on air-entraining recycled coarse aggregate concrete and non-air-entraining recycled coarse aggregate concrete with 0%, 25%, 50%, 75%, and 100% recycled coarse aggregate replacement. The test results showed that the incorporation of an air-entraining agent increased the proportion of medium pores (radius 0.01–0.05 μm) and large pores (radius 0.05–1 μm) in concrete by 13.31%–21.44%, which improved the frost resistance of concrete. The flexural strength of recycled aggregate concrete is significantly influenced by the total proportion of medium and large holes and is higher when the total proportion of medium and large holes is between 55% and 74%.

Wei et al. [19] conducted a study on the potential of recycled concrete as structural concrete aggregate by performing freeze-thaw cycles and 35% bending load tests on 100% replacement-rate C40 recycled concrete specimens. According to the test results, as the number of freeze-thaw cycles increased, the apparent density of the second-generation recycled coarse aggregate declined, but the crushing value, water absorption, and stability (mass loss) increased. With the accumulation of freeze-thaw cycle damage, the particle size of the second-generation recycled coarse aggregate gradually decreases, and the sieve curve of the recycled coarse aggregate after 400 freeze-thaw cycles cannot meet the requirement of continuous particle size. Bai et al. [20] conducted a study on the uniaxial compressive mechanical properties and damage evolution law of recycled concrete containing silica fume under the action of freeze-thaw cycles by performing 0, 50, 75, 100, and 125 freeze-thaw cycles and uniaxial compression tests on recycled concrete with 0% and 9% silica fume substitution, followed by observing the microstructure, microcrack extension process, and reaction products by acoustic emission and scanning electron microscopy. The test results showed that the peak stress and elastic modulus of the specimens tended to decrease, and the peak strain and mass loss rate gradually increased with the increase in the number of freeze-thaw cycles. The addition of silica fume produced an additional volcanic ash reaction that led to an increase in calcium alumina, which reduced the porosity of the recycled concrete and significantly increased the compressive strength at different numbers of freeze-thaw cycles. The compressive strength increased by 21.84% and 12.06% at the number of freeze-thaw cycles of 0 and 125, respectively. Xiao et al. [21] conducted freeze-thaw cycling tests on recycled concrete with a water-cement ratio of 0.45 and 30% recycled aggregate admixture in 3% Na_2SO_4, 5% Na_2SO_4, and 10% Na_2SO_4 (mass fraction) solutions and water by performing freeze-thaw cycling tests based on recycled aggregate concrete weight, relative dynamic elastic modulus, compressive strength loss, damage depth thickness, and concrete compressive strength of the damaged layer. The damage deterioration law of recycled aggregate concrete was studied based on the weight, relative dynamic elastic modulus, compressive strength loss, damage depth thickness, compressive strength of damaged concrete, etc. The composite damage mechanism of recycled aggregate concrete under the action of sulfate and freeze-thaw cycles was also analyzed using scanning electron microscopy. The test results indicate that as the number of freeze-thaw cycles increases, the amount of damage to recycled aggregate concrete increases, along with a decrease in ultrasonic velocity, an increase in thickness, a decrease in compressive strength, and a decrease in overall degree of damage. Recycled aggregate concrete in 5% Na_2SO_4 solution freeze-thaw damage is serious, the worst frost resistance is in 3% Na_2SO_4 solution, where the degree of deterioration of recycled concrete in the early freeze-thaw cycle is less than water, the degree of deterioration of recycled aggregate concrete after 200 freeze-thaw cycles is greater than water, and the degree of deterioration in 10% Na_2SO_4 solution is greater than 3% Na_2SO_4 solution and water. Pilehvar et al. [22] conducted a study on the effect of microencapsulated phase change materials on the physical and mechanical properties of two types of concrete under freeze-thaw cycle conditions by performing freeze-thaw cycle tests on geopolymer concrete (GPC)) and silicate cement concrete containing two different microencapsulated phase change materials (MPCM). The test results showed that the addition of microencapsulated phase change materials provided excellent frost resistance to concrete, and the reduction in compressive strength of concrete with the addition of microencapsulated phase change materials was smaller compared to concrete without the addition of microencapsulated phase change materials. When the temperature decreases to 0 °C, the initial setting time of the silicate cement paste becomes longer due to the low temperature and high viscosity. For geopolymer slurries, the initial setting time becomes shorter due to the phase separation of the alkali solution at low temperatures. The increase in the concentration of microencapsulated phase change material reduces the final setting time of silicate cement and geopolymer slurry. Li et al. [23] conducted a study of the stability and thermal properties of lightweight aggregates containing mortar-phase change materials by performing cyclic freeze-thaw mass loss, latent heat decay, Fourier infrared spectroscopy tests, and quasi-steady-state operation of a longitudinally protected comparative calorimeter. The test results show that when the lightweight aggregate containing mortar-phase change material is wrapped in a certain way, the lightweight aggregate can adsorb the mortar-phase change material stably without chemical reaction or large latent heat loss. The surface cladding of lightweight aggregates containing mortar-phase change materials using net cement paste is more compatible with the system than conventional material cladding. The ability of the mortar-phase change material to retard or prevent ice formation in the mortar and the release of heat from the mortar-phase change material are sufficient to inhibit ice formation when the carrier of the mortar-phase change material is replaced with a lightweight aggregate of up to 50% vol. Yeon et al. [24] conducted a feasibility study on the use of phase change materials to mitigate freeze-thaw damage in concrete pavements by conducting freeze-thaw cycle tests and mechanical properties tests on mortar specimens containing microencapsulated phase change materials. The test results show that low transition temperature phase change materials have the potential to extend the freeze-thaw life of concrete pavements with minimal effects even after prolonged exposure to ambient temperatures well below the transition temperature. The incorporation of microencapsulated phase change materials negatively affects the compressive and flexural strength of the mortar, but the early volume stability of the mortar is enhanced.

4 Current Problems and Future Prospects

In recent years, scholars at home and abroad have conducted extensive research on the mechanisms of freeze-thaw damage and the frost resistance of recycled aggregate concrete. The current problems in the study of the freeze-thaw resistance of recycled aggregate concrete are as follows:

(1) As the surface of recycled coarse aggregate has cracks as well as residual parts of the lower-strength mortar, the higher the replacement rate of recycled coarse aggregate, the worse the concrete frost resistance. The current research on the effect of recycled coarse aggregate replacement rate on the freezing resistance of recycled aggregate concrete is still in its infancy, and the research on the mechanism of freeze-thaw cycle breakage of recycled aggregate concrete and the establishment of the theoretical model are not perfect.
(2) Most of the current research on the frost resistance of recycled concrete focuses on the effect of recycled coarse aggregate replacement rate on the frost resistance of concrete, with less research on the frost resistance of recycled fine aggregates and recycled powders on concrete.
(3) Due to the existence of interfacial transition zone with high porosity and poor densification in recycled aggregate concrete, its frost resistance is poor. Phase-change materials can prevent the pore solution in recycled aggregate concrete from freezing

and improve the frost resistance of recycled aggregate concrete. Mineral admixtures, surface organic coating agents (inhibitors), air-entraining agents, fiber-reinforced materials, and other materials can also effectively improve the frost resistance of concrete. The current study of the frost resistance of recycled aggregate concrete after the addition of the above materials and the mechanism of impact is less thorough.

In response to the above problems, the following outlook is proposed for the study of the frost resistance of recycled concrete:

(1) Targeted research on the freeze-thaw damage mechanism of recycled coarse aggregate concrete is conducted to further improve the theoretical model by specifically studying and analyzing the matrix damage process when the recycled coarse aggregate concrete is subjected to freezing damage from a microscopic perspective.
(2) More research is needed on the frost resistance of recycled aggregate concrete with different replacement rates of recycled fine aggregates and recycled powders.
(3) Fibers, mineral admixtures, and phase change materials are added to recycled aggregate concrete to improve the frost resistance of recycled aggregate concrete, and their effects on the frost resistance of recycled aggregate concrete and the mechanism of the effects are studied.

5 Conclusion

Compared with ordinary concrete, the frost resistance of recycled aggregate concrete is poor, and its mechanical properties after freeze-thaw cycles decrease as the replacement rate of recycled aggregates increases. This phenomenon is due to the existence of microcracks in some of the recycled aggregates and the high porosity and poor density of the interface transition zone of the recycled aggregate concrete. The improvement of recycled aggregate surface properties, recycled aggregate concrete interfacial transition zone properties, and recycled aggregate concrete porosity can effectively improve the frost resistance of recycled aggregate concrete. The addition of silica fume to the recycled aggregate concrete creates an additional volcanic ash reaction that leads to an increase in ettringite, which reduces the porosity of the recycled aggregate concrete. Silane modification on the surface of recycled aggregate concrete makes the surface of recycled aggregate form a hydrophobic organic film to prevent the accumulation of water on the surface of recycled aggregate, and the interface transition zone is more fully hydrated to form a dense structure. The addition of phase-change materials to recycled aggregate concrete can inhibit ice formation due to their inherent ability to absorb and release heat. In addition, the addition of glazed hollow beads, air-entraining agents, and other materials to recycled aggregate concrete can also effectively improve its frost resistance. In the future, studies on the effect of recycled fine aggregates and recycled powders with different replacement rates on the frost resistance of recycled aggregate concrete will be conducted. To study the effect of fibers, mineral admixtures, and phase change materials on the frost resistance of recycled aggregate concrete and the mechanism of the effect. The matrix damage process of recycled concrete subjected to frost damage is studied from a microscopic perspective to further refine the theoretical model.

Acknowledgements. The authors would like to acknowledge the financial support from Liaoning Provincial Natural Science Foundation Doctoral Research Start-up Project (2023-BS-133) and Basic Research Project of Liaoning Provincial Department of Education (LJKM20220920) and Project Funding for China Testing & Certification International Group (JSKF202311).

References

1. Niu, H., Fan, Y., Zhang, X., et al.: Experimental study on the freeze-thaw resistance of recycled concrete with different replacement ratios. J. Mater. Sci. Eng. **36**(04), 615–620 (2018)
2. Bai, W., Wang, W., Yuan, C., et al.: Experimental study on stress-strain relationship of recycled concrete under freezing and thawing environment. Concrete (06), 9–12 (2021)
3. Zou, C., Fan, Y., Hu, Q.: Experimental study on the basic mechanical property of recycled concrete after freeze-thaw. Build. Struct. **40**(S1), 434–438 (2010)
4. Wang, C., Liu, L., Cao, F., et al.: Experimental study on mechanical properties of recycled concrete after freeze-thaw cycles. J. Build. Struct. **41**(12), 193–202 (2020)
5. Dai, J., Zhai, H., Huang, B., et al.: Study on mechanical properties of recycled concrete after freeze-thaw cycles. Concrete (05), 61–64+68 (2022)
6. Cao, W., Liang, M., Dong, H., et al.: Experimental study on basic mechanical properties of recycled concrete after freeze-thaw cycles. J. Nat. Disasters **21**(03), 184–190 (2012)
7. Chen, A., Zhang, Q., Wang, J., et al.: Freez-thaw cycle test and a damage mechanics model for recycled concrete. Eng. Mech. **26**(11), 102–107 (2009)
8. Wang, C., Zhang, D., Cao, F., et al.: Research on mechanical properties and damage of recycled concrete after being subjected to freeze-thaw cycles. Ind. Constr.
9. Fan, Y.: Experiment and study on influence of freeze-thaw cycle on the mechanics behavior of frost-resistant recycled aggregate concrete. Harbin Institute of Technology (2009)
10. Shang, Y.: The experimental study on the mechanical properties and frost resistance of recycled aggregate concrete. Harbin Institute of Technology (2010)
11. Qi, Z.: Study on constitutive relation of recycled aggregate concrete damaged by freeze-thaw under monotic and cyclic loading. Harbin Institute of Technology (2016)
12. Wang, W., Wang, Y., Chen, Q., et al.: Bond properties of basalt fiber reinforced polymer (BFRP) bars in recycled aggregate thermal insulation concrete under freeze–thaw cycles. Constr. Build. Mater. **329**, 127197 (2022)
13. Xiao, Q., Wu, Z., Qiu, J., et al.: Capillary water absorption performance and damage constitutive model of recycled concrete under freeze–thaw action. Constr. Build. Mater. **353**, 129120 (2022)
14. Hao, L., Liu, Y., Xiao, J.: Durability of recycled aggregate thermal insulation concrete under combined flexural loading and freeze–thaw cycles. Constr. Build. Mater. **272**, 121652 (2021)
15. Saberian, M., Li, J.: Effect of freeze–thaw cycles on the resilient moduli and unconfined compressive strength of rubberized recycled concrete aggregate as pavement base/subbase. Transp. Geotech. **27**, 100477 (2021)
16. Bao, J., Zheng, R., Yu, Z., et al.: Freeze-thaw resistance of recycled aggregate concrete incorporating ferronickel slag as fine aggregate. Constr. Build. Mater. **356**, 129178 (2022)
17. Zou, D., Wang, Z., Shen, M., et al.: Improvement in freeze-thaw durability of recycled aggregate permeable concrete with silane modification. Constr. Build. Mater. **268**, 121097 (2021)
18. Deng, X., Gao, X., Wang, R., et al.: Investigation of microstructural damage in air-entrained recycled concrete under a freeze–thaw environment. Constr. Build. Mater. **268**, 121219 (2021)

19. Wei, D., Zhu, P., Yan, X., et al.: Potential evaluation of waste recycled aggregate concrete for structural concrete aggregate from freeze-thaw environment. Constr. Build. Mater. **321**, 126291 (2022)
20. Bai, W., Song, Z., Yuan, C., et al.: Study on mechanical properties and damage mechanism of recycled concrete containing silica fume in freeze–thaw environment. Constr. Build. Mater. **375**, 130872 (2023)
21. Xiao, Q.H., Li, Q., Cao, Z.Y., et al.: The deterioration law of recycled concrete under the combined effects of freeze-thaw and sulfate attack. Constr. Build. Mater. **200**, 344–355 (2019)
22. Pilehvar, S., Szczotok, A.M., Rodríguez, J.F., et al.: Effect of freeze-thaw cycles on the mechanical behavior of geopolymer concrete and Portland cement concrete containing micro-encapsulated phase change materials. Constr. Build. Mater. **200**, 94–103 (2019)
23. Li, W., Ling, C., Jiang, Z., et al.: Evaluation of the potential use of form-stable phase change materials to improve the freeze-thaw resistance of concrete. Constr. Build. Mater. **203**, 621–632 (2019)
24. Yeon, J.H., Kim, K.K.: Potential applications of phase change materials to mitigate freeze-thaw deteriorations in concrete pavement. Constr. Build. Mater. **177**, 202–209 (2018)

Study on Thermal and Durability Properties of Shape-Stabilized Phase Change Particles

Yanhui Wang[1], Chonghui Zhou[2](✉), Yichao Zhang[2], and Yang Shi[1]

[1] China Construction Third Engineering Bureau Group (Zhejiang) Co., Ltd., Hangzhou, China
[2] Shenyang Jianzhu University, Shenyang, China
18809846820@163.com

Abstract. In this paper, the research status of shape-stabilized phase change particles at home and abroad was extensively investigated, and the composite phase change materials of decanoic acid and myristic acid were prepared by heating and fusion of decanoic acid and myristic acid, and the optimum mass ratio of decanoic acid and myristic acid was measured as 6:4 through the step-cooling curve and the DSC curve test, and the phase change temperature of decanoic acid-myristic acid composite phase change materials was 26 °C and the enthalpy of the phase change was 131.82 J/g. The adsorption performance and durability of shape-stabilized phase change particles were tested through vacuum adsorption and encapsulation. The phase change enthalpy was 131.82 J/g. The phase change particles were prepared by vacuum adsorption and encapsulation, and the adsorption performance and durability of the phase change particles were tested to find out the key problems affecting the performance, analyze the causes of the problems, and explore the solution paths to the causes, to reduce the energy consumption of heating and air conditioning during the operation phase of buildings, a new type of shaped phase change material that can buffer indoor temperature changes is prepared and its related parameters are measured to promote its application.

Keywords: shape-stabilized phase change particles · composite phase change materials · adsorption properties · durability properties

1 Introduction

With the rapid advancement of China's housing and urban-rural construction tasks, now easy to enter the housing and urban-rural construction business "14th Five-Year Plan" critical period, by 2025, the urban resident population of 770 million people, the urbanization rate of the resident population increased to 56.1% [1]. At the same time, with the improvement of people's living standards, people have more and more new needs for the living environment, but in order to improve the living comfort and maintain the optimal temperature of the human body indoors, resulting in increasing energy demand and carbon emissions, so in order to save energy and reduce carbon emissions we urgently need to find a new building energy efficiency methods.

Phase change material is a thermal functional material that undergoes a change in its physical state and is accompanied by a large amount of thermal energy absorption and

I. Tuns et al. (Eds.): CIBv 2023, LNCE 510, pp. 12–23, 2024.
https://doi.org/10.1007/978-3-031-60765-3_2

release when the temperature of the material reaches the phase change temperature. It has received widespread attention in recent years. Shaped phase change particles are particles obtained through adsorption and packaging processes on phase change materials. While maintaining the original thermal properties of phase change materials, they improve the problems of easy leakage and poor stability of phase change materials.

In the field of shape-stabilized phase change materials, foreign research started earlier. As early as in the 1980s, Lane et al. analyzed more than 200 kinds of phase change materials, compared the performance of shaped phase change particles made of various encapsulation methods, and applied them to buildings for research, all of which achieved better energy-saving effects [2]. China began to study shape change materials in the early 21st century. Qin Penghua's team from Tsinghua University encapsulated paraffin wax with different densities of polyethylene as a support material to make shape change particles, which significantly improved the heat storage per unit of body volume in practical applications [3]. At present, domestic and international research on shape-stabilized phase change materials mainly focuses on mechanical and thermal properties.

2 Experimental Materials for Shape-Stabilized Phase Change Particles

2.1 Composite Phase Change Materials

By reviewing the related research literature [4, 5], fatty acids were selected as phase change materials. The decanoic acid (CA)-myristic acid (MA) composite phase change material made by heating and melting of decanoic acid (CA) and myristic acid (MA) has a suitable phase change temperature, and the enthalpy of the phase change is also high, and the physical and chemical properties are stable. Both decanoic acid (CA) and myristic acid (MA) were produced by Shandong Yusuo Chemical Technology Co.

2.2 Adsorption, Encapsulation Materials

At present, in a large number of studies of shape-stabilized phase change particles, most of them choose expanded perlite as the adsorption carrier of phase change materials. The density of expanded perlite is 80–160 kg/m^3, and the thermal conductivity is between 0.034 and 0.04 W/(m-K) [6]. The expanded perlite used in this test was provided by Xingcheng Trench Musen Horticultural Development Co [7].

After the adsorption of expanded perlite is completed, encapsulation process is also needed. Through reading the literature, the choice of phenylpropylene emulsion [8] and cement slurry [9] as encapsulation materials have better encapsulation effect, the cement used P.O42.5 type cement, phenylpropylene emulsion samples produced by Shandong Yusuo Chemical Science and Technology Co.

3 Properties Study of Shape-Stabilized Phase Change Particles

3.1 Properties Study of Composite Phase Change Materials

3.2 Step-Cooling Curve Test

Based on the material quality, a ratio of 7 types of decanoic acid and myristic acid was designed. The mass ratio of decanoic acid (CA) to myristic acid (MA) was 2:8, 3:7, 4:6, 5:5, 6:4, 7:3, and 8:2, respectively.

(1) Prepare 50 g of each ratio sample according to the above ratio, weigh two types of fatty acids by mass using a small electronic scale, and mix them into a small beaker. Heat the water in the water bath heating pot to 80 °C, then put it into a small beaker containing two types of fatty acids, and continuously stir with a glass rod to ensure that the material is heated evenly.
(2) After observing that all the solids in the beaker have dissolved, continue heating for 30 min and stir with a glass every 3 min. When stirring, stir clockwise for 10 s first and then counterclockwise for 10 s to ensure sufficient co melting of the two fatty acids.
(3) After 30 min, remove the beaker and wait for the sample to cool down to obtain the composite phase change material at room temperature. The composite phase change material in solid state is shown in Fig. 1.

Fig. 1. Composite phase change materials in solid state

Perform the step cooling curve test according to the following steps, as shown in Fig. 2.

(1) Heat the water in the constant temperature water bath to 60 °C, place it in a small beaker containing 50 g of composite phase change material, insert a portable thermometer probe into the center of the beaker, immerse the probe about 2 cm into the center of the liquid surface and keep it still, and read the temperature at the center of the beaker.

(2) When the temperature of the thermometer rises to 60 °C, take the beaker out of the thermostatic water bath, place it on the foam cover and start timing. Record the temperature once every 20 s until the composite phase change material is completely solidified. Stop recording and pull out the portable thermometer probe.
(3) After wiping the portable thermometer probe clean, place it in the air to test the ambient temperature. After the temperature stops fluctuating, read the temperature at this time.

Fig. 2. Schematic diagram of step cooling curve test

After measuring the temperature data of seven sets of composite phase change materials with different ratios, the step-cooling curves were plotted, as shown in Fig. 3.

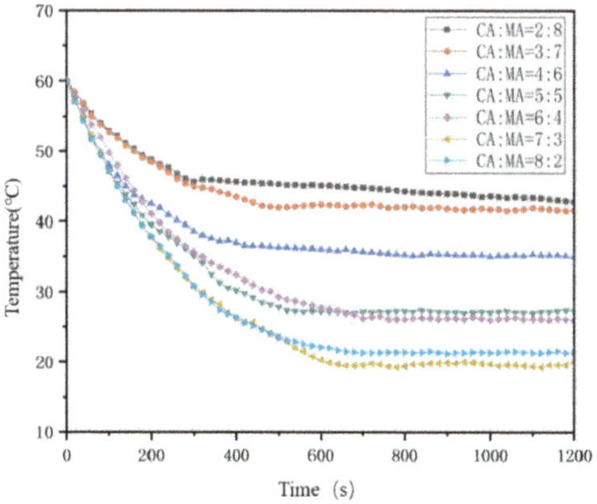

Fig. 3. Step-cooling curve of composite phase change material

As can be seen from Fig. 3, the temperature of the composite phase change material with seven different ratios enters a stage of smooth fluctuation after a rapid decrease, at which time the composite phase change material reaches the phase change temperature.

According to the plotted step-cooling curves Fig. 1, the phase change temperatures of the composite phase change materials with different ratios are shown in Table 1.

Table 1. Phase change temperature of composite phase change materials with different ratios

CA to MA mass ratio	Phase transition temperature (°C)
2:8	46
3:7	42.2
4:6	36
5:5	27.2
6:4	26
7:3	19.8
8:2	21.3

According to the research related to the dynamic thermal comfort theory, the indoor temperature that the human body feels comfortable is between 19 °C and 23 °C [10, 11]. In the practical application of composite phase change materials, the phase change temperature should be slightly higher than the human comfort temperature in order to play a role in the application. The phase change temperatures of the composite phase change materials with CA:MA = 2:8 and 3:7 were too high, so these two ratios of composite phase change materials were excluded, and the remaining five groups of composite phase change materials were subjected to the DSC test, and were further screened based on the enthalpy of phase change.

3.2.1 DSC Testing

DSC, also known as Differential Scanning Calorimetry in Chinese, is a thermal analysis method that can measure the relationship between temperature and heat flux of a substance, draw a heat flux temperature curve, and analyze and calculate many thermodynamic parameters based on this curve, such as specific heat capacity, reaction heat, and transition heat. The instrument that uses differential scanning calorimetry for testing is called a differential scanning calorimeter, and the recorded heat flux temperature curve is called the DSC curve. The data of temperature and heat flow rate are exported in the relevant data files and analyzed and processed, the DSC curve is plotted, and the enthalpy of phase change is calculated according to the mathematical principle, and the DSC curve is shown in Fig. 4, 5, 6, 7 and 8. The material components in Figs. 4, 5, 6, 7 and 8 are decanoic acid (CA) and myristic acid (MA). The mass ratio of decanoic acid (CA) to myristic acid (MA) is 2:8, 3:7, 4:6, 5:5, 6:4, 7:3, and 8:2, respectively.

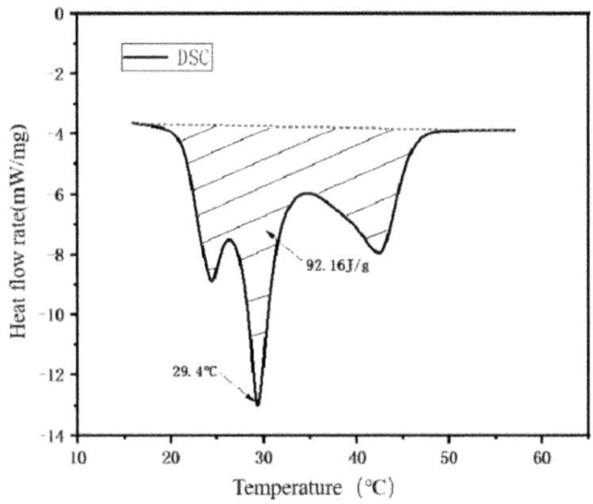

Fig. 4. DSC curve of composite phase change material with CA:MA = 4:6

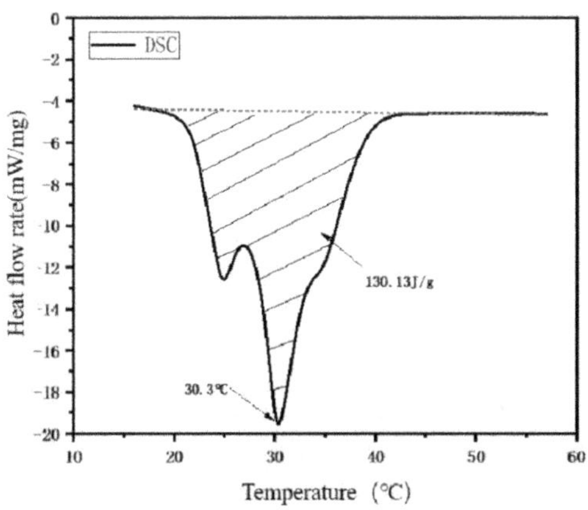

Fig. 5. DSC curve of composite phase change material with CA:MA = 5:5

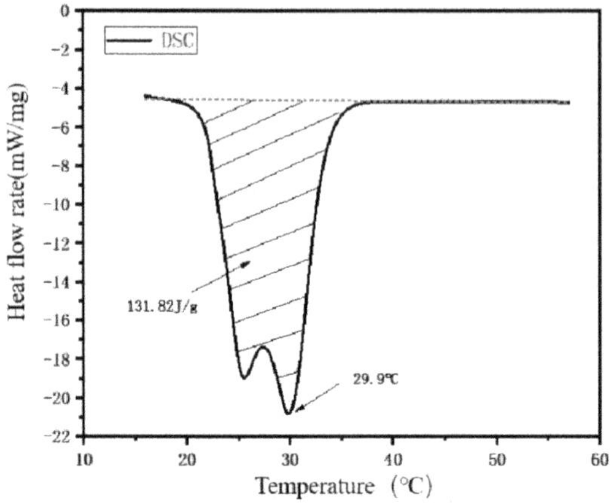

Fig. 6. DSC curve of composite phase change material with CA:MA = 6:4

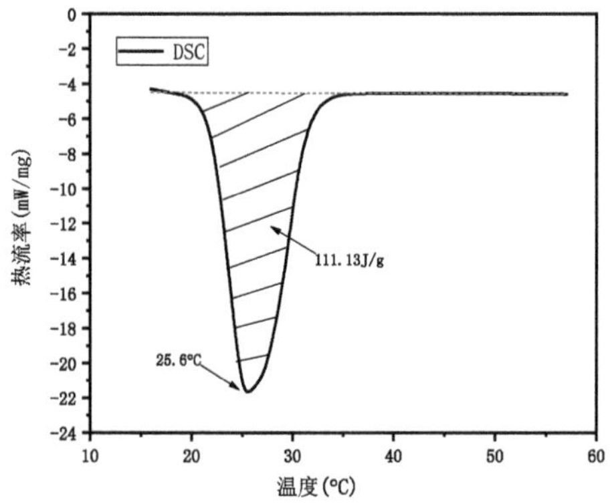

Fig. 7. DSC curve of composite phase change material with CA:MA = 7:3

Based on the DSC curves of the composite phase change materials, the phase change enthalpy of composite phase change materials was calculated for each ratio as shown in Table 2.

The CA:MA = 6:4 composite phase change material had the largest enthalpy of phase change, so it was decided that the CA:MA = 6:4 composite phase change material would be used for all subsequent tests.

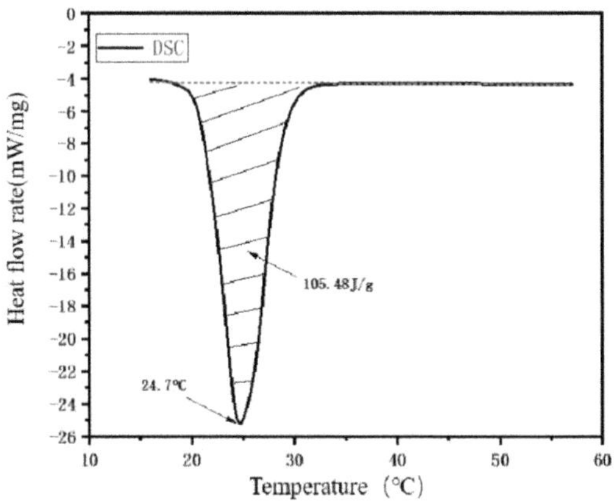

Fig. 8. DSC curve of composite phase change material with CA:MA = 8:2

Table 2. Phase change enthalpy of composite phase change materials with different ratios

CA to MA mass ratio	Enthalpy of phase change (J/g)
4:6	92.16
5:5	130.13
6:4	131.82
7:3	111.13
8:2	105.48

3.3 Expanded Perlite Adsorption Property Test

Expanded perlite was used in 1 group of particle gradation, a total of 8 copies were prepared, the mass of each copy was 20 g. The adsorption time was set to 0.5 min, 1 min, 1.5 min, 2 min, 2.5 min, 3 min, 4 min, 5 min, respectively, and the adsorbed mass and adsorption rate were calculated, and the adsorption rate-adsorption time curve was plotted. The adsorption performance test data are shown in Table 3, and the adsorption rate-adsorption time curve is shown in Fig. 9.

As determined by the adsorption rate-adsorption time curve in Fig. 7, a reasonable adsorption time for expanded perlite is 4 min, and the adsorption rate is around 600%.

Table 3. Adsorption performance test table

Time (min)	m_0 (g)	m_1 (g)	Adsorption capacity (g)	Adsorption rate (%)
0.5	20	108.6	88.6	443
1	20	121.1	101.4	507
1.5	20	127.2	107.2	535
2	20	131	111	555
2.5	20	135.1	115.1	576
3	20	137.1	117.1	586
4	20	140.4	120.4	602
5	20	139.2	119.2	596

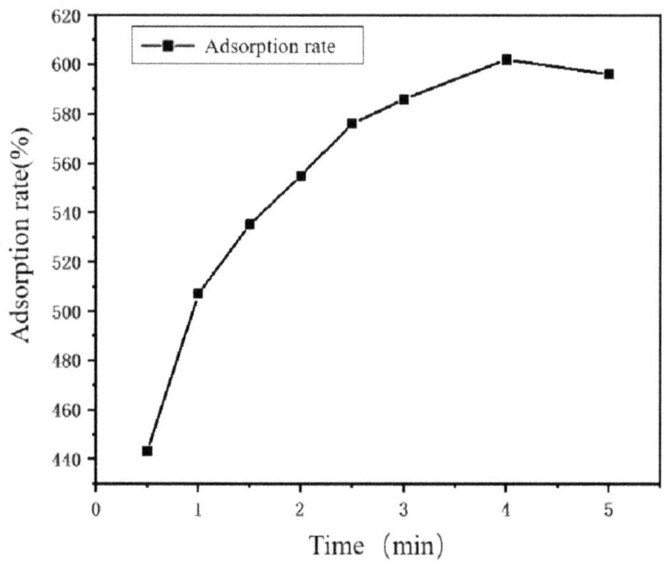

Fig. 9. Adsorption rate adsorption time curve

3.4 Durability Testing of Shape-Stabilized Phase Change Particles

As mentioned above, a composite phase change material with a mass ratio of 6:4 for decanoic acid to myristic acid was ultimately selected. In order to further compare the packaging effect of cement paste and styrene acrylic lotion, durability test is required after curing of stereotyped phase change particles of the two packaging methods. By repeatedly heating and cooling fixed phase change particles, simulating real usage situations, testing the quality loss of fixed phase change particles during phase change cycles, and evaluating the durability performance of different packaging materials. Due to the use environment of shaped phase change materials under cyclic temperature, durability tests should be conducted to ensure their service life under temperature cycling. The loss rate of phase change cycling can intuitively express the changes in material durability of shaped phase change materials after temperature cycling. Weigh three parts of adsorbed expanded perlite, each with a mass of 200 g, and prepare shaped phase change particles by means of non encapsulation, cement paste encapsulation and styrene acrylic lotion encapsulation respectively. After curing, put the shaped phase change particles into an electric thermostat at 50 °C for heating for 20 min, and then take them out to the natural environment for heat release. Repeat the above steps, measure the mass of the solidified phase change particles after 0, 10, 20, 30, 40, and 50 cycles of heat absorption and release, and calculate the phase change cycle loss rate. The phase change cycle mass loss rates of shaped-stabilized phase change particles are shown in Table 4, and the phase change cycle loss rate variations are shown in Fig. 10.

Table 4. Mass loss rate of shaped-stabilized phase change particles in phase change cycle

Number of phase change cycles (time)	Rate of loss of unencapsulated mass (%)	Rate of loss of cement encapsulated mass (%)	Styrene-acrylic emulsion encapsulated mass loss rate (%)
10	6.9	0.1	0
20	12.6	0.4	0.1
30	19.7	0.7	0.3
40	22.1	0.9	0.7
50	24	1.2	0.9

From Table 4 and Fig. 10, it can be found that the mass loss rate grows faster when the phase change cycles are in the range of 0–30 times, and there is a significant decrease in the growth rate of the mass loss rate in the 30th to 50th times.

Comparing the three curves in Fig. 8, the mass loss rates of the phenylpropylene emulsion encapsulated and cement net paste encapsulated shape-stabilized phase change particles at 50 cycles of phase change were 0.9% and 1.2%, respectively, with cement encapsulation reducing the loss rate of the composite phase change material at 50 cycles of phase change by 95% and phenylpropylene emulsion encapsulation reducing it by 96.25%.

From the perspective of the durability of the shape-stabilized phase change particles, phenylpropylene emulsion is more suitable for encapsulation material, but cement paste also has excellent encapsulation effect. Considering the practical application scenarios of shape-stabilized phase change particles, cement mortar is used to encapsulate the shape-stabilized phase change particles.

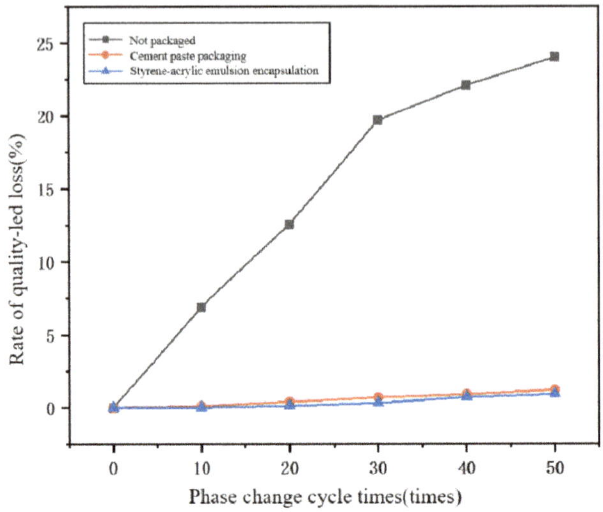

Fig. 10. Variation diagram of phase change cycle loss rate

4 Conclusion

In this paper, the material and process for preparing the shape-stabilized phase change particles were investigated to obtain the following conclusions:

When the mass ratio of decanoic acid to myristic acid is 6:4, the composite phase change material formed by heating and melting the two has a phase change temperature of 26 °C and a phase change enthalpy value of 131.82 J/g. Compared with other composite phase change materials, this ratio of composite phase change material can greatly alleviate temperature changes through phase change reactions.

In response to the difficulty in determining the completion time of adsorption in vacuum adsorption method, an adsorption rate adsorption time curve was drawn through adsorption performance testing. Based on the curve, a reasonable adsorption time of 4 min was determined, and the corresponding adsorption rate was about 600%. On this basis, the effectiveness of phase change materials can be ensured as much as possible.

The stereotyped phase change particles encapsulated with styrene acrylic lotion have serious clustering phenomenon and are not suitable for use in concrete, but their encapsulation effect is good. The mass loss rate after 50 phase change cycles is 0.9%. The stereotyped phase change particles encapsulated with cement paste have no clustering phenomenon, and the encapsulation effect is not different from that of styrene acrylic lotion. The mass loss rate after 50 phase change cycles is 1.2%. In general, cement paste is more suitable for encapsulation of stereotyped phase change particles.

References

1. Ministry of Housing and Urban-Rural Development. Outline of the 14th Five-Year Plan for Housing and Urban-Rural Construction, 10 September 2021
2. Lane, G.A.: Solar Heat Storage: Latent Heat Material. CRC Press, Florida (1986)
3. Qin, P., Yang, R., Zhang, Y., et al.: Thermal properties of conformal phase change materials. J. Tsinghua Univ. (Nat. Sci. Ed.) (06), 833–835 (2003)
4. Li, J.: Research on the Application of Shape-Stabilized Decanoic Acid/Stearic Acid Phase Change Material in Heat Storage Concrete. Wuhan University of Technology, Wuhan (2017)
5. Guo, Y., Zhu, Y., Wu, Q., et al.: Preparation and characterization of decanoic-myristic acid phase change heat storage microcapsules. New Build. Mater. **44**(10), 104–107 (2017)
6. Taicang Zhitang Commune Insulation Material Shake Factory. Expanded Perlite. Jiangsu Chem. Ind. (04), 42-45 (1977)
7. Zhang, Y., Wang, Y., Zhou, J., et al.: Preparation and characterization of desulfurized gypsum-based phase change energy storage wall materials. New Build. Mater. **47**(04), 135–138+146 (2020)
8. Zhang, Y., Wang, Y., Zhou, J., et al.: Preparation and characterization of phase change energy storage gypsum. Therm. Sci. **40**, 217 (2020)
9. Li, J., Zhang, J., Fu, M., et al.: Properties of mesoporous SiO2-loaded organic-based binary shape-stabilized composite phase change energy storage materials. Mater. Guide **35**(S2), 483–487 (2021)
10. Jin, Z.: Thermal Adaptation Behavior and Thermal Comfort Zone of Residents in Different Climate Regions. Chongqing University, Chongqing (2011)
11. Ning, H.: Research on Human Thermal Comfort and Thermal Adaptation of Heating Building Environment in Severe Cold Areas. Harbin Institute of Technology, Harbin (2017)

Mechanical Properties and Frost-Thaw Behavior in Road Concrete

B. I. Luca, A. Panțiru(✉), and M. Bărbuță

Technical University "Gheorghe Asachi" from Iassy, Faculty of Civil Engineering and Building Services, 1 Prof. Dimitrie Mangeron Blv., 700050 Iassy, Romania
{bogdan-ionel.luca,alexandru.pantiru}@student.tuiasi.ro

Abstract. This study aims to determine the effect of wastes on the mechanical and durability characteristics of road concrete. Steel slag was used as replacement of aggregates in different percentages. Silica fume was used as replacement of 10% of cement. The mechanical characteristics such as: flexural strength, split tensile strength and compressive strength were experimental determined and analyzed. The behavior to frost-thaw cycles of road concrete with wastes was also analyzed. The mechanical strengths presented high values in comparison with that of the control mix (without wastes) for reduced percentages of aggregate replacement. The behavior to frost-thaw cycles was better in the case of road concrete with waste in comparison with the control mix.

Keywords: Road concrete · Frost-Thaw · Silica Fume

1 Introduction

The network of road transport is permanently developing and new and rehabilitated roads are built every year. The road industry involves a huge consumption of raw materials and energy (electricity and fuels), and has a significant negative environmental impact, contributing to global warming [1].

The greenhouse gas emissions per meter per year associated with the construction of road infrastructure has been estimated at 14.7 kg CO2eq. [2]. The material extraction, production and built are the main contributor to carbon emissions and energy consumption [3, 4].

In the road industry significant amounts of raw materials from quarry are annually used. Waste materials are resulted as byproducts from all human industrial, commercial, building and demolition activities and their re-cycle is very important for the environment protection. In the last years significant researches have been made to investigate the possibilities of using different wastes as replacement of natural aggregates in construction industry. Fly ash, silica fume, rice husk, ground granulated blast furnace, are used as addition or replacement of cement [5–10], demolition wastes, steel slag, chopped polyethylene terephthalate, polystyrene granules, tires waste, agricultural waste, etc., are used as replacement of aggregates [11–15]. The wastes are influencing the mechanical

I. Tuns et al. (Eds.): CIBv 2023, LNCE 510, pp. 24–36, 2024.
https://doi.org/10.1007/978-3-031-60765-3_3

and durability characteristics of concrete. The behavior to frost-thaw cycles is a major property of road concrete which influences the service life [16–20].

In experimental program presented in the article, the road concrete containing silica fume as replacement for cement and steel slag as replacement of aggregates was studied. Flexural strength, compressive strength, split tensile strength were determined and discussed. The behavior of green road concrete subjected to freeze-thaw cycles was also analyzed.

2 Experimental Program

2.1 Materials

In the study, the mixtures of road concrete were prepared without wastes (control mix) and with wastes (green road concrete with silica fume and steel slag). The control mix contained the following materials:

- cement type ROMCIM II 52,5 – 380 Kg/m^3;
- quarry aggregates sand sort 0–4 mm – 639 Kg/m^3;
- quarry aggregates sort 4–8 mm – 317 Kg/m^3;
- quarry aggregates sort 8–16 mm – 402 Kg/m^3;
- quarry aggregates sort 16–22 mm – 426 Kg/m^3;
- water 190 l;
- Master Build Rheoubuild/water reducing additive in a dosage of 1.2% from cement's amount;
- Master Build Air 9060.

The green road concrete was prepared by replacing cement in a dosage of 10% with silica fume and steel slag as substitute of aggregates in dosages of 10%, 20% and 30% (noted G) (Table 1A, 1B and 1C).

Table 1A. Composition RECIPE - BCR (380 kg/mc) - road concrete with ballast types 4–8 mm (steel slag substitute) and cement content 380 kg/m^3 (silica fume substitute)

Component	BCRS (F1 = G0)	BCRS (G1)	BCRS (G2)	BCRS (G3)
Aggregate 0–4 mm [kg/m^3]	639	639	639	639
Aggregate 4–8 mm [kg/m^3]	317	**285,3**	**253,6**	**221,9**
Aggregate 8–16 mm [kg/m^3]	402	402	402	402
Aggregate 16–22,5 mm [kg/m^3]	426	426	426	426

(*continued*)

Table 1A. (*continued*)

Component	BCRS (F1 = G0)	BCRS (G1)	BCRS (G2)	BCRS (G3)
steel slag substitute for Aggregate 4–8 mm [kg/m^3]	—	**10%** **31,7**	**20%** **63,4**	**30%** **95,1**
Cement ROMCIM CEM II 52,5 R [kg/m^3]	**342**	**342**	**342**	**342**
Water [l/m^3] (A/C ratio = 0,5)	190	190	190	190
Plasticizer additive Master Build Rheoubuild 572 [l/m^3]	4,18	4,18	4,18	4,18
Air entrained Master Build Air 9060 [l/m^3]	2,28	2,28	2,28	2,28
Silica Fume 10% of cement	**38**			

Table 1B. Composition RECIPE - BCR (380 kg/mc) - road concrete with ballast types 8–16 mm (steel slag substitute) and cement content 380 kg/m^3 (silica fume substitute)

Component	BCRS (F1 = G0)	BCRS (G4)	BCRS (G5)	BCRS (G6)
Aggregate 0–4 mm [kg/m^3]	639	639	639	639
Aggregate 4–8 mm [kg/m^3]	317	317	317	317
Aggregate 8–16 mm [kg/m^3]	402	**361,8**	**321,6**	**281,4**
Aggregate 16–22,5 mm [kg/m^3]	426	426	426	426
steel slag substitute for Aggregate 8–16 mm [kg/m^3]	—	**10%** **40,2**	**20%** **80,4**	**30%** **120,6**
Cement ROMCIM CEM II 52,5 R [kg/m^3]	**342**	**342**	**342**	**342**
Water [l/m^3] (A/C ratio = 0,5)	190	190	190	190
Plasticizer additive Master Build Rheoubuild 572 [l/m^3]	4,18	4,18	4,18	4,18
Air entrained Master Build Air 9060 [l/m^3]	2,28	2,28	2,28	2,28
Silica Fume 10% of cement	**38**			

Micro-silica was MICROSIL 90 GRIGIO from AZICHEM Italian Group and respects the technical characteristics given by SREN 13263-1 2005 [21]. The chemical

Table 1C. Composition RECIPE - BCR (380 kg/mc) - road concrete with ballast types 16 – 22.5 mm (steel slag substitute) and cement content 380 kg/m^3 (silica fume substitute)

Component	BCRS (F1 = G0)	BCRS (G7)	BCRS (G8)	BCRS (G9)
Aggregate 0–4 mm [kg/m^3]	639	639	639	639
Aggregate 4–8 mm [kg/m^3]	317	317	317	317
Aggregate 8–16 mm [kg/m^3]	402	402	402	402
Aggregate 16–22,5 mm [kg/m^3]	426	**383,4**	**340,8**	**298,2**
steel slag substitute for Aggregate 16–22.5 mm [kg/m^3]	—	**10%** **42,6**	**20%** **85,2**	**30%** **127,8**
Cement ROMCIM CEM II 52,5 R [kg/m^3]	**342**	**342**	**342**	**342**
Water [l/m^3] (A/C ratio = 0,5)	190	190	190	190
Plasticizer additive Master Build Rheoubuild 572 [l/m^3]	4,18	4,18	4,18	4,18
Air entrained Master Build Air 9060 [l/m^3]	2,28	2,28	2,28	2,28
Silica Fume 10% of cement		**38**		

composition of this product, according to the manufacturer's specifications, is approximately 90% silicon (SiO_2), with minimal amounts of other oxides such as aluminum oxide (Al_2O_3), iron oxide (Fe_2O_3), calcium oxide (CaO), magnesium oxide (MgO), potassium oxide (K_2O), and sodium oxide (Na2O), with a maximum concentration of 0.5%. It is important to note that these values may slightly vary depending on the specific batch of the product.

2.2 Samples and Tests

In order to determine the mechanical characteristics and the loss of resistance in compression of road concrete subjected to freeze-thaw cycles, the following samples were prepared:

- 150 mm sizes cubes, 3 samples for compression test, according to [22, 23].
- 150 × 150 × 600 mm sizes prism for testing in flexure and splitting, according to [24, 25].
- 100 mm-side cubes were prepared with 6 samples of each concrete mix, 3 samples being used as control samples, while the other 3 were subjected to freeze-thaw cycles in the TWIN FEUTRON S:D-0795 climatic chamber [26] (Figs. 1 and 2).

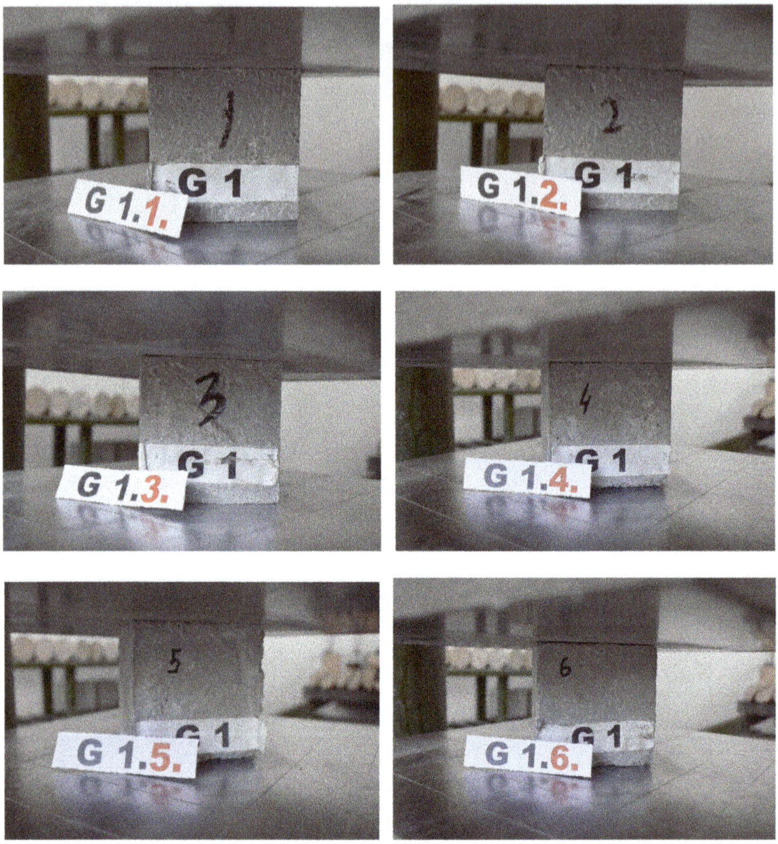

Fig. 1. Samples

For determining the loss of compressive strength to freeze-thaw cycles the test was carried out in accordance with SR EN 3518:2009 [26], by comparing the values of compressive strength. The concrete developed in this research was subjected to 150 freeze-thaw cycles. After the completion of 150 cycles of 4 h freeze - 4 h thaw, the samples were tested for compression (Fig. 3).

The variation of compressive strength (ŋ) of the samples subjected to cycles was calculated using the following relationship:

$$h = \frac{R_m - R_t}{R_m} * 100 \quad (1)$$

ŋ = (Rm−Rt)/Rm × 100, where

R(m) – the arithmetic means of the compressive strengths of the control samples, N/mm^2.

R(t) – the arithmetic means of the compressive strengths of samples subjected to freeze-thaw cycles, N/mm^2.

Fig. 2. TWIN FEUTRON S:D-0795 climatic chamber

2.3 Results and Their Interpretation

Mechanical Properties of Green Road Concrete. For determining mechanical characteristics (f_{ti},, f_c and f_{td}) according to [23–25], the following formworks were used: cubes of 150 mm sizes and prisms of 150 mm × 150 mm × 600 mm. The samples were kept according to standard conditions until testing at 28 days [22].

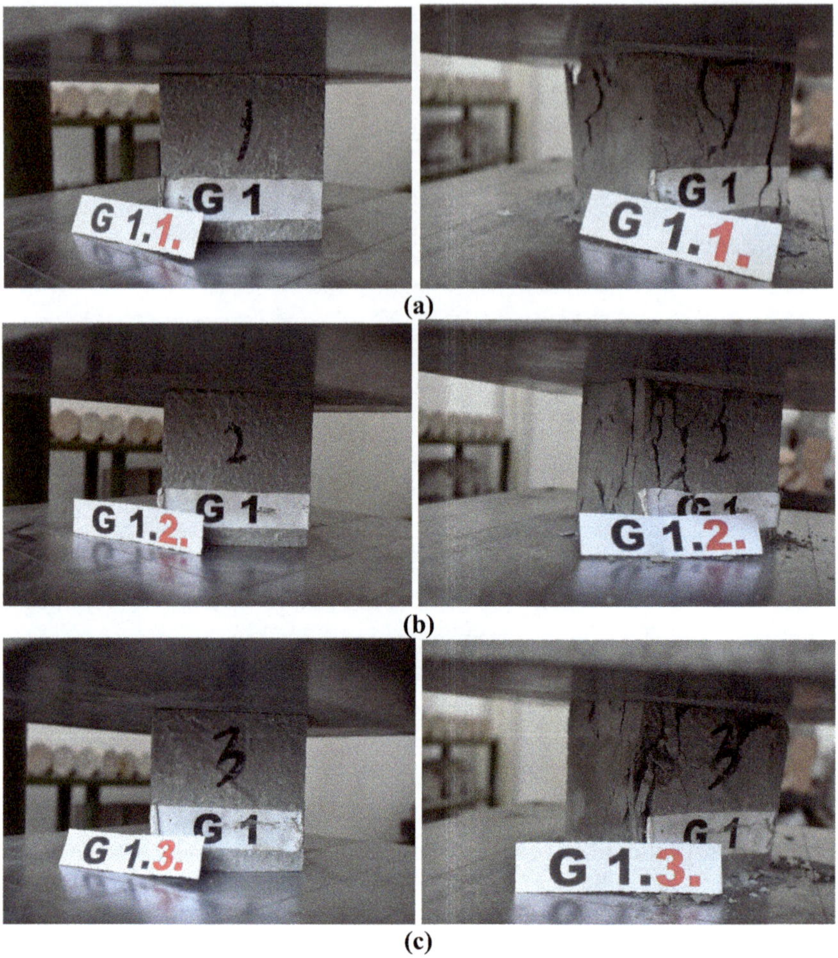

Fig. 3. Samples subjected to axial compression

The following results were obtained for green concrete road:

Flexural Strength

In Fig. 4 there are presented the values of flexural strength.

The highest value of flexural strength was obtained for mixture G0 which is the control mix. All samples presented values of f_{ti} smaller than the control mix. For sample with steel slag as substitute for aggregate 4–8 mm [kg/m^3] sample G3 with 30% steel slag obtained the highest value. For sample with steel slag as substitute for Aggregate 8–16 mm [kg/m^3] sample G6 with 30% steel slag obtained the highest value. For sample with steel slag as substitute for aggregate 8–22.5 mm [kg/m^3] sample G9 with 30% steel slag obtained the highest value.

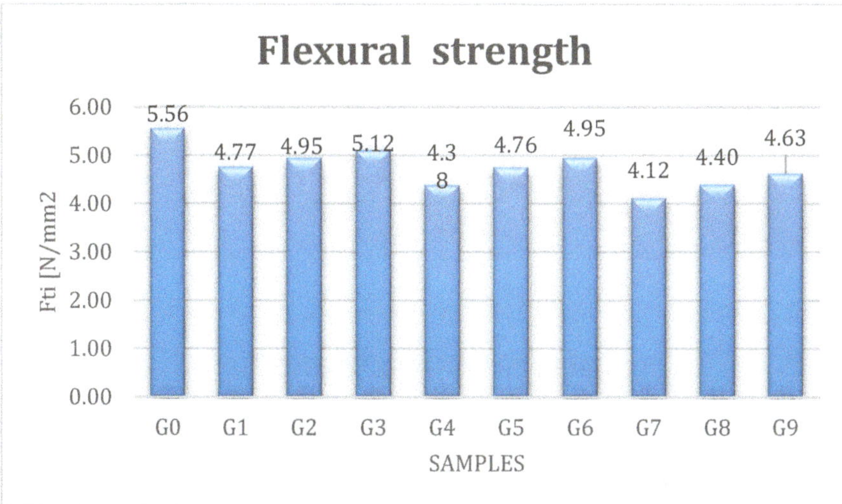

Fig. 4. Variation of flexural strength of green concrete

For the mix with 30% steel slag and 10% silica fume was obtained the best value of f_{ti}.

Split Tensile Strength

In Fig. 5 there are presented the values of split tensile strength.

The highest value of split tensile strength was obtained for mixture G0 which is the control mix. For sample with steel slag as substitute for aggregate 4–8 mm [kg/m^3] sample G1 with 10% steel slag obtained the highest value. For sample with steel slag as substitute for aggregate 8–16 mm [kg/m^3] sample G4 with 10% steel slag obtained the highest value. For sample with steel slag as substitute for aggregate 8–22.5 mm [kg/m^3] sample G7 with 10% steel slag obtained the highest value.

For the mix with 10% steel slag and 10%silica fume was obtained the best value of f_{td}.

Compressive Strength

In Fig. 6 there are presented the values of compressive strength.

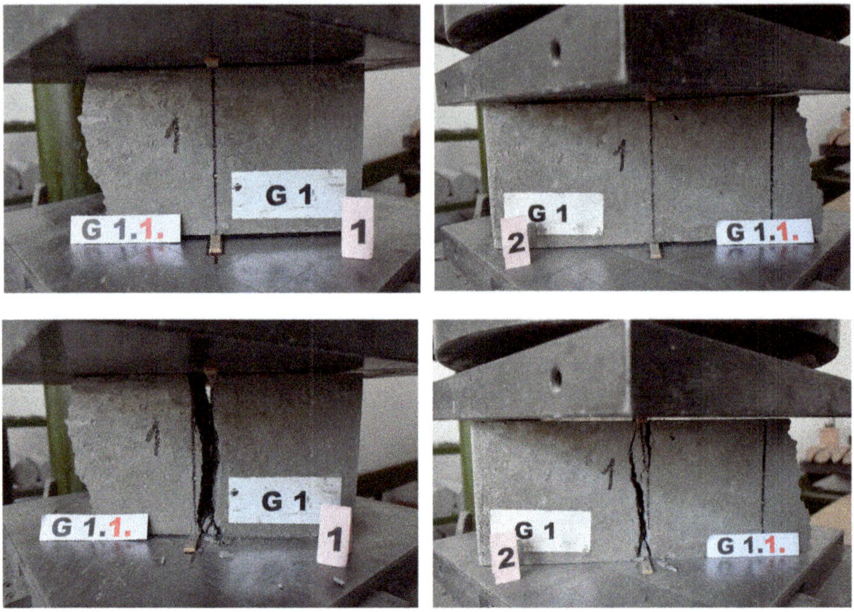

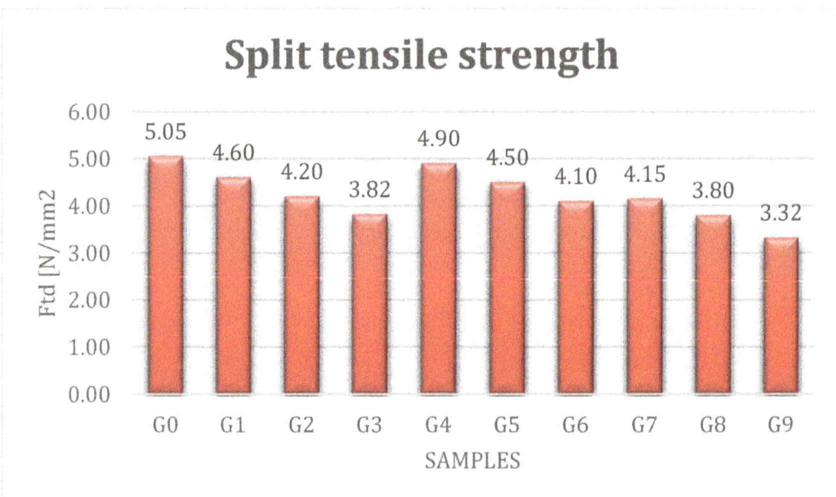

Fig. 5. Variation of split tensile strength of green concrete

The highest value of compressive strength was obtained for mixture G0 which is the control mix. For sample with steel slag as substitute for aggregate 4–8 mm [kg/m^3] sample G1 with 10% steel slag obtained the highest value. For sample with steel slag as substitute for aggregate 8–16 mm [kg/m^3] sample G4 with 10% steel slag obtained the highest value. For sample with steel slag as substitute for aggregate 8–22.5 mm [kg/m^3] sample G7 with 10% steel slag obtained the highest value.

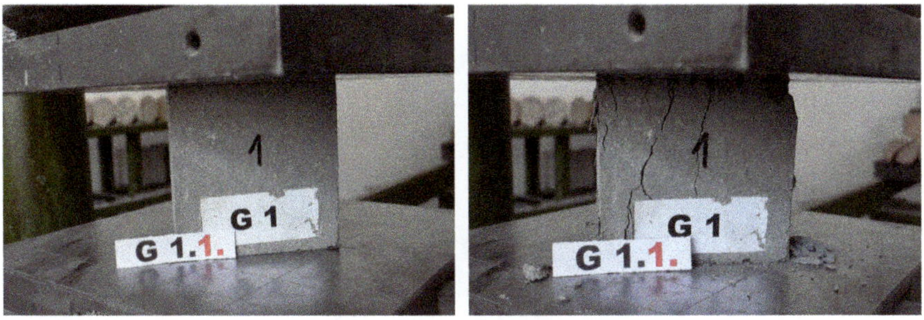

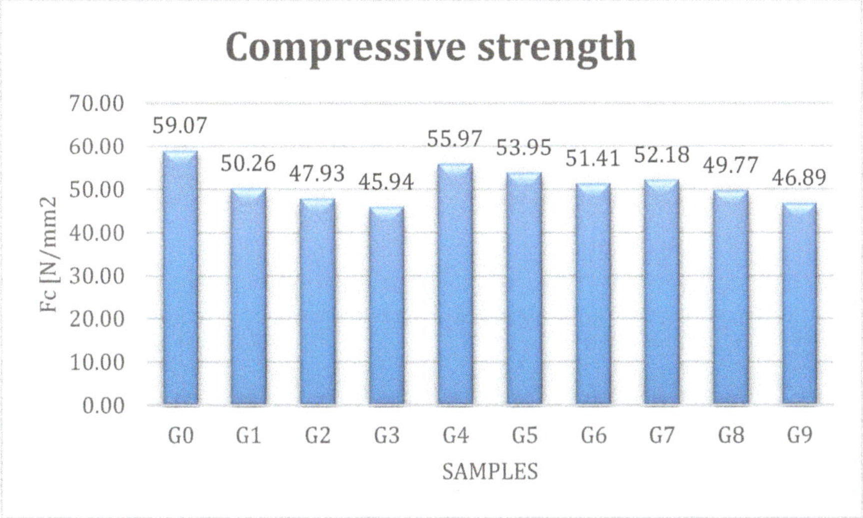

Fig. 6. Variation of compressive strength of green concrete

For the mix with 10% steel slag and 10%silica fume was obtained the best value of f_c.

Behavior to Frost-Thaw Cycles

(See Fig. 7 and Table 2).

The samples subjected to 150 freeze-thaw cycles were firstly analyzed as physical behavior, all the samples were without any deterioration of surfaces. The analyses of behavior in compression shown that after 150 cycles of freeze-thaw cycles the samples with steel slag and silica fume presented a higher value of f_c in comparison with the control mix.

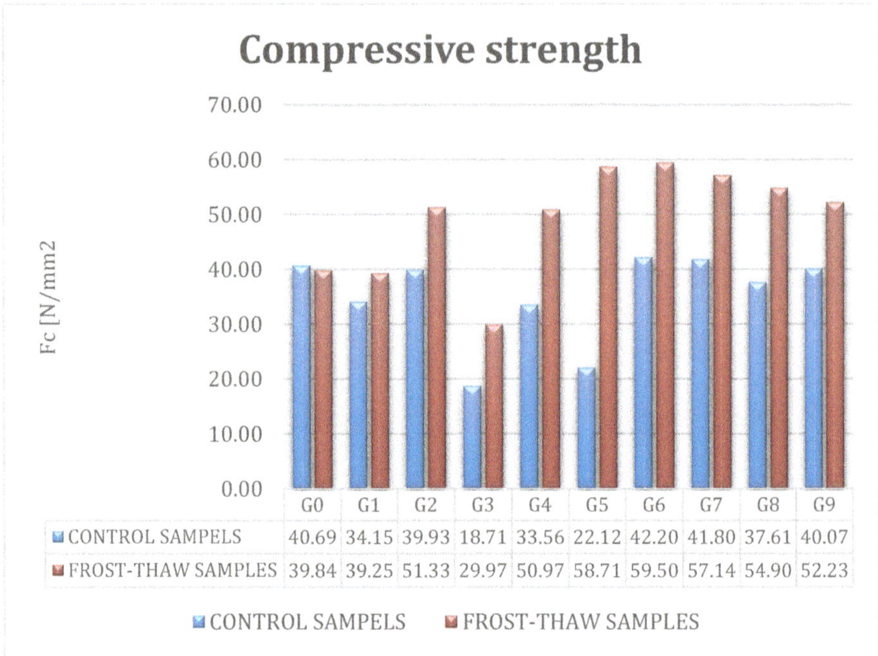

Fig. 7. Compressive strength of control concrete Rm and concrete subjected to freeze-thaw cycles

Table 2. Loss of compressive strength of concrete subjected to freeze-thaw cycles, η (%)

Sample	Control samples	Frost-thaw samples	η (%)
G0 [N/mm^2]	40,69	39,84	2,09
G1 [N/mm^2]	34,15	39,25	−14,93
G2 [N/mm^2]	39,93	51,33	−28,55
G3 [N/mm^2]	18,71	29,97	−60,18
G4 [N/mm^2]	33,56	50,97	−51,88
G5 [N/mm^2]	22,12	58,71	−165,42
G6 [N/mm^2]	42,2	59,5	−41
G7 [N/mm^2]	41,8	57,14	−36,7
G8 [N/mm^2]	37,61	54,9	−45,97
G9 [N/mm^2]	40,07	52,23	−30,35

3 Conclusions

The paper analyses the mechanical characteristics of road concrete prepared with steel slag as substitution of aggregate and silica fume as substitution of cement. All mechanical characteristics presented values smaller than that of the control mix. For flexural strength, the mix with replacement of sort 4–8 mm with 30% steel slag presented the best value. For splitting and compression, the mix with replacement of sort 8–16 mm with 10% steel slag presented the best value. To frost-thaw cycles al the mixes with steel slag and silica fume presented a better behavior under 150 cycles.

The green concrete road, prepared with steel slag as aggregate substitution and silica fume as replacement of cement is indicated to be use in road construction. Supplementary researches are necessary for establishing the other characteristics of green road concrete and economical efficiency. The green road concrete is a promising way to help the environment and to reduce the energy consuming and pollution.

Acknowledgements. This paper was supported by the Project "Network of excellence in applied research and innovation for doctoral and postdoctoral programs"/InoHubDoc, project co-funded by the European Social Fund financing agreement no. POCU/993/6/13/153437.

References

1. Cass, D., Mukherjee, A.: Calculation of greenhouse gas emissions for highway construction operations by using a hybrid life-cycle assessment approach: case study for pavement operations. J. Constr. Eng. Manag. **137**, 1015–1025 (2011)
2. Diaz Piloneta, M., Terrados-Cristos, M., Alvarez-Cabal, J.V., Vergara-Gonzales, E.: Comprehensive analysis of steel slag as aggregate for road construction: experimental testing and environmental impact assessment. Materials **14**(13), 3587 (2021). https://doi.org/10.3390/ma14133587
3. Jiang, R., Wu, P.: Estimation of environmental impacts of roads through life cycle assessment: a critical review and future directions. Transp. Res. Part Transp. Environ. **77**, 148–163 (2019)
4. Bonoli, A.; Degli Esposti, A., Magrini, C.: A case study of industrial symbiosis to reduce GHG emissions: performance analysis and LCA of asphalt concretes made with RAP aggregates and steel slags. Front. Mater. (2020)
5. Kannur, B., Chore, H.S.: Strength and durability study of low-fines self-consolidating concrete as a pavement material using fly ash and bagasse ash. Eur. J. Environ. Civ. Eng. **27**(11) (2023)
6. Bagaria, A., Juneja, D.: Experimental research on influence of marble powder, silica fume and polypropylene fiber on the porous concrete. Mater. Today Proc. (2023). https://doi.org/10.1016/j.matpr.03.240
7. Padhan, S.K., Behera, N.: Performance assessment of pervious concrete road on strength and permeability by using silica fume. Mater. Today **60**(Part 1), 559–568 (2022). https://doi.org/10.1016/j.matpr.2022.02.018
8. Dondi, G., Mazzotta, F., Lantieri, C., Cuppi, F., Vignali, V., Sangiovanni, C.: Use of steel slag as an alternative to aggregate and filler in road pavements. Materials **14**(2), 345 (2021). https://doi.org/10.3390/ma14020345
9. Chen, R., Cai, G., Dong, X., Pu, S., Dai, X., Duan, W.: Green utilization of modified biomass by-product rice husk ash: a novel eco-friendly binder for stabilizing waste clay as road material. J. Clean. Prod. **376**, 134303 (2022)
10. Toma, U.T., Podder, J., Sadia, K.N., Rahman, M.W.: Development of artificial aggregates using GGBS, fly ash and waste vegetable oil as sub-base material for road construction. In: 6th International Conference on Advances in Civil Engineering (ICACE-2022) CUET, Chattogram, Bangladesh (2022)

11. Loureiro, C.L.A., et al.: Steel slag and recycled concrete aggregates: replacing quarries to supply sustainable materials for the asphalt paving industry. Sustainability **14**(9), 5022 (2022). https://doi.org/10.3390/su14095022
12. Jiarui, G., Liu, X., Zhang, Z.: Road base materials prepared by multi-industrial solid wastes in China: a review. Constr. Build. Mater. **373**, 130860 (2023)
13. Dawodu, P.O., Apata, A.C., Imoh, U.U., Akintunde, F.: Comparative study of cement replacement with waste plastic in interlocking paving stone for highway construction in Nigeria. J. Eng. Res. Rep. **24**(1), 20–29 (2023). Article no. JERR.95278. ISSN: 2582-2926
14. Faldessai, K., Lawande, S., Kelekar, A., Gurav, R., Kakodkar, S.: Utilization of ceramic waste as a partial replacement for cement in concrete manufacturing. Mater. Proc. (2023). https://doi.org/10.1016/j.matpr.2023.06.453
15. Pandey, A., Kumar, B.: Utilization of agricultural and industrial waste as replacement of cement in pavement quality concrete: a review. Environ. Sci. Pollut. Res. **29**, 24504–24546 (2022)
16. Yu, H., Dong, B., Liu, Z.: Effects of fiber reinforcement on the frost resistance of cementitious composites containing fly ash and silica fume. Constr. Build. Mater. **273**, 121736 (2021)
17. Wang, Q., Li, Z., Li, C., Li, Y.: The effects of fly ash and silica fume on the durability of concrete subjected to freeze-thaw cycles. Materials **13**(16), 3547 (2020)
18. Wang, D., Zhou, X., Meng, Y., Chen, Z.: Durability of concrete containing fly ash and silica fume against combined freezing-thawing and sulfate attack (2023)
19. Wang, J., Cheng, Q., Liu, J., Liu, C., Zhang, X.: Effects of Ultrafine Silica Additives on Freeze-Thaw Durability of Concrete (2018)
20. Li, Q., Zhang, P., Qin, J., Ma, M., Zhang, J., Wang, Y.: Durability of Concrete with Ultrafine Silica Additives under Freeze-Thaw Action (2020)
21. SREN 13263-1:2009 Silica fume for concrete. Part 1: Definition, specifications and conformity criteria
22. SR EN 12390-2:2019 Romanian Standard Association, Testing hardened concrete. Part 2: Making and curing specimens for strength tests
23. SR EN 12390-3:2005 Romanian Standard Association, Testing hardened concrete. Part 3: Compressive strength of test specimens
24. SR EN 12390-5:2005 Romanian Standard Association, Testing hardened concrete. Part 5: Flexural strength of test specimens
25. SR EN 12390-6:2010 Romanian Standard Association, Testing hardened concrete. Part 6: Split tensile strength of test specimens
26. SR EN 3518:2009 Romanian Standard Association, Tests on concrete. Determination of freeze-thaw resistance by measuring the variation of compressive strength and/or dynamic modulus of elasticity, Bucharest, Romania

Lightweight Concrete with Wastes Used for Hollow Concrete Blocks

L. G. Hanuseac, G. Bejan, P. Mihai, M. Bărbuță, A. Panțiru(✉), and R. G. Mihai

Technical University "Gheorghe Asachi" from Iassy, Faculty of Civil Engineering and Building Services, 1 Prof. Dimitrie Mangeron Blv., 700050 Iassy, Romania
alexandru.pantiru@student.tuiasi.ro

Abstract. In the article are presented the experimental and numerical researches on concrete blocks realized of green lightweight concrete. For preparing the hollow blocks the light weight concrete with wastes type chopped plastic bottles (PET), sawdust and fly ash was used. Experimental study on physical-mechanical and deformation characteristics of green lightweight concrete were done. The hollow blocks were tested in compression. Failure of blocks was ductile for both types of concrete, the blocks of lightweight concrete with saw dust presented a major degradation. The numerical analyses performed with ATENA software indicated a similar type of failure as that from the tests that indicates a good correlation of results.

Keywords: Fly ash · Lightweight concrete · Hollow concrete block

1 Introduction

Pollution is constantly increasing. The construction industry is one of the most polluting industries. And the most polluting and used material is concrete, which has cement in its composition. This information is not new, but the methods and trials aimed at reducing CO_2 are always innovative and welcome [1–3]. The building materials industry has researched and developed new types of materials, aiming to be more sustainable and environmentally friendly [4–6]. Concrete products have diversified, in the concrete mix using different materials, as a replacement of cement and aggregates, or as additions. Materials of different natures have been replaced: natural, industrial products, waste, and the replacement of cement is impetuously necessary, since in the production of this material a huge amount of CO_2 is released into the atmosphere. Cement was replaced by various powdery waste, such as: silica smoke, fly ash, blast furnace, steel slag, glass powder, vegetable ash, etc. [7–9]. It was studied how natural aggregates can be replaced with waste, and whether the finished product is suitable for large-scale use, so in concrete recipes appeared compounds such as: slag, demolished waste, shredded plastic, glass, crushed tire, core, polystyrene granules, vegetable aggregates, etc. [10–16]. Studies have found that lightweight concrete is often used for non-structural elements of buildings. These concretes have lower production cost, better thermal and acoustic characteristics than standard concrete, low weight, easy handling, etc. [17, 18]. Moreover, they have

I. Tuns et al. (Eds.): CIBv 2023, LNCE 510, pp. 37–49, 2024.
https://doi.org/10.1007/978-3-031-60765-3_4

solved part of the big problem of waste storage, because some of these concretes have waste in their recipes [19, 20]. Lightweight concrete can be prepared with a lot of types of wastes which replace cement or aggregates [21–23].

The paper analyses the behavior of hollow blocks manufactured with lightweight concrete prepared by replacing natural aggregates with wastes of chopped plastic bottles and sawdust in a percentage of 70% and experimentally tested in axial compression.

2 Materials and Methods

2.1 Materials

For preparing the hollow blocks the light weight concrete with wastes type chopped plastic bottles (PET), sawdust and fly ash was used. The lightweight concrete was prepared using a mix of ordinary concrete of grade C20/25. The following components of concrete C20/25 were used:

- cement type CEM II 42,5 -- 360 kg/m^3;
- river aggregates: sand sort 0–4 mm – 803.16 kg/m^3;
- sand sort 4–8 mm – 384.12 kg/m^3;
- sand sort 8–16 mm – 558.72 kg/m^3;
- water 180 l;
- Master Glenium SKY 617 superplasticizer/water reducing additive in a dosage of 1% from cement's amount

The receipts are presented in Table 1.

Table 1. Receipts BCPET and BCRU with thermo power plant ash.

Receipt	BCPET	BCRU
Cement [kg/m^3]	324	324
Water [l/m^3]	180	180
Thermo power plant ash [kg/m^3]	36	36
Sand [kg/m^3]	241	241
Shredded plastic	562	0
Sawdust	0	562
Aggregate [kg/m^3]	384	384

(continued)

Table 1. *(continued)*

Receipt	BCPET	BCRU
Aggregate [kg/m^3]	558	558
Additive [kg/m^3]	3.24	3.24

The sort 0–4 mm was replaced 70% in volume with wastes type chopped plastic bottles (PET) and sawdust which had been presented in previous article [25]. The lightweight concrete was noted BCPET, in the case of replacement of sand with shredded plastic bottles and BCRU, in the case of replacement of sand with sawdust. The fly ash from Electric Power Plant Holboca Iasi, that was characterized in previous research [26] was used in a quantity of 36 kg/m^3, as 10% replacement of cement. The water was 180 l/m^3. Superplasticizer type Master Glenium Sky was used, 1% from the cement dosage.

2.2 Experimental Set up

The physical characteristics of lightweight concrete with shredded plastic and sawdust were determined according to European standards [27] and the experimental test shown a density of concretes smaller than 2000 kg/m^3, that characterizes a lightweight concrete. The compressive strength, f_c was determined according to European standard [28].

For pouring the lightweight concrete blocks the formwork of sizes L = 360 mm, l = 180 mm and H = 190 mm was realized, Fig. 1 and Fig. 2.

Fig. 1. The formwork

Fig. 2. Lightweight concrete pouring

The shape and sizes of blocks of lightweight concrete are presented in Fig. 3.

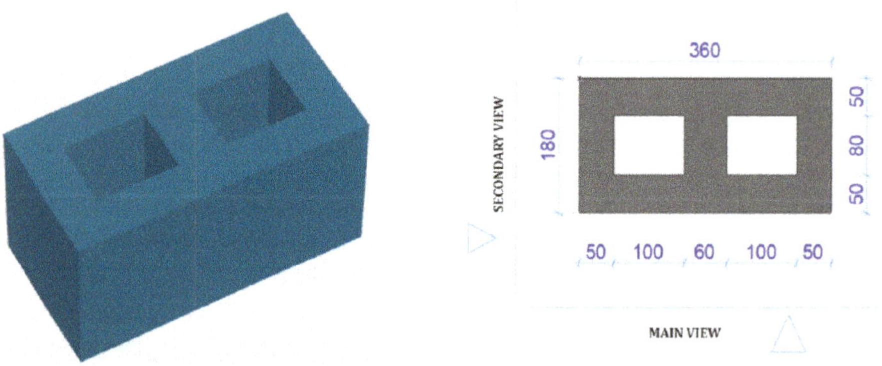

Fig. 3. Hollow block of lightweight concrete with wastes

3 Results and Discussion

3.1 Compressive Strength of Concrete with Wastes

The values of compressive strengths for lightweight concrete with shredded plastic bottles and saw dust are presented in Table 2.

Table 2. Experimental values of compressive strength and density

Sample	f_c, [MPa]	Density [kg/m^3]
BCPET	20.41	1997
BCRU	11.04	1894

According to the results observed in Table 2, the concrete with chopped PET (BCPET) presented a higher value of *compressive strength* than the concrete with sawdust (BCRU), for the same dosage of aggregate replacement.

The density of concrete with PET was bigger than that of concrete with sawdust for the same dosage of aggregate replacement. Both values were smaller than 2000 kg/m^3 that signifys a lightweight concrete.

3.2 Stress-Strain Behaviour of Concrete with Waste

The deformation and elasticity characteristics for lightweight concrete were determined according to [29], on 3 cylinders of 100 mm diameter and 200 mm length.

In Fig. 4 the force-displacement curves of lightweight concrete BCPET and BCRU are presented.

In Fig. 4 can be observed that the diagrams are similar to that of ordinary concrete. The curves are non-linear from the first stage of loading until the maximum value of the load. After that, was registered a decrease of force and non-elastic behavior for concrete with PET and a more elastic behavior for concrete with sawdust. The deformations at failure are approximately the same for both types of concretes with wastes, still a higher deformation can be seen in the concrete with chopped PET.

3.3 Modulus of Elasticity of Concrete with Wastes

For lightweight concrete BCPET and BCRU, the elasticity modulus was determined in accordance with [29] and presented the following values:

- for BCPET the value was $E_{c1} = 12893$ N/mm^2
- for BCRU the value was $E_{c2} = 14326$ N/mm^2

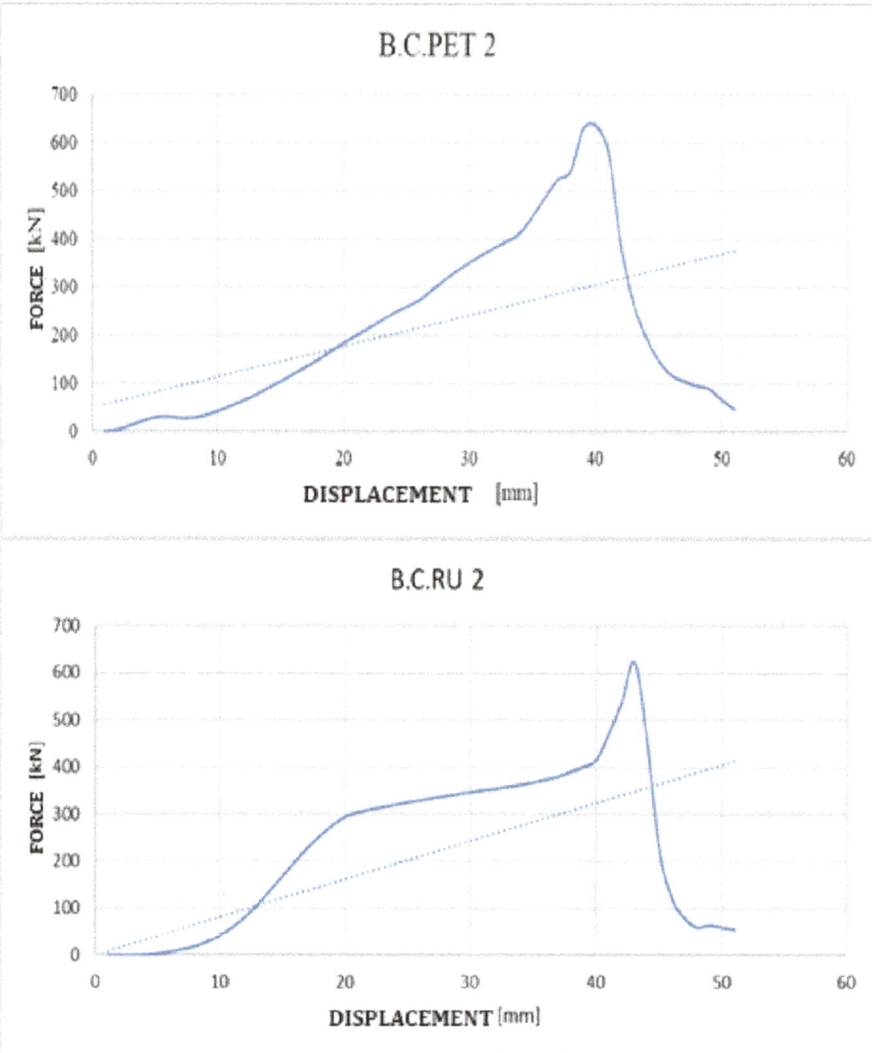

Fig. 4. Compressive force-displacement diagram of lightweight concrete with wastes

3.4 Hollow Blocks Experimental Test

The blocks from Fig. 5 were subjected to axial compression at the hydraulic machine of 3000 KN. The data registration was done with device ADLINK which was connected with a displacement traductor that measures a deformation between 0 to 100 mm.

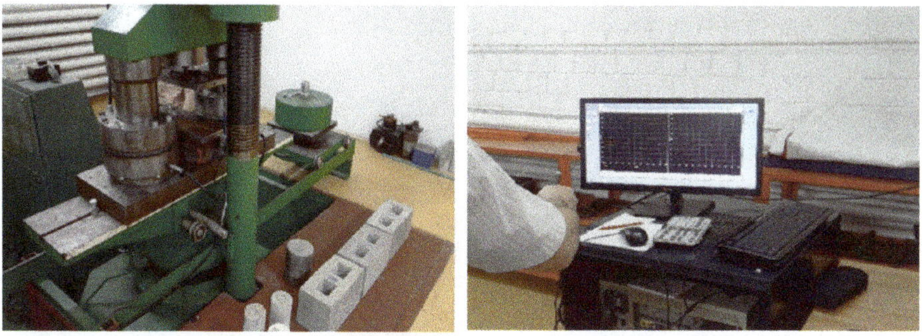

Fig. 5. Testing lightweight concrete blocks

Behavior of blocks of lightweight concrete with BCPET in compression is presented in Fig. 6.

Fig. 6. Experimental test of hollow blocks of lightweight concrete BCPET

The failure type of the block BCPET was gradual, by forming the first vertical cracks in central zone and near corners. With the increasing of the loads, the cracks continued to developed, until the failure, especially at the top of the block.

Behavior of lightweight concrete BCRU in compression is presented in Fig. 7.

Fig. 7. Experimental test of hollow blocks of lightweight concrete BCRU

The failure type of block BCRU was gradual, the first horizontal cracks occurred at the top and vertical cracks occurred near corners. The failure was gradual, with vertical cracks of high width which resulted in the totally degradation of edge lightweight concrete.

In Table 2 are presented the experimental values obtained for hollow blocks of lightweight concrete with waste after testing in axial compression. The maximum value of the load was divided by the gross contact area of the hollow block, including holes and was noted f_{chb1}, and by the net contact area, noted f_{chb2}.

The indirect tension test was calculated with relation (1):

$$f_{thb} = \frac{2P}{\pi LH} \tag{1}$$

where: P is the value of the maximum load, H is the height of the block (190 mm), L is the split length (160 mm) if the holes are not neglected, or 360 mm if is considered the total length (Table 3).

Analyzing the experimental results, it can observe that *compressive strength* f_{chb} of blocks of lightweight concrete is influenced by the type of waste, but the values are closed. A higher value of compressive strength was obtained in the case of blocks with shredded plastic bottles (BCPET). According to [30] the minimum compressive strength must be 7 N/mm^2 and all the blocks satisfy this condition for their use for masonry in seismic areas.

The split tensile strength f_{thb} of blocks of lightweight concrete is also influenced by the type of waste: the blocks type BCPET, for both tensile strength values presented higher values than that for blocks with sawdust, but also the values are very closed.

Table 3. Experimental results of the test on hollow blocks

No.	Sample	Sizes of Block mm	Sizes of holes mm	Maximum force kN	f_{chb1} N/mm^2	f_{chb2} N/mm^2	f_{thb1} N/mm^2	f_{thb2} N/mm^2
1	BCPET1	360 × 180 × 190 mm	100 × 80 × 190	637.2	9.83	13.06	5.93	13.35
2	BCPET2	360 × 180 × 190 mm	100 × 80 × 190	633.6	9.77	12.98	5.90	13.27
3	BCRU1	360 × 180 × 190 mm	100 × 80 × 190	625.2	9.65	12.81	5.82	13.10
4	BCRU2	360 × 180 × 190 mm	100 × 80 × 190	621.9	9.59	12.74	5.79	13.03

The mechanical characteristics experimentally determined are in concordance with those given in the literature for hollow blocks [22, 25, 30, 31].

3.5 Numerical Simulation of Blocks Behavior in Compression

A program ATENA was used for determining the stress distribution in the blocks during the loading. The cracks occurrence was observed during the simulation. The simulation of blocks and the stress distribution for the tested blocks models, Fig. 8 are presented in Fig. 9 and Fig. 10. The program used mechanical and deformation characteristics of lightweight concrete.

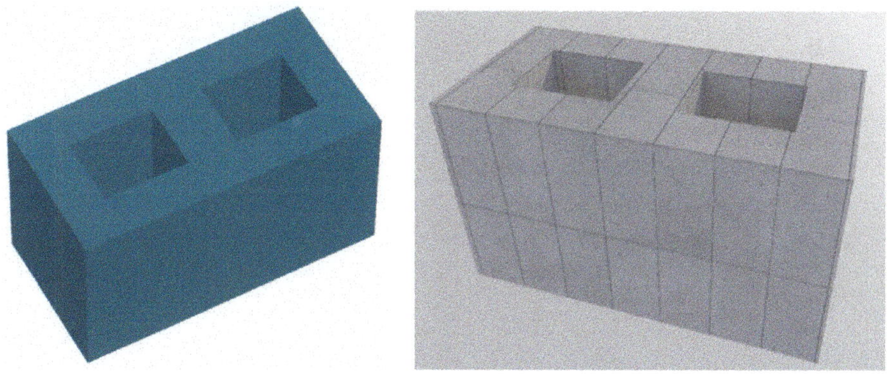

Fig. 8. FEM model in ATENA software

A good correlation was obtained between numerical simulation and experimental tests: the pattern of cracks and failure type were similar to that given by numerical simulation.

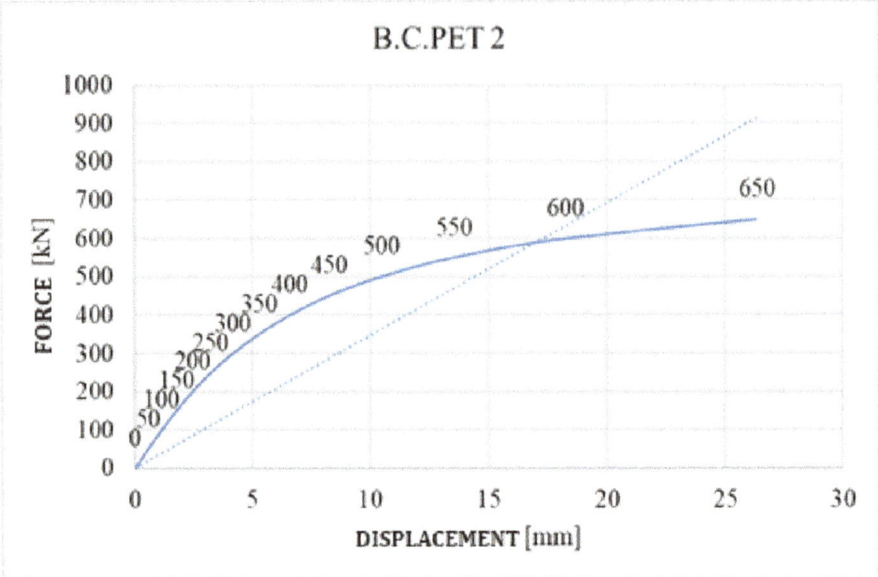

a)

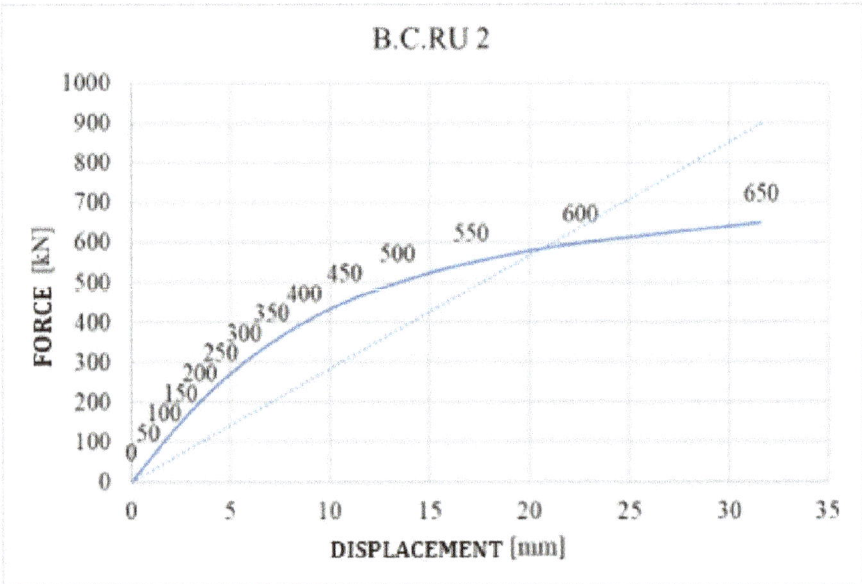

b)

Fig. 9. Simulation of behavior in compression of block BCPET (a) and BCRU (b)

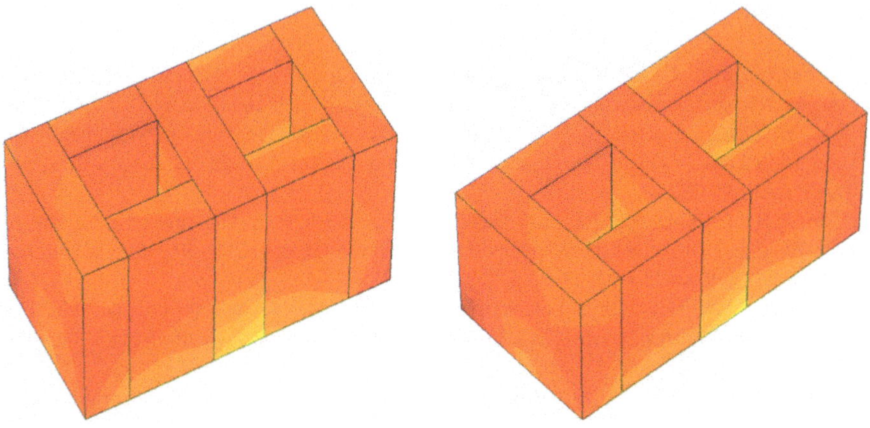

a) b)

Fig. 10. Stresses distribution in the block BCPET (a) and BCRU (b)

4 Conclusions

In the experimental research the authors studied blocks of green lightweight concrete that was prepared with fly ash, shredded plastic bottles and saw dust as replacement of aggregate 0–4 mm. Fly ash was used as replacement to 10% of cement. The wastes replaced 70% of the aggregates 0–4 mm in volume and so the natural resources of rocks are protected.

The experimental tests on blocks of lightweight concrete subjected to axial compression prepared with wastes indicated the following conclusions:

- The density of lightweight concrete BCRU was smaller than that of concrete BCPET.
- Both types of blocks were subjected to axial compression. The blocks of BCPET presented a higher value of compressive stress in comparison with the block of BCRU.
- The split tensile stress of blocks presented higher values in the case of blocks with BCPET than in the case of blocks of BCRU.
- The results of numerical simulation with ATENA software for both types of blocks are in good correlation with the experimental results, the errors are under 7%.

The values of mechanical characteristics of lightweight concrete blocks experimentally obtained shown that they satisfy the requirements necessary for realizing self-weight masonry walls. Further researches on blocks and masonry walls will continue to contribute to the knowledge of these green construction materials and their practical application.

Acknowledgements. This paper was supported by the Project "Network of excellence in applied research and innovation for doctoral and postdoctoral programs"/InoHubDoc, project co-funded by the European Social Fund financing agreement no. POCU/993/6/13/153437.

References

1. De Maeijer, P.K., Craeye, B., Snellings, R., Kazemi-Kamyab, H.: Effect of ultra-fine fly ash on concrete performance and durability. Constr. Build. Mater. **263** (2020)
2. Liu, Y., Zhang, Z., Hou, G., Yan, P.: Preparation of sustainable and green cement-based composite binders with high-volume steel slag powder and ultrafine blast furnace slag powder. J. Clean. Prod. **289** (2021)
3. Ganesh Babu, O., Danya, T.R., Sakthieswaran, N., Suresh, P., Muthuraman, U.: Sustainable characteristics of fly ash based geopolymer concrete incorporating alccofine, zeolite and rubber fibers. Roman. J. Mater. **51**(1), 17–24 (2021)
4. Bayraktar, O.Y.: Possibilities of disposing silica fume and waste glass powder, which are environmental wastes, by using as a substitute for Portland cement. Environ. Sci. Pollut. Res. **28** (2021)
5. Verma, M., Dev, N.: Effect of ground granulated blast furnace slag and fly ash ratio and the curing conditions on the mechanical properties of geopolymer concrete. Struct. Concr. 1–15 (2021)
6. Kolawole, T., Olusola, K.O., Babafemi, A.J., Olalusi, O.B., Fanijo, E.: Blended cement binders containing bamboo leaf ash and ground clay brick waste for sustainable concrete. Materialia **15** (2021)
7. Das Bheel, N., Jokhio, M.A., Abbasi, J.A., Lashari, H.B., Qureshi, M.I., Qureshi, A.S.: Rice husk ash and fly ash effects on the mechanical properties of concrete. Eng. Technol. Appl. Sci. Res. **10**(2), 5402–5405 (2020)
8. Dhage, B., Rathi, V., Kolase, P.: Experimental study on partial replacement of cement by banana leaves ash. Int. J. Anal. Exp. Modal Anal. **12**(9), 1869–1876 (2020)
9. Charitha, V., Athira, V.S., Jittin, V., Bahurudeen, A., Nanthagopalan, P.: Use of different agro-waste ashes in concrete for effective upcycling of locally available resources. Constr. Build. Mater. **285** (2021)
10. Papachristoforou, M., Anastasiou, E.K., Papayianni, I.: Durability of steel fiber reinforced concrete with coarse steel slag aggregates including performance at elevated temperatures. Constr. Build. Mater. **262** (2020)
11. Dong, Q., Wang, G., Chen, X., Tan, J., Gu, X.: Recycling of steel slag aggregate in Portland cement concrete: an overview. J. Clean. Prod. **282** (2021)
12. Hassan-Ghasemi, A., Nematzadeh, M.: Tensile and compressive behaviour of self-compacting concrete incorporating PET as fine aggregate substitution after thermal exposure: experiments and modelling. Constr. Build. Mater. **289** (2021)
13. Sharma, D.K., Kumar, N., Kumar, R.: A research paper on partial replacement of coarse aggregate by waste tires in cement concrete. EPRA Int. J. Multidiscip. Res. (IJMR) **7**(5), 19–28 (2021)
14. Ştefan (Oancea), I., Bărbuţă, M., Budescu, M., Mihai, P., Banu, O.M., Ţăranu, N.: Particularities regarding the mechanical behaviour of some types of sustainable concrete mixes with waste materials. Roman. J. Mater. **48**(2), 236–244 (2018)
15. Rattanachu, P., Toolkasikorn, P., Tangchirapat, W., Chindaprasirt, P., Jaturapitakkul, C.: Performance of recycled aggregate concrete with rice husk ash as cement binder. Cem. Concrete Compos. **108** (2020)
16. Gradinaru, C.M., Barbuta, M., Ciocan, V., Serbanoiu, A.A.: A study on the effects of the cement and mineral aggregates replacement with waste materials. IOP Conf. Ser.: Mater. Sci. Eng. **399**, 012020 (2018). https://doi.org/10.1088/1757-899X/399/1/012020
17. Zoum, K.S., Moon, J., Cho, J.-Y., Kim, J.J.: Experimental study on strength and durability of lightweight aggregate concrete containing silica fume. Constr. Build. Mater. **114**, 517–527 (2016)

18. Risdianto, Y.: Addition of fly ash and aluminium slag as cement substitute materials to cellular lightweight concrete. Int. J. Eng. Bus. Soc. Sci. **1**(03), 248–256 (2023)
19. Chen, P., Wang, Z., Cao, S., Rong, X., Shi, Z., Wang, H.: Study on axial compressive stress-strain relationship of alkali-activated slag lightweight aggregate concrete. Constr. Build. Mater. **364**, 129991 (2023)
20. Guo, Z., Wang, H., Sun, S., Feng, S., Shu, L., Tang, C.: Improvement mechanism of the mechanical properties and pore structure of rubber lightweight aggregate concrete with S95 slag. Sustainability **15**(2), 1008 (2023). https://doi.org/10.3390/su15021008
21. Šeputytė-Jucikė, J., Vėjelis, S., Kizinievič, V., Kairytė, A., Vaitkus, S.: The effect of expanded glass and crushed expanded polystyrene on the performance characteristics of lightweight concrete. Appl. Sci. **13**(7), 4188 (2023). https://doi.org/10.3390/app13074188
22. Hanuseac, L., Barbuta, M., Bejan, L., Rosu, R., Timu, A.: Experimental study on hollow blocks with wastes. In: Proceedings 14th International Conference on Interdisciplinarity in Engineering—INTER-ENG, Târgu Mureş, Romania, vol. 63, pp. 79–86 (2020)
23. Brito, L.B.Q., Agrawal, P., de Mélo, T.J.A., de Figueiredo Brito, G., da Silva Morais, C.R.: Recycling of waste glass for the production of hollow blocks using the kiln-casting process. Clean. Waste Syst. **4**, 100079 (2023)
24. SR EN 197-1:2011. Cement, Part 1: Composition, specifications and conformity criteria for common cements. Romanian Standard Association, Bucharest, Romania
25. Barbuţă, M., Mihai, P., Nicuţă, A.M., Bejan, L., Tăranu, N., Banu, O.M.: Hollow concrete blocks with wastes replacing the natural aggregates. Roman. J. Mater. **52**(1), 83–89 (2022)
26. Barbuta, M., Marin, E., Campeanu, S.M., Paraschiv, G., Lepadatu, D., Bucur, R.D.: Statistical analysis of the tensile strength of coal fly ash concrete with fibers using central composite design. Adv. Mater. Sci. Eng. **15**, 1–7 (2015)
27. SR EN 12390-7:2005. Testing hardened concrete. Part 2: Density of hardened concrete. Romanian Standard Association, Bucharest, Romania (2005)
28. SR EN 12390-3:2011. Testing hardened concrete. Part 3: Compressive strength of test specimens. Romanian Standard Association, Bucharest, Romania (2011)
29. SR EN 13412:2007. Determination of modulus of elasticity in compression. Romanian Standard Association, Bucharest, Romania (2007)
30. Marathe, S., Mithanthaya, I.R., Shetty, S.: Strength behaviour of masonry blocks produced using green concrete. In: Das, B., Neithalath, N. (eds.) Sustainable Construction and Building Materials. LNCE, vol. 25, pp. 33–40. Springer, Singapore (2019).https://doi.org/10.1007/978-981-13-3317-0_4
31. Qasim, M.F., Abbas, Z.K., Abed, S.K.: Production of load bearing concrete masonry units (blocks) from green concrete containing plastic waste and nano silica sand powder. J. Eng. **28**, 54–70 (2022)

Study of the Elastic Buckling Structural Behavior Using Finite Element Analysis

Ştefan Mocanu[1], Cristian Ştefan Dumitriu[1(✉)], and Ştefan Emil Ionescu[2]

[1] Faculty of Mechanical Engineering and Robotics for Construction, Technical University of Civil Engineering Bucharest, 59, Calea Plevnei, Bucharest, Romania
cristian.dumitriu@utcb.ro

[2] 'Ferdinand I' Technical Military Academy of Bucharest, 39 Bucureşti-Târgovişte str., Bucharest, Romania

Abstract. The study of slender structures in mechanical or civil/industrial constructions represents a continuous challenge from the view point of the risk factors identifying and offering sustainable solutions and methods to extend their life span. One of the most common and dangerous sources for the lifetime limitation of this type of structure is the buckling. The use of numerical algorithms in the applications for analyzing the structures, algorithms based on the finite element method, is a frequent solution in approaching the problem. Today, there are various software applications that have as their starting point the classic approach to the problem according to the Strength of Materials. The purpose of this article is to compare the classical approach with the output of running the working environments/platforms to establish the obtained solution convergence, highlighting the dispersion degree as well. The analyzed subject is the double joint section. Regardless the used software, the high degree of complexity of the problem requires a two steps algorithm: a buckling analysis in the linear domain followed by a nonlinear analysis. In addition, compared to other similar approached of the phenomenon, the use of the finite element with full options regarding the structure discretization (meshing) will be considered.

Keywords: Buckling · Finite Element Method · Simulation

1 Introduction

It is known that a structural element subjected to compression stress will have a bearing capacity directly proportional to the value of the geometric characteristics of the section (cross-sectional surface area, axial moment of inertia about the neutral axis of the section, etc.). This fact can lead to the tendency to create high and narrow profile sections within the design activity, which become unstable due to the loss of stability of the form of elastic equilibrium, i.e., buckling by compression.

The designers' decision to opt for slender structures is fully motivated taking into account both economic arguments, the emergence of new materials and technologies as well as the facilities offered by current computer systems. The realization of such

I. Tuns et al. (Eds.): CIBv 2023, LNCE 510, pp. 50–60, 2024.
https://doi.org/10.1007/978-3-031-60765-3_5

structures constitutes design themes in applications belonging to various industries, starting from civil constructions, the load-bearing elements of some industrial equipment, and up to the aerospace industry. The challenge consists in optimizing these structures under the conditions of their exploitation in safe conditions.

Given the diversity of the fields in which these structures are used, it is obvious that numerous researchers had approaches dedicated to the applications of the studies carried out. Starting from the treatment of stability phenomena [1], structures with thin walls [2–10], with solid sections [11], or with perforations [12, 13] are analyzed, and applications from different fields [14–19] as well as different materials (metallic or composite) [3, 20–23].

With the help of numerical algorithms based on the finite element method, this article proposes to approach the stability problems of slender structures subjected to compression. The aim is to obtain a high degree of confidence in the result (having as a reference the traditional analytical solutions) with decent computing resources. Regardless of the software application used [24, 25], finding solutions to an analysis problem using F.E.M. involves simultaneously solving several equations. Two large families of solving methods exist the direct and the iterative methods. The direct methods use inverse matrix type, LU decomposition, and Gauss-Jordon numerical algorithms.

The iterative methods are approximate calculation techniques for finding a solution, repeating the calculation sequence until the required degree of convergence is obtained. The high degree of complexity of the application also requires careful study regarding the set of initial conditions imposed as input sizes.

The primary purpose of the present article is to validate the analysis of linear and nonlinear simulation output using numerical and finite element methods against classical computational methods.

The solutions given by different authors are generally presented in a single situation, either theoretically or by applying FEM using specialized software. This article uses three computational solutions to assess the degree of convergence of the output simulation parameters covering linear and nonlinear approaches. Discussions and conclusions on the results are important observations of the theoretical and FEA techniques.

2 Problem Formulation

The aim of the study is to compute the safety factor to buckling phenomena using the analytical and the numerical approaches and to conclude about the dispersion of the obtained results. For this study has been chosen a simple slender structure represented by a double articulated structure of rectangular section (Fig. 1).

For the numerical example: P = 10 kN, l = 2 m and S235 JR steel grade material. The calculation method used is named *the coefficient buckling safety c* [1, 2].

3 Results and Discussion

3.1 Analytical Approach

The buckling domain for this example is determined by comparing the slender coefficient of the above structural beam λ with the slender nominal coefficient of the S235 material grade, λ_0^{S235}=105.

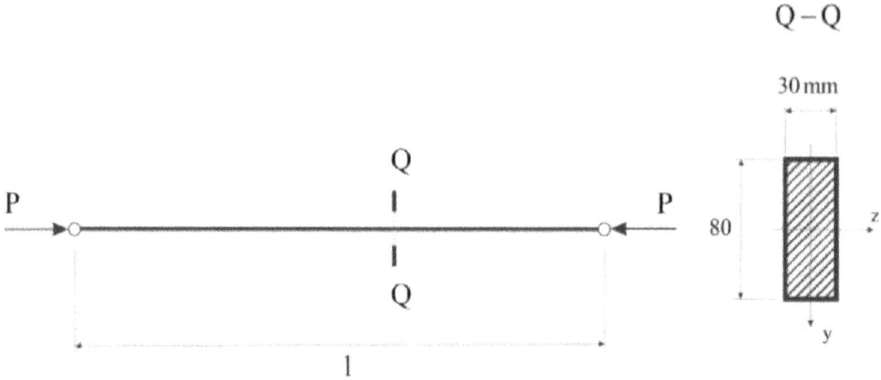

Fig. 1. Model of double articulated beam

The slenderness coefficient λ is determined as follows [3, 4].

$$\lambda = \frac{l_f}{i_{min}}$$

where:

l_f – buckling length. For a double articulated beam $l_f = l$.

i_{min} – minimum radius of inetia; $i_{min} = \sqrt{\frac{I_{min}}{A}}$

I_{min} – the minimum inertia momment of the bar section; A – sectional aria

$$\lambda = \frac{l}{\sqrt{\frac{I_{min}}{A}}} = 230.9$$

It is noted that$\lambda > \lambda_0^{S235}$,, so this example belongs to the elastic buckling domain (Euler buckling).

The effective buckling safety factor is calculated as:

$$c_{ef} = \frac{P_{cr}}{P} = \frac{\frac{\pi^2 \cdot E \cdot I_{min}}{l_f^2}}{P} = 9.33.$$

3.2 Numerical Approach

As specified in the first chapter, regardless of the exact type of software application or the platform/operating system used, the relatively high degree of complexity of the problem requires solving it in two steps. First is a buckling analysis in the linear domain, which will provide the estimated value of the critical buckling load. The second step is a nonlinear analysis, having as load the initial load multiplied by the previously established buckling safety coefficient [24].

This way of approach is because the linear buckling analysis estimates the critical buckling load (the buckling safety coefficient represents the load factor) for the undeformed shape of the structure. Against linear buckling analysis, the nonlinear buckling analysis will remake each linear calculation step considering the structures' deformation. Generic speaking, the nonlinear analysis will deal with the values of the structure's loading scheme, sequentially. For each loading step, the structure's stiffness matrix is recalculated at the corresponding state of deformation [25, 26].

In this study, the use of the finite element with complete options in terms of the discretization of the structure (meshing) will be considered. The simplified solutions of the beam-meshing type do not support nonlinear analyzes or any support schemes. The presented case, for example, is not being able to be modeled satisfactorily with the simplified beam-meshing solution.

The following present the solution approach using SolidWorks software. In the pre-processing step was built an assembly of two conjugate parts (Fig. 2) that allows modeling the desired buckling calculation scheme (the supporting scheme is the double articulated beam and the loading scheme is the centric compression).

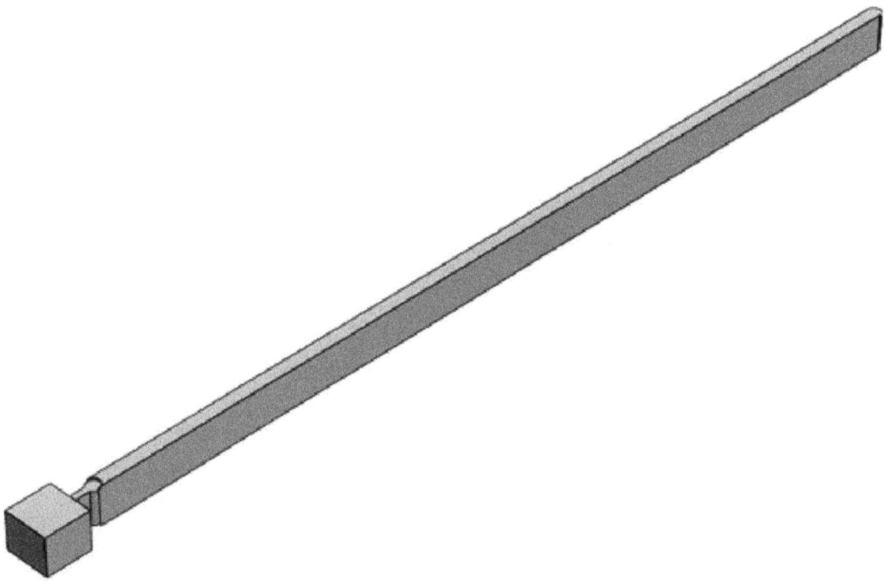

Fig. 2. Assembly of model in SolidWorks.

Carrying out the study itself involves making specific settings (Fig. 4) regarding the solving engine (solver-Direct Sparse) as well as the control method of the test steps (control-Arc Length), as well as the iteration method (iterative-Newton/Raphson), settings specific to the type of study desired [7, 8].

The analysis settings of the study will be completed by the type of connection between the two parts that build the assembly (Connectors – Pin Connector), allowing the degree of freedom in terms of rotation about the axis of the joint itself, as well as through the adequate choice of the support/loading scheme for the beam extremities as designed (Fig. 3).

Going through the successive stages of discretization (meshing generation - Fig. 5) and numerical analysis, the estimated values of the buckling safety coefficient are reached in the linear and nonlinear versions of the study (Figs. 6 and 7).

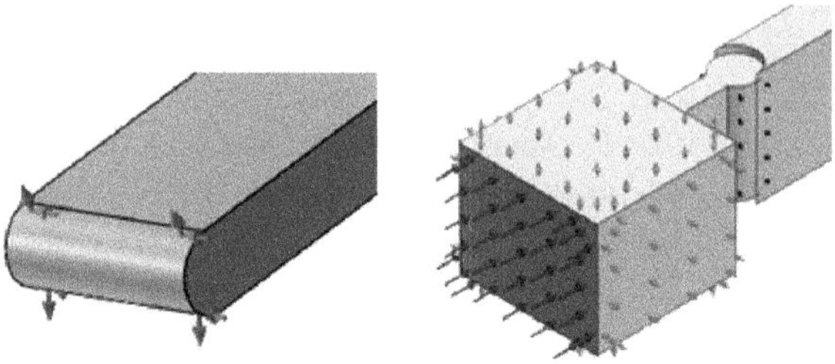

Fig. 3. Representation of connections and fixtures

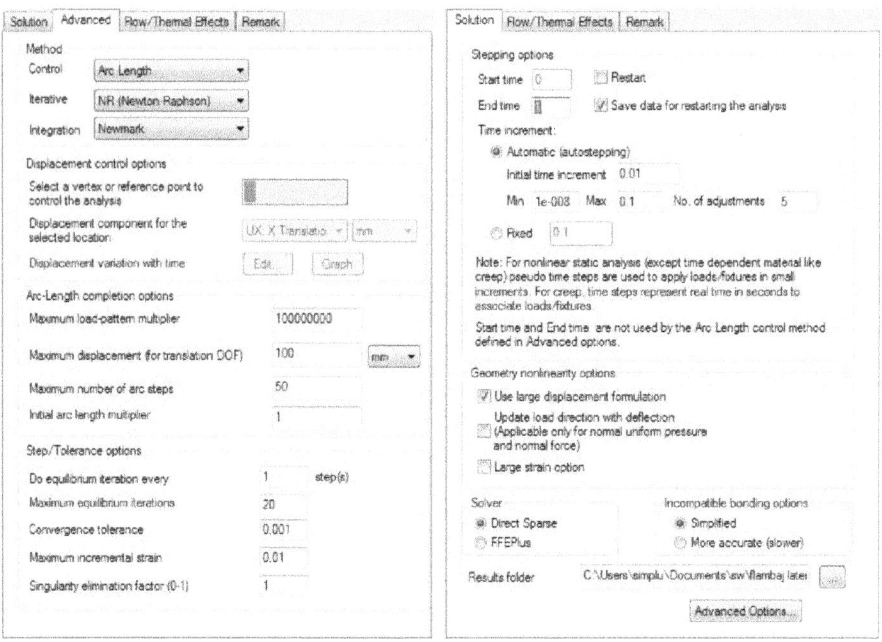

Fig. 4. Analysis settings parameters window

In this computing case, a direct solving algorithm is chosen, DSS (Direct Sparse Solver), against the default iterative solving algorithm FFEPlus (Fourier Finite-Element Plus) which is more suitable for linear static analyses.

After several simulation attempts with different algorithms, the maximum degree of accuracy of the solutions (including the absence of any error messages) is obtained for the Crisfield-Riks control method for numerical simulation (Incremental Arc Length Control Method), respectively for the Newton-Raphson iterative calculation method [27].

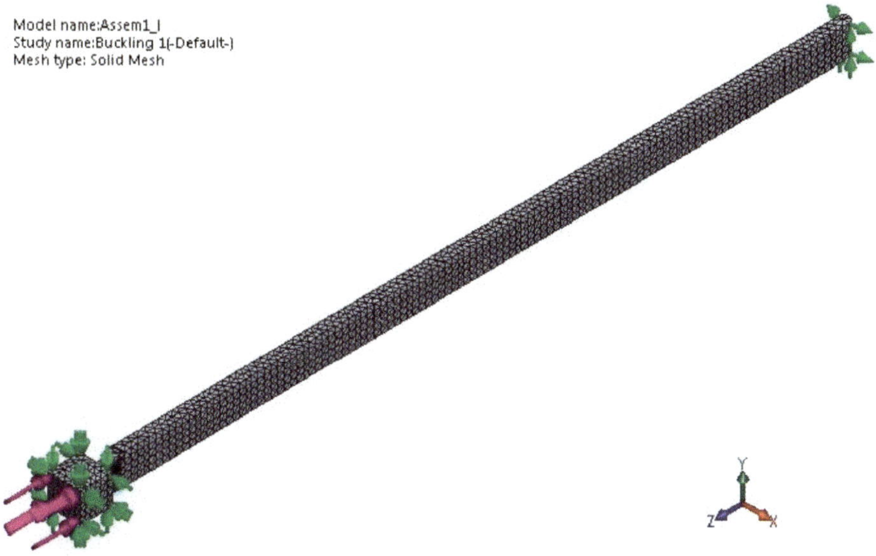

Fig. 5. Representation of discretized model and constraints

The use of the nonlinear solution algorithm is mainly due to the amplitude of the structural deformations, as the change of the stiffness matrix terms cannot be ignored as the load increases.

In the case of some state-of-the-art applications, the presence of a dedicated nonlinear calculation option (large displacement formulation) is noted, considering the large deformations of the studied structure, which is a valid variant in the case of using solid meshing or shell discretization [7].

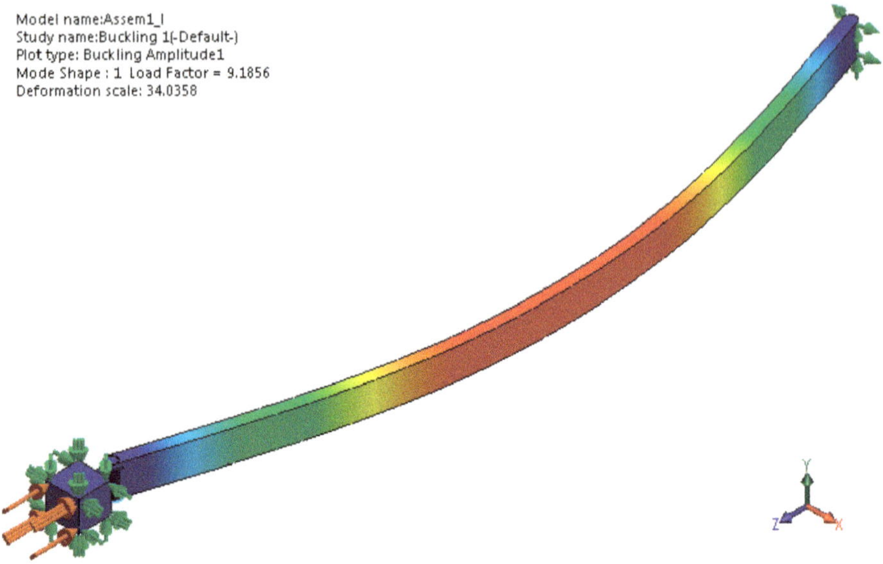

Fig. 6. Deformed representation and the safety coefficient for linear analysis (c = 9.18)

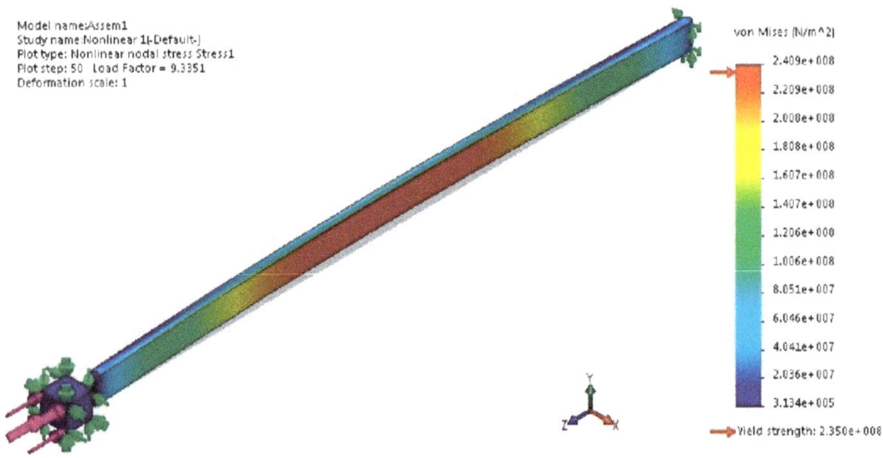

Fig. 7. Deformed representation and the safety coefficient of nonlinear analysis (c = 9.33)

3.3 Discussions

The convergence of the obtained solution can be ascertained from the set of output quantities presented in Figs. 6 and 7, set according to which the value of the buckling safety coefficient is near the value obtained by analytical calculation. Therefore, it is confirmed the critical value of buckling that was found by the previously performed linear analysis (Figs. 8 and 9) and the validity of the classical analytical approach.

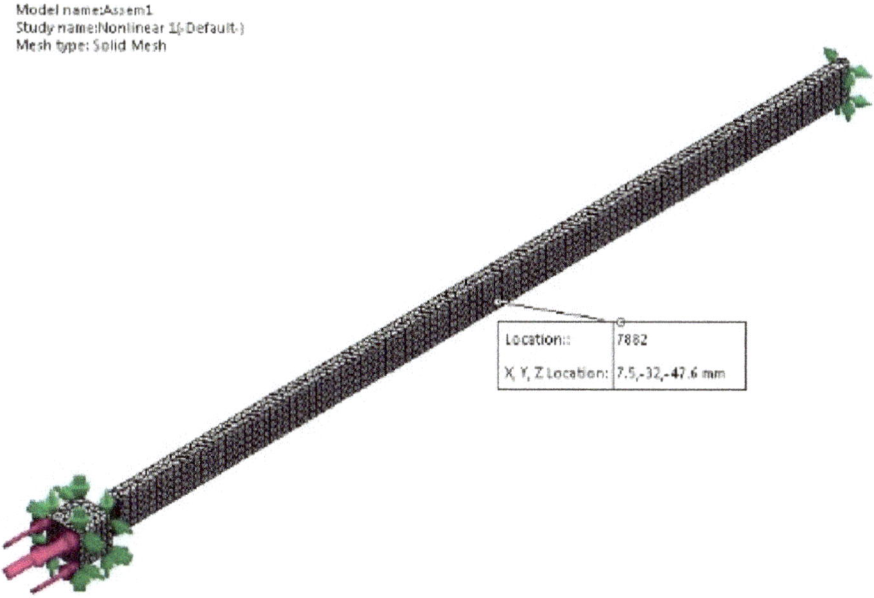

Fig. 8. The nonlinear response of the obtained solution – location of the Node 7882

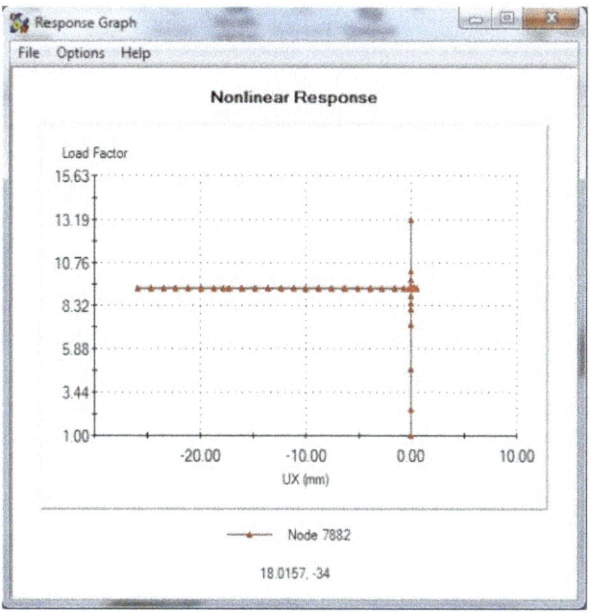

Fig. 9. The nonlinear response of the obtained solution – Response Graph

Following the study, a certain degree of dispersion of the results was also found in the case of the use of various software applications and/or working environments/platforms used (Ansys, SolidWorks/Windows, Salome-Meca, OpenFoam, Code Aster/Linux), remaining thus valid is the rule of using, by comparison, at least two approaches. The safety coefficient of nonlinear analysis performed using the SalomeMeca software is c = 9.323 which is very close to c = 9.33 obtained with SolidWorks and by analytical calculation, as presented in Table 1.

Table 1. Comparative bucking safety coefficients

Analytical approach	Numerical approach		
Euler buckling domain	*SolidWorks Linear analysis*	*SolidWorks Nonlinear analysis*	*SalomeMeca Nonlinear analysis*
9.33	9.18	9.33	9.323

4 Conclusions

Choosing the proper computation method is crucial when accurate results matter most. On the other hand, the software's FEA capabilities for meshing the model are essential factors in the accuracy and computation performance of the results, along with the versatility of the implemented solver algorithms.

The results obtained in this buckling study show a small dispersion of about 1.6% between the computation methods, linear (the buckling safety factor calculation) vs nonlinear. A minimal dispersion is noticed when identical computational methods are used regarding the software products.

The authors emphasize that the highest importance should be paid to the computational method that complies with the material mechanical response, deformation magnitude, fixture and loads state, and structure complexity in buckling analysis. In any case, if a high level of accuracy is required, the model discretization for the finite element analysis stage also becomes an important factor in design and research studies.

Among the studied cases, the highest level of result accuracy (versus traditional analytical solutions) was obtained for the latest generation of commercial software applications. However, it is noted significant steps also for the open-source family of programs in terms of regarding nonlinear analysis and the algorithms based on the finite element method, even in the absence of dedicated implementation programs (cumbersome support/loading schemes, advanced options regarding the settings of the solving engines, etc.).

References

1. Falzon, B.G.: Buckling and Post Buckling Structures: Experimental, Analytical and Numerical Studies. Imperial College Press, London (2008)

2. Norfaieqwan, M., et al.: Buckling analysis of a thin-walled structure using finite element method and design of experiments. Aerospace **9**(10), 541 (2022)
3. Rozylo, P., Debski, H.: Stability and load-carrying capacity of short composite Z-profiles under eccentric compression. Thin-Walled Struct **157**, 107019 (2020)
4. Einafshar, N., Lezgy-Nazargah, M., Beheshti-Aval, S.B.: Buckling, post-buckling and geometrically nonlinear analysis of thin walled beams using a hypothetical layered composite cross-sectional model. Acta Mech. **232**, 2733–2750 (2021)
5. Habtemariam, A.K., Bianco, M.J., Könke, C., Zabel, V.: Geometrically nonlinear formulation of generalized beam theory for the analysis of thin-walled circular pipes. Thin-Walled Struct. **173**, 109029 (2022)
6. Zhou, H.Q., An, Y., He, B., Qi, H.N.: Finite strip-Riccati transfer matrix method for buckling analysis of tree-branched cross-section thin-walled members. Adv. Mech. Eng. **14**, 1–15 (2022)
7. Carrera, E., Pagani, A., Augello, R.: Large deflection and post-buckling of thin-walled structures by finite elements with node dependent kinematics. Acta Mech. **232**, 591–617 (2021)
8. Shahbazian, A., Wang, Y.C.: Application of the direct strength method to local buckling resistance of thin-walled steel members with non-uniform elevated temperatures under axial compression. Thin-Walled Struct. **49**(12), 1573–1583 (2011)
9. Ajeesh, S.S., Jayachandran, S.A.: Spline finite strip analysis of thin-walled flexural members subjected to general loading with intermediate restraints. Thin-Walled Struct **158**, 107171 (2021)
10. Florescu, V., et al.: Use of a novel resistive strain sensor approach in an experimental and theoretical study concerning large spherical storage tank structure behavior during its operational life and pressure tests. Sensors **20**(2), 525 (2020)
11. Ruggerini, A., et al.: Post-buckling analysis based on the implicit corotational method. Int. J. Solids Struct. **163**, 40–60 (2019)
12. Pham, C.H.: Shear buckling of plates and thin-walled channel sections with holes. J. Constr. Steel Res. **128**, 800–811 (2017)
13. Kumari, E., Saxena, D.: Buckling analysis of folded structures. Mater. Today Proc. **43**, 1421–1430 (2020)
14. Aabid, A., et al.: Structural analysis of three-dimensional wings using finite element method. Aerosp. Syst. **5**, 47–63 (2022)
15. Juliano, A., et al.: Preliminary assessment of potential pile tip damage and extrusion buckling. Ocean Eng. **275**, 114071 (2023)
16. Florescu, V., et al.: Design contributions to the elaboration of new modeling schemes for the buckling assessment of hydraulic actuators. Metals **10**(9), 1143 (2022)
17. Mocanu, S., et al.: Novel procedures for sustainable design in structural rehabilitation on oversized metal structures. Metals **12**(7), 1107 (2023)
18. Florescu, V., et al.: Research on the parameters influencing the numerical analysis of the fatigue behaviour of a forklift-arm. Rom. J. Transp. Infrastr. **10**(1), 1–10 (2021)
19. Zakuan, M.A.M.B.M., Aabid, A., Khan, S.A.: Modelling and structural analysis of three-dimensional wing. Int. J. Eng. Adv. Technol. **9**, 6820–6828 (2019)
20. Zhang, G., et al.: Buckling analysis of thin-walled metal liner of cylindrical composite overwrapped pressure vessels with depressions after autofrettage processing. Sci. Eng. Compos. Mater. **28**, 540–554 (2021)
21. Doan, Q.H., Thai, D.: A numerical study of the effect of component dimensions on the critical buckling load of a GFRP composite strut under uniaxial compression. Materials **13**, 931 (2020)
22. Khazaal, D.S., Al-Khafaji, H.M.A., Abdulsahib, I.A.: Buckling behavior of aluminum alloy thin-walled beam with holes under compression loading. J. Eng. **26**, 137–154 (2020)

23. Kumar, P.R., Gupta, G., Shamili, G.K., Anitha, D.: Linear buckling analysis and comparative study of un-stiffened and stiffened composite plate. Mater. Today Proc. **5**, 6059–6071 (2018)
24. https://help.solidworks.com/2019/english/SolidWorks/cworks/c_Simulation_Studies_2.htm?id=b93bb7cb1a6e4ea59ebaba1d01931305#Pg0. Accessed 23 July 2023
25. https://www.caee.utexas.edu/prof/kallivokas/teaching/ANSYS_examples/ansys56theory.pdf. Accessed 23 July 2023
26. https://www.csc.fi/web/elmer/documentation. Accessed 23 July 2023
27. https://help.solidworks.com/2012/english/solidworks/cworks/Iterative_Solution_Methods__Newton-Raphson_(NR)_Scheme.htm

Structures' Global Stiffness, a Criterion for the Materials' Selection

Ştefan Mocanu, Cristian Ştefan Dumitriu(✉), and Anca Ruxandra Sasu

Faculty of Mechanical Engineering and Robotics for Construction, Technical University of Civil Engineering Bucharest, 59, Calea Plevnei, Bucharest, Romania
cristian.dumitriu@utcb.ro

Abstract. Structural and material optimization remains crucial in today's constructive design and machine building area. The importance of this topic relies on its applications for testing the behavior of different materials under the action of various loading and the evaluation of the computation error. When choosing the materials for a specific application, an important criterion is the global structure stiffness, which must be precisely determined. This paper presents a case study of a statically indeterminate, circular plane structure used in specific engineering devices. This structure's particular construction requires determining a dimensional parameter from the viewpoint of optimal rigidity (structure stiffness) to ensure minimal device size. A dual approach, the traditional calculus versus the Finite Element Method simulation, is used to solve the problem. The results of the two finite element software (Axis and SolidWorks) used in this study are also compared. Correlating the structural stiffness to the material properties is the best approach to the optimal design solution.

Keywords: Structure Stiffness · Material Optimization · Finite Element Method

1 Introduction

Optimization in the machinery industry has always been a three-point-of-view problem related to structural design, material selection, and fabrication cost optimization. Structural design goals include rational exploitation, economical maintenance, and safety operation. Industrial machinery design, and not only, is continuously updated to new research and improvements in the materials field. Finally, a good balance between performance and cost gives the products better survival live on the market.

The present work represents the design step, addressing structural optimization matters from a global stiffness point of view. A modern approach to damping systems combines the mechanical behavior of the springs with the wire rope isolators' capability [1]. While the coil springs are axially stressed, the wire rope isolators are stressed by compression in a transversal direction. In both cases, mechanical analysis with different materials should be performed for the specific load cases.

When choosing the materials for a specific application, an important criterion is the structure stiffness which should be precisely determined. In this context, the results

I. Tuns et al. (Eds.): CIBv 2023, LNCE 510, pp. 61–70, 2024.
https://doi.org/10.1007/978-3-031-60765-3_6

presented in this article are important for practical applications when mechanical constraints (sectional properties, deflection, shape of damper) are imposed, and optimal material must be selected based on its characteristics (Young's modulus, elongation, tensile strength, etc.).

Structural displacement field and beam internal forces assessment utilizing analytical and finite element methods (FEM) represent the subject of many research studies [2–6]. Not a few times in their studies, the researchers were concerned with validating the result by running a finite element software application with experimental measurements [7–10].

Kang et al. [8] provide extensive information concerning the effects of the thickness of collar stiffeners around the openings of wind turbine structures; the results are compared to the general criteria, by means of regression analysis being proposed a collar stiffener thickness equation for large diameter cylindrical shell. Buczkowski et al. [9] provided a guiding example of a FEA analysis developed to ascertain mechanical and structural performance of crane structures.

The study focuses on a damping isolator, similar to the wire rope isolator, placed in a particular supporting arrangement, typically used for self-leveling devices. Two beams laterally sustain the circular frame structure, and the force 2P is applied vertically, as presented in Fig. 1 (left side). The structure stiffness is given by material property, Young's modulus, and the sectional inertia moment. This approach proposes to find a modeling solution that can be used in different material applications for vibration-damping systems. Comparative analysis between the results of the classic analytical methods [11] and generally used FEM software [12] is done to investigate the output's convergence to confirm the results' accuracy.

2 Problem Formulation

Consider that a 2P magnitude payload is supported using a hybrid plane frame structure presented in Fig. 1. The mechanical arrangement is presented on the left side, and the equivalent force distribution is on the right. The goal is to determine the geometrical parameter "a" for obtaining maximized structural-related stiffness. The bending stiffness of the circular and linear frame components are identical.

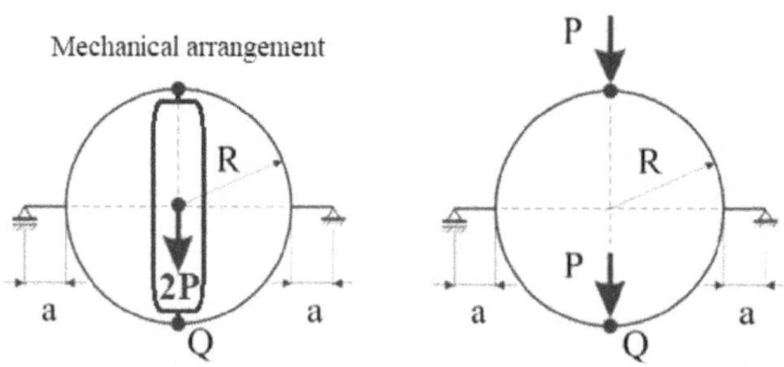

Fig. 1. The statically indeterminate hybrid frame structure

3 Results and Discussion

3.1 Traditional Approach

To solve the problem, the possible isostatic form of the studied three-times hyperstatic plane frame structure (with the use of symmetry) presented in Fig. 2 is used in the following.

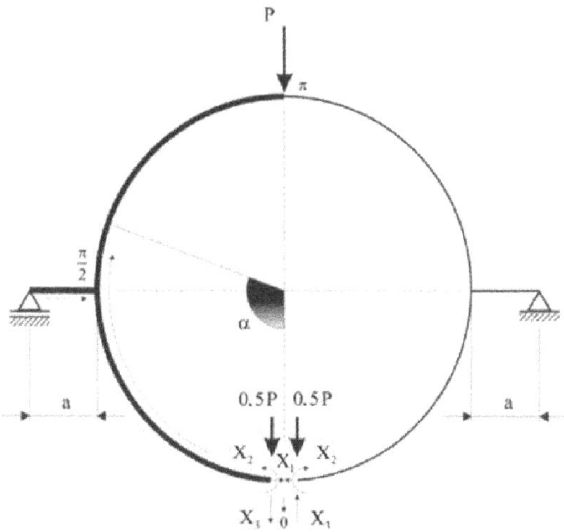

Fig. 2. Isostatic form of the hybrid frame. The angle α is the position variable of the current section, and X_1, X_2, X_3 are the three unknown internal forces that must be evaluated.

To find the problem solution, the system of continuity equations (expressed in the canonical form), Eq. (1), must be solved. This system is associated with the efforts method, one of the methods that remove the indeterminacy of an indeterminate static system (like that from Fig. 1) [11, 13].

To establish the characteristic terms of the continuity equations, one may use the Maxwell-Mohr energy relation [14] for the particular case of the exclusive use of the bending stress term. This simplifying is allowed by the type of the studied bar system - planar, loaded in its plane.

Due to the particular frame shape and loading scheme, only two of three unknown internal forces must be calculated ($X_1, X_2 \neq 0$; $X_3 = 0$) [7].

With the use of bending specific form only in Maxwell-Mohr energetic relation, one must solve a system of two first-degree equations (in two unknowns) expressed in canonical form:

$$\begin{cases} \delta_{11}X_1 + \delta_{12}X_2 + \Delta_{10} = 0 \\ \delta_{21}X_1 + \delta_{22}X_2 + \Delta_{20} = 0 \end{cases}, \tag{1}$$

where:

- δ_{ij} represents the displacement of section i due to the virtual force $\overline{1}$ loading scheme exclusively acting in section j, displacement concerning the chosen isostatic form,

- Δ_{i0} denotes the displacement of section i due to the initial loading scheme on the same chosen isostatic form.

In terms of the Maxwell-Mohr equation, the computational expressions for this particular plane frame structure - plane loading scheme, plane structure, use of symmetry - are:

$$\begin{cases} \delta_{ij} = \frac{1}{EI} \int m_i m_j ds \\ \Delta_{i0} = \frac{1}{EI} \int m_i M_0 ds \end{cases} \tag{2}$$

where:

- m_i is the polynomial expression of the bending moment internal force due to the unitary virtual force loading scheme exclusively present in the section i, on chosen isostatic form and.

- M_0 is the polynomial expression of the bending moment internal force due to the initial loading scheme, for the same isostatic form, respectively.

By taking into account the polynomial expressions of characteristic bending moments m_1, m_2 for $X_1 = \overline{1}, X_2 = \overline{1}$, one can observe that:

$$\begin{cases} \frac{1}{2}\delta_{11} = \int_0^{\pi} [-R(1 - cos\ \alpha)]^2\ R d\alpha \\ \frac{1}{2}\delta_{22} = \int_0^{\pi} (-1)^2 R d\alpha \\ \frac{1}{2}\delta_{12} = \frac{1}{2}\delta_{21} = \int_0^{\pi} [(-1)(-R(1 - cos\ \alpha))]\ R d\alpha \end{cases} \tag{3}$$

and

$$\frac{1}{2}\Delta_{10} = \int_0^{\frac{\pi}{2}} [(-R(1 - cos\ \alpha))(0.5PR\ sin\ \alpha)] R d\alpha + \int_{\frac{\pi}{2}}^{\pi} [(-R(1 - cos\ \alpha))(Pa + P - 0.5PR\ sin\ \alpha)] R d\alpha, \tag{4}$$

$$\frac{1}{2}\Delta_{20} = \int_0^{\frac{\pi}{2}} (-1)(0.5PR\ sin\ \alpha) R d\alpha + \int_{\frac{\pi}{2}}^{\pi} (-1)(Pa + PR - 0.5PR\ sin\ \alpha) R d\alpha \tag{5}$$

After elementary computation, the expressions (3)–(5) become:

$$\frac{1}{2}\delta_{11} = 1.5\pi R^3, \quad \frac{1}{2}\delta_{22} = \pi R, \quad \frac{1}{2}\delta_{12} = \frac{1}{2}\delta_{21} = \pi R^2 \tag{6}$$

$$\frac{1}{2}\Delta_{10} = -\left(0.5 + \frac{\pi}{2}\right) PR^3 - \left(1 + \frac{\pi}{2}\right) PR^2 a \tag{7}$$

$$\frac{1}{2}\Delta_{20} = -\frac{\pi}{2} PR^2 - \frac{\pi}{2} PRa \tag{8}$$

thus, the solutions of the system (1) are:

$$\begin{cases} X_1 = \frac{P}{\pi}\left(1 + 2\frac{a}{R}\right) \\ X_2 = \frac{PR}{\pi}\left[\left(\frac{\pi}{2} - 1\right) + \left(\frac{\pi}{2} - 2\right)\frac{a}{R}\right] \end{cases} \tag{9}$$

Choosing as performance criterion the vertical displacement v_Q of section, it results that:

$$v_Q = \frac{1}{EI} \int m_Q^0\, Mds \tag{10}$$

where:

- E denotes Young's modulus,
- I is the sectional moment of inertia,
- EI is the bending sectional stiffness,
- m_Q^0 is the polynomial variation expression for the bending diagram corresponding to the isostatic form with virtual unitary loading scheme exclusively,
- M is the polynomial expression corresponding to isostatic form with the initial loading scheme.

Concluding,

$$\begin{aligned} &\frac{EI}{2} v_Q = \frac{Pa \cdot a}{2} \cdot \frac{2}{3} \cdot 0.5a + \\ &\int_0^{\frac{\pi}{2}} 0.5R^2 sin\ \alpha \left[RP\left(\frac{cos\ \alpha}{\pi} - \frac{1}{2} + \frac{1}{2} sin\ \alpha \right) + aP\left(\frac{2cos\ \alpha}{\pi} - \frac{1}{2} \right) \right] d\alpha + \\ &\int_{\frac{\pi}{2}}^{\pi} 0.5R(a+R) \left[RP\left(\frac{cos\ \alpha}{\pi} + \frac{1}{2} - \frac{1}{2} sin\ \alpha \right) + aP\left(\frac{2cos\ \alpha}{\pi} + \frac{1}{2} \right) \right] d\alpha, \end{aligned} \tag{11}$$

or

$$\begin{aligned} &\frac{EI}{2} v_Q = PR^3 \left(-\frac{0.5}{2\pi} + \frac{0.75\pi}{4} - 0.5 \right) + PR^2 a \left(-\frac{1}{\pi} + \frac{\pi}{4} - 0.5 \right) \\ &+ a^2 PR \left(-\frac{1}{\pi} + \frac{0.25\pi}{2} \right) + \frac{1}{6} Pa^3. \end{aligned} \tag{12}$$

Denoting by $q = a/R$, relation (11) becomes:

$$v_Q = \frac{2PR^3}{EI} \left(\frac{1}{105.58} - \frac{1}{30.384} q + \frac{1}{13.443} q^2 + \frac{1}{6} q^3 \right). \tag{13}$$

Thus, the v_Q displacement takes the smallest value when the first derivative of the deflection (12) with respect to q is zero:

$$\frac{d\,(v_Q)}{dq} = 0 \Rightarrow -\frac{1}{30.384} + \frac{2}{13.443} q + \frac{3}{6} q^2 = 0 \Rightarrow \tag{14}$$

$$q_1 = 0.147801, q_2 = -0.443357. \tag{15}$$

We keep the positive value because the a and R are always positive.

3.2 Numerical Approach

The following numerical example is used for numerical analysis: $P = 150\text{daN}, R = 5\text{m}, E = 2.1 \times 10^5 MPa, I = 4.93 \times 10^6\text{mm}^4$, the main performance criteria parameter is "q", therefore the vertical deflection is considered as $v_Q = f(q)$.

The solution obtained by the classical analytical method was compared with the result provided by running Dassault Systèmes SolidWorks 2014 [15, 16] and AxisVM13. SolidWorks represents a complex software tool for design, and simulation purposes. It is used to "(…) *calculate displacements, reaction forces, strains, stresses, failure criterion, Factor of safety and error estimates. Available loading conditions include point, line, surface, and thermal loads, elastic orthotropic materials are also available. The SW Simulation software also offers several types of nonlinear studies*" [12]. The core feature of the AxisVM13 software – ideal for construction engineers – is the integrated "visual modeling" system with which the 3D model can be built. Visual modeling allows graphical tracking of the entire modeling and calculation process up to the phase of evaluating the results [17, 18].

The accuracy of output measurement depends on many variables, among which the relevant ones are the type and size of meshing elements (Table 1), the quality details of the fixture (nodal restraints, degrees of freedom (DOF), boundary conditions (B.C.)) scheme modeling. It is known that certain FEM algorithms involve several uncertainties and approximations that largely influence both accuracy and reliability of each processing stage [12].

Although both software are based on the same mathematical approach, as will be seen below, there are differences regarding the offered result. Determining the cause of the appearance of these differences as well as their level is a challenge for the authors. Thus, through their internal structure, the software forces the use of different reasoning regarding introducing the limit conditions in terms of the loading scheme, especially the support. Axis offers a simpler way of modeling supports, while SolidWorks forces the creation of assembly-type entities for each support (due to the discretization mode for the finite element category used). Figures 3 and 4 illustrate in detail the specific manner of each software in terms of imposing the boundary conditions in terms of the support scheme.

Fig. 3. Restrain D.O.F. modelling schema in AxisVM13 (left) vs. SolidWorks 2014 assembly (right). Left-hand simple support in Fig. 1

Besides the special aesthetics and somehow unexpected final deformed shape, despite the presumed differences between used methods to a certain extent, an excellent scatter

Fig. 4. Restrain D.O.F. modelling schema in AxisVM13 (left) vs. SolidWorks 2014 (right). Right-hand hinged support in Fig. 1.

of output measurements is obtained, especially in terms of the range of "q" parameter concerning the global stiffness aspect.

However, global boundary conditions induced by global support modeling schemes can cause, in some cases, large differences in accuracy and reliability.

Mesh parameters and support modeling scheme largely influence the output accuracy, even minimum changes in terms of initial analysis solver settings (pre-processing stages), being responsible for result scatter. Structure's so-called "singularities" can cause stress concentration that artificially raises local stress field, mesh element type can cause distorted line elements with poor "aspect-ratio" [14], material linear-elastic behavior, also loading scheme modeling with respect to Barre de Saint-Venant's principle.

Through the study topic, minimizing the vertical component of displacement, Q was pursued by varying the length of the rectilinear sections (of length a). The chosen performance criterion was evaluated about the fraction $q = a/R$, where R represents the average radius of the circular sector of the curve bar.

Table 1 compares the result obtained by using different types of discretization elements for the value of the vertical projection of the Q section displacement depending on the structure's geometric parameters.

Indeed, some differences can be remarked in terms of v_Q as deflection magnitude for the corresponding "q" parameter, mainly due to (simplified) beam type element meshing used in AxisVM13 ($q = 0.148$ at $v_Q = 2.443$) versus solid element mesh (where the number of mesh elements increased significantly) used in SolidWorks ($q = 0.151$ at $v_Q = 2.432$) and due to physical support modeling scheme.

On the other side, the traditional approach considers the bending terms only in the Maxwell-Mohr expression. This can explain the minimal differences between the left and right-hand side graphical representation of v_Q as a function of q in the FEM and classical approach, especially for the AxisVM13 software (see Fig. 5).

Therefore, special consideration is required when using different numerical analysis methods by considering the pros and cons of each approach.

The results presented in Table 1 were used for predicting the deflection value using and Gene Expression Programing algorithm [19].

Table 1. Vertical deflection with respect to various parameters

a (m)	$q = a/R$	v_Q (mm)	
		beam mesh	solid mesh
0.5	0.1	2.551	2.548
0.6	0.12	2.479	2.475
0.7	0.14	2.445	2.438
0.73	0.146	2.443	2.434
0.735	0.147	2.443	2.433
0.74	**0.148**	**2.443**	2.433
0.745	0.149	2.444	2.433
0.75	0.15	2.444	2.432
0.755	**0.151**	2.444	**2.432**
0.76	0.152	2.445	2.432
0.8	0.16	2.454	2.435
0.9	0.18	2.507	2.472
1	0.2	2.608	2.549

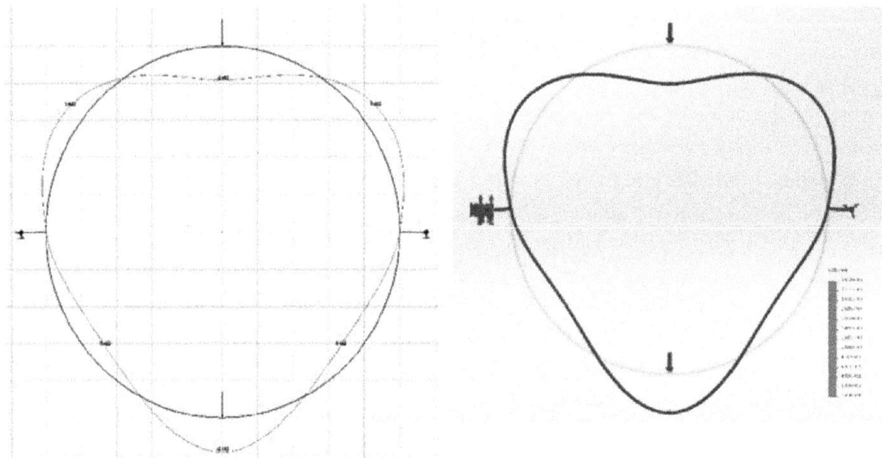

Fig. 5. Deformed shape and output scatter plot in AxisVM13 – left plot (q-axis of abscissae, v_Q - axis of ordinates) and in SolidWorks 2014 (q-axis of abscissae, v_Q - axis of ordinates) – right plot.

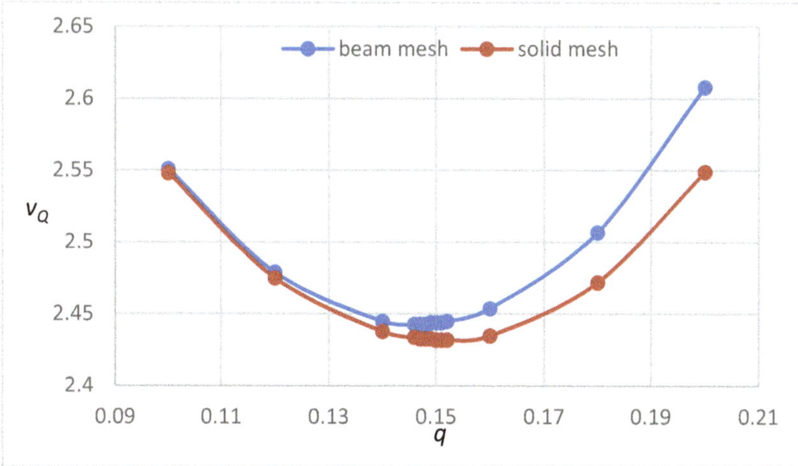

Fig. 6. Comparative research results in F.E.A. approach, beam mesh (AxisVM13) vs. solid mesh (SolidWorks 2014): v_Q evolution.

4 Conclusions

Calculating the *q* parameter leads to finding the optimal structure stiffness for a given material (mechanical properties and sectional parameter). The two curves presented in Fig. 6 show a different slope as the *q* parameter increases, so the convergence of the results decreases. This behavior has a numerical explication based on the finite element type. By increasing the *a* value, for a constant radius *R*, the frame volume increases and consequently the number of finite elements. The specific computational differences between beam mesh and solid mesh grow as the frame size increases, and the results will differ more.

One of the article's goals was to demonstrate, by converging the results for the optimal value of the *q* parameter, that using as simple meshing elements as possible (i.e., beam meshing elements) reduces solving time and saves computational resources. Further research will focus on finding the optimal material properties for an imposed deflection, v_Q, concerning the parameter *q*.

Least but not last, the presented numerical model gives a practical solution for finding the maximum structural stiffness for similar technical problems.

Modern-day structural optimization, based on finite element method principles rather than classical calculus methods, asks for straightforward studies with a certain degree of complexity to validate the final technical solutions. This article depicts such a study and emphasizes the importance of software algorithm accuracy in imposed boundary conditions, support and loading scheme degrees of freedom, and exact choice in terms of the finite element type.

References

1. ENIDINE Homepage. http://www.enidine.com/en-us. Accessed 11 Aug 2023

2. Zhu, Y., Liang, W., Zhao, X., Wang, X., Xia, J.: Strength and stability of spherical pressure hulls with different viewport structures. Int. J. Press. Vessel. Pip. **176**, 103951 (2019). https://doi.org/10.1016/j.ijpvp.2019.103951
3. Takezawa, A., Nishiwaki, S., Kitamura, M., Silva, E.C.N.: Topology optimization for designing strain-gauge load cells. Struct. Multidiscip. Optim. J. **42**, 387–402 (2010). https://doi.org/10.1007/s00158-010-0491-0
4. Girip, A., Calotă, R., Ilie, A.: Aspects regarding the optimal insulation thickness, the cost and energy savings for cold storage in Romania. Appl. Sci. **11**, 11455 (2021). https://doi.org/10.3390/app112311455
5. Florescu, V., Mocanu, S., Neagu, A., Sescu-Gal, C.: Research on the parameters influencing the numerical analysis of the fatigue behaviour of a forklift-arm. Rom. J. Transp. Infrastr. **10**(1), 1–10 (2021). https://doi.org/10.2478/rjti-2021-0001
6. Georgescu, M.R., Meslem, A., Nastase, I., Sandu, M.: Numerical and experimental study of the international space station crew quarters ventilation. J. Build. Eng. **41**, 102714 (2021)
7. Mocanu, S., Rece, L., Burlacu, A., Ronţescu, C., Modrea, A.: Novel procedures for sustainable design in structural rehabilitation on oversized metal structures. Metals **12**(7), 1107 (2023)
8. Kang, S.-Y., Won, D., Park, J.-S., Kang, Y.-J., Kim, S.: Structural behavior of large-diameter cylindrical shell with stiffened opening. Metals **11**(9), 1413 (2021). https://doi.org/10.3390/met11091413
9. Buczkowski, R., Żyliński, B.: Finite element fatigue analysis of unsupported crane. Polish Marit. Res. **28**(1), 127–135 (2021). https://doi.org/10.2478/pomr-2021-0012
10. Florescu, V., Mocanu, S., Rece, L., Ursache, R., Goga, N., Marian, C.V.: Use of a novel resistive strain sensor approach in an experimental and theoretical study concerning large spherical storage tank structure behavior during its operational life and pressure tests. Sensors **20**(2), 55 (2020). https://doi.org/10.3390/s20020525
11. Timoshenko, S., Gere, J.: Theory of Elastic Stability, 2nd edn. McGraw-Hill, New York (1963)
12. Akin, J.E.: Finite Element Analysis Concepts via SolidWorks. Rice University, Houston (2009)
13. Hibbler, R.C.: Structural Analysis, 8th edn. Prentince Hall, New Jersey (2021)
14. Maxwell – Mohr method. Available online. https://kmim.wm.pwr.edu.pl/kaleta/wp-content/uploads/sites/23/2020/04/Maxwell-Mohr-method_introduction.pdf. Accessed 11 Aug 2023
15. SOLIDWORKS Help. https://help.solidworks.com. Accessed 11 Aug 2023
16. https://blogs.solidworks.com/solidworksblog/2019/06/solidworks-simulation-accracy-report-by-afnor.html. Accessed 11 Aug 2023
17. CONSOFT. www.consoft.ro/axisvm. Accessed 11 Aug 2023
18. CADWORKS Homepage. www.cadworks.ro. Accessed 11 Aug 2023
19. Dumitriu, C.S., Mocanu, S., Panaitescu, R., Sasu, A.R., Tonciu, O.: Genetic Algorithm approach for modeling the structural global stiffness. Eng. Proc. **39**(1), 32 (2023). https://doi.org/10.3390/engproc2023039032

Seismic Performance of Replaceable Friction Energy Consuming Steel Joints for Assembled RC Beams and Columns

Weining Sui(✉), Yu Hou, Lin Zhu, and Atulinda Kato Aganyira

Civil Engineering College, Shenyang Jianzhu University, No.25, Hunnan Middle Road, Shenyang 110000, China
weiningsui@sjzu.edu.cn

Abstract. In response to issues such as dispersed damage and difficult repair of assembled RC beam-column joints after earthquakes, this paper proposes an assembled (Reinforced Concrete) RC beam-column replaceable friction energy-consuming steel joint. The steel joint consists of a friction energy dissipation system and a U-shaped shear-resistant steel plate system. Numerical simulations using ABAQUS software were conducted on one cast-in-place joint and six steel joints under low-cycle reciprocating loading. The simulation results comparing the cast-in-place joint with the steel joints show that the steel joints have comparable initial stiffness and ultimate bearing capacity, but stronger energy dissipation capacity. After seismic action, the main structure remains in an elastic working state, achieving toughness and seismic resistance of the structure. The simulation analysis of the effect of the design bearing capacity coefficient μ on the functional recoverability and concentrated energy dissipation performance of the structure shows that energy dissipation is mainly concentrated in the steel joints. With the increase of the bearing capacity coefficient, the total energy dissipated in the steel joints initially increases and then decreases, but the proportion of energy consumption of steel joints will continue to decrease. Under the premise of meeting the demand of bearing capacity, selecting a reasonable bearing capacity coefficient can make the energy consumption of the steel joints reach over 95%, thereby achieving the purpose of centralized energy dissipation of the structure and rapid recovery of the functional use after earthquakes.

Keywords: Assembled RC beam-column joints · Numerical simulations · Toughness against earthquakes · load capacity coefficient · Steel joints

1 Introduction

The prefabricated concrete frame structure has advantages such as low environmental impact, high component quality, short construction period, and low labor requirements, making it the direction of future development in the construction industry. Previous seismic damage investigations have shown that some prefabricated concrete structures suffered severe damage during earthquakes [1, 2] mainly due to the failure of the connection joints between prefabricated components, leading to the collapse of the entire structure. Therefore, the beam-column connection remains a key research subject.

I. Tuns et al. (Eds.): CIBv 2023, LNCE 510, pp. 71–83, 2024.
https://doi.org/10.1007/978-3-031-60765-3_7

Dry assembly beam-column connections, during installation, do not require on-site wet work, resulting in a higher assembly rate and better alignment with the current industrialized production requirements. Several researchers have conducted extensive experimental and finite element simulation studies to enhance the seismic performance of prefabricated frame beam-column connections. Englekirk et al. [3, 4] and Li Xiangmin et al. [5] proposed a ductile connection system to improve the seismic performance of structures. This system involves embedding ductile links within prefabricated columns, connecting them to prefabricated beams using bolts. During seismic events, the ductile links undergo plastic deformation, dissipating energy and reducing seismic forces, thus preventing damage to other structural components. However, this approach involves the casting of ductile connectors within the column, and after seismic action, the core region of the connection exhibits shear failure characteristics, which can hinder post-earthquake connection repair. Zhao Bin et al. [6] suggested embedding screws at the column node and casting concrete beams with I-shaped steel joints as an integral component. Subsequently, they used screws and rubber bearings for assembly, creating a semi-rigid connection. Experimental results showed that while bolted connections exhibited improved load-carrying capacity compared to cast-in-place connections, their energy dissipation performance was poorer, and significant cracks developed in the concrete core region in the later stages of loading. In summary, although certain existing structural approaches have enhanced the seismic performance of prefabricated beam-column connections, they still suffer from drawbacks such as dispersed damage under rare seismic events, uncontrollable locations for plastic deformation development, and the risk of damage to the core region of column ends and critical connection points. Once structural damage occurs, it becomes challenging to achieve repairs.

In recent years, seismic disaster investigations have revealed [7] that while the overall collapse of buildings due to seismic forces has been effectively controlled, the economic losses and societal impact caused by earthquakes are still significant. The severe damage sustained by buildings after an earthquake makes the repair process challenging, leading to prolonged interruptions in the normal functioning of buildings and hindering daily life and production. To address the need for structures that do not require extensive repair or maintenance after earthquakes, or that can quickly recover their normal functionality with minimal repairs, the concept of resilient structures has emerged [8]. In pursuit of this concept, scholars have proposed structural systems where damage to connecting nodes is controllable, and components are replaceable. This is achieved by placing replaceable connecting components at plastic hinge locations, and inducing damage concentration in these replaceable components. The goal is to achieve either no damage or minimal damage in concrete beam-column structures. After replacement of the damaged components, the structure can rapidly restore its functional use.

Ye Jianfeng et al. [9] proposed a beam-column connection node based on replaceable energy-dissipating hinges. After an earthquake, the seismic performance remains essentially unchanged by replacing the metal damper components, achieving functional recovery. Huang Wei et al. [10] introduced a beam-column node connected by multi-slot energy-dissipating devices. Experimental results indicate that this node can achieve damage concentration, facilitate rapid seismic damage recovery, and exhibit comparable load-bearing capacity and stiffness to conventionally cast nodes. Additionally, it possesses higher ductility and energy-dissipating capability.

Ma Zhehao et al. [11] proposed an assembly-type RC beam-column node with artificial energy-dissipating plastic hinges. By utilizing plastic perforated energy-dissipating steel plates for connection, the structure exhibits higher load-bearing capacity and lower damage, effectively enhancing seismic resistance. Song et al. [12] introduced a self-centering concrete beam-column connection with a web friction device. Research indicates that this structure functions to reduce residual deformations and minimize structural damage. Aninthaneni et al. [13] investigated the hysteric performance of prefabricated frame sub-assemblies with end plate connections. The study results demonstrate that this connection can be used for high-rigidity anti-bending connections and developed a detachable sub-component energy-dissipating system. Wang Meng et al. [14] proposed the use of low-yield-point steel as connecting components, enabling ductile energy dissipation resembling a fuse function. This alteration in the failure mode of the node helps prevent the main structure from entering the plastic phase prematurely.

In summary, with the increasing demands on building structures, research on prefabricated nodes has evolved from seismic resistance and isolation to the current development of recoverable functional structures. Recoverable functional structures mainly encompass prestressed systems and replaceable connection component systems. Recent studies have found that prestressed steel bars in prestressed systems suffer from severe elastic losses and complex construction procedures [15]. On the other hand, in replaceable connection component systems, small-scale BRB structures are commonly used as replacement devices [9, 16]. However, this system still faces challenges such as high component replacement costs and intricate manufacturing processes.

In light of the problems encountered in the development of prefabricated concrete structures today, this paper proposes a steel node with replaceable friction energy dissipation capability. Different from conventional prefabricated nodes, this node allows artificial control of strength by adjusting pre-tension at the beam end. Moreover, a friction energy dissipation damper is installed at the connection location. During seismic events, the damper's contact interface begins to slide, exerting energy dissipation effects to safeguard the entire structure from damage. Additionally, it helps confine damage locations within the replaceable steel construction area. This node exhibits excellent energy dissipation capabilities and controllable plastic development, making it of significant research value and importance.

2 Node Structure and Working Principle

The key components of the replaceable friction energy consuming steel joints for assembled RC beams and columns (hereinafter referred to as the steel node) are the friction energy dissipation system and the U-shaped anti-shear steel plate system. The force transmission path of this node is clear, with the friction energy dissipation system primarily bearing the bending moment at the beam end, while the U-shaped anti-shear steel plate system takes on the entire shear force. As illustrated in Fig. 1, during installation, the steel node components are first connected to the precast beams with bolt holes using high-strength bolt assemblies. Subsequently, the precast beam with ductile joints is hoisted to the column end for alignment, and the ductile joints are connected to the core area of the joints through high-strength bolt pairs and T-section steel with flange extension as backing plates to complete the assembly work.

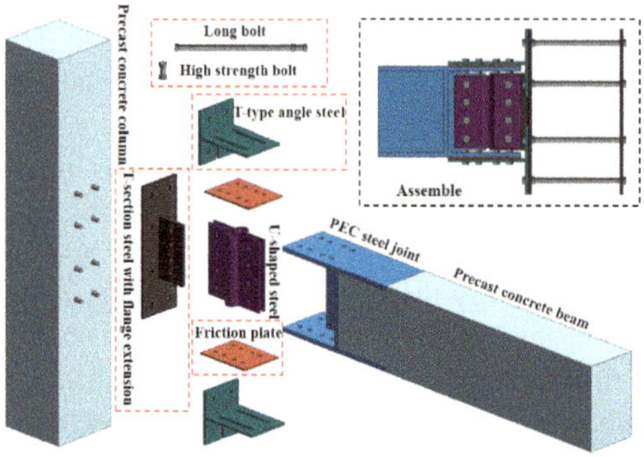

Fig. 1. Replaceable friction energy dissipation steel joint structure of assembled RC beam-column.

3 Seismic Performance Analysis of the Replaceable Friction Energy Dissipation Steel Node

3.1 Node Parameter Design

To exploit the damage control and concentrated energy dissipation characteristics of the steel node under seismic loads, a design load capacity coefficient μ between the steel node and the concrete beam has been defined, as shown in Eq. (1). When μ is greater than 1, the reinforcement in the beam-column tends to yield earlier than the steel node, leading to damage in the primary structural elements and a reduction in the recoverable functionality of the node. When μ is less than 1, the steel node can yield and dissipate energy earlier under seismic actions, playing a protective role for non-energy dissipating components such as the reinforcement in the beam-column.

$$\mu = \frac{M_f}{M_{b,r}} \tag{1}$$

$$M_{b,r} = n \cdot f_y \cdot A_s \cdot h_s \tag{2}$$

In the equation, M_f represents the initial slip moment of the friction damper; $M_{b,r}$ is the elastic resistance moment of longitudinal steel bars in the concrete beam; n is the number of tension or compression zone steel bars; f_y is the yield strength of longitudinal steel bars in the beam; A_s is the cross-sectional area of a single steel bar; hs is the distance between the upper and lower longitudinal steel bar force couples.

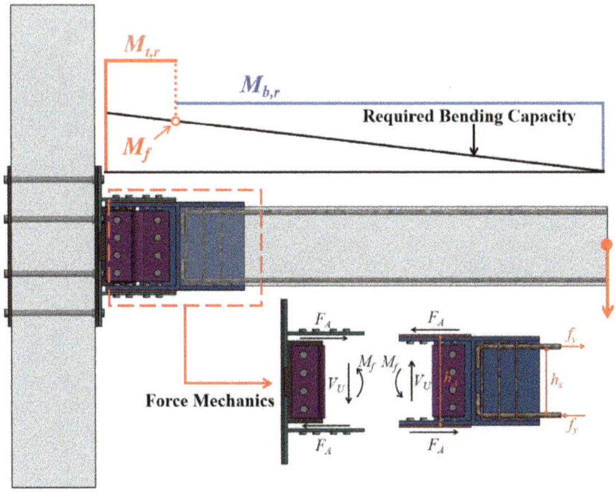

Fig. 2. Force mechanism of beam-column joint.

The initial slip moment M_f of the friction damper is obtained by the equivalent force couple formed by the tension and compression of the upper and lower friction dampers, as shown in the force analysis in Fig. 2. Additionally, at this node, the U-shaped shear-resistant steel plate bears the majority of the shear force at the beam end. Therefore, the friction damper can be approximately considered to only bear axial tensile and compressive forces, that is:

$$M_f = F_A \cdot h_A \tag{3}$$

In the equation, F_A represents the initial sliding force of the friction damper, and h_A is the distance between the upper and lower force couples of the friction damper.

For structures with post-earthquake recovery functionality, their basic working principle under seismic actions involves the concentrated energy dissipation of connectors, while main components such as beams and columns remain essentially in an elastic state. After an earthquake, structural functionality can be restored by simply replacing damaged connectors. To validate the ability of nodes to dissipate concentrated energy and achieve post-earthquake replaceability, six sets of node specimens were designed by adjusting the preload force of high-strength bolts to change the load capacity coefficient. The model numbers are XJ, PT, PT-1 to PT-7, as shown in Table 1. Single-parameter analysis was conducted using the finite element method. Through computation and comparison, the concentrated damage mechanism of steel nodes was further analyzed, and the rapid recovery capability of their post-earthquake functionality was assessed.

It is noteworthy that in the previous context, the bearing capacity factor μ was defined by the ratio of the moment resistance at the node at the initial sliding moment of the friction damper to the elastic moment resistance of the longitudinal steel bars in the concrete beam. However, as displacement is gradually applied, the U-shaped steel plate shear-resistant system also begins to contribute to the moment resistance of the node, and these contributions cannot be ignored. This results in the actual moment resistance M_r being greater than the initial sliding moment of the friction damper M_f in Eq. (3) thereby causing the bearing capacity factor μ to increase. To account for the impact of

Table 1. Model number and node parameters.

Model number	Pre-tightening force of high strength bolts/kN	U-shaped steel material	Bearing capacity coefficient	Lifting ratio
XJ	-	-	-	-
PT	34	Q235	0.47	0.19
PT-1	28	Q235	0.38	0.19
PT-2	40	Q235	0.55	0.19
PT-3	45	Q235	0.62	0.19
PT-4	50	Q235	0.69	0.19
PT-5	60	Q235	0.81	0.19
PT-6	70	Q235	0.96	0.19
PT-7	78	Q235	1.06	0.19

Note: The lifting ratio is the increase ratio of bearing capacity coefficient caused by the strengthening of U-shaped steel plate in the later period.

this situation, the concept of "enhancement ratio-$\mu_{1,r}$" is introduced, namely:

$$\mu_{l,r} = \frac{M_r - M_f}{M_{b,r}} \tag{4}$$

As only the U-shaped steel plate provides additional moment resistance after the sliding of the friction damper at this node, Eq. (4) can also be expressed as:

$$\mu_{l,r} = \frac{M_U}{M_{b,r}} \tag{5}$$

In the equation, M_U represents the moment resistance provided by the U-shaped steel plate after the sliding of the friction damper.

The beam-column section dimensions were designed according to the requirements of reference [11], with PT serving as the standard model. The specific parameters for PT and XJ can be found in Fig. 3. PT-1 to PT-7 only differ in the preload force applied to the high-strength bolts, while the other characteristics of the models remain consistent with PT.

In the connection components, the material constitutive model for the 10.9-grade M22 high-strength bolts adopts a two-linear model, with a yield strength (f_y) of 995 MPa and an ultimate strength (f_u) of 1160 MPa. The bolt hole sizes, positions, and spacing comply with the design specifications for high-strength bolts. T-shaped connectors and U-shaped shear-resistant steel plates are made of Q235 steel (yield strength of 235 MPa, and post-yield stiffness is 0.01 times the initial stiffness). The remaining steel components are constructed from Q345 steel (yield strength of 345 MPa). The elastic modulus (E) for all steel materials is 21000 MPa, following the kinematic hardening criterion. The concrete constitutive model employs the C30 plastic damage model, and specific parameter relationships are referenced from experimental results in literature [11].

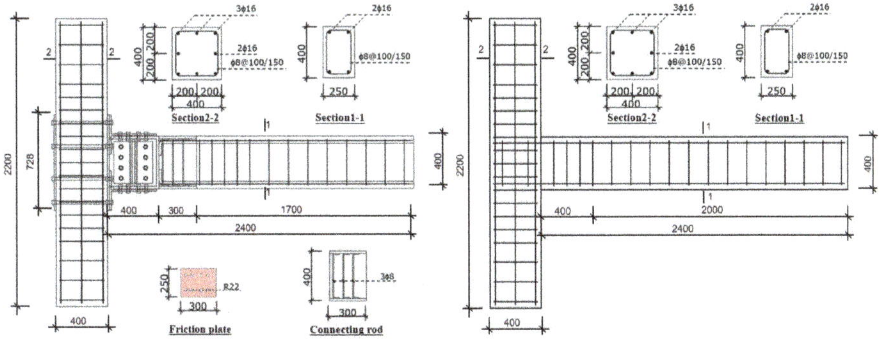

(a) Dimensions of steel joints and detail drawings of members(PT)

(b) Detail drawing of steel bar size of cast-in-place joint(XJ)

Fig. 3. Reinforcement size and component details of beam-column joints

Seismic data is referenced to a specific engineering project, with a seismic design intensity of 7° in the strong motion zone. The design basic seismic acceleration is 0.10 g, and the foundation is of Type II soil without liquefiable layers. The design seismic grouping is the first group with a damping ratio of 0.05. To facilitate quasi-static loading, the model adopts the following loading Protocol: Prior to component yielding, each load increment is 5 kN, with 1 cycle per increment. After specimen yielding, increments are multiples of the yield displacement (Δy), with 2 cycles per increment. All models follow the same loading protocol, as shown in Fig. 4. The total nodal rotation is defined as the displacement at the loading point divided by the total beam length. The maximum applied displacement is 80 mm, corresponding to a rotation of 0.035 radians.

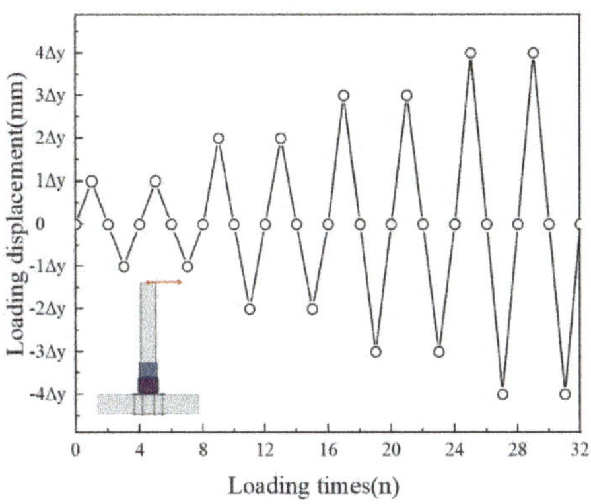

Fig. 4. Loading regime

3.2 Finite Element Model

A refined beam-column joint model was established using Abaqus finite element analysis software. To ensure the accuracy of the computational results, 8-node linear reduced integration solid elements (C3D8R) were employed for the concrete beams, columns, bolts, and connection components. Local mesh refinement and thickness direction refinement were applied to critical force-bearing regions. The reinforcing bars were modeled using T3D2 elements, and the Embed command was used to incorporate them into the concrete without considering bond slip between the reinforcement and concrete.

The contact interactions between components were simulated using the "Face to Face" command, with Coulomb friction employed in the tangential direction. The friction coefficient between brass plates and steel components was set at 0.2, while for other components, it was set at 0.3 [17]. The normal direction was defined as "Hard Contact". The contact between steel components and concrete beams utilized the Tie command. The Bolt Load command was used to apply pre-tension to the bolt shank's mid-plane. Subsequent analysis steps simulated the loss of bolt pre-tension by setting "Fixed Current Length."

In order to prevent the impact of stress concentration on the results, reference points (RP1, RP2, RP3) were established at the column and beam ends. The column-end face and the loading point region were coupled to these reference points.

3.3 Simulation Results Comparison and Analysis

Stress Comparison and Analysis between Cast-in-Place Nodes and Steel Nodes
In the final loading stage ($\Delta = 80$ mm), the stress distribution comparison between the cast-in-place nodes (XJ) and the steel nodes (PT) is illustrated in Fig. 5. From the figure, it can be observed that after loading completion, the XJ node experiences severe tensile damage at the intersection of the beam and column, extending towards the core region of the concrete node. The compressive damage extends from the beam end in an approximately 60° triangular region towards the loading end, and at the bottom of the beam root, the damage has penetrated the entire cross-section. This situation corresponds to the compressive failure of the concrete at the bottom of the beam observed in the experiment [11], accompanied by the phenomenon of longitudinal and shear reinforcement leakage. The PT node exhibits horizontal tensile damage along the beam direction, presenting a pattern of fine and short cracks with a relatively uniform distribution. Compressive damage is only slightly evident at the casting end of the PEC joint where it connects with the concrete.

Examining the stress contour of the reinforcement in Fig. 5(a,d) reveals that after loading completion, all tensile reinforcements in the core region and at the beam-column intersection in the XJ node have yielded. The maximum stress, σ_{max}, is 620 MPa, corresponding to a strain ε_{max} of 0.13375, approximately 60 times the elastic strain. This indicates severe structural damage to the building under seismic action. In contrast, the PT node exhibits its maximum reinforcement stress at the casting end of the PEC, σ_{max} = 350 MPa, where both longitudinal and transverse reinforcements in the beam-column joint remain in the elastic working stage. Consequently, compared to the cast-in-place node, this steel node can effectively form a plastic hinge mechanism at the beam end,

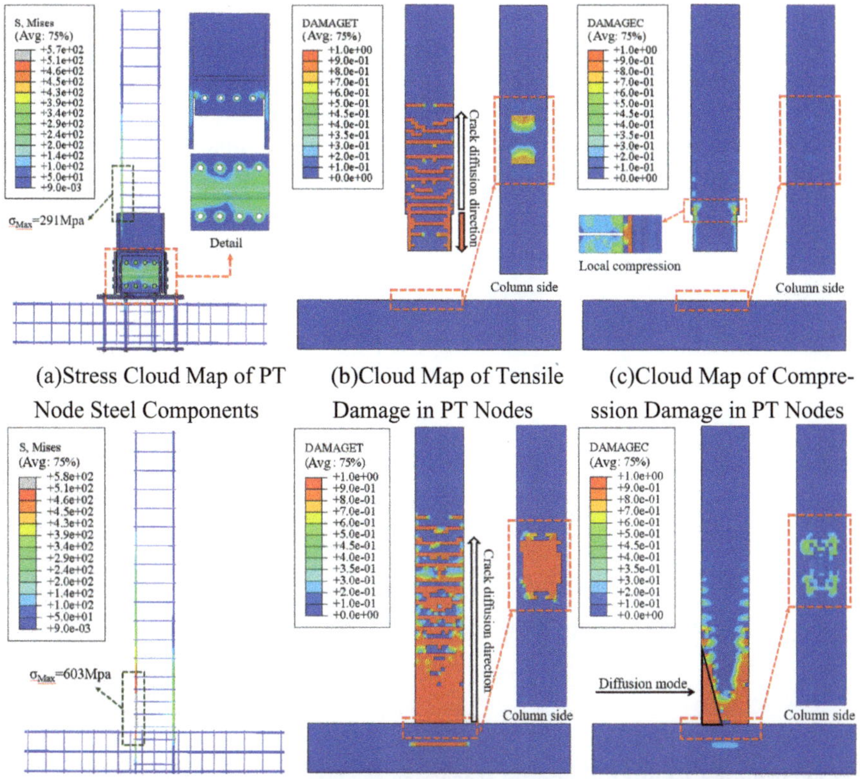

(a)Stress Cloud Map of PT Node Steel Components (b)Cloud Map of Tensile Damage in PT Nodes (c)Cloud Map of Compression Damage in PT Nodes

(d) Stress Cloud Map of Reinforcement in XJ Node (e) Tensile Damage Cloud Map of XJ Node (f) Cloud Map of Compression Damage in XJ Nodes

Fig. 5. Cast-in-place node and steel node cloud diagram

inducing damage concentration in the steel components. After an earthquake, the primary structure remains within the elastic range, safeguarding the main components and achieving a "fuse" effect.

Comparative Analysis of Hysteresis Energy Dissipation Performance

As shown in Fig. 6, comparing the hysteresis curves and skeleton curves of nodes XJ and PT, it can be observed that the hysteresis curve of XJ node exhibits a "spindle shape," while the hysteresis curve of PT node shows a "parallelogram," indicating that the steel PT node possesses a stronger energy dissipation capability. In the early loading stage, both the load and initial stiffness of XJ node are higher than those of PT node. XJ node undergoes yielding at a positive displacement of 15.9 mm (yield displacement selected based on the method proposed by Park [18]), and the component's load reaches its peak value at a displacement of 23.8 mm. Subsequently, the load continuously decreases until the end of the loading process.

For the PT node in this study, the moment of initiation of sliding in the frictional energy dissipation system is defined as the yielding point, with a yielding load of 20.2

kN. In the subsequent loading phase, due to the yielding of the shear-resistant system with U-shaped steel plates, the skeleton curve shows a slow initial increase in load, followed by a trend towards a stable state, providing sufficient strength for the structure in the later stages of loading. Clearly, for this steel node, when the load coefficient μ = 0.47, the peak load capacity reaches 31.9 kN, which is comparable to the peak load capacity of a cast-in-place node of the same size (F_{max} = 32.6 kN).

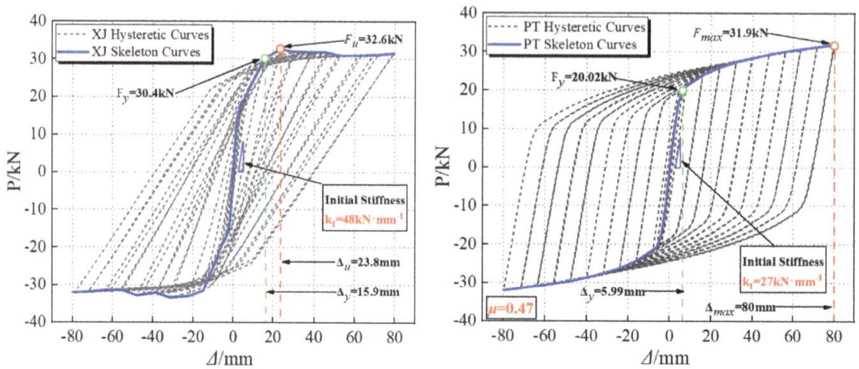

(a) Hysteretic Curves of XJ node (b) Hysteretic Curves of PT node

Fig. 6. Comparison of hysteretic curves between XJ node and PT node

Post-earthquake Rapid Functional Recovery Performance Analysis

To verify the post-earthquake functional recovery performance of the structure, this study utilized the energy balance functionality in ABAQUS to extract the energy dissipation ratios of various components and beams-columns at the prefabricated node. In this node, the overall model energy dissipation can be divided into two parts: the first part is frictional energy dissipation (ALLFD), and the second part is plastic energy dissipation (ALLPD), as illustrated in Fig. 7 The left vertical axis in the figure compares the energy dissipation of different components, including the total energy dissipation of the frictional energy system (F-eds), the U-shaped shear-resistant steel plate system(U-sps), and the concrete beam-column construction system (C-b&c). The right vertical axis represents the ratio of the energy dissipated by the steel components to the total energy dissipation of the entire node (Te-r), with the horizontal axis being the load-bearing capacity coefficient.

As shown in Fig. 7, a comparison of the total energy dissipation of different components under various load coefficient values reveals that when the load coefficient is less than 0.81, the total energy dissipation of the node exhibits an increasing trend with the increase of the load coefficient. However, when the load coefficient exceeds 0.81, the total energy dissipation of the node shows a negative growth phenomenon. The total energy dissipation proportion of the node consistently decreases with an increase in the load coefficient. When μ is less than 0.81, the total energy dissipation proportion gradually decreases, and after μ exceeds 0.81, the energy dissipation proportion decreases rapidly. When μ = 1.06, compared to μ = 0.96, the energy dissipation proportion decreases

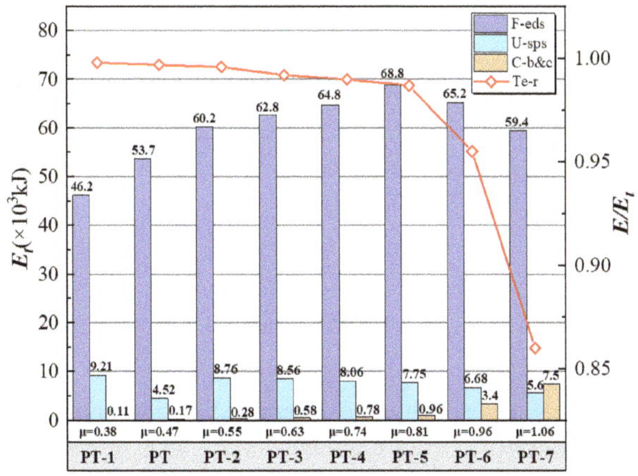

Fig. 7. Total energy consumption of each node component

from 96% to 86%. This decline in energy dissipation proportion is primarily attributed to the fact that, with a load coefficient greater than $\mu > 0.81$, main structural components such as beams and columns start to yield and participate in energy dissipation.

It is evident that for the concrete partial frame nodes selected in this paper, the critical point where the performance of the steel nodes undergoes a significant change is when the load-bearing capacity coefficient $\mu = 0.81$. This differs from the consideration in previous studies, which used $\mu = 1$ as the dividing point. In reference [19], when $\mu < 1$, the replaceable components of the steel nodes yield earlier than non-dissipative components, thereby achieving the design objective of concentrated energy dissipation. However, in this paper, to maintain the main structure in an elastic phase throughout the entire stage, the later increase in the load-bearing capacity coefficient caused by the strengthening effect of U-shaped steel plates should be considered, as shown in Table 1. As all U-shaped steel plates use the same material and dimensions, the consistent increase ratio is 0.19. Hence, when the load-bearing capacity coefficient $\mu + 0.19 > 1$, the device's ability to transmit bending moments has already exceeded the yield load of the steel reinforcement, resulting in the transfer of energy dissipation from the connection components to the main structure. At this point, the connection components are unable to protect the main structure, which also elucidates that $\mu = 0.81$ is a specific parameter for the steel nodes in this paper.

As indicated in the preceding text, under the action of low-frequency cyclic loads, the steel node effectively dissipates energy and induces plastic hinges to appear in the steel node region. When the load coefficient is in the range of $\mu = 0.47$ to 0.81, it ensures that the steel node has sufficient bearing capacity and energy dissipation performance while also ensuring minimal or low damage to the main structural components. After an earthquake, the structure can quickly regain its functional use by simply replacing friction pads, bolt assemblies, and U-shaped steel plate components as needed.

For such resilient connection nodes, the load coefficient is a critical parameter influencing both seismic performance and recoverability of the structure. The minimum load coefficient can be determined by designing the resilient node to have a bearing capacity comparable to that of cast-in-place concrete nodes. The maximum load coefficient is determined based on the increase in bearing capacity due to the strengthening of some components after node yield and the deformation-induced rise in structural capacity. Finally, values within this range can be selected based on actual engineering requirements to meet varying seismic demands.

4 Conclusion

This paper, through numerical simulation and parametric analysis, draws the following conclusions:

(1) Under the same bearing capacity, the assembled beam-column joints proposed in this paper are significantly better than the cast-in-place joints in terms of ductility and energy dissipation capacity, showing excellent seismic performance.
(2) The structural load-bearing coefficient significantly affects the seismic and recoverable performance of this steel node. When the load-bearing coefficient is not greater than 0.81, the total energy dissipation of the node increases with an increase in the load-bearing coefficient. Within the range of coefficients from 0.47 to 0.81, the node exhibits the best energy dissipation capability and recoverability.
(3) This structural system departs from the traditional concept of 'equivalent cast-in-place.' Under moderate or severe seismic effects, structural damage and seismic energy are concentrated and dissipated within the steel nodes, thereby safeguarding the main structures such as concrete beams, columns, and the core area of the joints. Post-earthquake, only a partial replacement of detachable components is required for the restoration of structural functionality, forming a resilient structural system.

Acknowledgments. The research was supported by Foundation of Liaoning Province Education Administration (Grant No. JYTZD2023163) and Shenyang Bureau of Science and Technology (Grant No. 23-407-3-18). The writers wish to express gratefully acknowledged.

References

1. Korkmaz, H.H., Tankut, T.: Performance of a precast concrete beam-to-beam connection subject to reversed cyclic loading. Eng. Struct. **27**(9), 1392–1407 (2005)
2. Savoia, M., Buratti, N., Vincenzi, L.: Damage and collapses in industrial precast buildings after the 2012 Emilia earthquake. Eng. Struct. **137**, 162–180 (2017)
3. Englekirk, R.E.: Development and testing of a ductile connector for assembling precast concrete beams and columns. PCI J. **40**(2), 36–51 (1995)
4. Nakaki, S.D., Englekirk, R.E., Plaehn, J.L.: Ductile connectors for a precast concrete frame. PCI J. **39**(5), 46–59 (1994)
5. Xiangmin, L., Rundong, G., Qingfeng, X.: R&D of low yield and high ductile rods and its application in precast concrete joint. Earthq. Resistant Eng. Retrofitting **34**(4), 42–46 (2012). (in Chinese)
6. Zhao, B., Lv, X.L., Liu, L.Z.: Experimental study on seismic behavior of precast concrete beam-column subassemblage with fully assembled joint. Earthq. Eng. Eng. Vib. **25**(1), 81–87 (2005). (in Chinese)

7. Lv, X.L., Quan, L.M., Jiang, H.J.: Research trend of earthquake resilient structures seen from 16WCEE. Earthq. Eng. Eng. Dyn. **37**(3), 1–9 (2017). (in Chinese)
8. Lv, X.L., Chen, Y., Mao, Y.J.: New concept of structural seismic design: earthquake resilient structures. J. Tongji Univ. **39**(7), 941–947 (2011). (in Chinese)
9. Ye, J.F., Zheng, L.Q., Yan, G.Y., et al.: Experimental study on hysteretic performance of replaceable energy-dissipating prefabricated hinges. Eng. Mech. **38**(8), 42–54 (2021). (in Chinese)
10. Hu, G., Huang, W., Xie, H.: Mechanical behavior of a replaceable energy dissipation device for precast concrete beam-column connections. J. Constr. Steel Res. **164**, 105816 (2020)
11. Ma, Z.H., Zhang, J.G., Liang, H.Z., et al.: Numerical research on prefabricated frame joint based on artificial dissipative plastic hinge under low-reversed loading. China Civ. Eng. J. **53**(S2), 162–168 (2020). (in Chinese)
12. Song, L.L., Guo, T., Cao, Z.L.: Seismic response of self-centering prestressed concrete moment resisting frames with web friction devices. Solid Dyn. Earthq. Eng. **71**, 151–162 (2015)
13. Aninthaneni, P.K., Dhakal, R.P., Marshall, J., et al.: Nonlinear cyclic behavior of precast concrete frame sub-assemblies with "dry" end plate connection. Structures **14**, 124–136 (2018)
14. Guo, Y., Wang, M.: Experimental study on mechanical behavior and resilient performance of steel frame connection with low-yield-point steel fuses. Eng. Struct. **266**, 114599 (2022). https://doi.org/10.1016/j.engstruct.2022.114599
15. Song, L.L., Guo, T.: Study on seismic performance of self-centering prestressed concrete frames with web friction devices. Eng. Mech. **31**(12), 47–56 (2014). (in Chinese)
16. Hoveidae, N., Tremblay, R., Rafezy, B., et al.: Numerical investigation of seismic behavior of short-core all-steel buckling restrained braces. J. Constr. Steel Res. **114**, 89–99 (2015)
17. Latour, M., Piluso, V., Rizzano, G.: Experimental analysis on friction materials for supplemental damping devices. J. Constr. Build. Mater. **65**, 159–176 (2014)
18. Park, R.: Evaluation of ductility of structures and structural assemblages from laboratory testing. Bull. New Zealand Nat. Soc. Earthq. Eng. **22**(3), 155–166 (1989)
19. Men, J.J., Li, T., Zhang, H.H.: Study on seismic behavior and on earthquake-resilient performance of beam-column joint with replaceable t-stubs. Eng. Mech. 1–17 (2023). http://kns.cnki.net/kcms/detail/11.2595.03.20230303.1504.028.html. Accessed 05 July 2023, (in Chinese)

Research on Mechanical Properties of Square Steel Tubular Column-H-Shaped Steel Beam Connected with External Diaphragm

Weining Sui(✉), Lin Zhu, and Yu Hou

Civil Engineering College, Shenyang Jianzhu University, No.25, Hunnan Middle Road, Shenyang 110000, China
weiningsui@sjzu.edu.cn

Abstract. The mechanical properties of beam-to-column joints of steel frames are an important factor in determining whether steel frames can perform normally under seismic action, but the current research mainly focuses on the mechanical properties of semi-rigid connection such as cruciform or end-plate types, and the research on the performance and seismic performance of semi-rigid connection in steel frames is not sufficient. In view of this, finite element simulations of single-storey single-span experimental steel frames are carried out using finite element analysis software ABAQUS and verified by tests, and multi-parameter analysis is carried out on the width-to-thickness ratio of the column and the height-to-width ratio of the panel zones, to obtain the effects of different parameters on the mechanical properties of the steel frames with semi-rigid connection, and to provide support for the promotion of the use of semi-rigid connection in steel frame design. The research results indicated that: the initial stiffness, yield load and ultimate load of the steel frame increase with the increase of column width-to-thickness ratio; the structural load and initial stiffness increase with the increase of the height of the beam web, and the stress concentration phenomenon at the connection and beam flanges can be improved by reducing the height of the beam web, but the change of the aspect ratio of the panel zones has a small effect on the structural ductility.

Keywords: Panel zone · Semi-rigid connection · Square steel tubular column-H-beam steel frame · Mechanical properties · Finite element analysis

1 Introduction

In recent years, with the rapid development of the national economy, the social and industrial structure has been optimised. The traditional reinforced concrete structure has gradually withdrawn from the stage of history, and the steel structure has been widely used in various building structures because of its good anti-seismic performance, convenient construction, green environmental protection and other advantages. In steel frame structural systems, the mechanical properties of beam-column connection determine the

I. Tuns et al. (Eds.): CIBv 2023, LNCE 510, pp. 84–97, 2024.
https://doi.org/10.1007/978-3-031-60765-3_8

performance of buildings under seismic effects, such as in the 1994 Northridge earthquake in the U.S.A. and the 1995 Hanshin earthquake in Japan, which resulted in different degrees of brittle damage of connections due to the effects of welding residual stresses and weld defects at the welded connections [1, 2]. The results of the seismic investigations showed that the rigid connection had poor ductility and were not sufficient to dissipate the energy generated by the earthquakes, after which the study of beam-column joints in steel structures became a research priority [3, 4].

Eurocode 3 [5] firstly initiated the study of semi-rigid connection, expecting the seismic energy to be dissipated through the rotational deformation of the connection. Semi-rigid connection has various structural forms, and there are many variable parameters of members, and the force relationship between members is complicated, which makes it extremely difficult to solve the force performance of semi-rigid connection. However, with the large number of steel structure buildings, researchers have carried out a large number of studies on semi-rigid connection through theoretical analyses, experimental studies and numerical modelling methods to gain a more in-depth understanding of the seismic performance of the connection. Shi et al. [6] carried out a numerical simulation study on the relationship between the bolt diameter and the thickness of the end plate, and obtained the main mechanical property indexes such as the connection bearing capacity, initial stiffness and hysteresis curve by analysis the law of influence of the size of the beam cross-section on the force performance and damage characteristics of the semi-rigid connection. Li et al. [7] carried out an experimental study on weak axis end plate connection and the results showed that components such as end plate thickness and end plate stiffening ribs affect the initial stiffness and ultimate bearing capacity of the connections. Simplified calculation method of bending moment angle was also proposed and suggestions were made for design and engineering. Fanaie et al.[8] used the method of arranging external stiffeners to strengthen CFT beam-to-column joints and carried out experimental research work. The results showed that the damage mode of the new connection in the stiffened channel was the weld stress concentration damage, the steel beam flange arc weakening method could reduce the stress concentration of the beam flange, and the special bending-resistant steel frames could be applied with this connection. Alhendi et al. [9] discussed the law of influence of bolt strength and angle on connection forces under monotonic loading and evaluated its applicability by comparing it with the results of existing moment angle model calculations. Bian et al. [10] conducted low circumferential reciprocating loading tests for 14 steel framed with thin-walled square steel tubular column. The results showed that the welded connection had poor load carrying capacity and their stiffness, load carrying capacity and seismic performance could be improved by arranging stiffening ribs. Wang et al. [11] proposed a new scheme of replaceable angle connectors for steel frame joints, which achieves structural damage control and ensures elasticity of beams and columns by arranging angle connectors with high ductility, high energy dissipation capacity and low yield point. The influence of the weakening of the connectors and the size of the energy-consuming beam sections on the seismic performance of the connection is investigated. Sharma and Wang et al. used semi-rigid-jointed steel frames to study their seismic vulnerability and to evaluate the structural model's probability of exceedance and collapse reserve factor in the post-earthquake limit state, as well as to give recommendations for the design of semi-rigid connection stiffnesses and ultimate load carrying capacity [12, 13].

In the actual project, for multi-storey high-rise steel frame structure in both directions of the force situation is close to, should give priority to the selection of closed-section columns, closed-section columns and H-beam connections are usually used through diaphragm connection, the internal diaphragm connection or the external diaphragm. In this regard, Sui et al. [14] completed the hysteresis performance test on the steel frame with through diaphragm connection. Wang et al. [15] completed the static tests and numerical simulations of two types of internal diaphragm and through diaphragm connection, and the results showed that the initial damage of the connections were located in the connecting cross-section of the lower flange of the beams, and the through diaphragm configuration made the lower flange of the beams more prone to fracture compared to the internal diaphragm configuration, and the size of the deformation of the connection substructures corresponding to the fracture of the lower flange of the beams mainly had an effect on the development of the bending mechanism, which then influenced the maximum dynamic loads that could be borne by the joint substructures. The maximum dynamic load that can be carried by the substructure is affected by the deformation of the joint substructure corresponding to the fracture of the lower flange of the beam.

The external diaphragm is simple to construct, does not need to cut the steel tubular column, ensures the integrity of the joint, avoids the eccentricity of the column that may be caused by the accumulation of construction errors, and the connection has a high rate of factory prefabrication, and in the form of the state vigorously advocating the assembly building, the external diaphragm connection will be more widely promoted and applied in the actual project. However, for the external diaphragm, according to the classification of joints in Eurocode 3, it can be seen that it generally belongs to the semi-rigid connection, so it is necessary to evaluate the mechanical properties of the connection. In this regard, Mou et al. [16] designed seven types of low circumferential reciprocating load tests of beam-column connection with external diaphragm with unequal height. The results showed that the damage mode of the specimens was dominated by shear damage in the whole or part of the panel zone, and the connection load-bearing deformation capacity was significantly affected by factors such as the height of the steel beam cross-section and the width-to-thickness ratio of the steel column cross-section, and the connections showed good stress performance. Yang et al. [17] completed four low-circumference reciprocating load tests based on flange-connected steel tubular concrete column-steel beam external diaphragm, and the results show that the specimens have better seismic performance, and the width of the external diaphragm is an important factor affecting its force performance. On the basis of finite element analysis and statistical regression, an optimal design method for the external diaphragm applicable to this connection form is proposed. Tong et al. [18] carried out eight wide flat rectangular steel tubular concrete columns - H-shaped steel beams small-scale external diaphragm connection proposed static test, the results show that the connection presents the beam flange and external diaphragm welding parts of the local buckling, cracking and connection domain of the column web shear damage in two modes, the small-scale external diaphragm still has a high load carrying capacity, in the engineering application, can be used to stretch the width of the external diaphragm than the current relevant design codes can be used.

In summary, researchers have carried out a large number of experimental and numerical simulation studies on common semi-rigid connections and semi-rigid connected steel frames, and have obtained the connection angle and stiffness calculation formula. However, the existing research mainly focuses on the cross-shaped connection or end plate type, and the influence of the performance of the external diaphragm connection in steel frames is not sufficiently discussed, and there is a lack of clear calculation methods for the theoretical calculations of this type of steel frames at this stage, so it is necessary to carry out a research on steel frames.

Therefore, the aim of this paper is to carry out a multi-parameter analysis of steel frames to obtain the effects of different parameters on the bearing capacity of steel frames as well as the damage modes, to enrich the relevant theoretical studies and to provide support for the promotion of the use of semi-rigid connection in steel frame.

2 Finite Element Modelling and Validation

2.1 Finite Element Modelling

In order to study the mechanical properties of steel frames containing semi-rigid connections under low-cycle reciprocating loading process, a finite element model establishes based on the low-cycle reciprocating loading test specimen SR-0 of cold-formed, thick-walled square steel tubular column-H shaped steel beam connected with external diaphragm. The test specimen is a 1:2 scaled-down single-storey single-span planar frame specimen, with semi-rigid connections at all connections. Steel frame storey height is 1975 mm, width is 2700 mm, frame beam is H type steel beam with cross-section size 250 × 150 × 8 × 16, frame column is cold-formed thick-walled square steel tubular column with cross-section size 250 × 8, beam and column steel is Q355B, and bolts' specification is 10.9 grade M20 high-tensile bolts (see Fig. 1).

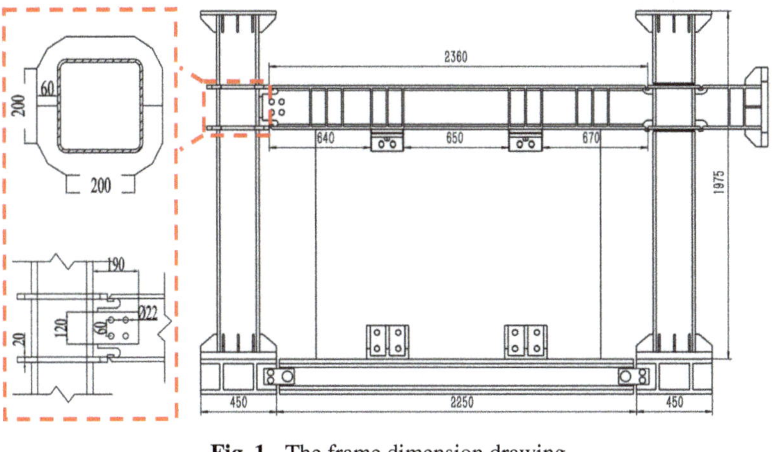

Fig. 1. The frame dimension drawing

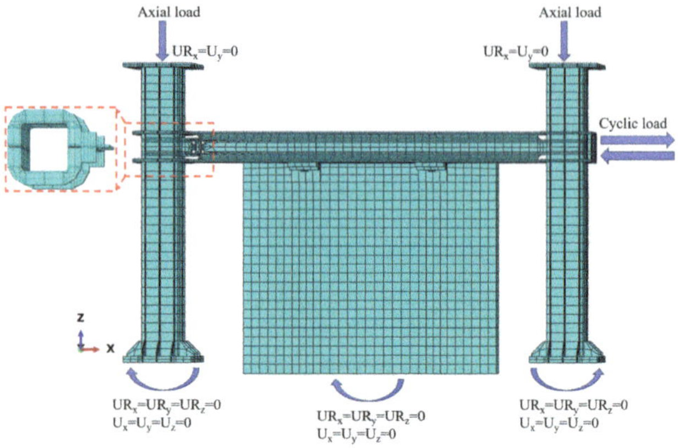

Fig. 2. The numerical modelling of H-beam external diaphragm steel frames

The finite element model is shown in Fig. 2. Considering the principle of model simplification, the three-dimensional eight-connection linear solid unit C3D8R with reduced integration is used, except for the high-strength bolts. The high-strength bolts are then modelled using the B31 cell. The model integrates material and geometric nonlinearities. In the test, the column foot was fixed with the support, the wall plate was fixed with the foundation beam, the foundation beam was hingedly connected with the support, and the foundation beam and the support were fixed relative to the ground. The foundation beam did not rotate significantly during the test loading process. Therefore, in the finite element model, considering the principle of model simplification, the column foot and wall plate are set as fully cemented. The model was built with full reference to the previous tests [14], including the test specimen size, material properties, boundary conditions and other factors. The material properties are taken according to the pre-test, as shown in Table 1. Welding-induced defects, welding residual stresses and residual deformations are treated using Merge. In the finite element modelling is simplified to lateral loading using cross-section coupling.

Table 1. Finite element material parameters

Test piece part	f_y/MP$_a$	f_u/MP$_a$	E/MPa	E_{st}/MPa	v
Beam flange	429	514	2.79×10^5	2.79×10^4	0.3
Beam web	357	510	1.81×10^5	1.81×10^4	0.3
Column	419.1	548.4	1.76×10^5	1.76×10^4	0.3
External diaphragm	257.1	430.2	1.73×10^5	1.73×10^4	0.3

Note: f_y is the yield strength; f_u is the tensile strength; E is the modulus of elasticity; E_{st} is the reinforcement modulus; v is the Poisson's ratio

2.2 Finite Element Model Validation

In this section, the hysteresis curve and damage modes of specimen SR-0 are compared with the FEA results to verify the accuracy and feasibility of finite element modelling of external diaphragm beam-column connections with ABAQUS software, and on this basis, the finite element analysis of column width-to-thickness ratio and panel zones aspect ratio is carried out for external diaphragm connections of square steel tubular column-H-beam, along with the same modelling method. The finite element simulation uses the same loading protocol as the test, as shown in Fig. 3. Figure 4 shows the comparison graph between the test and the finite element hysteresis curve. It can be seen from the graph that the hysteresis curve obtained from the finite element simulation analysis and the hysteresis curve obtained from the test are basically the same in shape, with a high degree of agreement, and in the elasticity stage, the two hysteresis curves basically overlap, both of which are in the form of a full, which shows a good ability to dissipate energy and mechanical properties; The symmetry of the hysteresis curve obtained is higher than the hysteresis curve obtained from the test because the boundary conditions, welding and other conditions in the finite element simulation are ideal. Figure 5 shows the stress cloud of specimen SR-0 finite element simulation loaded to a displacement of 49 mm, from which it can be seen that the plastic hinge was generated at the beam end of the frame when it was loaded to 49 mm, and plastic damage occurred, and the foot of the column also produced the local buckling failure phenomenon of the foot of the column at the stage of ultimate load. The finite element analysis results are in good agreement with the test results, thus verifying the correctness of the finite element analysis model.

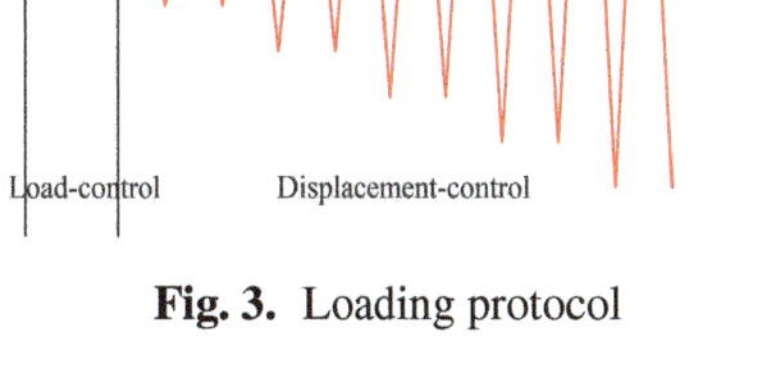

Fig. 3. Loading protocol

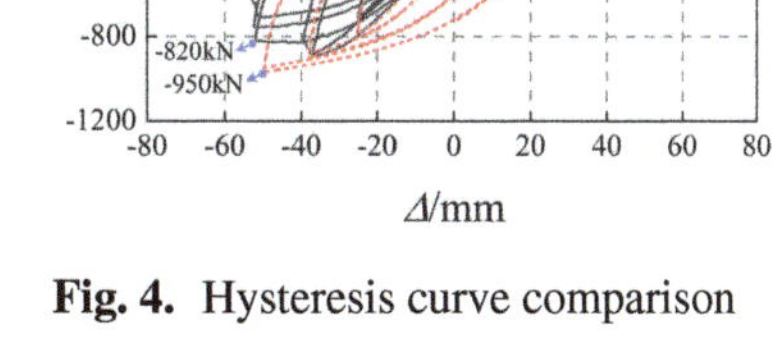

Fig. 4. Hysteresis curve comparison

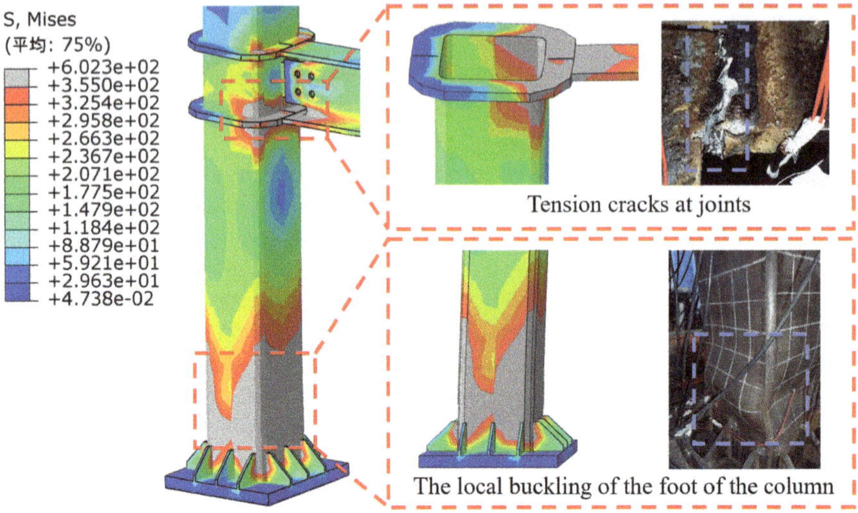

Fig. 5. Comparison between finite element simulation and experimental phenomena of specimen SR-0

3 Analysis of Finite Element Simulation Results

3.1 Influence of Width-to-Thickness Ratio *D/t* of the Column

In order to study the effect of column width-to-thickness ratio *D/t* on the mechanical properties of square steel tubular column-H-beam with external diaphragm in steel frame, three model specimens, SR-1, SR-2 and SR-3, were designed for finite element simulation and analysis of structural system under different *D/t* by using single-parameter analysis method. The geometrical parameters of the structural system with different *D/t* ratio are listed in Table 2.

Table 2. Geometrical parameters of the structural system

Specimen number	Section specification		*D/t*
	Column $D \times t$ (mm)	Beam $H \times B_f \times t_w \times t_f$ (mm)	
SR-1	250 × 6	250 × 150 × 8 × 16	42
SR-2	250 × 8	250 × 150 × 8 × 16	31
SR-3	250 × 10	250 × 150 × 8 × 16	25

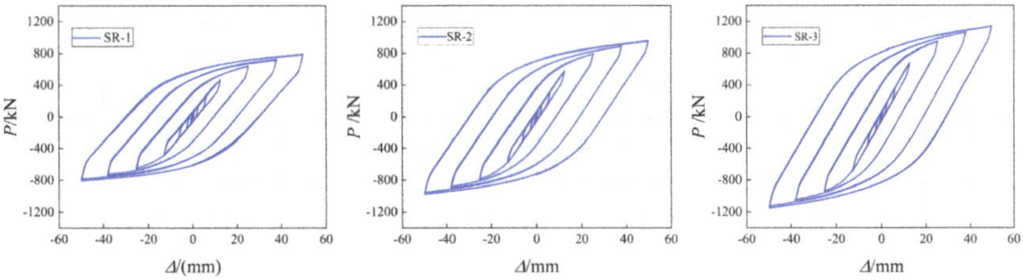

(a) SR-1 (b) SR-2 (c) SR-3

Fig. 6. The hysteresis curve of specimens

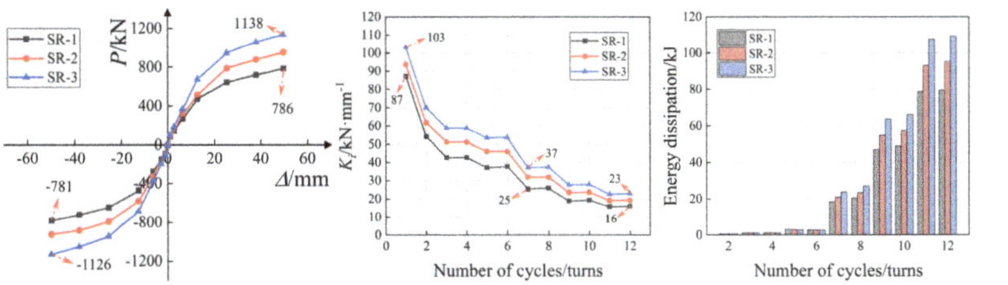

(a) Skeleton curve (b) Stiffness degradation curve (c) Energy consumption capacity

Fig. 7. Comparison of specimen curves for three column width-to-thickness ratios

Figure 6 shows the comparison of the hysteresis curves of the structural system for the three specimens, and the skeleton curve, stiffness degradation, and energy dissipation of a single turn of the structural system are listed in Fig. 7. As can be seen from the figures, the hysteresis curves of the models are full, which shows good energy dissipation capability. The energy dissipated by the specimen at the early stage of loading is very small and basically negligible. When the lateral loading continues to increase, the energy dissipation of the specimen begins to increase significantly, and the local buckling deformation of the frame column footing and the local buckling of the frame beam absorb and dissipate part of the seismic energy. As the width-to-thickness ratio decreases, the energy dissipation capacity of the structure continues to increase. From the skeleton curves, it can be seen that the initial stiffness and ultimate load carrying capacity of the structural system increase continuously with the increase of column wall thickness. In this paper, the method as shown in Fig. 8. The method shown to determine the yield load and yield displacement of the specimen. When the lateral horizontal displacement is less than or equal to the yield displacement (the displacement is about 0–20 mm), the specimen is in the elastic working state, and when the lateral horizontal displacement exceeds the yield displacement, the specimen enters the elastic-plastic working state. The lateral horizontal displacement and load are linearly related at the beginning of loading, and the specimen is in elastic working state. As the lateral displacement continues to increase, the skeleton curves begin to demonstrate non-linear behaviour. The specimen bearing capacity is significantly affected by the change in column width-to-thickness ratio, resulting in a gradual decrease in the magnitude of the load increase. The decrease in the slope of the curve indicates that the specimen stiffness has been degraded to different degrees and has entered the elastic-plastic working stage. The degradation trend of the stiffness of each specimen is basically the same. The initial stiffness of specimen SR-1 was 87 kN/mm, and the initial stiffness of SR-3 was increased by 18% relative to SR-1.

The yield point in the structural skeleton curve is determined by the universal bending moment method to obtain the ductility capacity of the ductile square steel tubular column-H-beam external diaphragm steel structure system. The characteristic points of the skeleton curves of the structural system for the three column width-to-thickness ratios are listed in Table 3, from which it can be seen that the smaller the width-to-thickness ratio of the columns, the higher the yield load of the specimen, the higher the yield displacement, and the decrease in the coefficient of ductility. Compared with specimen SR-1, the yield load of specimen SR-3 is increased by 50%, the ultimate load capacity is increased by 45% and the ductility coefficient is decreased by 8%. At a column wall thickness of 10 mm (width-to-thickness ratio of 25), the ductility factor is 2.42, which is greater than the *Specification for Structural Steel Buildings* (AISC 360–10) and the general steel structure ductility coefficient criterion 2 specified in *Seismic Provisions for Structural Steel Buildings* (AISC 341–10) indicate that the specimen ductility is still in a desirable state.

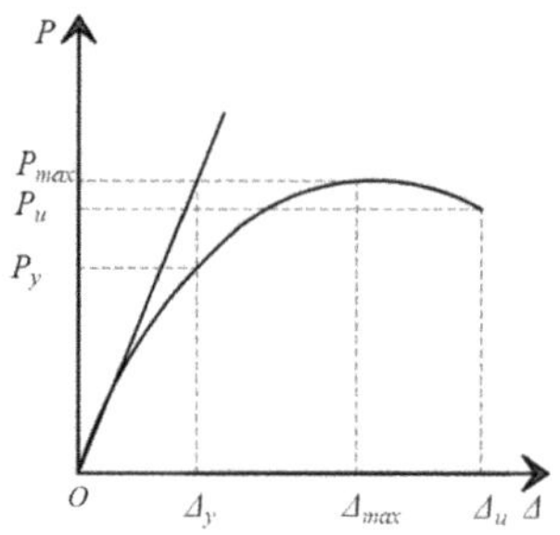

Fig. 8. The universal bending moment method

Table 3. Characteristic points of variation of column width-to-thickness ratio parameters

Specimen number	Yield load point		Ultimate load point		μ
	P_y/kN	Δ_y/mm	P_u/kN	Δ_u/mm	
SR-1	561.17	19.08	786.58	49.9	2.62
SR-2	664.85	19.81	956.36	49.91	2.52
SR-3	839.13	20.61	1138.56	49.89	2.42

3.2 Influence of Aspect Ratio *H*/*D* of Panel Zone

When the semi-rigid connection is subjected to lateral load, the beam web is the main shear force transfer member and will transfer the shear force of the frame beam end to the column wall, in order to study the influence of the beam web height on the connection bearing capacity and stiffness, the finite element model is established, keeping other parameters unchanged and only changing the height of the beam web, the specific model parameters are shown in Table 4.

Table 4. Geometrical parameters of the structural system

Specimen number	Section specification		H/D
	Column $D \times t$ (mm)	Beam $H \times B_f \times t_w \times t_f$ (mm)	
SR-2	250×8	$250 \times 150 \times 8 \times 16$	1
SR-4	250×8	$350 \times 150 \times 8 \times 16$	1.4
SR-5	250×8	$450 \times 150 \times 8 \times 16$	1.8

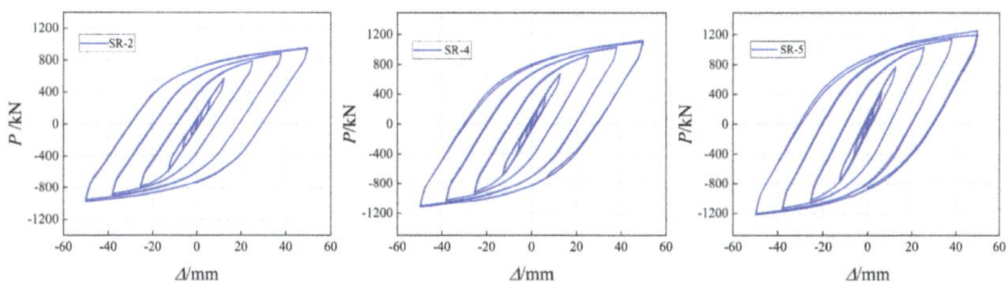

(a) SR-2 (b) SR-4 (c) SR-5

Fig. 9. The hysteresis curve of specimens

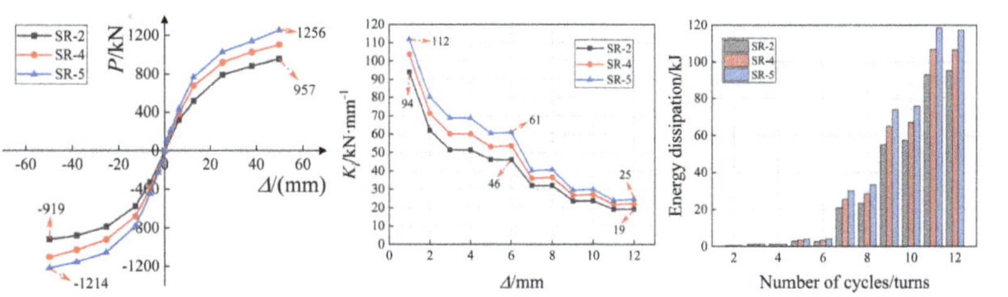

(a) Skeleton curve (b) Stiffness degradation curve (c) Energy consumption capacity

Fig. 10. Comparison of model curves for three panel zone height-to-width ratios

Figure 9 shows the comparison of the hysteresis curves of the specimens under different panel zone aspect ratios. It can be seen from the figures that the panel zone aspect ratio has little influence on the hysteresis performance of the structural system, and the hysteresis curves of the specimens have the same shape, all of them are in the shape of full "shuttle", which show good energy dissipation capacity and mechanical properties; At the beginning of loading, the hysteresis curves of the specimens basically overlapped and showed similar mechanical properties, and no residual deformation was produced when the load was reduced to zero. Figure 10 shows the skeleton curves, stiffness degradation, and comparative curves of energy dissipation in a single turn for each specimen with different aspect ratio in the panel zone. From the figures, it can be seen that different panel zone height-to-width ratios have a greater effect on the P-Δ skeleton curve stiffness, and the stiffness of the elastic phase increases with the increasing H/D. The initial stiffness of SR-5 is improved by 19% compared with that of SR-2. The stiffness of each specimen continued to decrease with model loading, while the stiffness of specimens with larger initial stiffness decreased faster. With the growth of plastic deformation, its single-turn hysteresis back dissipation energy tends to increase.

The yield loads, ultimate loads, etc. of the three panel zones aspect ratio specimens extracted from the skeleton curves are listed in Table 5. As can be seen from the table, the increase in the height-to-width ratio of the panel zone has a significant effect on the ultimate load carrying capacity of the specimens, which is increased by 16% and 31% for specimens SR-4 and SR-5, respectively, compared with specimen SR-2. The change in the height of the beam web has a small effect on the overall frame ductility coefficient, and the ductility coefficients of each specimen are greater than 2, which indicates that the structural energy dissipation performance is good.

Table 5. Panel zone aspect ratio parameter variation characteristic points

Specimen number	Yield load point		Ultimate load point		μ
	P_y/kN	Δ_y/mm	P_u/kN	Δ_u/mm	
SR-2	664.85	19.81	956.36	49.91	2.52
SR-4	811.14	19.91	1107.86	49.95	2.51
SR-5	892.75	19.11	1256.43	49.9	2.61

The effect of increasing the beam web height on the stress-strain and deflection diagrams of the frame containing steel frames with semi-rigid connection at ultimate displacement state is shown in Fig. 11. Figure 12 shows the equivalent plastic strain of the specimen loaded to the ultimate displacement state, which refers to the total amount of deformation in the region subjected to the action of all-directional stress components. When $0 < \text{PEEQ} \leq 1$, the test specimen is in the elastic working state, and when PEEQ > 1, it indicates that the test specimen enters into the plastic working state. It can be seen from the graph that increasing the height of the beam web has a greater impact on the stress and deformation of the steel frame, and the connection between the external diaphragm and the beam flange produces an obvious stress concentration phenomenon, the maximum plastic deformation in the deformation region increased from 0.988 to 1.786, indicating that the specimen enters the plastic working state. In practice, the brittle damage of the weld is prone to occur here to make the structural failure, and the place enters into the yielding stage earlier. The stress concentration at the beam flange connection is due to the fact that part of the load at the beam end is transferred to the external diaphragm through the beam flange, and the other part of the shear force is transferred to the column wall through the beam web. As the height of the beam web continues to decrease, the degree of stress concentration in the beam flange connection and the column foot gradually decreases, and the seismic performance is improved, the web of the panel zone will be unstable with the increase in the height of the beam web, and the height of the beam web should be taken into account in the design of the panel zone.

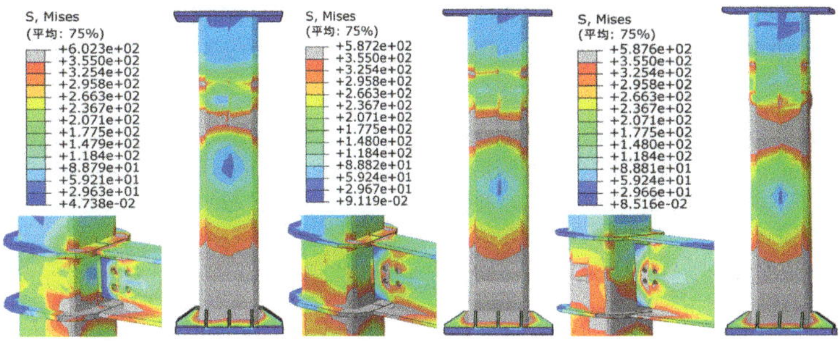

(a) SR-2 (b) SR-4 (c) SR-5

Fig. 11. Stress clouds of the structural system for three panel zone aspect ratios

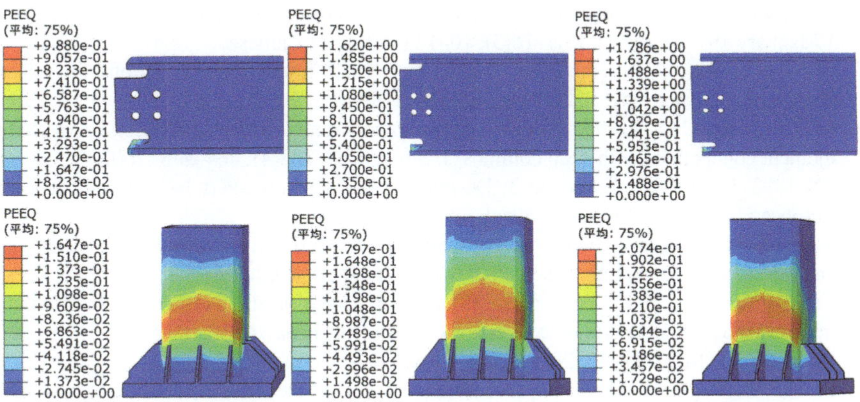

(a) SR-2 (b) SR-4 (c) SR-5

Fig. 12. PEEQ of the structural system for three panel zone aspect ratios

4 Conclusion

In this paper, the following conclusions are obtained from the finite element parametric analysis of the frame column width-to-thickness ratio and the panel zone height-to-width ratio:

(1) The frame column width-to-thickness ratio *D*/*t* has a greater effect on the specimen's bearing capacity and stiffness, with the increase of column width-to-thickness ratio, the specimen's bearing capacity and stiffness are in the trend of increasing, but the specimen's ductility will be continuously reduced with the increase of column width-to-thickness ratio.
(2) The variation of the aspect ratio *H*/*D* in the panel zone has an obviously effect on yield load and ultimate load of the structure, with the increase of the aspect ratio the ultimate load increased. The initial stiffness of the structure also increases with the increase in the height of the beam web, and the rate of increase in stiffness is also increasing. Reducing the height of the beam web improves the stress concentration at the connections and beam flanges as well as at the column footing. The changing in the aspect ratio of the panel zone has a small effect on the structural ductility.

Acknowledgments. The research was supported by Foundation of Liaoning Province Education Administration (Grant No. JYTZD2023163) and Shenyang Bureau of Science and Technology (Grant No. 23-407-3-18). The writers wish to express grate-fully acknowledged.

References

1. Watanabe, E., Sugiura, K., Nagata, K., Kitane, Y.: Performances and damages to steel structures during the 1995 Hyogoken-Nanbu earthquake. Eng. Struct. **20**(4–6), 282–290 (1998)
2. Qingchang, H.: Structural damage to houses in Kobe city in the great Hanshin earthquake of 17 January 1995. J. Struct. Eng. **16**(3), 10–12 (1995). (in Chinese)
3. Chen, S.J., Yeh, C.H., Chu, J.M.: Ductile steel beam-to-column connections for seismic resistance. J. Struct. Eng. **122**(11), 1292–1299 (1996)
4. Chi, B., Uang, C.M.: Cyclic response and design recommendations of reduced beam section moment connections with deep columns. J. Struct. Eng. **128**(4), 464–473 (2002)
5. EN 1993–1–8 (2005), Eurocode 3: Design of steel structures-Part 1–8: Design of joints. European Committee for Standardization
6. Yongjiu, S., Meng, W., Yuanqing, W., Gang, S.: Analysis on the behavior of steel frame end-plate connections. Eng. Mech. **28**(9), 51–58 (2011). (in Chinese)
7. Fengyu, L., Xiaoyan, Z., Yuzhou, S., Sifan, D.: Research on static tests of beam-to-column minor-axis extended end-plate connections and seismic behavior of steel frames. Prog. Steel Build. Struct. **18**(2), 41–47 (2016). (in Chinese)
8. Fanaie, N., Moghadam, H.S.: Experimental study of rigid connection of drilled beam to CFT column with external stiffeners. J. Constr. Steel Res. **153**, 209–221 (2019)
9. AlHendi, H., Mahmoud, M., Celikag, M.: Numerical study of top-and seat-angle with double web-angle connection with high-strength steel. J. Constr. Steel Res. **174**, 106297 (2020)
10. Bian, J., Cao, W., Zhang, Z., Qiao, Q.: Cyclic loading tests of thin-walled square steel tube beam-column joint with different joint details. Structures **25**, 386–397 (2020)
11. Meng, W., Yi, S., Lu, Y.: Damage control and optimal design of steel frame connection with low yield point angle steel components. J. Struct. Eng. **42**(12), 76–89 (2021). (in Chinese)
12. Sharma, V., Shrimali, M.K., Bharti, S.D., Datta, T.K.: Seismic fragility evaluation of semi-rigid frames subjected to near-field earthquakes. J. Constr. Steel Res. **176**, 106384 (2021)
13. Wei, W., Shuling, H., Chao, Z.: Seismic fragility analysis of beam – through steel braced frames based on IDA method. J. Struct. Eng. **42**(4), 42–49 (2021). (in Chinese)

14. Sui, W., Dong, Z., Zhang, X., Wang, Z.: Experimental study of the in-fluence of the shear strength of panel zones on the mechanical properties of steel frame structures. Structures **44**, 42–57 (2022)
15. Wei, W., Xi, Q., Junjie, W.: Comparison of progressive collapse behavior between inner-diaphragm stiffened and through-diaphragm stiffened concrete filled SHS column-H beam joints. J. Struct. Eng. **38**(S1), 362–368 (2017). (in Chinese)
16. Ben, M., Houkai, J., Chunwei, Z.: Experimental investigation on seismic behavior of unequal-depth H-shaped steel beam to square steel column connection with external-diaphragm. J. Struct. Eng. **39**(8), 80–89 (2018). (in Chinese)
17. Hong, Y., Yijin, Y., Le, K., Jinfeng, C., Yang, L.: Study on mechanical behavior of flange-connected CFST column-steel beam joint with external stiffening ring plate. J. Struct. Eng. **43**(2), 157–172 (2022). (in Chinese)
18. Lewei, T., et al.: Hysteretic behavior of small-size outer ring joints between concrete-filled rectangular tubular columns and H-shaped beams. J. Struct. Eng. **44**(2), 142–153 (2023). (in Chinese)

Estimation of FRP-To-Concrete Bond Strength of Carbon and Basalt Fabric Under Moisture Conditions: A Neural Network-Based Approach

Aman Kumar[1,2](✉), Harish Chandra Arora[1,2], Prashant Kumar[1,2], and Nishant Raj Kapoor[1]

[1] Academy of Scientific and Innovative Research (AcSIR), Ghaziabad 201002, India
aman.civil17@gmail.com

[2] Department of Structural Engineering, CSIR-Central Building Research Institute, Roorkee 247667, India
hcarora@cbri.res.in

Abstract. The deterioration of reinforced concrete (RC) structures is the main issue throughout the globe. The researchers are persistently dedicated to resolving the issue of deterioration in RC structures. In the last few decades, fiber-reinforced polymer (FRP) has been continuously used to retrofit deteriorated structural members. The various advantages of FRP composites such as high-strength-to-low-weight, ease of use and transport, etc. increase the use of FRP composites in the retrofitting industry. The FRP-to-concrete bond strength is one of the main contributing parameters that decide the efficacy of the FRP composite with the concrete surface. Various analytical models are available to predict the FRP-to-concrete bond strength, but there is no analytical model that can estimate the FRP-to-concrete bond strength under moisture conditions. This study aims to evaluate the FRP-to-concrete bond strength under moisture conditions with an artificial neural network and analyse the durability of FRP composite under moisture conditions. The considered dependent parameters that affect the bond strength under moisture conditions were the width of the concrete block, bonded length of FRP composite, thickness of FRP composite, width of FRP composite, exposure type, temperature, relative humidity, duration, elastic modulus of FRP composite and compressive strength of concrete. The developed machine-learning model was accurate, fast, reliable, easy to use, and economical to predict the FRP-to-concrete bond strength under moisture conditions.

Keywords: Deterioration · Bond strength · FRP-to-concrete bond · Moisture conditions · Machine learning · ANN

1 Introduction

Nowadays, the strengthening of deteriorated reinforced concrete (RC) structures with externally bonded fiber-reinforced polymer (FRP) is very common. Carbon, glass, and basalt composites are the most commonly used FRP materials. The carbon composites are available in both the wraps as well as in the form of laminates/plates [1].

I. Tuns et al. (Eds.): CIBv 2023, LNCE 510, pp. 98–110, 2024.
https://doi.org/10.1007/978-3-031-60765-3_9

The bond between the FRP-to-concrete surface is the key parameter for transferring the stresses from the concrete surface to the bonded fabric materials. Most of the observed failures in the research studies performed have happened due to the weakening of the bond between the composites and concrete [2]. Therefore, it is important to estimate the bond strength to delay or even avoid the bond failure. In scientific literature, standard guidelines as well as analytical models are accessible to estimate the bond strength between FRP composite and concrete substrate [3]. As, the bond is affected by several parameters like concrete grade, surface condition (surface preparation), mechanical properties of adhesive material and environmental factors, etc. Available standard guidelines as well as analytical models are capable of only estimating the FRP-to-concrete bond strength without external exposures. As per the author's knowledge, there is no standard guideline or analytical model, that can consider the effect of external exposure on the FRP-to-concrete bond strength.

Machine learning (ML) is currently being used extensively to address issues in structural engineering, concrete technology, and other fields. There are several ML models available in the published literature to predict the FRP-to-concrete bond [4–10]. However, only a few studies are available in the scientific literature to estimate the FRP-to-concrete bond strength under environmental exposures with ML techniques [11].

Yazdani et al. (2023) [12] evaluated the FRP-to-concrete bond strength with three design guidelines and twenty-eight analytical models and the collected dataset contains 345 test specimens. Among all the models, the performance of the Serbescu et al. (2013) [13] model was good and holds the first rank in the ranking system. The aim of this study is to modify the existing analytical model provided by Serbescu et al. (2013) [13] and develop an ANN-based ML model to estimate the FRP-to-concrete bond strength under different environmental exposures.

This manuscript is organised into four sections; in the first section, the introduction with research significance is described. The second section provides an overview of the collected dataset, analytical models, and development of the ANN model. The results of the analytical and developed ANN models are expressed in section third. The last section summarises the performance of the modified analytical model and the developed ANN model.

1.1 Research Significance

FRP-to-concrete bond is a crucial parameter that defines the overall efficiency of the strengthening system. Estimating the FRP-to-concrete bond strength under different environmental exposures is essential because external bonded strengthening composite systems are directly in contact with the outside environment. The available analytical and design guidelines did not consider the effect of environmental exposure on the FRP-to-concrete bond strength. This study aims to provide an analytical as well as computational-based model to estimate the FRP-to-concrete bond strength under different environmental exposures. Both the developed models are easy to use, provide good accuracy, and are reliable in estimating the FRP-to-concrete bond strength under different environmental exposures.

2 Methodology

The methodology section was categorized into three categories (i) details of the collected dataset and input parameters that affect the FRP-to-concrete bond strength under different environmental exposures, (ii) details of the analytical models, and overview of the ANN model and development of novel ANN-based model to predict the FRP-to-concrete strength under different exposures.

2.1 Details of Collected Dataset

The dataset, which includes 291 experimental values, was compiled from the published literature [14–24]. The considered input parameters were the compressive strength of concrete (f'_c), width of concrete block (b_c), bonded length (L_f), width of FRP composite (b_f), elastic modulus of FRP composite (E_f), tensile strength of FRP composite (f_f), exposure type (E_t), exposure condition (E_c), temperature (T), relative humidity (RH) and duration (D) and the bond strength (P_u) was the output parameter. To reduce the number of input parameters, the variables such as E_t, E_c, T, and RH were converted to E_t/E_c and T/RH. The correlation coefficient of the input and output parameters is shown in Fig. 1.

According to the Pearson correlation coefficient plot, the input parameters b_c, L_f, t_f, and D show a good relationship with FRP-to-concrete bond strength.

2.2 Performance Indices

To assess the performance of the ANN and analytical models, the six performance indices were used in this study namely: correlation coefficient (R), root mean square error (RMSE), mean absolute error (MAE), mean absolute percentage error (MAPE), Nash–Sutcliffe (NS) efficiency and a20-index. The mathematical expression of above mentioned performance indices is available in the research articles of Kapoor et al. [25] and Kumar and Saini [26].

2.3 Analytical Models

Serbescu Model

In 2013, Serbescu et al. [13] developed an FRP-to-concrete bond strength model. The expression to calculate the bond strength is expressed as:

$$P_u = \frac{2}{3}\beta \frac{455}{b_f + 350} b_f \left(0.8\sqrt{f_{cu}}\right) L_e \quad (1)$$

where, b_f and L_e are the width of the FRP composite and the effective length of the bond, respectively. β is the surface preparation factor and values are 0.55, 0.85, and 1.00 for minimal, recommended, and superior surface preparations, respectively, whereas f_{cu} is the compressive strength of concrete.

$$L_e = \sqrt{\frac{E_f t_f}{2.8 f_t}} \quad (2)$$

where, E_f and t_f are the elastic modulus, and thickness of FRP composite, respectively, while f_t is the tensile strength of concrete.

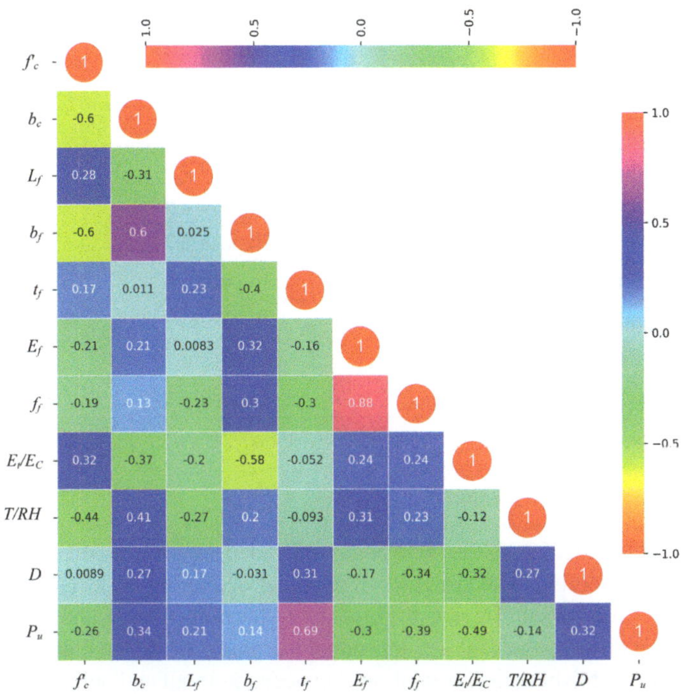

Fig. 1. Pearson correlation coefficient plot.

Modified Serbescu Model

The modified equation to estimate the bond strength is expressed in Eq. 3.

$$\left(0.55\frac{455}{b_f + 350}b_f\left(0.8\sqrt{f_{cu}}\right)L_e\right) \times E_p \times E_{factor} \tag{3}$$

From Eq. 2, the effective bond length may be computed.

where, f_t is the tensile strength of concrete and can be taken as 10% of the compressive strength (f_{cu}), and E_p represents the exposure factor, and the values of the exposure factor are given in Table 1. E_{factor} is the factor only valid for the elastic modulus of FRP composite less than 60 GPa. Otherwise, its value should be equal to one. The value of E_{factor} is equal to 3.5 for FRP composite whose E_f is less than 60 GPa.

Table 1. Different exposure conditions with suggested factors.

Exposure	Exposure Name	E_p
Water immersion	Exposure 1	1.1
270 thermal cycle in air	Exposure 2	1.0
5% NaCl salt water	Exposure 3	1.3
pH 4.0 H_2SO_4	Exposure 4	1.2
Aging in environmental chamber	Exposure 5	1.6
Water soaking + Freeze-thaw 100	Exposure 6	2.9
Freeze-thaw	Exposure 7	1.3
Elevated water temperatures	Exposure 8	0.8
Cyclic temperature	Exposure 9	1.35
Cyclic Wet–Dry Cycles in potable water	Exposure 10	1.6
Cyclic Wet–Dry Cycles in salt solution (5%)	Exposure 11	1.4

2.4 Artificial Neural Network

Artificial neural networks are the most commonly used supervised ML technique in the area of civil engineering. The structure of artificial neural networks (ANNs) typically consists of three fundamental layers: the input layer, the hidden layer, and the output layer. The input layer encompasses all the dependent variables, while the hidden layer serves the purpose of establishing the correlations between the input and output variables [27]. Consequently, the predicted value produced by the network tends to converge towards the desired output values.

Selection of Best Neuron

To enhance the performance of the ANN algorithm, the dataset was normalized in the range of -1 to + 1. This normalisation method was used to lessen the effect of different data scales and increase the effectiveness of the neural network learning process. Subsequently, the pre-processed dataset was partitioned into three distinct categories: training, testing, and validation. The training set constituted 70% (203 dataset), of the data, serving as the basis for training the ANN model. The testing set included 15% (44 dataset) of the data and was employed to assess the generalization and predictive capabilities of the trained network. Finally, the validation set, also comprising 15% (44 dataset) of the data, was utilized to fine-tune the model and validate its performance on unseen data instances.

The processed dataset went through the ANN algorithm and the neurons in the hidden layer were set to three to twenty neurons and a total of eighteen ANN models were developed. The selection of the best neuron was based on the correlation coefficient (R) and mean square error (MSE) values. Based on the ranking system of R and MSE values, the performance of the ANN model at the 14th neuron was good and shows the highest R-value and lowest MSE value among all the ANN models (3 to 20 neurons). The Fig. 2 also validated the best performance of the selected neuron. In Fig. 2(a) parrot colour circle shows the highest R-value in the training, testing, and validation set. Similarly, in Fig. 2(b), the lowest MSE-value selected by the best neuron for training, testing, and validation set is shown by a parrot colour circle.

The "*tangent Sigmoid*" and "*purelin*" activation function was used in the hidden and output layer respectively. In Subsect. 3.1, the formulation of the developed ANN model is presented.

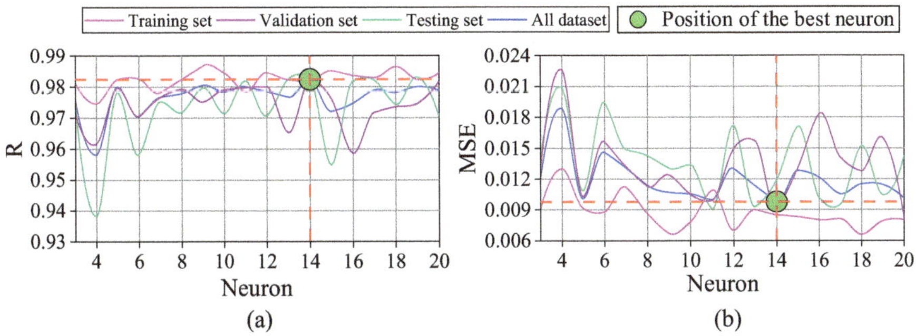

Fig. 2. Selection of optimum neuron based on (a) R and (b) MSE-value.

3 Results and Discussions

In this section, the results of the developed ANN model and analytical models have been explained. Table 2 presents an overview of the performance achieved by both the ANN model and analytical models. This table provides a comparative analysis of the effectiveness and efficiency achieved by these respective modeling approaches in addressing the research objectives. In the ANN model, the R-value of the training, testing, validation, and all datasets was 0.9844, 0.9755, 0.9710, and, 0.9814, respectively. The achieved R-value of the proposed modified model was 0.9683, which was 26.08% higher than Serbescu's [13] model. Similarly, the R-value of the evolved ANN model was 1.35% higher than the proposed analytical model. The NS and a20-index of proposed analytical model and ANN model were also greater than Serbescu's model.

MAE, MAPE, and RMSE metrics demonstrate how well the ANN and analytical models perform in terms of errors. Among all the models, the MAE, MAPE and RMSE values of the ANN model (for all datasets) were the lowest with values of 1.54 kN, 8.91%, and 2.05 kN, respectively, as shown in Table 2.

Table 2. Result of ANN and analytical models.

Model		R	MAE (kN)	MAPE (%)	RMSE (kN)	NS	a20-index	SD
ANN	Training	0.9844	1.44	8.03	1.92	0.9689	0.9655	10.51
	Validation	0.9755	1.92	11.08	2.34	0.9516	0.8409	
	Testing	0.9710	1.63	10.79	2.29	0.9383	0.8182	
	All	0.9814	1.54	8.91	2.05	0.9630	0.9244	
Proposed modified model		0.9683	2.11	11.97	2.77	0.9671	0.8144	10.40
Serbescu et al. (2013)		0.7680	10.98	49.92	7.83	0.2388	0.0825	4.87

The performance of the ANN and analytical models is shown graphically using a scatter plot, absolute error plot, and error frequency distribution plot. In Fig. 3 and Fig. 4, the first, middle, and third plot shows the scatter plot, absolute error plot, and frequency distribution plot, respectively. According to Fig. 3a, Fig. 3b and Fig. 3c (absolute error plot (middle), all the absolute error values were below 6 kN, in the training, and testing datasets, except the validation dataset.

In the validation dataset, only one value goes beyond the absolute error range of 6 kN. As per Fig. 3a, Fig. 3b, and Fig. 3c (frequency distribution of errors (third plot)), the range of the errors for training, validation, and testing dataset was −4.90 to 5.91 kN, -4.35 to 6.04 kN, and −4.83 to 5.97, respectively, while the error range of whole dataset was −4.90 to 6.04 kN.

In analytical models, the 35.73% and 95.88% values of the Serbescu's [13] and modified Serbescu's models was lies less than 6 kN error ranges as shown in Fig. 4a and Fig. 4b (middle plot), respectively. The error range of Serbescu's [13] and modified Serbescu's models was −0.20 to 31.27 kN and −7.20 to 9.48 kN, respectively (Fig. 4a and Fig. 4b (third plot)). The distribution of errors in the developed model (ANN) and analytical models is also presented using a violin with a box plot and distribution area plot (Fig. 5).

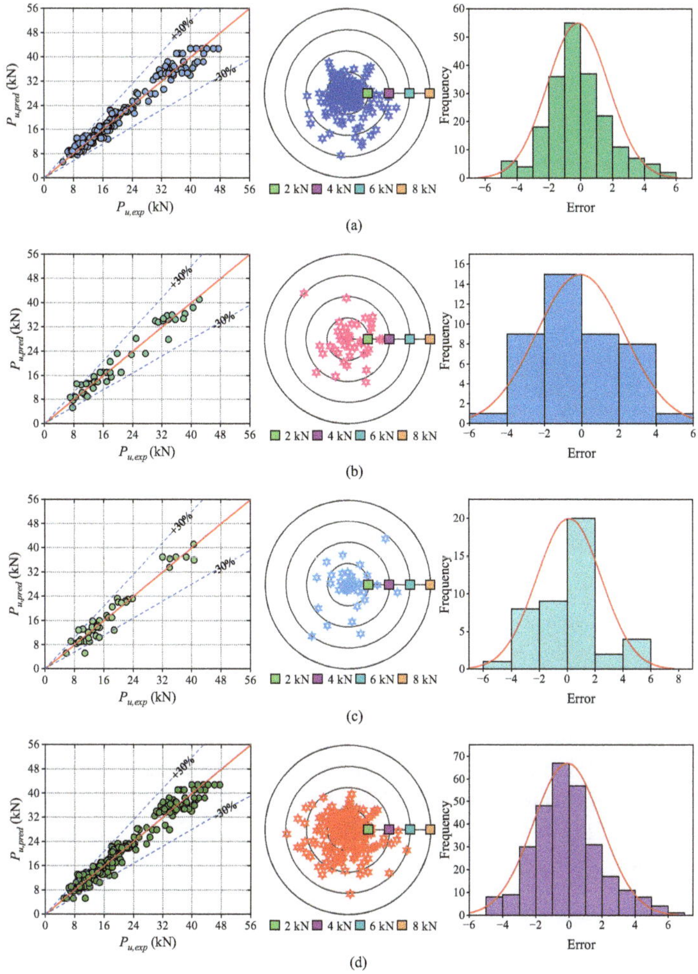

Fig. 3. Results of developed ANN model (a) training set, (b) validation set, (c) testing set, and (d) All set.

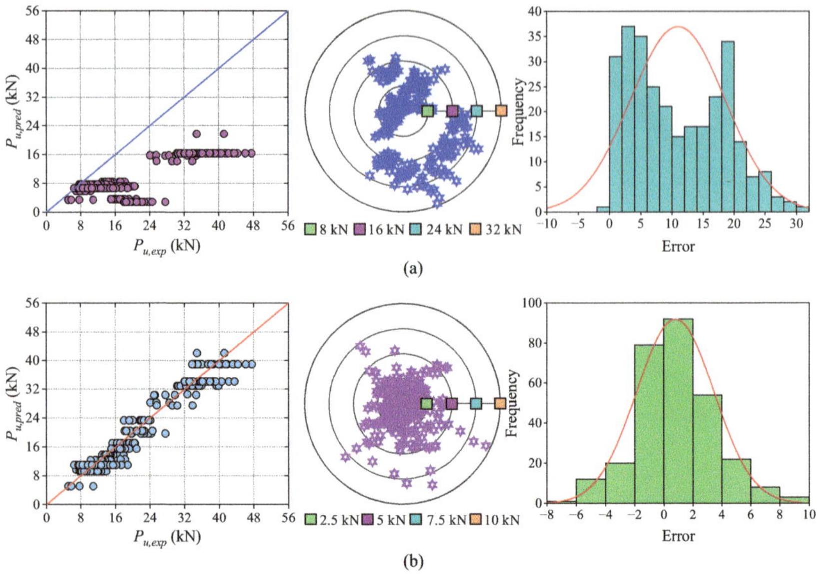

Fig. 4. Results of analytical models (a) Serbescu's model and (b) proposed modified model.

Taylor plot is also used to validate the performance of the developed (ANN) and analytical models as shown in Fig. 6. The *x* and *y*-axis represent the standard deviation (SD), the blue dotted line shows the SD of the experimental dataset, while the dark green line shows the RMSE values. The curved line shows the R-value of the analysed models. The performance of the ANN and modified Serbescu's model was very near to the dotted blue line and shows good performance.

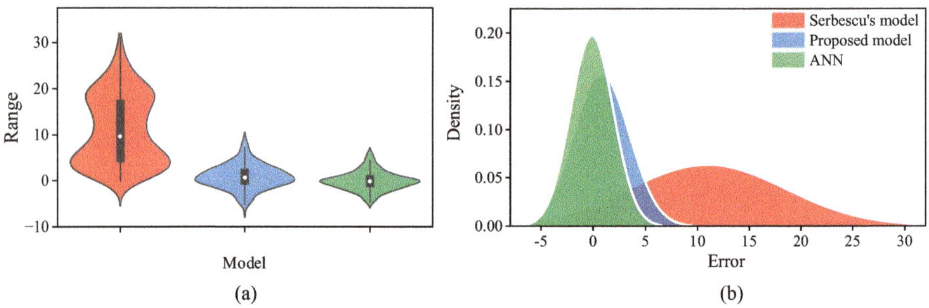

Fig. 5. Performance comparison of analytical models (a) Serbescu's model and (b) proposed modified model.

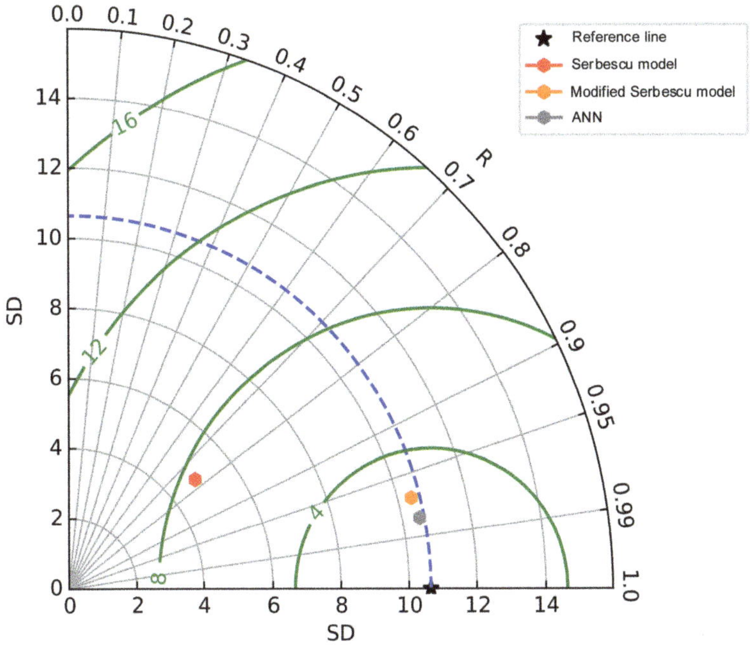

Fig. 6. Taylor plot.

3.1 Proposed ANN Formulation

To estimate the FRP-to-concrete bond strength under different exposure conditions the ANN formulation is expressed in Eq. 4.

$$\begin{aligned} & 0.4844A_1 - 0.6685A_2 + 0.6413A_3 + 0.4986A_4 + 1.0204A_5 - 0.6689A_6 + 0.8578A_7 \\ & \quad + 0.6642A_8 + 0.3686A_9 - 0.9173A_{10} + 0.4894A_{11} + 0.8359A_{12} \\ & \quad + 0.2317A_{13} + 0.0403A_{14} - 0.0677 \end{aligned} \tag{4}$$

The values A_1 to A_{14} coefficient can be evaluated from Eq. 5. The values of input parameters ($f'_c, b_c, L_f, b_f, E_f, f_f, E_t, E_c, T, RH$, and D) should be normalized and in the range of − 1 to + 1. The solved example to estimate the FRP-to-concrete bond strength under one exposure condition is available in Appendix I.

$$\begin{pmatrix} A_1 \\ A_2 \\ A_3 \\ A_4 \\ A_5 \\ A_6 \\ A_7 \\ A_8 \\ A_9 \\ A_{10} \\ A_{11} \\ A_{12} \\ A_{13} \\ A_{14} \end{pmatrix} = tansig \begin{bmatrix} -0.4148 & -0.2895 & 0.1368 & 0.8797 & -0.7015 & 0.8039 & -0.3462 & 0.6169 & 0.5726 & -0.4924 \\ 0.9332 & -0.3461 & -0.3908 & 0.0335 & 0.2762 & 0.3724 & 1.0030 & 1.3426 & -0.4838 & 0.7297 \\ 1.0314 & 0.4836 & -0.5836 & 0.4682 & -0.0660 & -0.2495 & 0.9417 & -0.3243 & 1.3545 & -0.4379 \\ 1.0282 & -0.1272 & -0.2445 & 0.0689 & 0.8713 & 0.4435 & 0.7724 & -0.4678 & 0.1353 & 0.6210 \\ -0.1186 & 0.2035 & -0.1940 & -0.6939 & 1.3109 & -0.6229 & 0.0060 & 0.2743 & -0.8721 & 0.2433 \\ -0.4287 & -0.2517 & -0.9981 & 0.6409 & -1.2150 & -0.4322 & -0.7316 & -0.2149 & -0.9185 & -0.3657 \\ -0.3708 & -0.4109 & -0.2186 & 0.8737 & -0.9105 & 0.0483 & 0.4415 & 1.2100 & -0.3050 & -0.0504 \\ -0.0032 & -1.0052 & -0.5519 & 0.4959 & 1.0435 & -0.5066 & 0.2078 & 0.9643 & 0.9207 & 0.0234 \\ 0.7246 & 0.3022 & -0.8216 & -0.7918 & 0.1079 & -0.3258 & -0.8519 & 0.3203 & -0.8532 & 0.1337 \\ 0.1407 & -0.8293 & -0.6083 & 0.4900 & -0.0361 & -0.4229 & 1.2153 & 0.5309 & -0.3012 & -0.4132 \\ 0.8722 & 0.3583 & -0.6467 & 0.4011 & -0.3742 & -0.2216 & -0.5190 & -1.0487 & -1.0875 & -0.1508 \\ -0.6906 & -0.8551 & 0.5989 & 1.0614 & 0.1864 & 0.2076 & 0.7533 & -0.5440 & 0.5136 & -0.5207 \\ -0.3387 & 0.2790 & 0.0834 & 0.0956 & -0.8698 & 0.7813 & -0.7756 & -0.9466 & -0.5115 & -0.4609 \\ 0.4985 & -0.2233 & 0.1020 & -0.4195 & -0.2689 & 0.7401 & -1.0913 & -0.8913 & 0.5125 & 0.3365 \end{bmatrix} \times \begin{bmatrix} f'_c \\ b_c \\ L_f \\ b_f \\ t_f \\ E_f \\ f_f \\ \frac{E_t}{E_c} \\ \frac{T}{RH} \\ D \end{bmatrix} + \begin{bmatrix} 1.9346 \\ -1.3910 \\ -0.7749 \\ -1.0929 \\ 1.2827 \\ 0.6457 \\ -0.1089 \\ 0.0397 \\ 0.5559 \\ 0.3037 \\ 1.0323 \\ -1.1543 \\ -1.5117 \\ 1.7724 \end{bmatrix} \tag{5}$$

4 Conclusions

In this study, the FRP-to-concrete bond strength under different exposure conditions was predicted using developed ANN model and analytical models. In addition to that, a modified Serbescu model had also been proposed to estimate the bond strength under different exposures. Based on the analysed results, the following conclusions were been drawn:

The performance of the Serbescu model was not much effective in the exposure conditions and shows R and MAE values of 0.7680 and 10.98 kN, respectively. On the other hand, the modified Serbescu model shows better performance as compared with the Serbescu model. The R, MAE, MAPE, and RMSE values of the modified Serbescu model were 0.9683, 2.11 kN, 11.97%, and 2.77 kN, respectively. The developed ANN model shows good accuracy as compared to Serbescu [13] and modified Serbescu model. The R-value of the ANN model was 1.35% and 27.78% higher than the modified Serbescu and Serbescu [13] model. Even the MAPE of the developed ANN model was 25.56% lower than the modified Serbescu model. The performance of the ANN model and modified Serbescu's model was also verified by the Taylor and error plots.

Appendix I

Considered input parameters are $f'_c = 0.3922$, $b_c = -1$, $L_f = 0.2$, $b_f = -1$, $t_f = 0.8534$, $E_f = -0.3244$, $f_f = -0.4157$, $\frac{E_t}{E_t} = 1$, $\frac{T}{RH} = -0.0256$, and $D = 0.3332$

$$\begin{pmatrix} A_1 \\ A_2 \\ A_3 \\ A_4 \\ A_5 \\ A_6 \\ A_7 \\ A_8 \\ A_9 \\ A_{10} \\ A_{11} \\ A_{12} \\ A_{13} \\ A_{14} \end{pmatrix} = tansig \begin{bmatrix} -0.4148 & -0.2895 & 0.1368 & 0.8797 & -0.7015 & 0.8039 & -0.3462 & 0.6169 & 0.5726 & -0.4924 \\ 0.9332 & -0.3461 & -0.3908 & 0.0335 & 0.2762 & 0.3724 & 1.0030 & 1.3426 & -0.4838 & 0.7297 \\ 1.0314 & 0.4836 & -0.5836 & 0.4682 & -0.0660 & -0.2495 & 0.9417 & -0.3243 & 1.3545 & -0.4379 \\ 1.0282 & -0.1272 & -0.2445 & 0.0689 & 0.8713 & 0.4435 & 0.7724 & -0.4678 & 0.1353 & 0.6210 \\ -0.1186 & 0.2035 & -0.1940 & -0.6939 & 1.3109 & -0.6229 & 0.0060 & 0.2743 & -0.8721 & 0.2433 \\ -0.4287 & -0.2517 & -0.9981 & 0.6409 & -1.2150 & -0.4322 & -0.7316 & -0.2149 & -0.9185 & -0.3657 \\ -0.3708 & -0.4109 & -0.2186 & 0.8737 & -0.9105 & 0.0483 & 0.4415 & 1.2100 & -0.3050 & -0.0504 \\ -0.0032 & -1.0052 & -0.5519 & 0.4959 & 1.0435 & -0.5066 & 0.2078 & 0.9643 & 0.9207 & 0.0234 \\ 0.7246 & 0.3022 & -0.8216 & -0.7918 & 0.1079 & -0.3258 & -0.8519 & 0.3203 & -0.8532 & 0.1337 \\ 0.1407 & -0.8293 & -0.6083 & 0.4900 & -0.0361 & -0.4229 & 1.2153 & 0.5309 & -0.3012 & -0.4132 \\ 0.8722 & 0.3583 & -0.6467 & 0.4011 & -0.3742 & -0.2216 & -0.5190 & -1.0487 & -1.0875 & -0.1508 \\ -0.6906 & -0.8551 & 0.5989 & 1.0614 & 0.1864 & 0.2076 & 0.7533 & -0.5440 & 0.5136 & -0.5207 \\ -0.3387 & 0.2790 & 0.0834 & 0.0956 & -0.8698 & 0.7813 & -0.7756 & -0.9466 & -0.5115 & -0.4609 \\ 0.4985 & -0.2233 & 0.1020 & -0.4195 & -0.2689 & 0.7401 & -1.0913 & -0.8913 & 0.5125 & 0.3365 \end{bmatrix} \times \begin{bmatrix} 0.3922 \\ -1 \\ 0.2 \\ -1 \\ 0.8534 \\ -0.3244 \\ -0.4157 \\ 1 \\ -0.0256 \\ 0.3332 \end{bmatrix} + \begin{bmatrix} 1.9346 \\ -1.3910 \\ -0.7749 \\ -1.0929 \\ 1.2827 \\ 0.6457 \\ -0.1089 \\ 0.0397 \\ 0.5559 \\ 0.3037 \\ 1.0323 \\ -1.1543 \\ -1.5117 \\ 1.7724 \end{bmatrix} \tag{A1}$$

$$\begin{aligned} &0.4844 \times (0.7314) - 0.6685 \times (0.4664) + 0.6413 \times (-0.9805) + 0.4986 \times (-0.5824) + \\ &1.0204 \times (0.9977) - 0.6689 \times (-0.7686) + 0.8578 \times (-0.4900) + 0.6642 \times (0.9821) + \\ &0.3686 \times (0.9707) - 0.9173 \times (0.5217) + 0.4894 \times (-0.5491) + 0.8359 \times (-0.9856) + \\ &0.2317 \times (-0.9989) + 0.0403 \times (-0.9491) - 0.0677 - \mathbf{0.5868} \end{aligned} \tag{A2}$$

$$P_{u,un-normalized} = \frac{-0.5868 + 1}{2} \times 42.5 + 5.1 \tag{A3}$$

The predicted FRP-to-concrete bond strength was 13.88 kN, however, the experimental value was14.80 kN.

References

1. Derkowski, W., Walczak, R.: Possibilities of increasing effectiveness of RC structure strengthening with FRP materials. Materials **14**(6), 1387 (2021). https://doi.org/10.3390/ma14061387
2. Yao, J., Teng, J.G., Chen, J.F.: Experimental study on FRP-to-concrete bonded joints. Compos. B Eng. **36**(2), 99–113 (2005). https://doi.org/10.1016/j.compositesb.2004.06.001
3. Monaldo, E., Nerilli, F., Vairo, G.: Effectiveness of some technical standards for debonding analysis in FRP-concrete systems. Compos. B Eng. **160**, 254–267 (2019). https://doi.org/10.1016/j.compositesb.2018.10.022
4. Chen, S.Z., Zhang, S.Y., Han, W.S., Wu, G.: Ensemble learning based approach for FRP-concrete bond strength prediction. Constr. Build. Mater. **302**, 124230 (2021). https://doi.org/10.1016/j.conbuildmat.2021.124230
5. Zhang, J., Wang, Y.: Evaluating the bond strength of FRP-to-concrete composite joints using metaheuristic-optimized least-squares support vector regression. Neural Comput. Appl. **33**(8), 3621–3635 (2021). https://doi.org/10.1007/s00521-020-05191-0
6. Yuan, C., He, C., Xu, J., Liao, L., Kong, Q.: Bayesian optimization for selecting efficient machine learning regressors to determine bond-slip model of FRP-to-concrete interface. Structures **39**, 351–364 (2022). https://doi.org/10.1016/j.istruc.2022.03.043
7. Huang, H.-B., Zhang, W., Sun, Z.-G., Wang, D.-S.: Probabilistic models for predicting bond strength of externally bonded FRP-to-concrete joints based on Bayesian inference. Constr. Build. Mater. **329**, 127194 (2022). https://doi.org/10.1016/j.conbuildmat.2022.127194
8. Zhang, F., et al.: Prediction of FRP-concrete interfacial bond strength based on machine learning. Eng. Struct. **274**, 115156 (2023). https://doi.org/10.1016/j.engstruct.2022.115156
9. Pei, Z., Wei, Y.: Prediction of the bond strength of FRP-to-concrete under direct tension by ACO-based ANFIS approach. Compos. Struct. **282**, 115070 (2022). https://doi.org/10.1016/j.compstruct.2021.115070
10. Su, M., Zhong, Q., Peng, H., Li, S.: Selected machine learning approaches for predicting the interfacial bond strength between FRPs and concrete. Constr. Build. Mater. **270**, 121456 (2021). https://doi.org/10.1016/j.conbuildmat.2020.121456
11. Aghabalaei Baghaei, K., Hadigheh, S.A.: Durability assessment of FRP-to-concrete bonded connections under moisture condition using data-driven machine learning-based approaches. Compos. Struct. 114576 (2021). https://doi.org/10.1016/j.compstruct.2021.114576
12. Yazdani, A., Sanginabadi, K., Shahidzadeh, M.-S., Razmyan, S., Salimi, M.-R., Mostofinejad, D.: Evaluation of existing FRP-to-concrete bond strength models using data envelopment analysis. J. Compos. Constr. **27**(3), 04023026 (2023). https://doi.org/10.1061/JCCOF2.CCENG-4103
13. Serbescu, A., Guadagnini, M., Pilakoutas, K.: Standardised double-shear test for determining bond of FRP to concrete and corresponding model development. Compos. B Eng. **55**, 277–297 (2013). https://doi.org/10.1016/j.compositesb.2013.06.019
14. Pan, Y., Shi, J., Xian, G.: Experimental and numerical study of the CFRP-to-concrete bonded joints after water immersion. Compos. Struct. **218**, 95–106 (2019). https://doi.org/10.1016/j.compstruct.2019.03.043
15. Al-Lami, K., Colombi, P., D'Antino, T.: Influence of hygrothermal ageing on the mechanical properties of CFRP-concrete joints and of their components. Compos. Struct. **238**, 111947 (2020). https://doi.org/10.1016/j.compstruct.2020.111947
16. Liu, S., Pan, Y., Li, H., Xian, G.: Durability of the bond between CFRP and concrete exposed to thermal cycles. Materials **12**(3), 515 (2019). https://doi.org/10.3390/ma12030515
17. Gravina, R.J., Aydin, H., Visintin, P.: Extraction and analysis of bond-slip characteristics in deteriorated FRP-to-concrete joints using a mechanics-based approach. J. Mater. Civ. Eng. **29**(6), 04017013 (2017). https://doi.org/10.1061/(ASCE)MT.1943-5533.0001881

18. Zheng, X.H., Huang, P.Y., Guo, X.Y., Huang, J.L.: Experimental study on bond behavior of FRP-concrete interface in hygrothermal environment. Int. J. Polym. Sci. **2016**, 5832130 (2016). https://doi.org/10.1155/2016/5832130
19. Pan, Y., Xian, G., Silva, M.A.G.: Effects of water immersion on the bond behavior between CFRP plates and concrete substrate. Constr. Build. Mater. **101**, 326–337 (2015). https://doi.org/10.1016/j.conbuildmat.2015.10.129
20. Shi, J., Zhu, H., Wu, Z., Seracino, R., Wu, G.: Bond behavior between basalt fiber–reinforced polymer sheet and concrete substrate under the coupled effects of freeze-thaw cycling and sustained load. J. Compos. Constr. **17**(4), 530–542 (2013). https://doi.org/10.1061/(ASCE)CC.1943-5614.0000349
21. Colombi, P., Fava, G., Poggi, C.: Bond strength of CFRP–concrete elements under freeze–thaw cycles. Compos. Struct. **92**(4), 973–983 (2010). https://doi.org/10.1016/j.compstruct.2009.09.044
22. Lai, W.L., Kou, S.C., Poon, C.S., Tsang, W.F., Lai, C.C.: Effects of elevated water temperatures on interfacial delaminations, failure modes and shear strength in externally-bonded CFRP-concrete beams using infrared thermography, gray-scale images and direct shear test. Constr. Build. Mater. **23**(10), 3152–3160 (2009). https://doi.org/10.1016/j.conbuildmat.2009.06.012
23. Ceroni, F., Bonati, A., Galimberti, V., Occhiuzzi, A.: Effects of environmental conditioning on the bond behavior of FRP and FRCM systems applied to concrete elements. J. Eng. Mech. **144**(1), 04017144 (2018). https://doi.org/10.1061/(ASCE)EM.1943-7889.0001375
24. Li, J., Gravina, R., Visintin, P., Smith, S.T.: Durability and long-term performance of FRP-to-concrete joints under environmental conditioning: experimental and analytical study. J. Compos. Constr. **24**(4), 04020021 (2020). https://doi.org/10.1061/(ASCE)CC.1943-5614.0001023
25. Kapoor, N.R., Kumar, A., Kumar, A., Kumar, A., Kumar, K.: Transmission probability of SARS-CoV-2 in office environment using artificial neural network. IEEE Access **10**, 121204–121229 (2022). https://doi.org/10.1109/ACCESS.2022.3222795
26. Kumar, K., Saini, R.P.: Development of correlation to predict the efficiency of a hydro machine under different operating conditions. Sustain. Energy Technol. Assess. **50**, 101859 (2022). https://doi.org/10.1016/j.seta.2021.101859
27. Rafiq, M.Y., Bugmann, G., Easterbrook, D.J.: Neural network design for engineering applications. Comput. Struct. **79**(17), 1541–1552 (2001). https://doi.org/10.1016/S0045-7949(01)00039-6

Study on Flame Spread in the High-Rise Building with Different Window Aspect Ratios

Jiaming Liu(✉) and Yu Wang

School of Civil Engineering, Shenyang Jianzhu University, Shenyang, China
641351953@qq.com

Abstract. This paper studies the effect of window height to aspect ratios on the flame spread of high-rise buildings, and provides reference for the engineering and fire protection design of high-rise buildings. PyroSim is used to establish high-rise building models under six different window aspect ratios, and the temperature distribution isotherms of the model under different window aspect ratios are analyzed. The flame fusion height of longitudinal continuous four windows is increased by 0.3–0.7 m at 540 °C compared with that of three windows.It is concluded that the flame spread of the exterior walls of high-rise buildings is affected by the window aspect ratio and the number of Windows. The smaller the window aspect ratio is, the higher the flame fusion height is. Therefore, under the premise of meeting the building design specifications, the flame fusion height can be reduced by controlling the window aspect ratio in high-rise buildings, thus reducing the fire risk and gaining time for rescue.

Keywords: High-rise Building · Flame Fusion,Numerical Simulation · Aspect Ratio · Number of Windows · Dangerous Temperature

1 Introduction

High rise buildings are the most common type of building in Chinese cities today, and their fires have the characteristics of extremely fast spread, difficulty in extinguishing, and difficulty in rescuing trapped people. In response to the problem of vertical spread of high-rise building fires [1–6], scholars at home and abroad have conducted extensive research. Jeffs etal. [7] explored the impact of window overflow flames caused by indoor fires on outdoor insulation materials through experiments, and the results showed that: the power of the fire source The form of building structure and ventilation conditions will directly affect the degree of fire spread; Oleszkiewicz [8] studied the fire spread under different opening shapes using a full-scale test bench constructed with concrete slabs covered with non combustible materials. Experiments have shown that when the shape of the window approaches short width, most of the flames will flow out and spread towards the upper part of the wall; Galea et al. focused on studying the effect of window size on the propagation of fire along the exterior walls of buildings using computer numerical simulation methods. The research results indicate that a wide window is more likely to bring the plume flame closer to the outer facade than a narrow window; Qu Zhipeng

I. Tuns et al. (Eds.): CIBv 2023, LNCE 510, pp. 111–119, 2024.
https://doi.org/10.1007/978-3-031-60765-3_10

[10] used FDS to build an eighteen story concave flat residential building. Through numerical simulation, it was found that the smaller the aspect ratio of the window, the higher the temperature of the flames sprayed from the window on the exterior facade of the building; Fu Jiajia et al. [11]. Established a three story building facade fire overflow barrier simulation test bench. The experiment found that when the window height to width ratio is small, the width and thickness of the overflow fire plume are larger, making the lateral suction of the overflow plume weaker than the forward suction. The fire plume is prone to adhere to the wall after overflow.

In the above research, the actual size of high-rise buildings was taken as the research object, and there were relatively few specific forms and numerical analysis results of flame melting height.

On this basis, this article takes an ordinary high-rise building as the research object, with 540 °C as the dangerous temperature. By changing the aspect ratio of vertical windows and the number of vertical combustion windows, the influence of the above two factors on the height of multi window flame fusion is analyzed, and the law of multi window flame fusion is revealed, providing reference for the prevention and control of high-rise building fires.

2 FDS Simulation Settings

Research methods: simulation and comparative.

Research steps: Model establishment → Parameter setting → Combustion simulation → Changing variables → Comparative analysis.

2.1 Computation Model

The model building parameters:Build a model using a high-rise civil building. The total height of the building is 75 m, with a total of 25 floors, a floor height of 3 m, a wall thickness of 0.2 m, and a floor thickness of 0.1 m. The fire source is located on the 10th floor, with a heat release rate of 6 MW. The combustion process is an ultra fast fire, and the area of the fire source room is 70 m^2. Install thermocouple T1–T25 at the center of each layer's window, with a total of 25 measurement points and a grid size of 0.3 m × 0.3 m × 0.3 m. The high-rise building model is shown in Fig. 1.

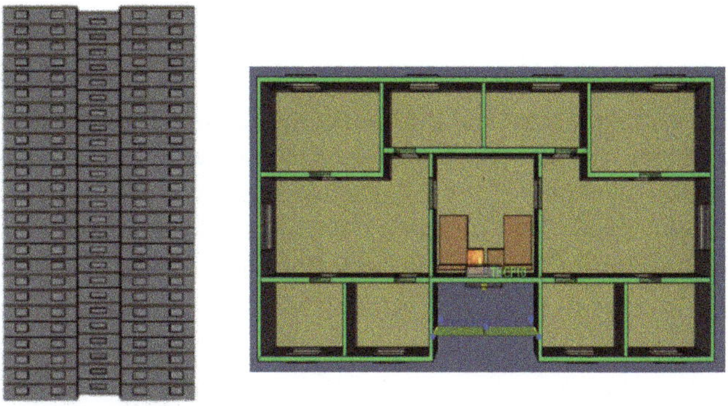

a Indoor layout **b** Building model

Fig. 1. The high rise building model

2.2 Working Condition Settings

Simulation duration: 1200 s. Environmental temperature: 20 °C.Heat release rate per unit area of fire source: 6 MW.The heat release rate curve adopts a non-stationary t that can truly reflect the development process of a fire ^{2} model.

To study the effect of height to width ratio of vertical windows on flame propagation in high-rise buildings, six window sizes were set as shown in Table 1 with approximately equal window areas, and then six height to width ratios were set. The selected window sizes comply with the design specifications for doors and windows [12].

Table 1. Working condition settings

Working condition	Window height H/m	Window Width W/m	aspect ratio λ/m	Area A/m^2
1	1.0	2.2	0.45	2.2
2	1.1	2.0	0.57	2.2
3	1.2	1.8	0.64	2.28
4	1.3	1.7	0.72	2.21
5	1.4	1.6	0.88	2.24
6	1.5	1.5	1	2.25

2.3 Dangerous Temperature

The author use a dangerous temperature of 540 °C (i.e. the outer wall window glass reaches the fire resistance limit and the temperature at which flame fusion occurs), at which point the flame will fuse. The author used the isothermal surface at the dangerous temperature as the outer contour of the flame to determine the flame fusion height. The flame fusion height reaches 540 °C, which is the height of the highest point of the flame minus the height of the fire source when the temperature is dangerous.

2.4 Fire Source Heat Release Rate (HRR)

According to the "Technical Standards for Building Smoke Control and Exhaust Systems" [13], the indoor fire heat release rate of high-rise buildings is 6 MW, and the fire load density is 0.51 MW/m^2. A non steady state t2 model is set, and ultra fast fire is selected for combustion process simulation. The heat release rate reaches 6 MW, and the time taken is 179 s.

3 Simulated Running Results

Figures 2, 3 and 4 show the isotherms during longitudinal continuous combustion for windows 2, 3, and 4. In the figure, Z represents the height of the high-rise building model along the Z-axis direction, and X represents the distance of the high-rise building model along the X-axis direction. The isotherm reflects the temperature distribution during the simulated combustion process. Analysis shows that when the window aspect ratio is 0.45–1 for 2 windows, the flame fusion height is 31.7–33.1 m when reaching 540 °C; When the window height to width ratio is 0.45–1, the flame fusion height is 32.3–33.8 m at 540 °C when the window height to width ratio is 3 windows; When the window aspect ratio is 0.45–1 and the flame fusion height reaches 540 °C, the flame fusion heights are 32.7–34.8 m, respectively, when the window aspect ratio is 4 windows.

It can be seen that under the conditions of longitudinal continuous 2, 3, and 4 windows, the flame fusion height decreases with the increase of window aspect ratio.

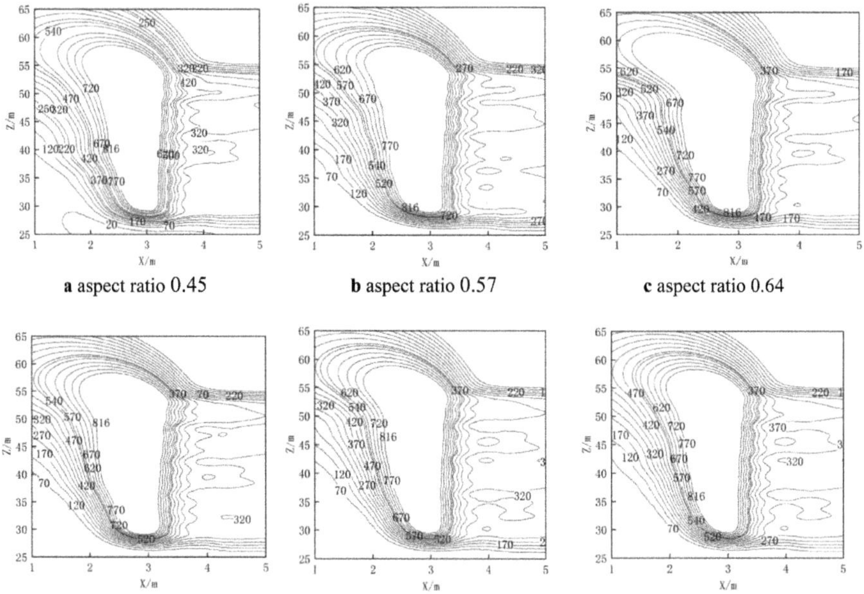

a aspect ratio 0.45 **b** aspect ratio 0.57 **c** aspect ratio 0.64

d aspect ratio 0.72 **e** aspect ratio 0.88 **f** aspect ratio 1.00

Fig. 2. Temperature distribution isotherm of longitudinal continuous two windows

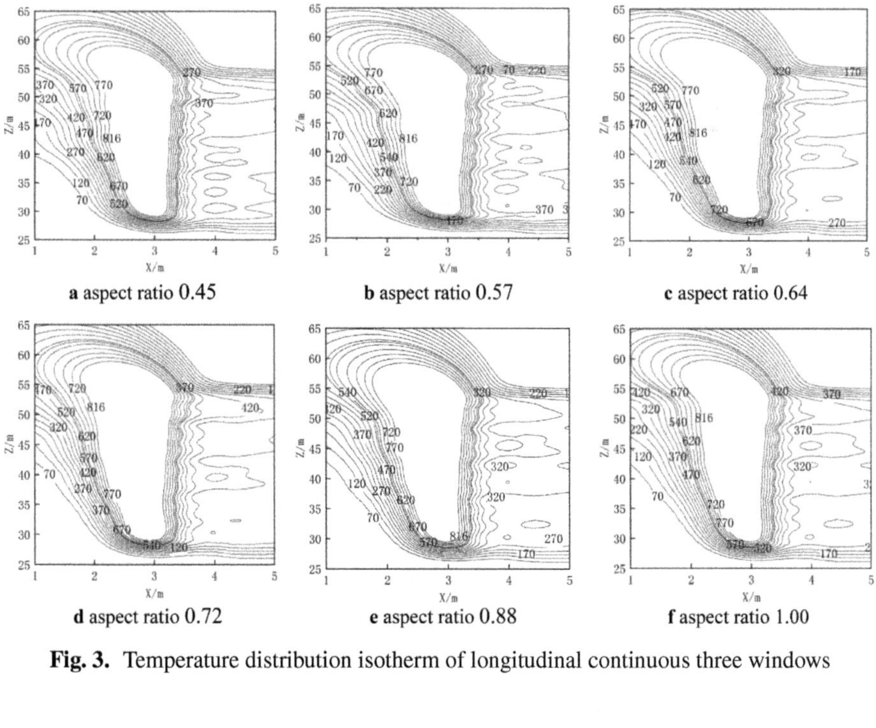

a aspect ratio 0.45 **b** aspect ratio 0.57 **c** aspect ratio 0.64

d aspect ratio 0.72 **e** aspect ratio 0.88 **f** aspect ratio 1.00

Fig. 3. Temperature distribution isotherm of longitudinal continuous three windows

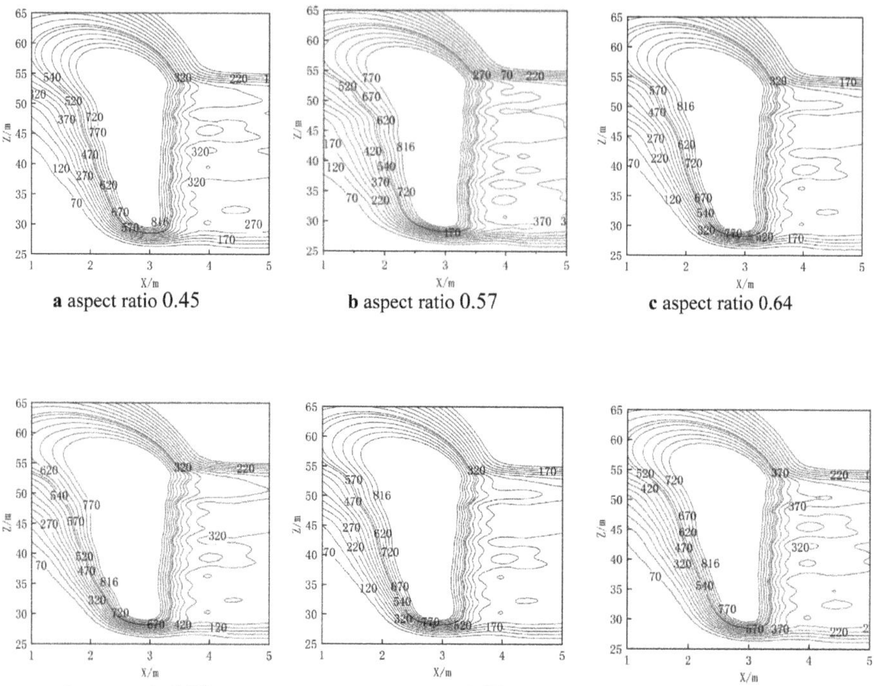

a aspect ratio 0.45 **b** aspect ratio 0.57 **c** aspect ratio 0.64

d aspect ratio 0.72 **e** aspect ratio 0.88 **f** aspect ratio 1.00

Fig. 4. Temperature distribution isotherm of longitudinal continuous four windows

4 The Results of Numerical Simulation

Table 2 shows the flame fusion height at dangerous temperatures when the window aspect ratio is 0.45–1 under different window numbers. Figure 5a shows the comparison of flame fusion height and flame fusion height with aspect ratio and number of windows when reaching dangerous temperatures. According to Table 2, when reaching 540 °C, under two window conditions, as the window aspect ratio increases, the flame fusion height decreases by 0.4 m–1.0 m; Under the three window condition, as the window aspect ratio increases, the flame fusion height decreases by 0.3m to 1.1 m; Under the four window condition, as the window aspect ratio increases, the flame fusion height decreases by 0.69–2.42 m.

As shown in Fig. 5b, with the increase of the number of windows, the flame fusion height reaching 540 °C gradually increases, and the flame fusion height of three windows increases by 0.5–1.1 m compared to two windows; The flame fusion height of four windows increases by 0.3–0.7 m compared to three windows. On the basis of four consecutive windows vertically, the number of windows continues to increase, and the growth rate of flame fusion height significantly decreases when reaching 540 °C. When the number of vertical continuous windows is 8, the flame fusion height stabilizes at 38.1 m.

The above results indicate that as the window aspect ratio increases sequentially, when reaching 540 °C, the flame fusion height decreases by 1.27% to 9.21% under the condition of two consecutive windows in the longitudinal direction; Under the condition of 3 windows, the flame fusion height decreased by 0.9% to 7.58%; Under the condition of 4 windows, the flame fusion height decreased by 0.6% to 6.65%. It can be seen that the flame fusion height at 540 °C decreases with the increase of window aspect ratio.

1. For high-rise buildings, the larger the window aspect ratio, the smaller the flame fusion height when reaching a dangerous temperature of 540 °C. In condition 6, when the window aspect ratio is 1, the flame fusion height is the smallest.
2. As the window aspect ratio increases sequentially, the flame fusion height decreases by 0.6%–9.21% when reaching 540 °C. Under continuous longitudinal two, three, and four window combustion conditions, the maximum flame fusion height decreases by 3.1 m, 2.6 m, and 2.3 m, respectively.
3. Under the same window aspect ratio conditions, the flame fusion height reaching 540 °C increases with the number of windows. The flame fusion height of vertically continuous three windows increases by 0.5–1.1 m compared to two windows at 540 °C; The flame fusion height of vertically continuous four windows increases by 0.3–0.7 m compared to three windows at 540 °C.
4. Select operating condition 6, with a window aspect ratio of 1, and continue to increase the number of windows for combustion simulation. When the number of longitudinal continuous windows is 8, the flame fusion height stabilizes at 38.1 m.

Table 2. Flame fusion height when reaching temperature 540 °C

aspect ratio	The height of flame fusion between two windows/m	The height of flame fusion between three windows/m	The height of flame fusion between four windows/m
0.45	33.7	34.3	34.6
0.57	32.7	33.2	33.7
0.64	32.1	32.9	33.3
0.72	31.4	32.4	33.1
0.88	31.0	32.0	32.7
1	30.6	31.7	32.3

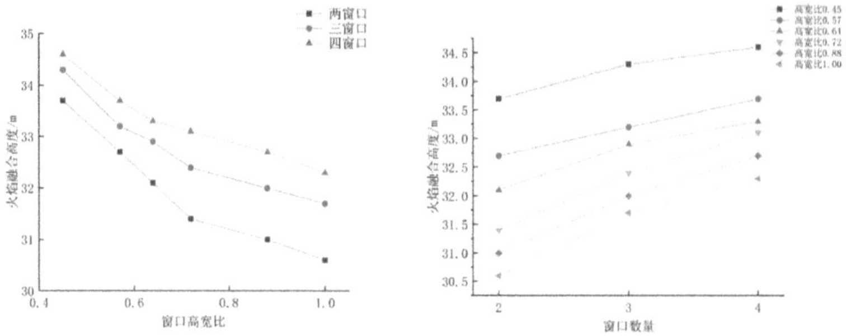

a Different aspect ratios **b** Different number of windows

Fig. 5. Comparison of flame fusion height when reaches temperature 540 °C

5 Conclusion

1. The flame fusion height of high-rise building exterior wall combustion decreases with the increase of window aspect ratio. Therefore, in practical engineering design, the harm caused by fires can be reduced by controlling the window size.
2. In this study, the window form used was a conventional square, and it remains to be studied whether there are different combustion patterns for some special window shapes in real life.
3. In actual engineering design, it is necessary to balance cost and safety at the same time. The author is not very familiar with engineering cost, and it is recommended that future research can use this as an improvement direction.

References

1. Ikeda, K.: Fire resistive design for preventing upward fire spread. J. Disaster Res. **6**(6), 558–567 (2011)
2. Kim, M., Kim, T., Yeo, I.H., Lee, D., Cho, H., Kang, K.I.: Improvement of standards on fire safety performance of externally insulated high-rise buildings: focusing on the case in Korea. J. Build. Eng. **35**, 101990 (2021)
3. Iii, J.H.M., Battaglia, F.: A computational study on the use of balconies to reduce flame spread in high-rise apartment fires. Fire Saf. J. Int. J. Devoted Res. Fire Saf. Sci. Eng. **39**(4), 277–296 (2004)
4. SFPE Task Group (2012) Guidelines for designing fire safety in very tall buildings, Public Review Draft (2012)
5. Ohlemiller, T.J., Shields, J.R.: The effect of surface coatings on fire growth over composite materials in a corner configuration. Fire Saf. J. **32**(2), 173–193 (1999)
6. Tsai, K.: Upward flame spread on a flat surface in a corner and between two parallel surface. J. Chinese Soc. Mech. Eng. **28**(3), 341–348 (2007)
7. Jeffs, G.M.F., Klingelhofer, H.G., Prager, F.H., et al.: Fire-performance of a ventilated facade insulated with a B2-classified rigid polyurethane foam. Fire Mater. **10**(2), 79–89 (1986)
8. Oleszkiewicz, I.: Fire exposure to exterior walls and flame spread on combustible cladding. Fire Technol. **26**, 357–375 (1990)

9. Galea, E.R., Berhane, D., Hofmann, N.A.: CFD analysis of fire plumes emerging from windows with external protrusions in high rise buildings. In: International Fire Science and Engineering Conference, Interflam96, Cambridge, UK (1996)
10. Qu, Z.: Numerical simulation of fire vertical spread of exterior wall of high-rise buildings with corridor[D].Shenyang, Shenyang Jianzhu University (2020)
11. Fu, J., Li, E., Zhang, J.: The influence of window size on the characteristics of fire overflow and its mathematical model characterization. Fire Sci. Technol. **35**(3), 316–320 (2016)
12. Ministry of Housing and Urban Rural Development of the People's Republic of China. GB/T 5824–2008 Series of Dimensions for Door and Window Openings in Buildings [S]. Beijing: China Planning Press (2008)
13. Sichuan Fire Research Institute of the Ministry of Public Security, Shanghai Public Security Fire Brigade. Technical Standard for Building Smoke Control and Exhaust Systems. GB 51251–2017 [S] Beijing: China Planning Press (2017)

Research on Surface Settlement Characteristics of Continuous Small Radius Shield Tunnels Based on Different Routes

Wei Sun, Tai Ge(✉), and HaoJie Chen

Shenyang Jianzhu University, Shenyang, China
gtvvnb1@126.com

Abstract. This paper takes the continuous small radius curve tunnel around Wanghua Street Station of Shenyang Metro Line 4 as the background, aiming to explore the impact of different line types on surface settlement. By establishing three different shield construction finite element models of straight line, small radius curve and continuous small radius curve, the comparison and analysis of shield tunneling processes of various line types are carried out, and the basic rules of surface settlement caused by continuous small radius shield construction are summarized. The correctness of the conclusion is verified by using empirical formulas. The results show that the surface settlement caused by continuous small radius shield construction conforms to the Gaussian distribution law, and the maximum value of surface settlement is located at the starting end of shield tunneling, and the amplitude and range of soil disturbance caused also continue to increase. Under the same conditions, the surface settlement caused by continuous small radius shield construction increases by about 16.4% and 33.3% respectively compared with small radius shield construction and straight section shield construction.

Keywords: Metro · Settlement · Continuous Small Radius · Different Routes

1 Introduction

When constructing shield tunnels in cities, it is often necessary to use small radius curved tunnels or continuous small radius curved tunnels to cope with terrain challenges. The main difference between continuous small radius shield tunneling and small radius shield tunneling is the complexity of the excavation environment. Continuous small radius shield tunneling involves the excavation of multiple small radius curved tunnels, which requires higher directional control and stability of the shield machine and requires frequent correction. Small radius shield tunneling, on the other hand, targets a single small radius tunnel with relatively small technical challenges, and the small radius and continuous turning characteristics of continuous small radius shield tunneling can cause significant surface settlement. Although shield tunneling construction technology has made significant progress in recent years, there are still certain technical bottlenecks when dealing with complex lines such as continuous curves. Therefore, in-depth research

I. Tuns et al. (Eds.): CIBv 2023, LNCE 510, pp. 120–130, 2024.
https://doi.org/10.1007/978-3-031-60765-3_11

on shield tunneling technology for continuous small radius curved tunnels is a necessary step before engineering excavation.

At present, the research on shield construction technology for straight tunnels is very mature, and the construction technology for small radius curves is also being gradually improved, while there is little research at home and abroad on the control of surface settlement in continuous small radius curved tunnels. The empirical formula method proposed by Peck R.B. [1] has been widely used by researchers at home and abroad, and the formula can estimate lateral surface settlement and maximum surface settlement with some authority. Xuan H et al. [2] conducted research and analysis on a large number of surface settlement data from actual engineering projects in China to demonstrate the applicability of empirical formulas in tunnel construction and provided applicable values for some excavation parameters. Jacobsz et al. [3] conducted centrifugal experiments to study the impact of tunnel excavation on nearby soil under different soil conditions, discussed the loss of surrounding soil during tunnel excavation, and recorded changes in surface settlement. Xianfeng M et al. [4] used a centrifugal model to simulate the formation loss and analyzed the relationship between the formation loss and surface settlement before and after construction. Chakeri H et al. [5] used the three-dimensional finite element software FLAC3D to numerically simulate the influence of different parameters on the inflection point value, and established a new inflection point estimation relationship. The new equation provides results similar to actual cases. Yafeng L [6] used MIDAS GTS finite element software to simulate the surface settlement around a subway foundation pit in Shenyang and compared it with on-site monitoring data to verify the usability of the model. Zhi D et al. [7] explained that in the construction of curved shield tunneling, when the shield posture changes, the construction load also changes and has an impact on the force acting on the segments. They also summarized the control measures and corrective technical measures for shield tunneling.

This paper takes the continuous small radius curve tunnel around Wanghua Street Station of Shenyang Metro Line 4 as the background and establishes a three-dimensional finite element numerical model for analysis by using MIDAS GTS software. By comparing the differences in surface settlement impact between continuous small radius shield construction and conventional shield construction, the basic rules of surface settlement caused by continuous small radius shield construction are summarized based on actual projects, and the correctness of the conclusion is verified using empirical formulas.

2 Establishment of Finite Element Model

2.1 Basic Information of Finite Element Model

Assuming that the soil elements around the tunnel in the shield tunneling construction model simulation are uniformly distributed and isotropic, the constitutive relationship adopts the modified Mohr Coulomb criterion, ignoring the soil friction force. The cross-section of the continuous small radius curved shield tunnel is circular, with an outer diameter of 6.4 m and an inner diameter of 6 m for soil excavation. The thickness of the grouting pores is 0.2 m, and the thickness of the pipe segments is 0.3 m. The buried depth of the tunnel excavation center is 15.2 m, and the thickness of the upper soil layer is 12 m.

2.2 Parameter Selection of Finite Element Model

Soil Parameters

The depth of influence of rock layers in the model of continuous small radius shield tunneling is taken as 50 m. Divide the soil in the model into 5 layers, with a thickness of 3 m, 7 m, 10 m, 10 m, and 20 m from the first layer to the fifth layer. During the simulation, the soil layer where the continuous small radius shield tunnel is located is round gravel soil, which makes the calculation results of this model more accurate. The parameters of the soil layer in the finite element model are shown in Table 1.

Table 1. Main material parameters.

Serial number	Stratum name	Volumetric weight (kN/m^3)	Poisson's ratio	Cohesive forces (kPa)	The angle of internal friction (°)	Elastic Modulus (MPa)
1	Composite fill	19.0	0.35	20	10	3.5
2	Gravelly sand 1	19.5	0.35	20	15	22.0
3	Round gravel 1	19.5	0.31	20	20	26.2
4	Gravelly sand 2	21.0	0.35	30	30	30.0
5	Round gravel 2	20.8	0.35	30	36	29.0

Material Parameters

The segments are made of prefabricated reinforced concrete lining blocks and assembled with bolts, and the shield machine uses an earth pressure balance shield machine. The material parameters are shown in Table 2.

Table 2. Main material parameters.

Name	Thickness (m)	Volumetric weight (kN/m^3)	Poisson's ratio	Elastic Modulus (MPa)
Segment	0.3	25	0.3	34500
Shield shell	0.06	0.2	78.5	208000
Grouting	0.2	20	0.25	1

Mesh Division of Finite Element Model

The three different routes selected for this simulation are linear section, small radius curve section, and "S" type continuous small radius curve section. The selected route is 60 m long, with a curve section radius of 325 m. The overall size of the shield tunneling model is 50 m long horizontally, 60 m long in the excavation direction, and 50 m deep in the soil layer. The overall grid division of the continuous small radius shield construction model is shown in Fig. 1.

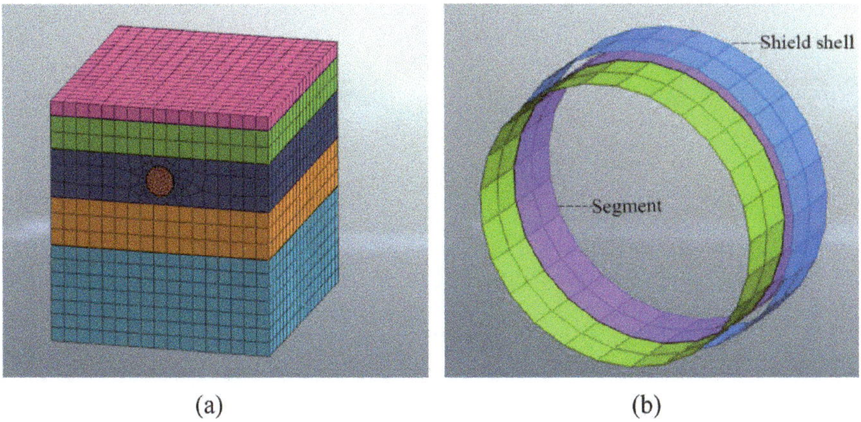

Fig. 1. Mesh division of continuous small radius shield tunneling construction model.

3 Analysis of Simulation Results

3.1 Analysis of Surface Settlement at Different Construction Stages

There are a total of 24 construction stages in this simulation of shield tunneling construction for three different routes, namely the initial stage S1 of shield tunneling, S10 at the midpoint of the excavation route when the shield tunneling machine has just arrived, S13 from the segment lining to the midpoint of the excavation route, and S24 at the end stage of shield tunneling. The magnitude and range of soil disturbance can be intuitively reflected through the surface settlement displacement cloud map. The displacement cloud maps of different stages of different routes are shown in Fig. 2, 3, 4 and 5(a), (b), and (c) are sequentially straight-line segments, small radius curve segments, and "S" type continuous small radius curve segments.

From the displacement cloud map of S1 stage, when the shield machine starts excavation along the tunnel axis direction, the surface settlement displacement cloud map phenomenon of continuous small radius curve section, small radius curve section, and straight section shield construction simulation is basically consistent. The settlement value decreases as the distance from the starting end of the tunnel increases; Along the excavation direction, the upper soil layer in front of the tunnel also experiences settlement due to the disturbance caused by shield tunneling; In the displacement cloud map, the uplift value decreases as the distance from the starting end of the tunnel increases.

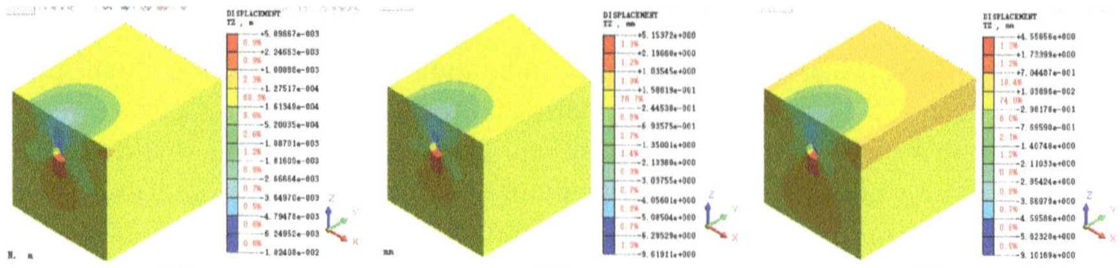

(a) (b) (c)

Fig. 2. Starting stage of excavation S1.

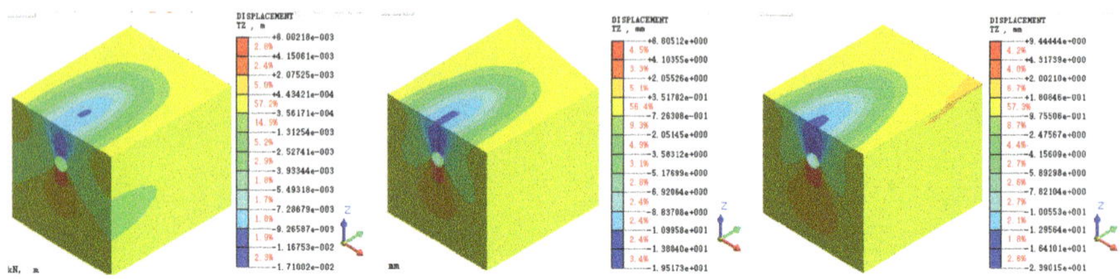

(a) (b) (c)

Fig. 3. Middle point of excavation route S10.

From the displacement cloud map of stage S10, when the shield machine starts excavation along the tunnel axis direction to half of the tunnel line, the surface settlement displacement cloud map phenomenon simulated by continuous small radius curve segments, small radius curve segments, and straight segment shield construction is basically the same as that of starting excavation, but the impact range is larger than that of stage S1. In continuous small radius shield tunneling, the settlement value at the beginning of excavation increases as the shield machine deepens into the excavation and is greater than the settlement value at that position when the shield machine is excavating. This is because during the excavation of curved tunnels, it is necessary to continuously correct the shield and segment attitudes. The soil layer around the tunnel at the beginning of the excavation is frequently corrected by equipment, resulting in consolidation settlement, and increasing its total surface settlement value.

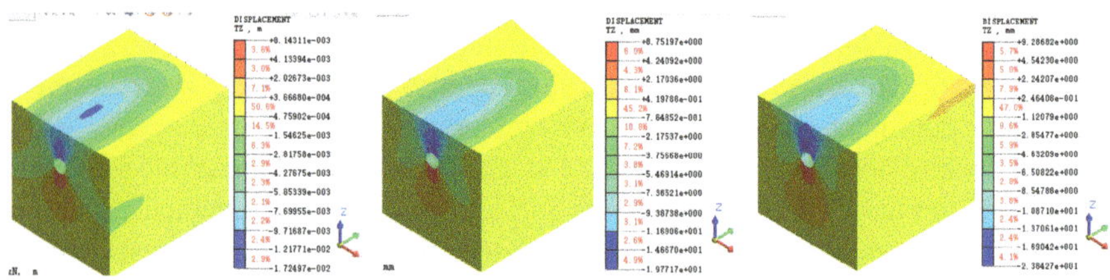

(a) (b) (c)

Fig. 4. Mid-point of segment lining S13.

From the displacement cloud map of S13 stage, when the segment lining in the shield tunneling machine starts to excavate along the tunnel axis direction to half of the tunnel route, the continuous small radius curve segment is not completely consistent with the surface settlement displacement cloud map phenomenon simulated by the small radius curve segment and straight segment shield tunneling construction. The settlement phenomenon caused by shield tunneling in small radius curved and straight sections is consistent with that in S10 stage and has not changed; During the construction of continuous small radius curved section shield tunneling, the surface settlement at the turning point of the excavation line is greater than the settlement in the front and rear directions of the line. This is because the opposite change in excavation direction occurs at the turning point, which increases the amount of over excavation of the surrounding rock and soil of the tunnel, thereby increasing the surface settlement at this stage.

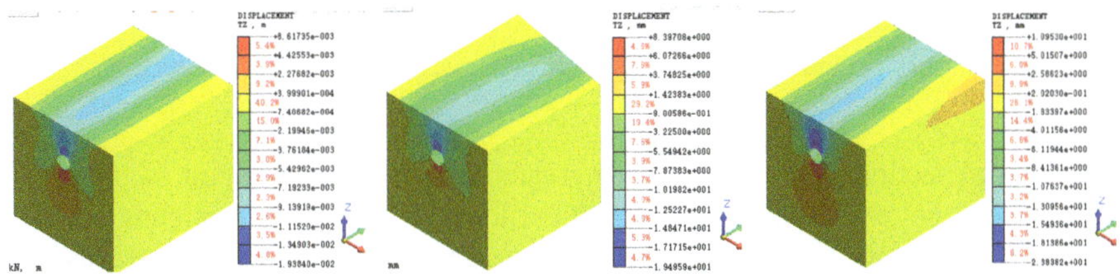

(a) (b) (c)

Fig. 5. End of excavation stage S24.

From the displacement cloud map of stage S24, after the shield tunneling machine completes excavation along the tunnel axis direction, the surface settlement displacement simulated by the continuous small radius curve section, small radius curve section, and straight section shield tunneling has reached a stable state, and the settlement value has reached its maximum.

3.2 Analysis of Settlement Trough Curve

In order to study the change rule of surface settlement caused by the disturbance to the surrounding soil caused by shield construction in a continuous small radius curve, a section is selected at Y = 30 m in the middle section along the boring direction of the model tunnel to analyze the law of surface settlement in the process of boring in the section and compare it with the straight line section and the small radius curve section, as shown in Fig. 6.

The surface settlement curves simulated by three different shield tunneling models, namely continuous small radius curve segment, small radius curve segment, and straight segment, all follow normal distribution curves. Among them, the surface settlement curve of straight segment shield tunneling is symmetrical about the central axis of the tunnel, while the surface settlement curve of small radius curve segment shield tunneling and continuous small radius curve segment shield tunneling deviates. With the completion of shield tunneling, the disturbance range of soil layer caused by continuous small

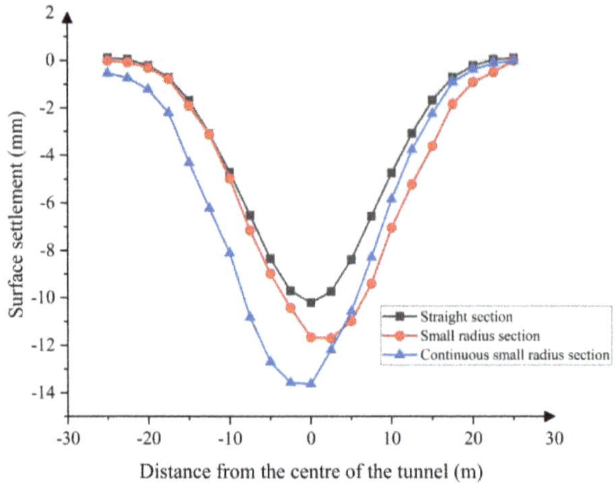

Fig. 6. Surface settlement curve of shield construction in different line sections.

radius curved section shield tunneling is greater than both, resulting in the maximum surface settlement of continuous small radius curved section shield tunneling, reaching 13.61 mm. Under the same conditions, the maximum surface settlement caused by continuous small radius shield tunneling increases by 1.92 mm compared to the small radius section and 3.40 mm compared to the straight section.

Compared with the surface settlement groove generated after excavation, the maximum surface settlement caused by the three different routes when the shield tunneling equipment cutter-head reaches the selected section accounts for approximately 41% to 57% of the final surface settlement when the cutter-head reaches the selected section; The proportion of maximum surface settlement in straight section construction is 56.41%, while the proportion of continuous small radius shield construction is 41.15%. The proportion of soil loss caused by continuous small radius shield construction is smaller than that of straight section shield construction. However, the proportion of consolidation settlement caused by continuous small radius shield tunneling and continuous correction of deviation is greater than that of straight section shield tunneling in three different routes when the cutter-head tunneling equipment reaches the selected section until the end of tunneling.

3.3 Time History Curve Analysis

In this subsection, the longitudinal surface settlement curves along the tunnel boring direction at the midpoint of the tunnel boring direction (Y = 30 m) in the shield construction model under three different routes are compared and analyzed in order to study the disturbance law in the selected section during shield boring with continuous small radius curves. The longitudinal surface settlement curves in the same section of the shield construction model under three different lines are shown in Fig. 7.

Under three different routes, namely continuous small radius curve section, small radius curve section, and straight section, the slope of the surface settlement curve

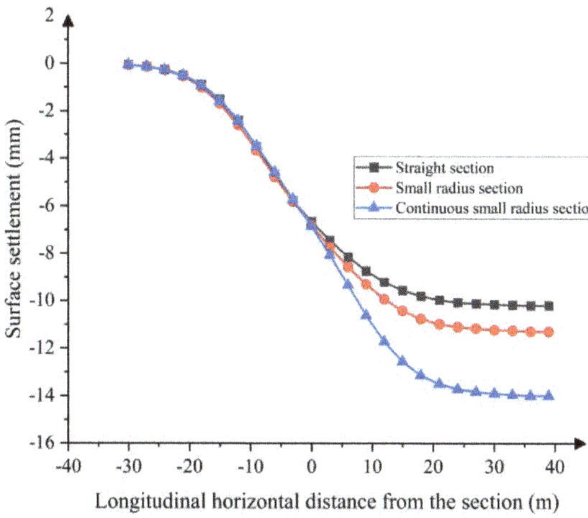

Fig. 7. Time history curve of longitudinal surface settlement during shield construction in different line sections.

of the straight section shield construction model is the smallest, indicating that the development of surface deformation in the straight section is slower. The slope of the surface settlement curve of the continuous small radius shield tunneling model is the highest, and after selecting the cross-section, the resulting surface settlement rate remains at a relatively high value, indicating that the surface deformation of the curved section develops rapidly; Until the excavation reaches a distance of 15 m away from the selected section, the rate begins to slow down. After about 21 m away from the selected section, the settlement change rate approaches 0, indicating that the impact of consolidation settlement begins to weaken and has a certain distance range on the selected section. Under the same conditions, the surface settlement effects generated by the three types of lines before reaching the selected section are basically the same. After the excavation face passes through the selected section, the surface settlement caused by consolidation settlement has the greatest impact in continuous small radius shield tunneling.

3.4 Comparison and Analysis of Peck's Empirical Formula and Simulation Results

According to the construction parameters of the subway shield tunnel, the empirical formula Peck in small radius curve shield construction was modified by Xiaolong H [8], and the formula is as follows.

$$S_C(x) = \frac{RD^2V_1}{4\sqrt{2/\pi}KZ_0(R-x)}\exp(\frac{-x^2}{2K^2Z_0^2}) \tag{1}$$

"R" is the size of the radius curve; "D" is the diameter of the pipe segment; "V_1" is the formation loss rate, taking a value of 0.9%; "K" is the width coefficient of the settlement tank, with a value of 0.45; "x" is the distance from the center of the tunnel; "Z_0" represents the depth of tunnel burial. The comparison of the calculation results for the four stages is shown in Table 3.

Table 3. Comparison of calculation results.

Stage	Numerical simulation	Peck's formula
S1	−0.53	−0.02
S10	−13.55	−13.77
S13	−13.61	−14.84
S44	0	−0.02

The radius of the curved tunnel is 325 m. Based on the value of simulation result x, calculate the surface settlement value of continuous small radius shield tunneling under the Peck formula and draw the surface settlement curve, as shown in Fig. 8.

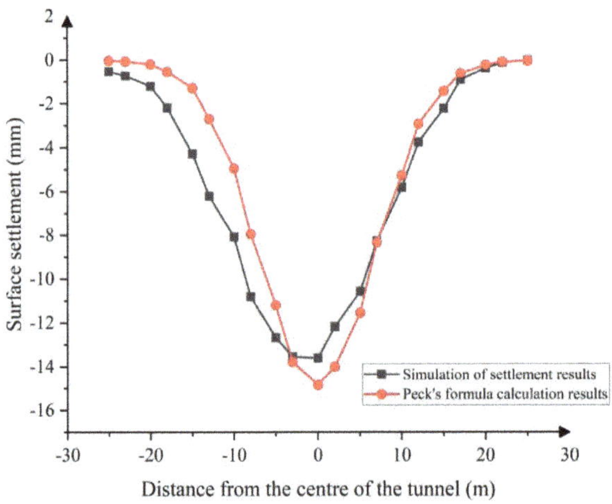

Fig. 8. Comparison between simulated surface settlement of continuous small radius shield construction and calculation results of Peck formula.

The deviation between the settlement simulation data and the small radius curve modified Peck formula calculation results is small, and the overall distribution shape is similar. The settlement curve predicted by the empirical formula and the curve change trend and maximum value obtained by finite element simulation are very similar, fully verifying the reliability of finite element software simulation analysis.

4 Conclusions

In the displacement cloud map of surface settlement at different construction stages, the surface settlement caused by continuous small radius shield tunneling follows a Gaussian distribution law, and the amplitude and range of soil disturbance caused by it also continue to increase. Unlike other lines, during the excavation process of continuous small radius shield tunneling, the maximum longitudinal surface settlement is always at the starting end of the shield tunneling. This is mainly due to frequent adjustment of direction and high resistance and friction to curved tunnels, which increases the disturbance of the shield tunneling machine to the soil. Therefore, it is recommended to enhance the turning flexibility of the shield machine, optimize the cutterhead design, and ensure accurate excavation direction to reduce excavation resistance and friction.

In continuous small radius shield tunneling, the settlement groove curve is asymmetric distribution about the center of the tunnel axis, and the maximum lateral surface settlement position shifts towards the inner side of the curve. Under the same conditions, the proportion of soil loss caused by continuous small radius shield construction is 15% less than that of straight section shield construction, from the beginning of excavation to the end of equipment cutter-head surface reaching the selected section stage. From the arrival of shield equipment at the selected section to the end of excavation stage, the consolidation settlement caused by continuous small radius shield construction deviation correction is more important than that of straight section shield construction, which is 15% more significant. In continuous small radius shield tunneling construction, the shield machine adapts to curved tunnel excavation in the initial stage, with lower excavation speed and disturbance; Subsequent correction operations increase disturbance to the soil, resulting in an increase in the proportion of consolidation settlement.

From the settlement trough curve and time history curve, under the same conditions, the increase in surface settlement caused by continuous small radius shield construction is about 16.4% compared to that caused by small radius shield construction, and the increase in surface settlement caused by straight section shield construction is about 33.3%. This data significantly demonstrates the challenges faced by continuous small radius shield tunneling in controlling surface settlement. Therefore, in-depth research on this construction technology is not only of great significance for construction safety, cost savings, and industry technological progress, but also fills the gap in related fields. However, given the complex and ever-changing situations in practical engineering, such as parallel excavation of hyperbolic tunnels, and insufficient research on other key parameters that affect settlement processes, it is recommended to further expand future research to provide more comprehensive construction guidance.

References

1. Peck, R.B.: Deep excavation and tunnelling in soft ground. 225–290 (1969)
2. Xuan, H., Ning, L., Standing, J.R.: Analysis of the applicability of Peck's formula in the prediction of ground deformation of tunnel construction in China. (01), 23–28+35 (2007)
3. Jacobsz, S.W., Standing, J.R., Mair, R.J., et al.: Centrifuge modelling of tunnelling near driven piles. Soils Found. **44**(01), 49–56 (2004)
4. Xianfeng, M., Junsong, W., Jianyun, L., et al.: Centrifugal model experimental study on ground loss and surface settlement caused by shield tunneling. **34**(05), 942–947 (2012)
5. Chakeri, H., Unver, B., Ozcelik, Y.: A novel relationship for predicting the point of inflexion value in the surface settlement curve. Tunn. Undergr. Space Technol. **43**, 266–275 (2014)
6. Yafeng, L.: Numerical analysis of deformation monitoring and supporting structure of subway deep foundation pit excavation. Shenyang University of Architecture (2018)

7. Zhi, D., Yuching, D., Xiao, Z., Shaojun, M.: Research and prospect on the influence and control of shield attitude change on tube sheet. **21**(21), 8745–8756 (2021)
8. Xiaolong, H.: Analysis of ground settlement during shield tunneling of small radius curved subway. Shijiazhuang Railway University (2020)

Study on Mechanical Properties of Hollow Sandwich Concrete Filled Steel Tubular Column-Steel Beam Assembled Joint System

Wei Sun, YongQi Zhang(✉), and JunShan Yang

Shenyang Jianzhu University, Shenyang, China
zyq_1816@126.com

Abstract. This paper studies the bearing capacity and seismic performance of a new type of assembled joint with tenon. The finite element model of hollow sandwich concrete-filled steel tubular column-steel beam assembled joint system is established by ABAQUS software, and its mechanical mechanism is analyzed. The static performance and hysteretic performance of bolted and tenon assembled joint system with different ring plate width, embedded depth, axial compression ratio and concrete strength are compared. Through the comprehensive comparative analysis of the stress nephogram, hysteretic behavior and skeleton curve of the members, the results show that the assembled joint system of hollow sandwich concrete filled steel tubular column and steel beam has reliable bearing capacity and good energy dissipation performance. The assembled joints based on tenons form fully assembled non-permanent beam-column joints, which makes the space layout more flexible and solves the problem of limited middle height in temporary buildings.

Keywords: hollow sandwich concrete filled steel tube · Finite element analysis · Bolt-tenon assembled joint · Seismic performance

1 Introduction

Hollow sandwich concrete-filled steel tube is a new type of compound concrete-filled steel tube developed on the basis of ordinary concrete-filled steel tube, which is formed by placing two steel tubes concentrically and pouring concrete between the interlayers [1]. Compared with ordinary concrete filled steel tube, it has the advantages of light weight, high bearing capacity, high bending stiffness and good fire resistance. Although the bearing capacity of hollow sandwich concrete filled steel tubular members has been widely studied [2–5], its application in engineering field has not been expanded. The main reasons [6] include the lack of high-precision positioning conditions in the construction site and the lack of simple and reasonable joints to connect columns and beams. At present, the research on the joints of concrete-filled steel tubular members with hollow sandwich is still in the initial stage, and its application in assembled structures is still blank. Therefore, this paper puts forward a connection form between steel beam and circular hollow sandwich concrete filled steel tube in order to promote its application in engineering field, especially in assembly field.

I. Tuns et al. (Eds.): CIBv 2023, LNCE 510, pp. 131–143, 2024.
https://doi.org/10.1007/978-3-031-60765-3_12

2 Introduction of Hollow Sandwich Concrete Filled Steel Tubular Column-Steel Beam-"tenon" Joint Assembly System

2.1 "tenon" Fabricated Node Structure

The structural diagram of the "tenon" assembled joint is shown in Fig. 1, and the whole joint consists of three parts: pier, screw and splicing hoop plate. Pier is the "link" of structural assembly, which realizes the connection between hollow sandwich concrete filled steel tubular columns and steel beams. Its inner cavity is the same as that of the adjacent column ends, and it is a circular inner steel pipe with threads. The threads are arranged to realize the connection of screws, and the pier column and the screw are connected in series to form an integral whole. Splicing hoop plate is specially made of grooved steel plate, and its flange thickness corresponds to that of steel beam, and the web thickness is half of that of steel beam. It is temporarily fixed to the joint pier by embedding, and the steel beam and splicing hoop plate are spliced into a whole by fixing channel steel and upper cover plate with bolts as the medium.

Fig. 1. Node construction diagram

2.2 Structural System of Assembled Joints Based on "tenon"

Frame Stress System Under the Constraint of "tenon" Joints
Frame structure is the most widely used structural system in multi-storey buildings, and it is widely used because of its flexible layout, adaptability to the requirements of different use functions and economy [7]. Based on the concept of "assembly" and "modularization", this paper puts forward a joint form of factory prefabrication, on-site direct assembly and connectable hollow sandwich concrete filled steel tubular column-steel beam-steel brace. Taking the joint as the connecting carrier and the hollow sandwich concrete filled steel tubular column-steel beam as the main stress component, a frame system which can realize the complete assembly of the structural stress system is formed, as shown in Fig. 2. Static performance analysis of hollow sandwich concrete filled steel tubular joints (a).

Support Stress System Under the Constraint of "tenon" Joint

At present, most of the commonly used supporting systems in engineering use circular steel pipes. After repeated use, the bearing capacity of steel is greatly reduced due to repeated welding, which is prone to brittle failure. In this paper, according to the particularity of the supporting system, the splicing hoop plate is designed in the form of web bolt connection tie rod [8, 9], and the vertical members and transverse members are connected in series through the bolted joints to form the supporting system. Concrete-filled steel tubular columns in the support system constrained by tenon joints meet the

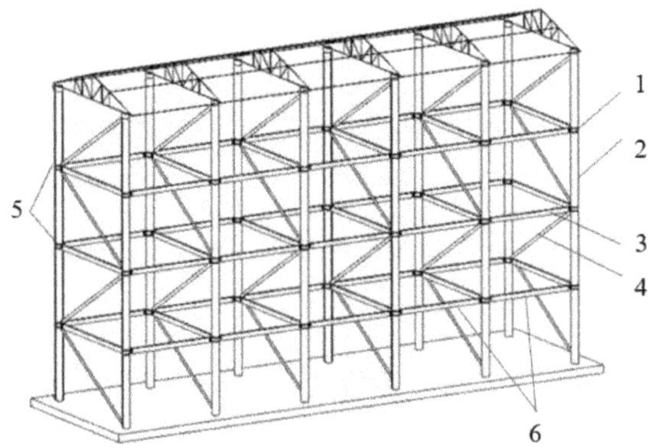

1- Splicing hoop plate
2- Hollow Sandwich Concrete Filled Steel Tubular Column
3-H steel beam
4- Pull rod
5- Column End Bearing Module
6- Beam End Bearing Module

(a) Frame stress system under joint constraints

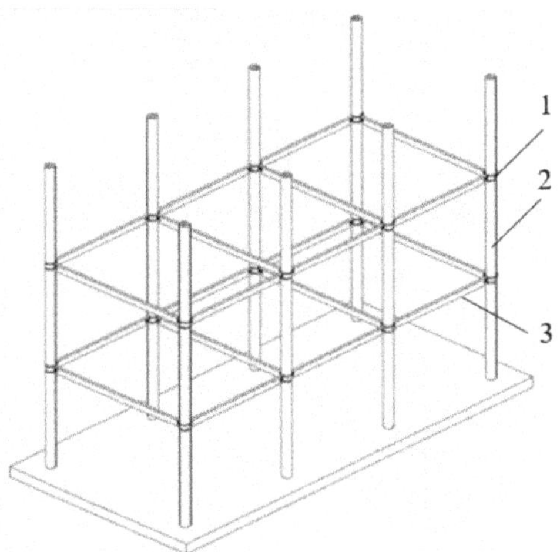

(b) Support system under node constraints

Fig. 2. Application of structural system constrained by prefabricated screw joints demand of bearing capacity and avoid the defect that individual steel tubes are easy to buckle. Due to the structural measures of bolt connection, the supporting system connected in series with tenon joints can be rotated for many times, which can avoid the disadvantage that the existing system is prone to brittle failure, as shown in Fig. 2(b).

3 Static Performance Analysis of Hollow Sandwich Concrete Filled Steel Tubular Joints

3.1 Establishment of Numerical Model

All components are solid elements of eight-node reduced integral, and the finite element model and component mesh division are shown in Fig. 3.

All the steels involved in the model are Q235 grade steels, and the five-line model is selected as the constitutive relation. The standard value of cube compressive strength of core concrete is 40 Mpa, The value of elastic modulus is $E_c = 4730\sqrt{f_c'}$, Poisson's ratio $\mu_c = 0.2$. The plastic damage model in ABAQUS is adopted for concrete, and the values of plastic parameters are shown in Table 1. The calculation formula of the selected concrete stress-strain model is as follows [10].

$$y = \begin{cases} 2x - x^2 & (x \leq 1) \\ \frac{x}{\beta_0 \times (x-1)^2 + x} & (x > 1) \end{cases} \tag{1}$$

其中：

$$x = \frac{\varepsilon}{\varepsilon_0}; \quad y = \frac{\sigma}{\sigma_0} \tag{2}$$

$$\sigma_0 = f_c \tag{3}$$

$$\varepsilon_0 = \varepsilon_c + 800\xi^{0.2} \cdot 10^{-6} \tag{4}$$

$$\varepsilon_c = (1300 + 12.5 f_c) \cdot 10^{-6} \quad \beta_0 = \left(2.36 \times 10^{-5}\right)^{[0.25+(\xi-0.5)^7]} \cdot f_c^{0.5} \cdot 0.5 \geq 0.12 \tag{5}$$

Table 1. Plastic parameters of concrete

expansion angle	eccentricity	$fb0/fc0$	Yield constant (K)	viscosity factor
30°	0.1	1.16	0.667	0.005

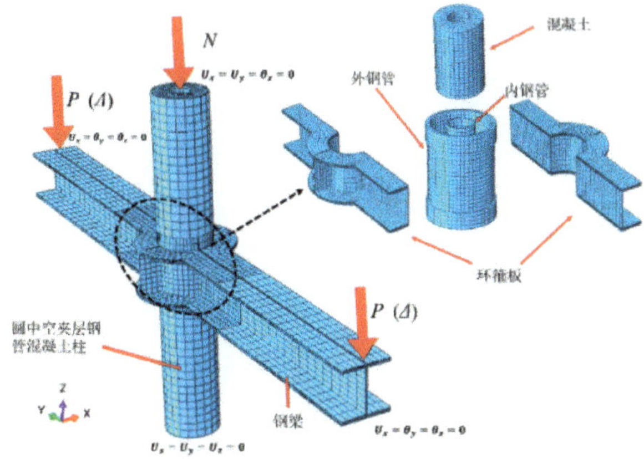

Fig. 3. Node computing model

In the model, the contact in the normal direction is simulated by "hard contact", and the tangential direction is simulated by Coulomb friction. The friction coefficient between the core concrete and the steel tube in the pier column is 0.25, the friction coefficient between the core concrete and the steel tube in the hollow sandwich concrete filled steel tube column is 0.6, and the friction coefficient between the steel tubes is 0.45. The screw and the inner steel tube perform Boolean operation, so that they can work together to bear the internal force [11].

3.2 Finite Element Verification

The finite element modeling analysis of CDSJ3–3 in reference is carried out [12]. Figures 4 and 5 show the comparison results of P-δ curve and failure mode between test and finite element simulation respectively. It can be seen that the experimental results are in good agreement with the simulation results, which verifies the accuracy of the model.

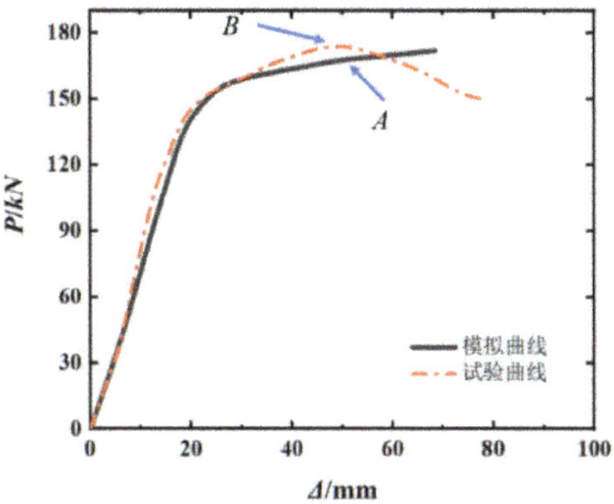

Fig. 4. Comparison of finite element and experimental P-Δ curves

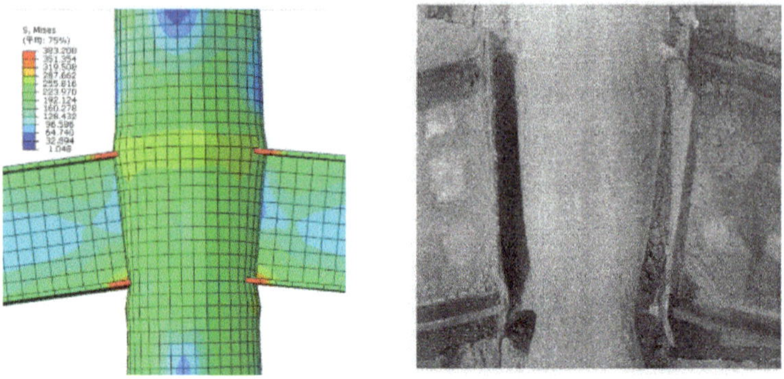

Fig. 5. Comparison of failure patterns between finite element and test

3.3 Parameter Impact Analysis

Outer Width of Hoop Plate

Figure 6 shows the influence of the outer width of the ring plate on the P-δ relationship of the node. With the increase of the outer width L of the hoop plate, the flexural capacity of the joint increases. When the outer width of ring plate is increased from 39 mm to 59 mm, 79 mm and 99mm, the ratio and ultimate flexural capacity of joints are increased by 29.28%, 31.80% and 4.36% and 9.11%, 14.60% and 18.81% respectively. The increase of the outer width of the ring plate is equal to the increase of the area of the outer ring of the hoop plate, which strengthens the bending bearing capacity of the joint area.

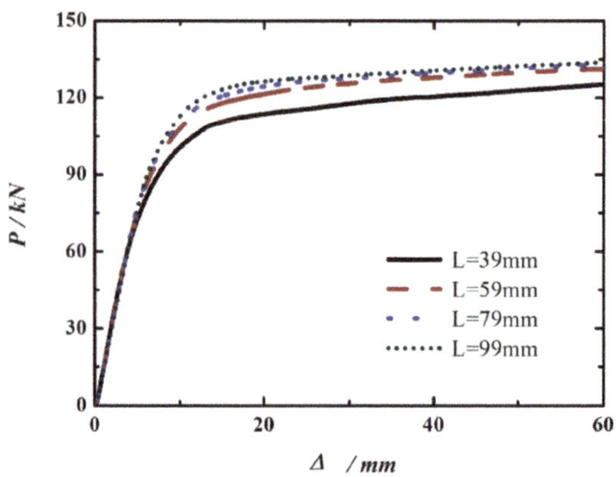

Fig. 6. The influence of annular plate width (L) on P-Δ relation curve

Embedding Depth

Figure 7 shows the influence of the embedding depth (b) on the P-δ relationship curve of the node. Within the parameters studied, with the increase of the depth of hoop plate embedded in the steel pipe outside the pier, the bending capacity of the joint is significantly enhanced. With the increase of embedding depth from 6 mm to 12 mm, 18 mm and 24 mm, the proportional ultimate flexural capacity of joints increased by 11.96%, 22.93% and 31.30%, and the ultimate flexural capacity increased by 23.91%, 44.73% and 55.67%, respectively. Increasing the embedding depth of hoop plate can effectively transfer the load at the beam end to the joint area, and the participation of pier column can effectively enhance the bending bearing capacity of the joint.

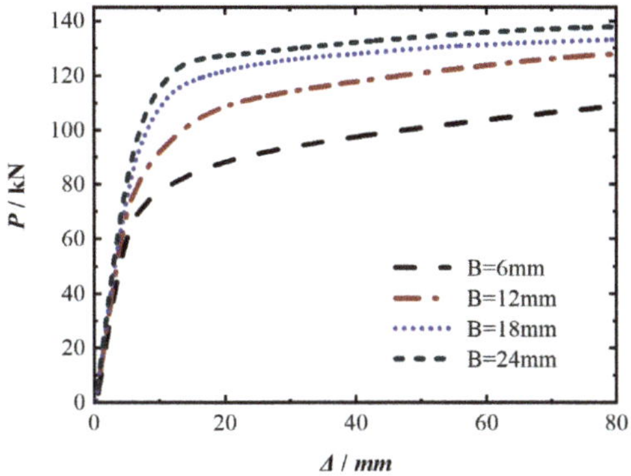

Fig. 7. The influence of embedding depth (B) on P-Δ relation curve

4 Hysteretic Behavior Analysis of 3 Circular Hollow Sandwich Concrete Filled Steel Tubular Joints

4.1 Force Mechanism Analysis

The failure mode of the joint is shown in Fig. 8, which shows that there is a large lateral displacement and the concrete failure section reaches the plane of the screw. The reason is that the bending moment generated by the reverse load at the beam ends on both sides is converted into positive and negative tensile stress, which makes the joint pier rotate. The screw drives the concrete at the column end to extrude, and the column end in the joint area resists the bending moment. The bending moment transmission path is shown in Fig. 9 below. With the increase of load, the concrete on the compression side of the column is gradually destroyed. When the concrete at the root of the column is damaged to the screw interface, the joint loses its bearing capacity, resulting in a large lateral displacement and the whole joint is destroyed. The screw transfers the stress generated by the bending moment to the concrete. When the concrete fails to reach the upper interface of the screw, the bending moment load from the beam end cannot be effectively transferred to the column end, which leads to the increase of the displacement in the joint area and reaches the failure state.

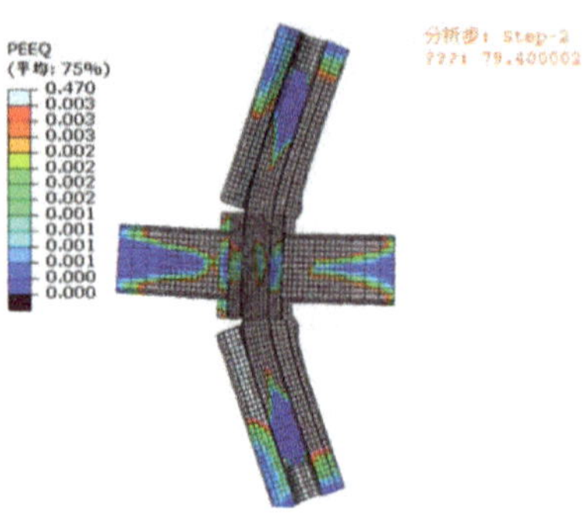

Fig. 8. Node failure mode

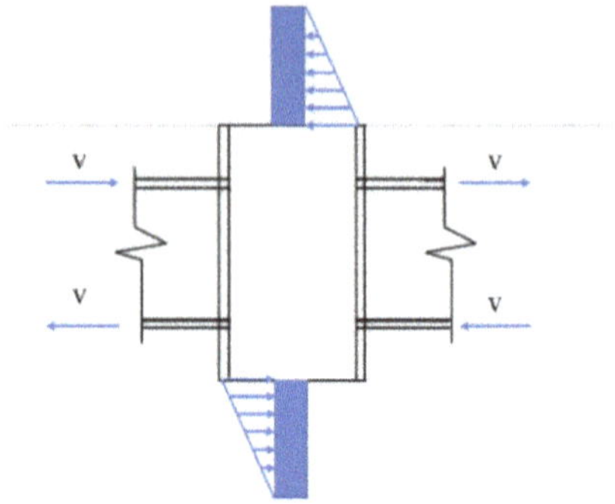

Fig. 9. Node force diagram

4.2 Parameter Impact Analysis

Axial Compression Ratio

Figure 10 shows the P-δ hysteretic curves of joints under different axial compression ratios. It can be seen from the figure that the stiffness of the joint gradually degrades under the low-cycle reciprocating load at the beam end, and the greater the axial compression ratio, the more obvious the degradation is. When the axial compression ratio is lower than 0.6, the "pinching" phenomenon of joints caused by slip is obvious, and it bears S-type and Z-type. When the axial compression ratio reaches 0.6 or above, the hysteretic curve is full. The skeleton curve extracted from the hysteretic curve is shown in Fig. 11. It can be seen from the figure that the axial compression ratio can delay the yield of the joint, and the ductility of the joint decreases with the increase of the axial compression ratio.

Figure 12 shows the relationship curve of energy dissipation coefficient e-/of joints under different axial compression ratios. It can be seen from the figure that the greater the axial compression ratio, the better the energy dissipation performance, but the development of joint displacement is limited under large axial compression ratio. When the axial compression ratio is less than 0.6, the joint shows better ductility, and the smaller the axial compression ratio, the stronger the ductility of the joint.

Concrete Strength

The P-δ hysteretic curves of concrete filled steel tubular joints with circular hollow sandwich under different concrete strengths are shown in Fig. 13. It can be concluded from the figure that when C30 concrete is used as the member, the shape of the hysteretic curve is full, and the constraint effect coefficient ξ of steel tube on concrete is the largest [13]. When C30 and C35 strength concrete is used, the hysteretic curve bears S-type and Z-type, and the slip phenomenon is obvious.

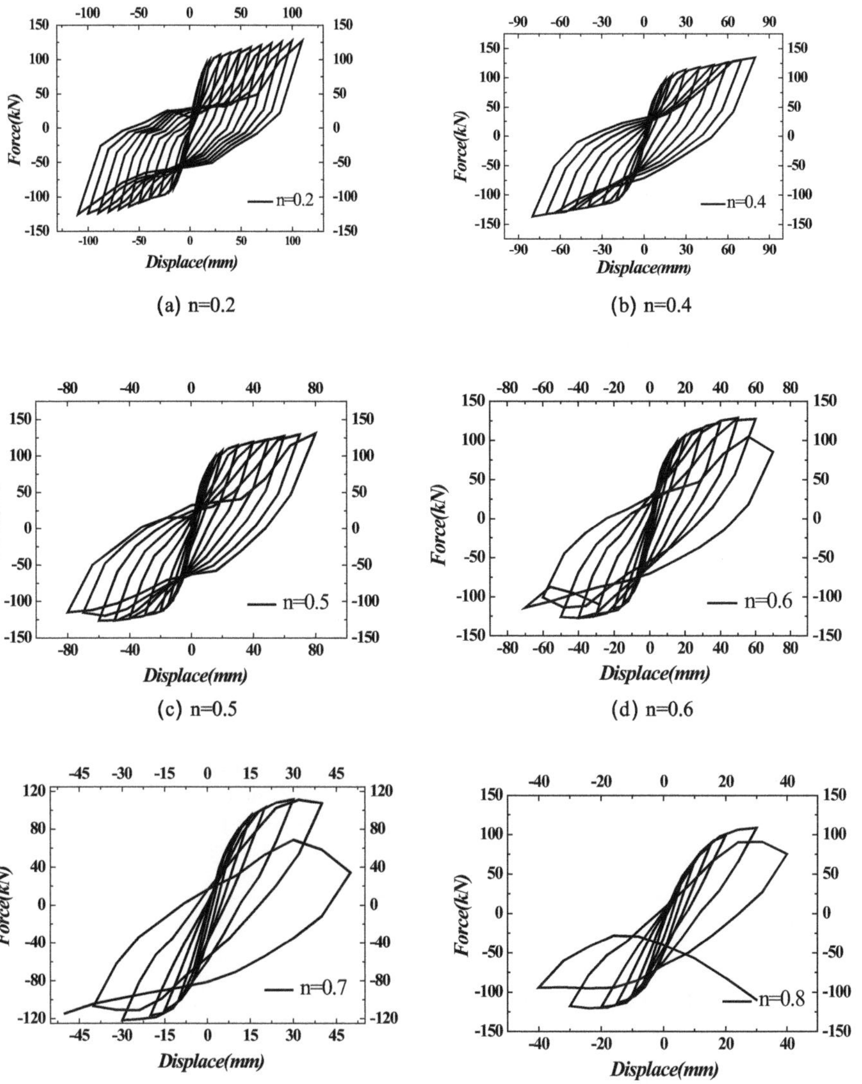

(a) n=0.2 (b) n=0.4

(c) n=0.5 (d) n=0.6

(e) n=0.7 (f) n=0.8

Fig. 10. P-Δ hysteresis curves of nodes under different coaxial pressure ratios

The skeleton curve extracted from the hysteretic curve is shown in Fig. 14. It can be seen that the strength of concrete is improved and the bearing capacity of joints is also improved, but the improvement is very small. When the strength grade of concrete is increased from C30 to C35, C40, C45 and C50, the ultimate bearing capacity of the joint is increased by 3.8%, 8.3%, 10.37% and 10.85% respectively.

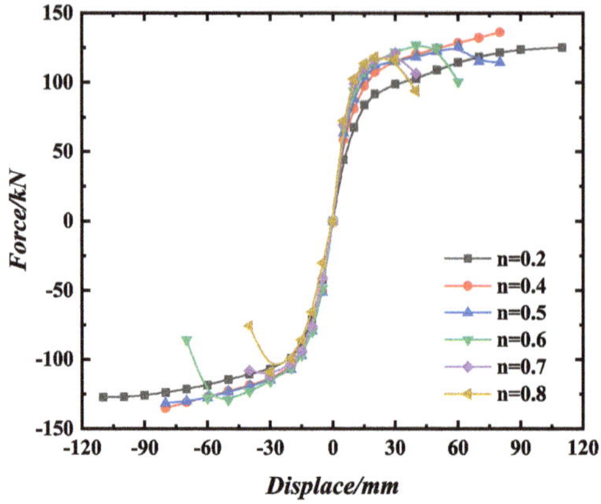

Fig. 11. Skeleton curves at different coaxial pressure ratios

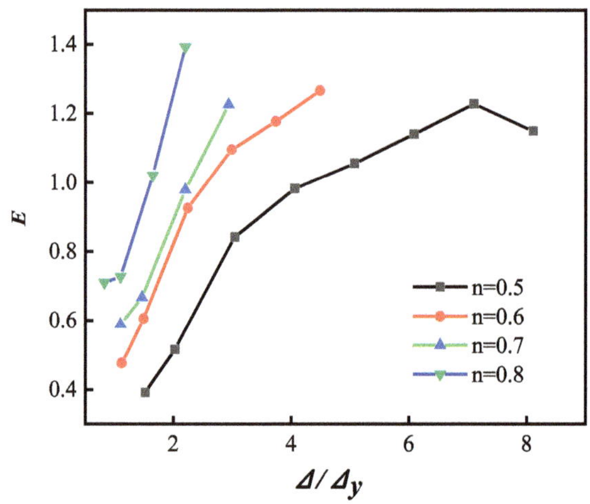

Fig. 12. Don't under the axial compression ratio of E $-$ Δ/Δ y relation curve

Figure 15 shows the relationship curve of energy dissipation coefficient e-/of joints under different concrete strengths. As can be seen from the figure, when higher strength concrete is used in the steel pipe, the better the energy dissipation performance of the joint is. When/is less than 2.5, the concrete strength has little influence on the energy dissipation capacity of the joint. When the load displacement is more than 2.5 times/, the energy dissipation capacity of the joint can be improved by using concrete with higher strength.

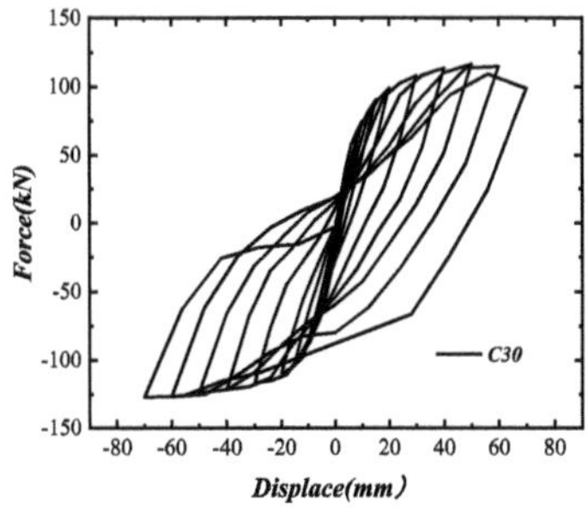

(a) f_{cu}=30MPa

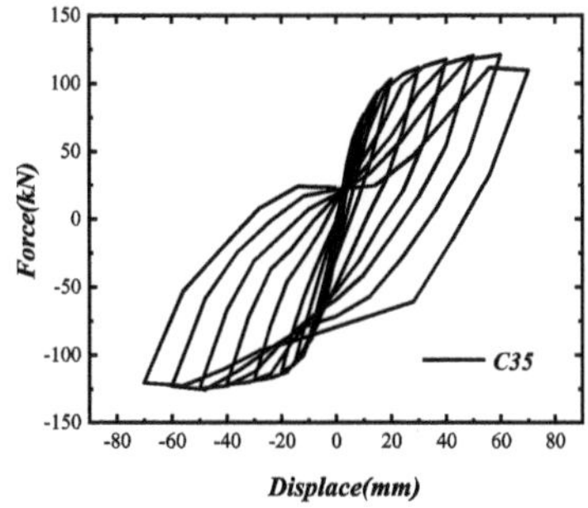

(b) f_{cu}=35MPa

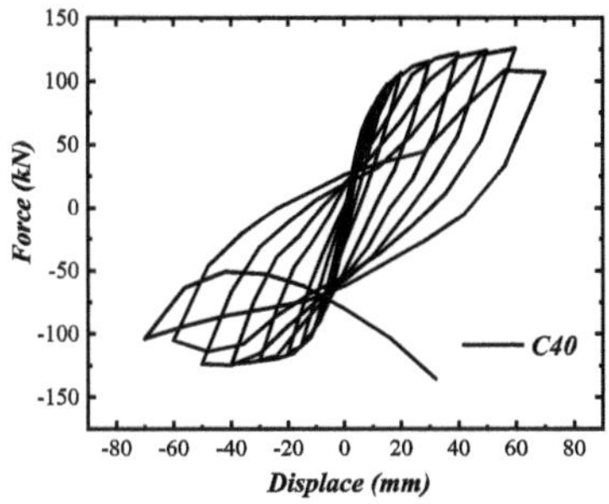

(c) f_{cu}=40MPa

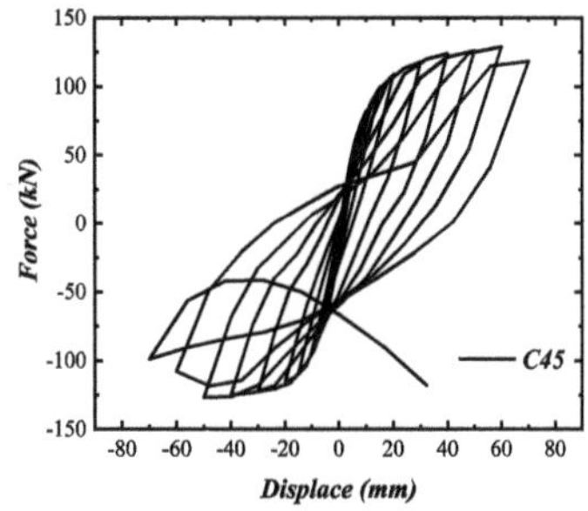

(d) f_{cu}=45MPa

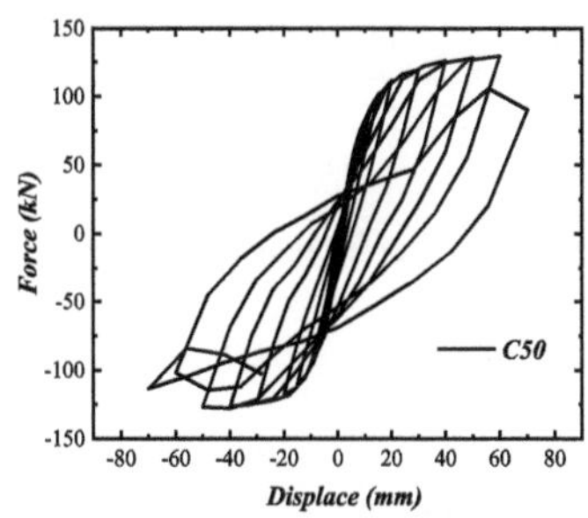

(e) f_{cu}=50MPa

Fig. 13. The P-Δ hysteresis curves of joints under different concrete strengths

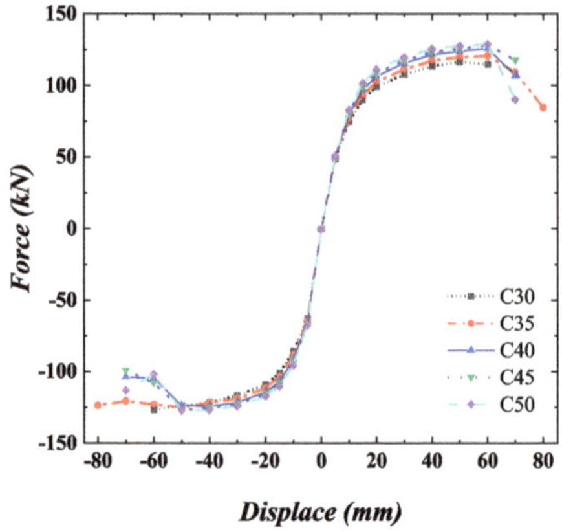

Fig. 14. Skeleton curves at different coaxial pressure ratios

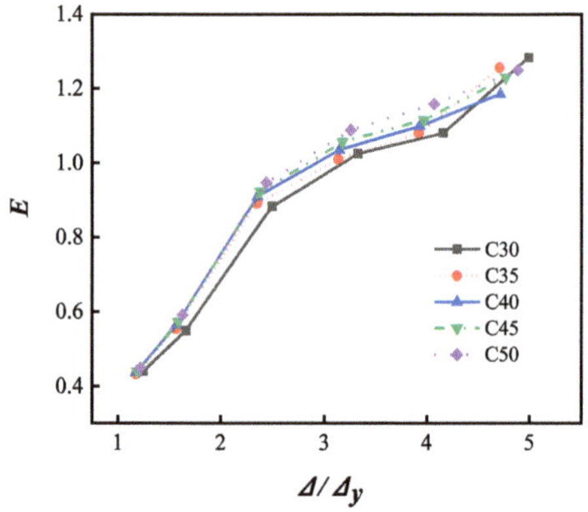

Fig. 15. Don't under the axial compression ratio of E − Δ/Δ y relation curve

5 Conclusion

(1) The outer width of the ring plate and the embedding depth of the spliced hoop plate are the key parameters affecting the static performance of the joint. The flexural capacity of the joint increases with the increase of the outer width and embedded depth of the ring plate.

(2) When the axial compression ratio is small, the ductility of the joint is stronger, and the stiffness degradation is less obvious. In addition, the greater the strength of concrete, the stronger the bearing capacity and ductility of the joints, and with the increase of loading displacement, the joints with higher strength concrete show better energy dissipation capacity.

(3) The assembled joint system of hollow sandwich concrete filled steel tubular column and steel beam proposed in this paper can be applied to embedded modular structure, and it is suitable for solving the problems of limited storey height in temporary buildings.

References

1. Han, L.H.: Concrete-Filled Steel Tube Structures – Theory and Practice, 2nd edn. Science Press, Beijing (2007)
2. Wei, S., Mau, S.T., Vipulananadan, C., Mantrala, S.K.: Performance of new sandwich tube under axial loading: experiment. J. Struct. Eng. **121**(12), 1806–1814 (1995)
3. Nakanishi, K., Kitada, T., Nakai, H.: Experimental study on ultimate strength and ductility of concrete filled steel columns under strong earthquake. J. Constr. Steel Res. **51**(3), 297–319 (1999)
4. Elchalakani, M., Xiao-Ling, Z., Grzebieta, R.: Concrete-filled steel circular tubes subjected to constant amplitude cyclic pure bending. Eng. Struct. **26**(14), 2125–2135 (2004)
5. Tao, Z., Han, L., Huang, H.: Study on mechanical properties of concrete-filled steel tube with square hollow sandwich under eccentric compression. J. Civ. Eng. (02), 33–40+51 (2003). (in Chinese)
6. Guo, Z., Liang, T.: Discussion on the characteristics of prefabricated architecture. Value Eng. (2), 233–235 (2017)
7. Sun, T.: Experimental study on static properties of concrete-filled steel tube (CFST) column-beam joint with circular hollow sandwich. Shenyang Jianzhu University (2011)
8. Li, Z., Xu, S., Liu, H., Jiao, A.: Shaking table test of a new prefabricated concrete-filled steel tube column and reinforced concrete beam frame support system. J. Civ. Eng. **49**(02), 22–30 (2016)
9. Zhang, M., Guo, X., Ma, D., Huang, L., Wang, W.: Design method of fabricated concrete-filled steel tube internal support system for foundation pit engineering. J. Beijing Univ. Technol. **42**(12), 88–96 (2016)
10. Gu, F.: Research on mechanical properties of hollow sandwich concrete-filled steel tubular column with tenon joint constraints. Shenyang Jianzhu University (2020)
11. Zhang, N.: Seismic test and theoretical analysis of semi-rigid joints of circular hollow sandwich concrete-filled steel tubular columns. Hefei University of Technology (2017)
12. Wu, Y., Cai, J., Yang, C., Zhu, C.: Test on joint of inner clapboard square concrete-filled steel tubular column and reinforced concrete beam. Build. Struct. **40**(07), 88–91 (2010)
13. Sun, W., Zhao, Y., Yan, S.: Analysis of mechanical properties of compound hollow concrete-filled steel tubular column under eccentric compression with new mortise and tenon joint constraints. Concr. Cem. Prod. (02), 46–49 (2015)

Study on Fire Resistance Limit of Bolted Prefabricated Beam-Beam Joints

Rui Ma(✉) and Yu Wang

School of Civil Engineering, Shenyang Jianzhu University, Shenyang, China
marui13842667222@163.com

Abstract. In this paper, the fire resistance limit state of a new bolted prefabricated beam-to-beam joint exposed to fire, and the influencing factors of its fire resistance limit were studied. Based on the ABAQUS finite element software, a numerical simulation of bolted prefabricated beam-beam joints under temperature load and static load is carried out, and the mid-span deflection curve is drawn according to the fire exposure time for different components. The mid-span deflection of the bolted prefabricated beam-beam joints gradually increases with the increase of fire exposure time, and breaks away from the linear growth when approaching the fire resistance limit. With the increase of fire exposure time, the growth rate of the mid-span deflection of the components is constantly increasing, and the growth rate of the mid-span deflection of specimens with different fire resistance is also different, the greater the load ratio of the components, the shorter their fire resistance limit; the higher the concrete strength grade, the thicker the concrete protective layer, and the longer the fire resistance limit. The size of the bolt pretightening force and the bolt strength grade have no effect on the fire resistance rating in the present model design.

Keywords: fabricated joint · beam-beam connection · temperature field · fire resistance limit · finite element analysis

1 Introduction

The reinforced concrete frame joints connect the frame beams and columns to form the structure, which plays the role of connecting each component into a whole and transferring load. The research on the fire resistance of the joints is the basis of the research on the fire resistance of the frame structures. In the past, the fire resistance of reinforced concrete members has been studied quite a few reports. Han et al. [1] presents a theoretical model for the deformation and strength of RC columns in fire, which, in conjunction with a finite element method for the temperature field analysis of RC columns, can be used to predict the fire resistance of RC columns and FRP-confined RC columns with or without fire protection. Wu et al. [2] conducted parameters for reinforced concrete beams constrained by axial and rotational constraints under fire

I. Tuns et al. (Eds.): CIBv 2023, LNCE 510, pp. 144–156, 2024.
https://doi.org/10.1007/978-3-031-60765-3_13

conditions and analyzes and studies the law of the influence of constraint stiffness on the internal force and deformation of reinforced concrete beams. There has been a lot of research work on the fire resistance of beam-column joints in building structures. Al-Jabri, J [3] focused on the experimental research, finite element analysis and simplified analysis methods of steel structure beam-column joints under fire are summarized. Yu et al. [4, 6] conducted experimental studies on the rotational deformation properties and tensile knot resistance of several common steel joints at high temperature, and proposed numerical simulation methods and simplified calculation models on the basis of the tests. Wang et al. [7] focused on the influence of different parameters such as the temperature field distribution of frame joints, beam deformation, axial compression ratio, beam layer thickness, and steel bar-equiping ratio on the fire resistance of frame joints in fire disasters. Based theoretical researches, Kong et al. [8] conducted a fire resistance test on the basic joint unit composed of supporting beams and supporting columns under the coupling effect of fire disaster and dead load, and further studied the temperature field distribution, deformation law, fire resistance performance and failure mode of joint unit specimens. Wu et al. [9] studied the lateral displacement of beam-column joints of concrete frame structures under the action of fire through numerical simulation analysis. Han et al. [10] conducted an experimental study of shaped steel concrete beam-column joints, and studied the temperature field distribution, fire resistance limit, failure form and practical calculation method of joint fire resistance. At present, there have been many studies on the static and seismic performance of prefabricated building joints and fire resistance. Most of the researches on fire resistance of structural joints has focused on the frame joints connecting beams and columns [11]. The temperature field distribution and fire resistance limit are studied by using full-scale or reduced-scale specimens for testing or simulation via computer software, but there is almost no literature analyzing of fire resistance performance of prefabricated joints for beam-to-beam connections. And there are very few literature on the fire resistance performance and fire resistance limit of beam-to-beam joints at high temperature, whether it is experimental or simulation.

In this paper, composite beam-beam joints composed of steel sections, bolts, steel plates, steel bars and concrete are studied. Their mechanical properties were studied and proved to have good earthquake resistance, but their fire resistance is still unclear. Therefore, this paper used ABAQUS to establish a finite element model under the ISO-834 standard temperature curve simulating the condition of fire disaster to study the temperature field distribution and temperature characteristics of the joints, as well as the influence of steel plate thickness and concrete protective layer thickness on its temperature field. Then, the model is thermally coupled by applying a static load to obtain the span deflection curve of the joint in fire disaster. The growth characteristics of the mid-span deflection, and the influence of anchor bolt preload, concrete strength grade, load ratio, anchor bolt strength grade, concrete cover thickness and other parameters on the fire resistance of the joints were analyzed.

2 Joint Structure

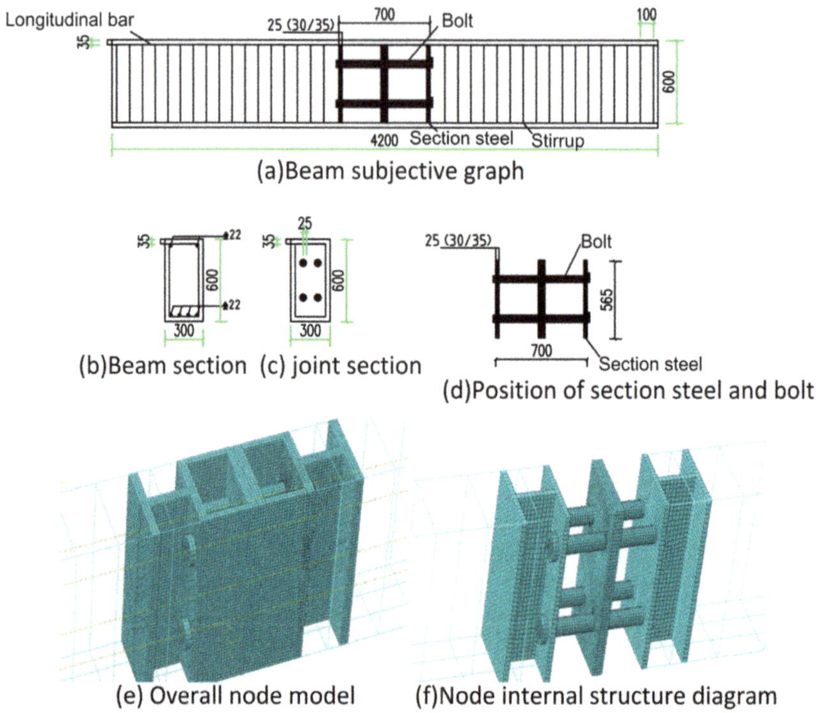

Fig. 1. Schematic diagram, three-dimensional diagram and internal structure diagram of bolted prefabricated beam-beam joint

The overall length of the beam $l = 4.2$ m, the beam section height $h = 600$ mm, width $b = 300$ mm. There are two upper longitudinal steel bars, and four lower longitudinal steel bars, and all their model number is HRB335, and the diameter of a longitudinal steel bar is 22 mm. The stirrup spacing is 200 mm, and the model number is HPB300; the concrete cover thickness is 30 mm, the bolt grade is M8.9, and the diameter is 50 mm. Both the section steel and the steel plate are made of Q235 grade steel with a thickness of 30 mm. In terms of assembly, to make the downward load act on the upper surface of the beam more uniformly and simulate the loading of force more realistically, two pads are created to attach to the beam and bind to the upper surface of the beam, with steel bars, sections, steel plates and bolts all embedding in concrete. In terms of boundary conditions and loads, both ends of the beam are set as hinge constraints to limit their displacement and release their corners. The four bolts are preloaded with the load applied at the center of the block. By finding contact pairs, the frictional contact properties between steel plates, steel plates and bolts, and section steel and bolts are defined. Three sides of the structure are exposed to fire, so that the state of joints in fire disaster can be more realistically simulated (Fig. 1).

2.1 Finite Element Model Building

Using ABAQUS simulation software to model the appeal joint model, the temperature field under simulated fire disaster is analyzed, and the fire resistance limit is obtained. As the temperature of concrete rises, the physical and mechanical properties of the internal components will change. Meanwhile, there may be indefinite chemical reactions between the components at high temperatures. It further highlights the importance of studying the mechanical properties of concrete at high temperatures.

2.1.1 Mechanical Properties of Concrete at High Temperature

This paper adopts the formula proposed by Guo et al. [12] for calculating the concrete compressive strength f at high temperature, peak strain of concrete at high temperature, calculation formula of elastic modulus and the Constitutive (stress-strain) relationship of concrete at high temperature. For the concrete tensile strength at high temperature, this paper uses the formula proposed by Guo et al. [14].

2.1.2 Mechanical Properties of Steel at High Temperature

The specific heat capacity formula given by the European standard EC3 [16] and EC4 [13] is:

$$C_s = \begin{cases} 425 + 7.73^{-1}T_s - 1.69 \times 10^{-3}T^2 + 2.22 \times 10^{-6}T^3 & 20\ °\text{C} \le T \le 600\ °\text{C} \\ 666 - \frac{13002}{T-738} & 600\ °\text{C} < T \le 735\ °\text{C} \\ 545 + \frac{17820}{T-731} & 735\ °\text{C} < T \le 900\ °\text{C} \\ 650 & 900\ °\text{C} < T \le 1200\ °\text{C} \end{cases} \quad (1)$$

where the unit of C_s is J/(kg·°C).

In this paper, the coefficient of thermal expansion and thermal conductivity formula given by European Standard EC3 [16] and EC4 [13], stress-strain relationship, given by the European code EC3 [16] is selected for steel bars reinforcement.

3 Research on Fire Resistance of Bolted Prefabricated Beam-Beam Joints

When the load-bearing structure is subjected to both temperature load and static load, the bearing capacity will be greatly reduced. Two kinds of limit states of structures are usually defined: one is the damage limit state, i.e., the bearing capacity of load-bearing structures and components is reduced to the degree of load corresponding to the design under fire disaster conditions; the other is the deformation limit state, i.e., the deformation of the load-bearing structure under temperature load and static load gradually exceeds a certain design size. To ensure personal and property safety, it is necessary to obtain the time for the bolted prefabricated beam to reach the limit state when it is subjected to fire, that is, the fire resistance limit, and the factors that affect the fire resistance limit. In this paper, based on the ABAQUS finite element software, the numerical simulation of bolted prefabricated beam-beam joints under temperature load and static load is carried out, and the fire resistance limit is obtained. The load ratio, bolt pre-tightening force, concrete strength, the influence of concrete cover thickness, bolt strength and other factors on its fire resistance limit are analyzed to provide a reference for the fire resistance of joints (Fig. 2).

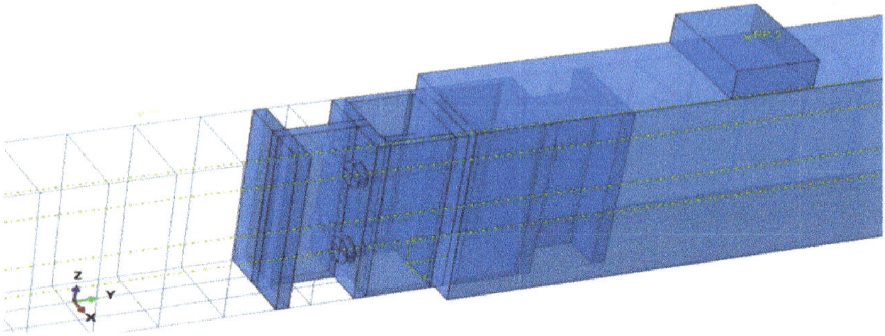

Fig. 2. Bolted prefabricated beam - internal structure diagram and joint diagram of beam

3.1 Judgment of Limit State of Structural Bearing Capacity in Fire Disaster

In this paper, the deformation limit state is used to determine whether the component has reached the limit state of bearing capacity, i.e., the deformation of the load-bearing structure under the temperature and static load gradually exceeds a certain design limit.

Referring to *Concrete Structure Fire Protection Technology*, the component is not suitable for continuous bearing deformation, and the specific judgment formula can be expressed as:

$$\delta \geq \frac{l}{20} \tag{2}$$

where δ is the maximum deflection of the member, mm;
l is the length of the member, mm.

Specimen condition setting.
The working conditions of bolted prefabricated beam-beam joints under different parameter variables are shown in the Table 1.

Table 1. Specific parameters and naming numbers of each test piece

Number	Load ratio (n)	Bolt preload(Pa)	Concrete strength class	Bolt strength class (Mpa)	Stirrup spacing (mm)	Thickness of concrete protective layer (mm)
L1	0.5	5 × 107	C30	900	200	30
L2	0.6	5×10^7	C30	900	200	30
L3	0.7	5×10^7	C30	900	200	30
L4	0.5	5×10^7	C25	900	200	30
L5	0.6	5×10^7	C25	900	200	30
L6	0.7	5×10^7	C25	900	200	30
L7	0.5	1 × 108	C30	900	200	30
L8	0.6	1 × 108	C30	900	200	30
L9	0.7	1 × 108	C30	900	200	30
L10	0.6	5×10^7	C30	900	200	30
L11	0.6	108	C30	900	200	30
L12	0.6	1.5 × 108	C30	900	200	30
L13	0.6	5×10^7	C30	900	150	30
L14	0.6	10^8	C30	900	150	30
L15	0.6	1.5×10^8	C30	900	150	30
L16	0.5	5×10^7	C30	900	200	30
L17	0.5	10^8	C30	900	200	30
L18	0.5	1.5×10^8	C30	900	200	30
L19	0.6	5×10^7	C25	900	200	30
L20	0.6	5×10^7	C30	900	200	30
L21	0.6	5×10^7	C40	900	200	30
L22	0.6	1.5 × 108	C25	900	200	30
L23	0.6	1.5 × 108	C30	900	200	30
L24	0.6	1.5 × 108	C40	900	200	30
L25	0.5	5×10^7	C25	900	150	30
L26	0.5	5×10^7	C30	900	150	30
L27	0.5	5×10^7	C40	900	150	30
L28	0.6	5×10^7	C30	600	200	30
L29	0.6	5×10^7	C30	900	200	30
L30	0.6	5×10^7	C30	1200	200	30

(*continued*)

Table 1. (*continued*)

Number	Load ratio (n)	Bolt preload(Pa)	Concrete strength class	Bolt strength class (Mpa)	Stirrup spacing (mm)	Thickness of concrete protective layer (mm)
L31	0.7	5×10^7	C30	600	200	30
L32	0.7	5×10^7	C30	900	200	30
L33	0.7	5×10^7	C30	1200	200	30
L34	0.6	5×10^7	C30	600	150	30
L35	0.6	5×10^7	C30	900	150	30
L36	0.6	5×10^7	C30	1200	150	30
L37	0.6	5×10^7	C30	900	200	30
L38	0.6	5×10^7	C30	900	200	35
L39	0.6	5×10^7	C30	900	200	40
L40	0.6	5×10^7	C30	900	150	30
L41	0.6	5×10^7	C30	900	150	35
L42	0.6	5×10^7	C30	900	150	40
L43	0.6	1.5×10^8	C30	900	200	30
L44	0.6	1.5×10^8	C30	900	200	35
L45	0.6	1.5×10^8	C30	900	200	40

3.2 Factors Affecting the Fire Resistance of Bolted Prefabricated Beam-Beam Joints

After knowing the basic situation of the fire resistance limit of the bolted prefabricated beam-beam joints, this paper selects six parameter variables to explore their influence on the fire resistance limit, which are load ratio, bolt preload, concrete strength grade, bolt strength grades and concrete cover thicknesses, to give recommendations for structural design in practice. Therefore, 45 specimens as indicated above were set up.

3.2.1 Effects of Different Load Ratios on the Fire Resistance of Bolted Prefabricated Beam-to-Beam Joints

Three groups of components (L1, L2, L3), (L4, L5, L6), (L7, L8, L9) were set to explore the effects of different load ratios on the fire resistance of bolted prefabricated beam-to-beam joints. The load ratio is defined as the ratio of the static load on the structure under fire to its ultimate bearing capacity at normal temperature, and the results are shown in Fig. 3.

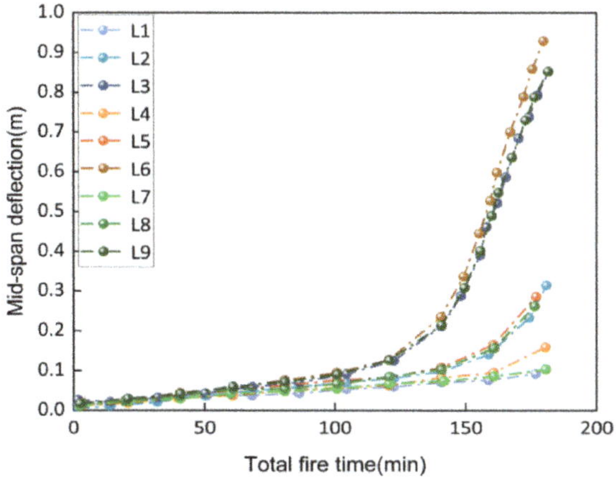

Fig. 3. Effects of different load ratios on the fire resistance of bolted prefabricated beam-to-beam joints

Figure 3 shows that the fire exposure time-mid-span deflection curves of the three groups of specimens have a relatively large degree of dispersion. In the first and third groups of specimens, the specimens of $n = 0.6$ and $n = 0.7$ have already left the linear displacement stage, while the specimens with a load ratio of $n = 0.5$ are still in the linear displacement stage, and the specimens with a load ratio of $n = 0.5$ in the second group have just finished the linear displacement stage when the fire exposure time reaches about 170 min. The fire resistance limits of specimens L1, L2 and L3 are ≥180 min, 172 min and 142 min respectively; The fire resistance limits of test pieces L4, L5, and L6 are ≥180 min, 171 min, and 140 min respectively; the fire resistance limits of L7, L8, and L9 are ≥180 min, 173 min, and 143 min respectively. The change of 0.1 times the load ratio caused the least impact on the fire resistance limit of 30 min, which shows that the load ratio has a great influence on the fire resistance limit of bolted prefabricated beam-beam joints. The reason should be that when the structural load is relatively large, each component bears a greater stress than usual, and has to generate a large strain in advance. With the continuous application of temperature loads, the mechanical parameters of the components are constantly changing, and soon reach their yield strength, and enter the plastic stage and failure stage ahead of time, resulting in a decrease in their fire resistance.

3.2.2 Effects of Different Bolt Pretightening Forces on the Fire Resistance of Bolted Prefabricated Beam-to-Beam Joints

Three groups of components (L10, L11, L12), (L13, L14, L15), (L16, L17, L18) were set to explore the influence of different bolt pretightening forces on the fire resistance of bolted prefabricated beam-to-beam joints, the result is shown in Fig. 4.

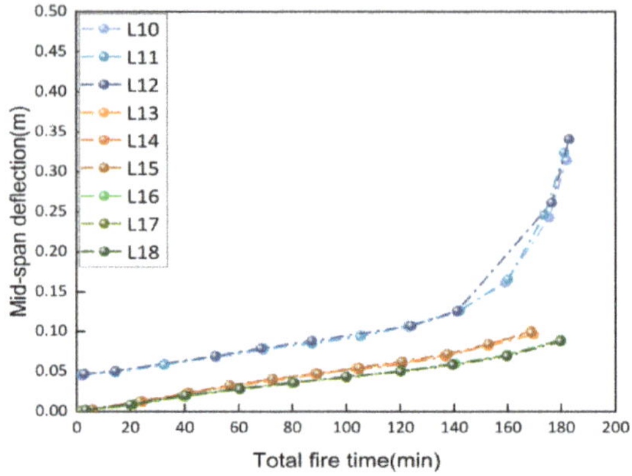

Fig. 4. Effects of different bolt pretightening forces on the fire resistance of bolted prefabricated beam-beam joints

Figure 4 above shows that no matter which group of specimens, the fire time-mid-span deflection curves under different bolt preloads are very close, and there is basically no difference in the mid-span deflection within the same fire exposure time. Among them, the fire resistance limits of specimens L10, L11, and L12 are 171 min, 172 min, and 172 min respectively; the fire limit states of specimens L13, L14, L15, L16, L17 and L18 have not reached, and the maximum mid-span deflection at 180 min of fire exposure is: 0.125 m, 0.122 m, 0.12 m, 0.092 m, 0.09 m, 0.089 m. This indicates that the size of the bolt pre-tightening force in the joint has no effect on the fire resistance rating under the existing design of the model. There are two reasons: firstly, under present design conditions, no matter whether the bolt pretightening force is set or not, the stress in the bolt is always small and cannot reach the yield state and fail; secondly, due to the continuous application of temperature loads, the material properties of the bolts change suddenly, and the applied bolt pretightening force can no longer compensate for the damage of the thermal and mechanical parameters of the bolts caused by the temperature.

3.2.3 Effects of Different Concrete Strength Grades on the Fire Resistance of Bolted Prefabricated Beam-Beam Joints

Three groups of components (L19, L20, L21), (L22, L23, L24), (L25, L26, L27) were set to explore the influence of different concrete strength levels on the fire resistance limit of bolted prefabricated beam-beam joints. The structure is shown in the Fig. 5.

Figure 5 shows that in each group of specimens, as the concrete strength grade changes, the time-deflection curves of the specimens also change significantly, and the curves of the three groups of specimens have the same trend. Among them, the fire resistance limits of specimens L19, L20, and L21 are 154 min, 173 min, and $\geq$180 min respectively; the fire resistance limits of specimens L22, L23, and L24 are 155 min, 173 min, and $\geq$180 min, respectively; the fire resistance limits of specimens L25, L26, and L27 are 164 min, 179 min, and $\geq$180 min, respectively. The concrete grade was upgraded from C25 to C30, and the fire resistance limit was increased by nearly 20 min. Though

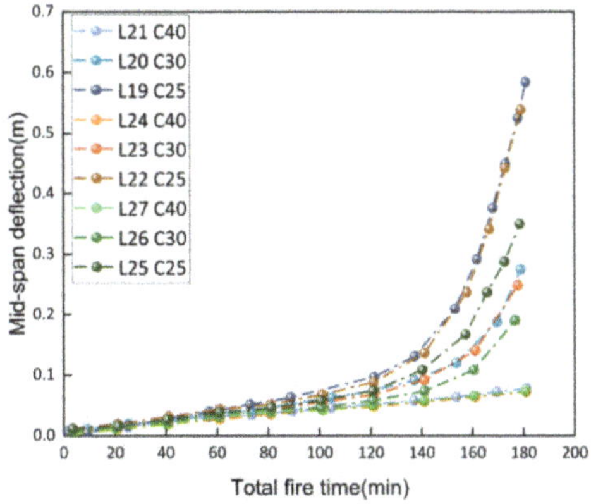

Fig. 5. Effect of different concrete strength grades on the fire resistance limit of bolted prefabricated beam-beam joints

the increase in fire resistance limit from C30 to C40 has not been determined due to the limited fire exposure time, it can be inferred from the fact that L21, L24, and L27 have very small mid-span displacements within 180 min of fire exposure time and are still in a linear change stage, and the fire resistance limit has increased larger. In summary, the strength grade of concrete has a great influence on the fire resistance limit of bolted prefabricated beam-beam joints, and the higher the strength grade of concrete, the longer the fire resistance limit. The reason is that the yield strength of high-strength concrete is greater than that of low-grade concrete when the temperature is high, so the strain generated is less, and the time to enter plastic deformation and destruction is slowed down, thereby reducing mid-span displacement and improving the fire resistance of the structure.

3.2.4 Influence of Different Bolt Strength Grades on the Fire Resistance Limit of Bolted Prefabricated Beam-Beam Joints

Three groups of components (L28, L29, L30), (L31, L32, L33), (L34, L35, L36) were set to explore the influence of different bolt strength grades on the fire resistance limit of bolted prefabricated beam-beam joints. The structure is shown in the Fig. 6.

Figure 6 shows that no matter which group of specimens, the fire time-mid-span deflection curves of different bolt strength grades are very close, and there is basically no difference in the mid-span deflection under the same fire time. Among them, the fire resistance limits of specimens L28, L29, and L30 are all about 173 min, and the fire resistance limits of specimens L31, L32, and L33 are 142 min, 142 min, and 143 min, while the specimens L34, L35, and L36 have not reached the limit state of the structure. The deflection at the 180th minute is 0.121 m, 0.12 m, 0.119 m respectively. This shows that the bolt strength rating has no effect on the fire resistance of existing bolted fabricated beam-to-beam joint designs. There are two reasons: the concentrated load and temperature load are not so large that the bolts of these three strengths yield, so that increasing the bolt strength has no effect; or the bolt strength drops very obviously

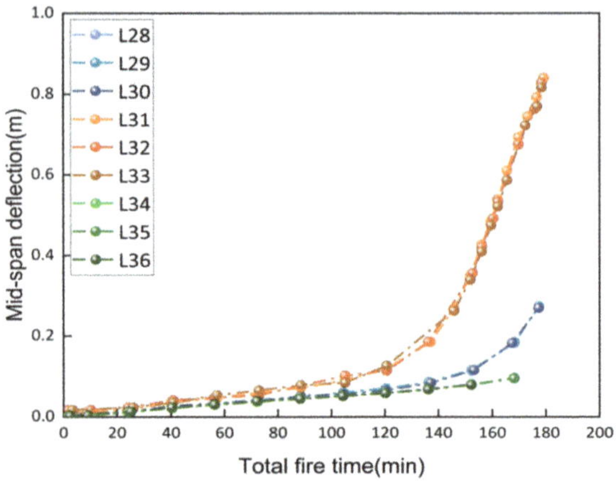

Fig. 6. Effect of different bolt strength grades on the fire resistance of bolted prefabricated beam-beam joints

at high temperature, and the increased bolt strength is still reduced and disappears as the temperature rises.

3.2.5 Influence of the Thickness of Concrete Cover on the Fire Resistance of Bolted Prefabricated Beam-Beam Joints

Three groups of components (L37, L38, L39), (L40, L41, L42), (L43, L44, L45) were set to explore the influence of different concrete cover thicknesses on the fire resistance of bolted prefabricated beam-beam joints. The results are as follows Fig. 7 shows.

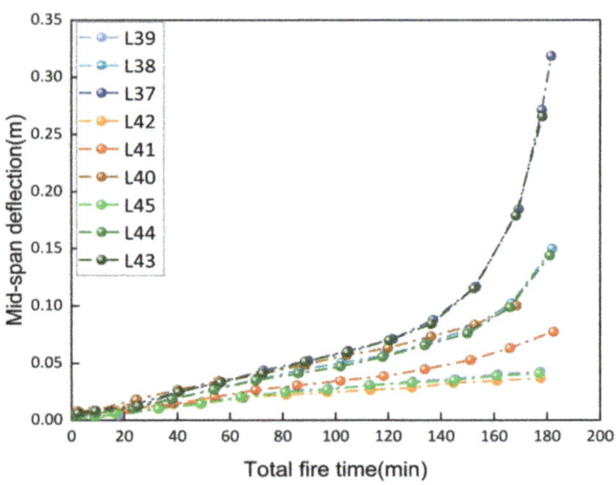

Fig. 7. Effect of different concrete cover thicknesses on the fire resistance of bolted prefabricated beam-beam joints

Figure 7 shows that in each group of specimens, as the thickness of the concrete cover changes, the time-deflection curves of the specimens also change significantly, and the curves of the three groups of specimens have the same trend. Among them, only specimens L37 and L43 have reached the fire resistance limit, respectively: 172 min and 173 min. The rest of the specimens did not reach the limit state of fire resistance within 180 min of fire exposure time. The maximum mid-span deflection—L38, L39: 0.146 m, 0.04 m; L40, L41, L42: 0.12 m, 0.075 m, 00.034 m; L44, L45: 0.143 m, 0.04 m. It shows that the thickness of the concrete cover has a great influence on the fire resistance of bolted prefabricated beam-beam joints. The greater the thickness of the concrete protective layer, the greater the fire resistance limit of the structure and the stronger the fire resistance. The reasons are analyzed: from the conclusion of Part 3, it can be known that increasing the thickness of the concrete cover has a significant effect on improving the temperature field of bolted prefabricated beam-beam joints. This leads to a certain degree of reduction in the temperature values of various parts of the structure with the same fire exposure time, and then according to the mechanical performance formulas of various materials at high temperatures in Part 2, the yield strength reduction coefficient of each component is optimized, so that the fire resistance limit of the overall structure is higher and the fire resistance performance is better.

4 Conclusions

The temperature field and mechanical model under the ISO-834 international standard temperature rise curve are established by ABAQUS software, and the temperature rise characteristics of the joint model, the influence of the thickness of the steel plate and the thickness of the concrete protective layer on the joint, the growth characteristics of the mid-span deflection of the joint under the thermal-mechanical coupling state, and the influence of parameters such as bolt preload, concrete strength grade, load ratio, bolt strength grade, and concrete cover thickness on the fire resistance limit of the joint are obtained using the method of controlling variables. The conclusions are as follows.

(1) As the thickness of the steel plate increases, the temperature field of the bolted prefabricated beam-to-beam joint decreases, but the overall impact is not significant. As the thickness of the concrete cover increases, the temperature of the bolt center of the bolted prefabricated beam-beam joint decreases significantly. Therefore, in actual engineering, it is feasible to reduce the temperature of the center of the joint by increasing the thickness of the concrete cover.
(2) As the fire exposure time increases, the mid-span deflection of nodes gradually changes from an initial slow linear growth to a parabolic growth with increasing growth rate. When approaching the fire resistance limit, there is obvious deformation that becomes larger and larger, and then the limit state of the structure is reached, and the damage is predictable to a certain extent.
(3) The pretightening force of bolts and the strength grade of bolts have no effect on the fire resistance of joints. Therefore, it is meaningless to improve the fire resistance of joints by increasing the pre-tightening force of bolts or increasing the strength grade of bolts.

(4) Concrete strength grade, concrete cover thickness and load ratio have great influence on the fire resistance limit of bolted prefabricated beam-beam joints. The higher the concrete strength grade, the greater the thickness of the protective layer, the smaller the load ratio, the longer the fire resistance limit of the joints, and the better the fire resistance. Therefore, with economy factors taken into account, improving the concrete strength grade, increasing the thickness of the concrete cover layer and the design value of the specimen bearing capacity have a significant effect on improving the fire resistance of the joints.

References

1. Han, L.H., Zheng, Y.Q., Teng, J.G.: Fire resistance of RC and FRP-confined RC columns. Mag. Concr. Res. **58**(8), 533–546 (2006)
2. Wu, B., Lu, J.Z.: A numerical study of the behavior of restrained RC beams at elevated temperatures. Fire Saf. J. **44**(4), 522–531 (2009)
3. Al-Jabri, K.S., Davison, J.B., Burgess, I.W.: Performance of beam-to-column joints in fire-a review. Fire Saf. J. Int. J. Devoted Res. Fire Saf. Sci. Eng. **43**(1), 50–62 (2008)
4. Yu, H.X., Burgess, I.W., Davison, J.B.: Tying capacity of web cleat connections in fire, part 1: test and finite element simulation. Eng. Struct. **31**(3), 651–663 (2009)
5. Yu, H.X., Burgess, I.W., Davison, J.B.: Tying capacity of web cleat connections in fire, part 2: development of component-based model. Eng. Struct. **31**(3), 691–708 (2009)
6. Yu, H.X., Burgess, I.W., Davison, J.B.: Experimental investigation of the behavior of fin plate connections in fire. J. Constr. Steel Res. **65**(3), 723–736 (2009)
7. Wang, Y.Z., Fu, C.G.: Nonlinear analysis on reinforced concrete frame joints under thermal-stress coupling. Adv. Mater. Res. **163**, 4559–4563 (2010)
8. Kong, W., Fu, C., Liu, W.: Fire resistance test of a reinforced concrete beam-supporting column transfer structure joint unit. Arab. J. Sci. Eng. **44**(5), 5117–5129 (2019)
9. Wu, B., Liu, J., Chen, X.: Numerical analysis of lateral displacement of beam-column joints in concrete frame structures subjected to fire. Adv. Struct. Eng. **21**(10), 1495–1509 (2018)
10. Han, L.H., Zheng, Y.Q., Tao, Z.: Fire performance of steel-reinforced concrete beam-column joints. Mag. Concr. Res. **61**(7), 499–518 (2009)
11. Song, T.-Y., Han, L.-H., Tao, Z.: Performance of steel-reinforced concrete beam-to-column joints after exposure to fire. J. Struct. Eng. **142**(10), 04016070 (2016)
12. Li, W., Guo, Z.: Experimental investigation of strength and deformation of concrete at elevated temperature. J. Build. Struct. **14**(01), 8–16 (1993). https://doi.org/10.14006/j.jzjgxb.1993.01.002
13. European Committee for Standardization, ENV 2007-1-2, Eurocode 4, Design of Composite and Concrete Structures, Part 1.2: Structural Fire Design, British Standards Institution, London (2007)
14. Guo, Z., Li, W.: Deformation testing and constitutive relationship of concrete under different stress-temperature paths. China Civ. Eng. J. **26**(05), 58–69 (1993)
15. Lie, T.T., Denham, E.M.A.: Factors affecting the fier resistance of circular hollow steel columns filled with bar-reinforced concrete. NRC-CNRC internal report, No. 651, Ottawa, Canada (1993)
16. European Committee for Standardization (CEN), Drafter EN 2007-1-2, Eurocode 3: Design of Steel Structures, Part 1.2: General Rules, Structural Fire Design. British Standards Institution, London (2007)
17. ISO: Fire-resistance tests-elements of building construction. ISO-834, Geneva (1980)
18. ISO: Fire-resistance tests-elements of building construction. Part 1: general requirements. ISO-834-1, Geneva (1999)

Study on the Boundary Slenderness Ratio Between Short Column and Medium-Long Column Under Axial Compression of Spiral Stirrups

Xin Chen(✉), Yun Iong Bai, and MaoSheng Li

Shengyang Jianzhu University, Shenyang 110168, Liaoning, China
15071500580@163.com

Abstract. In GB50010-2010 'Code for design of concrete structures', there is no clear stipulation on how to distinguish the spiral stirrup column from short column or medium-long column, which causes confusion to designers in practical engineering design. In order to quantitatively analyze the limit slenderness ratio of spiral stirrup axial compression short column and medium-long column, this paper establishes and verifies the finite element model of spiral stirrup axial compression column based on ABAQUS software. Based on the data obtained by finite element method, the load-displacement curve and ultimate bearing capacity of the component are analyzed, and the judgment method of the limit slenderness ratio of short column and medium-long column is given. Based on this judgment method, the finite element analysis of spiral stirrup axial compression column with different parameters is carried out. The results show that the limit slenderness ratio of the spiral stirrup axial compression short column and the middle and long column is 4.9 to 7.3, which is mainly related to the concrete strength and pitch of the component, and is negatively correlated with the two. The influence of concrete strength on the limit value is affected by the pitch. As the pitch increases, the influence of concrete strength on the limit value gradually decreases. On the basis of calculation and analysis, the proposed formula for calculating the boundary slenderness ratio of short columns and medium-long columns under axial compression of spiral stirrups is proposed.

Keywords: Boundary slenderness ratio · Finite element analysis · Spiral stirrup column

1 Introduction

The ideal concrete axial compression member does not exist in practical engineering. Due to the influence of factors such as inaccurate size, uneven material and load deviation, it is inevitable that there is a certain initial eccentricity, which leads to the generation of additional bending moment. For short columns with small slenderness ratio, the additional bending moment has little effect on the bearing capacity, which can be ignored [1].

I. Tuns et al. (Eds.): CIBv 2023, LNCE 510, pp. 157–170, 2024.
https://doi.org/10.1007/978-3-031-60765-3_14

As the slenderness ratio increases to medium-long columns or even slender columns, the additional bending moment and the corresponding lateral deflection are also increasing, and the bearing capacity is reduced more.

For the concrete column with spiral stirrups, under the axial pressure, the bearing capacity and ductility can be significantly improved due to the confinement effect of the spiral stirrups on the concrete in the core area. For the mechanical properties of spiral stirrup confined concrete, domestic and foreign scholars have carried out a large number of theoretical, experimental research and numerical simulation. In 1929, Richart [2] first studied the strength and deformation of concrete confined by circular hoop and the strength and deformation of concrete cylinder under transverse water pressure, and put forward the view that the ultimate strength and ultimate deformation of concrete confined by circular hoop were greatly improved. This view was also confirmed by later researchers Iyengar [3]. Scholars such as Chan [4] and Muzhir [5] also obtained similar conclusions.In order to obtain the accurate strength of confined concrete, Skeikh and Uzumeri [6] fully considered the influence of stirrup spacing and longitudinal reinforcement arrangement, and studied and analyzed the effective constraint area in the core concrete area. Based on this, the stress-strain expression of confined concrete is obtained. Through the monotonic axial compression test of high strength concrete columns confined by composite stirrups, Guan [7] compared and analyzed the test results. It is concluded that the ultimate bearing capacity and deformation performance of the specimens can be effectively enhanced by appropriately increasing the volume stirrup ratio of confined concrete columns and reasonably arranging stirrups. Xiao Yan [8] found in the experimental study on the seismic performance of high-strength concrete that the increase of stirrup spacing will cause the ductility of the specimen to decrease, but this ductility attenuation can be compensated by enhancing the strength of the stirrup, and the effect is obvious. Based on this conclusion, the stirrup design formula is developed, which takes into account the relative displacement and ductility parameters of the column end. Li Qingning [9] and Zhang Xinghu [10] carried out the low cyclic loading test of concrete columns confined by high-strength spiral stirrups. The results show that under the condition of high axial compression ratio, the improvement effect of high-strength spiral stirrups on the ductility and energy dissipation capacity of frame columns is very obvious. However, most of the existing studies are based on the study of mechanical properties of short columns, and there are few studies on spiral stirrup columns with large slenderness ratio. For the design of spiral stirrup columns, the Chinese code GB50010-2010 "Code for Design of Concrete Structures" [11] has corresponding provisions, but there is no clear stipulation on how to distinguish whether spiral stirrup columns belong to short columns or medium-long columns, which causes confusion to designers in practical engineering design.

In order to quantitatively analyze the boundary slenderness ratio of short column and medium-long column of spiral stirrup axial compression members under different parameters, a refined finite element analysis model of spiral stirrup column is established based on ABAQUS finite element analysis software in this paper, and the existing test results are used to verify the model. On this basis, the finite element analysis of the axial compression failure process of spiral stirrup columns with different concrete strength, pitch, longitudinal reinforcement ratio, yield strength of steel bar and aspect ratio is carried out. The load-displacement curves and ultimate bearing capacity of components with different slenderness ratios are compared and studied. The variation law of the boundary slenderness ratio of short columns and medium-long columns under axial compression of spiral stirrups under different parameters is explored, and the proposed calculation formula of the boundary slenderness ratio of short columns and medium-long columns under axial compression of spiral stirrups is given quantitatively.

2 Establishment of Finite Element Model

In order to study the influence of different parameters on the limit slenderness ratio of spiral stirrup axial compression members, this paper uses ABAQUS software to establish the elastoplastic finite element analysis model of circular spiral stirrup concrete columns, and uses the existing test results to verify it.

2.1 Unit Type Selection and Meshing

The finite element model of spiral stirrup column in this paper is composed of four parts: cushion plate, spiral stirrup, longitudinal reinforcement and concrete column. The longitudinal reinforcement and spiral stirrups are merged into the whole reinforcement cage and embedded in the concrete column. Without considering the slip between the reinforcement cage and the concrete, the top surface of the finite element model of the spiral stirrup column is bound with the bottom surface of the rigid pad. The displacement load and the change constraint are applied to the reference point of the outer surface of the rigid pad to simulate the boundary conditions of the hinged support at both ends.

The three-dimensional solid element (C3D8R) with eight-node reduction integration is used for the concrete and the upper and lower plates in the model. The two-node linear three-dimensional truss element (T3D2) [12] is used for the spiral stirrups and longitudinal bars. The mesh size of the steel cage and the concrete column is 20 mm (Fig. 1).

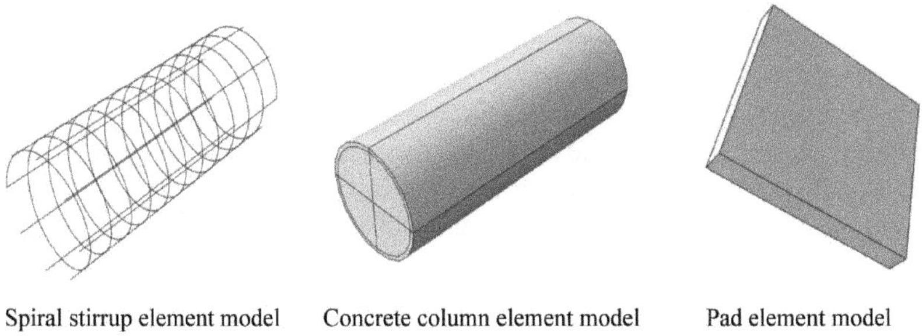

Fig. 1. Finite element component model

2.2 Constitutive of Steel and Concrete

In the concrete column with spiral stirrups, the core concrete is in a triaxial compression state under axial pressure, and the restraint effect of spiral stirrups on the core concrete should be considered. The constitutive data of concrete adopts the model of Han Hailin [13], which fully considers the influence of lateral restraint effect, has a wide range of parameters and good versatility. The specific expression of the model is as follows:

$$\begin{cases} y = 2x - x^2 & \mathrm{x} \le 1 \\ \mathrm{y} = \frac{\mathrm{x}}{\beta(\mathrm{x}-1)^2 + x} & \mathrm{x} > 1 \end{cases} \tag{1}$$

$$\begin{cases} x = \dfrac{\sigma_c}{\sigma_o} \\ y = \dfrac{\varepsilon_c}{\varepsilon_o} \end{cases} \tag{2}$$

where σ0 and ε0 are the peak value of confined concrete and the sum of peak stress, respectively. Strain; β is the curve shape control parameter, depending on the concrete.

The steel bar adopts a bilinear strengthening model, and the specific expression of the model is as follows [14]:

$$\sigma = \begin{cases} E_s\varepsilon, & \varepsilon < \varepsilon_y \\ f_y, & \varepsilon = \varepsilon_y \\ f_y + E_h(\varepsilon - \varepsilon_y), & \varepsilon > \varepsilon_y \end{cases} \tag{3}$$

In the formula, Es and fy are the elastic modulus and yield strength of steel bar respectively.

2.3 Verification of Finite Element Model

In order to verify the accuracy of the finite element model and simulation method of the spiral stirrup column in this paper, the specimens numbered HSC-A2 and HSC-B2 in Reference [15] and numbered S-1 and S-2 in Reference [16] were selected respectively, and the corresponding finite element model of the spiral stirrup column was established and numbered M-1, M-2, M-3 and M-4 respectively for analysis. The experimental design parameters and test results of the four spiral stirrup columns are shown in Table 1.

The finite element model of the existing experimental specimens is used to simulate the axial compression. The ultimate bearing capacity and stress-strain curve of the components are compared with the experimental results of the corresponding specimens in literature [15] and literature [16]. The ultimate bearing capacity test values and finite element simulation results of each group of specimens for example, as shown in Table 1, the finite element simulation results are slightly higher than the test results. The maximum error is 4.04%, the minimum error is 0.96%, and the average error is only 2.43%. It can be explained that the axial compression bearing capacity obtained by finite element simulation is in good agreement with the experimental measured bearing capacity.

The stress-strain curve of the spiral stirrup column obtained by the finite element simulation is compared with the experimental results of Reference [15] and Reference

[16], as shown in Fig. 2. It can be seen from Fig. 2 that the finite element analysis results in this paper are generally in good agreement with the experimental results in literature [15] and literature [16]. In summary, the elastic-plastic finite element model used in this paper to simulate the axial compression mechanical behavior of spiral stirrup columns has certain accuracy.

Table 1. Comparison of ABAQUS simulated results and experimental results

Specimen	f_{cu}/MPa	f_y/MPa	f_{sy}/MPa	d_y/mm	d_{sy}/mm	s/mm	N_{u1}/KN	N_{u2}/KN	$e/\%$
M-1	90.7	550	541	14	8	60	5588.7	5642.6	0.96%
M-2	103.5	525	541	16	8	50	7372.7	7507.3	1.83%
M-3	42.3	355	412	8	8	30	2721.8	2831.8	4.04%
M-4	42.3	355	630	8	6	30	2209.1	2274.1	2.91%

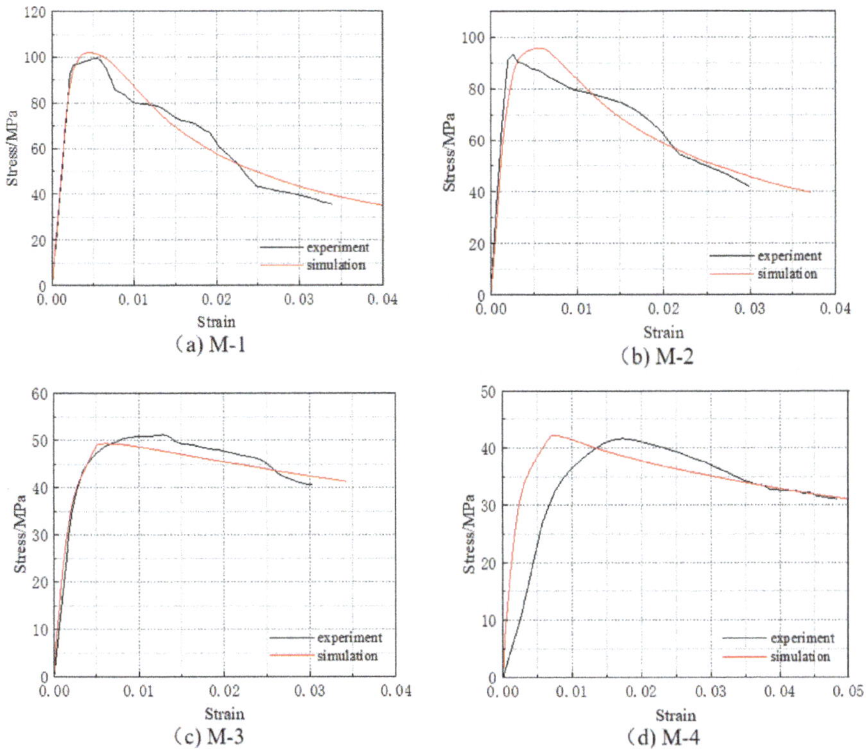

Fig. 2. Comparison of finite element and experimental stress-strain curve

2.4 Analysis of Failure Mode of Specimen

Figure 3 show the comparison between the finite element simulation failure mode and the test results of the circular spiral stirrup concrete axial compression short column. It can be seen from the figure that the finite element results are basically consistent with the test results, showing a waist drum type failure, and the finite element can better simulate the failure mode of the specimen. In summary, the finite element model of circular spiral stirrup concrete column established in this paper can better reflect the mechanical properties of the specimen.

(a) S-1 (b) S-2

Fig. 3. Comparison of experimental and finite element failure mode

3 Judgment Standard of Boundary Slenderness Ratio

Through the finite element modeling method of the spiral stirrup column in the first section, the finite element analysis models of three groups of spiral stirrup columns with different pitches and different concrete strength grades were established. The slenderness ratio was used as the control variable to simulate the failure process of the spiral stirrup column under axial pressure. The load-displacement curve, bearing capacity and stability coefficient were analyzed, and the judgment method of the boundary slenderness ratio between the short column and the medium-long column under axial compression of spiral stirrups was determined.

3.1 Determination of Basic Analysis Parameters

The diameter of the concrete in the core area of the spiral stirrup column is dcor = 400 mm, the thickness of the protective layer is 25 mm, the standard value of the yield strength of the spiral stirrup and the longitudinal reinforcement in the column is 400 MPa, and the longitudinal reinforcement is evenly arranged around the spiral stirrup. The specific parameters of the model are as follows: 1) The concrete strength grade is

C35, the spiral stirrup is C10@40, and the longitudinal reinforcement is 8C18; 2) The concrete strength grade is C55, the spiral stirrup is C10@40, and the longitudinal steel bar is 8C18; 3) The strength grade of concrete is C40, the spiral stirrup is C10@50, and the longitudinal reinforcement is 8C18, which are named L40-35, L40-55 and L50-40 respectively. The length-to-diameter ratio lo/dcor of the core area of each group of components increases from 3 to enough to determine its boundary value.

3.2 Load-Displacement Curve

The load-displacement curve reflects the corresponding relationship between the total load and the total longitudinal compression of the column, which is the macroscopic manifestation of the longitudinal stress and longitudinal average strain of the column [17]. Figure 4 is the load-displacement curve of three groups of finite element models of spiral stirrup columns with the change of length-diameter ratio.

It can be seen from Fig. 4(a) that when the aspect ratio of the component is within 5, the peak load of the spiral stirrup column remains basically unchanged. At this time, the second-order effect of the component is small, and the additional bending moment has little effect on its mechanical performance, which can be ignored. As the aspect ratio of the component increases from 5 to 6, the peak load begins to decrease significantly, indicating that the second-order effect of the component is more obvious at this time, and the impact on the bearing capacity cannot be ignored. As the aspect ratio of the component continues to increase to 8, the peak load of the spiral stirrup column gradually decreases, the influence of the additional bending moment caused by the initial eccentricity is significant, and the influence of material nonlinearity and geometric nonlinearity is more obvious.

It can be inferred that the limit slenderness ratio of component 2–1 is between 5 and 6. Similarly, it can be seen from Fig. 4(b) (c) that the boundary slenderness ratios of the corresponding components are between 6 and 7, 4 and 5, respectively. In summary, through the load-displacement curve, the boundary slenderness ratio range of the three groups of short columns and medium-long columns can be preliminarily inferred.

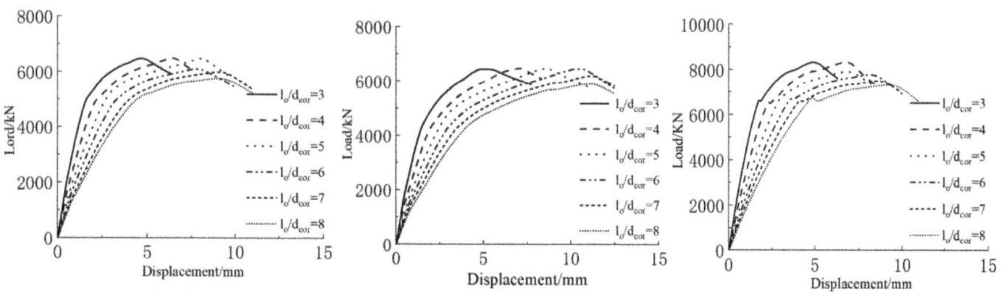

(a) L50 -40 (b) L40-35 (c) L40-55

Fig. 4. Load-displacement curves of different slenderness ratios

3.3 Peak Load

In order to obtain the limit slenderness ratio of different components more accurately, the length-diameter ratio of the components is changed in the range of the existing limit values in a unit of one tenth, and the corresponding curves of the peak load-length-diameter ratio of the three groups of components are drawn as shown in Fig. 5. It can be seen from (a) in Fig. 5 that when the aspect ratio of the component reaches 5.3, the peak load remains basically unchanged, and the bearing capacity of the component does not change greatly due to the second-order effect. The slenderness ratio is still within the slenderness ratio range of the short column. When the aspect ratio of the component increases from 5.3 to 5.4, the peak load changes significantly, from the original 6458.26 KN to 6150.71 KN, a decrease of 4.7%. Combined with Fig. 4, it can be seen that after the aspect ratio of the component reaches 5.4, as the aspect ratio continues to increase to 9, the peak load generally shows a gentle downward trend, and the influence of the second-order effect on the bearing capacity of the component is also gradually increasing, which cannot be ignored. It can be considered that when the aspect ratio of component 2–1 is 5.4, it is the limit slenderness ratio of short column and medium-long column under axial compression. The limit slenderness ratios of short column and medium-long column under axial compression of three groups of spiral stirrups are 5.4,6.9 and 4.9 respectively. It can be seen that with the change of concrete strength and pitch, the limit values of aspect ratio corresponding to the change of peak load are not the same.

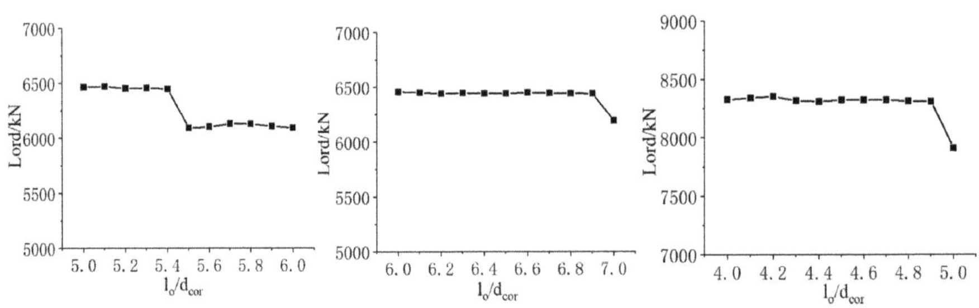

(a) L50 -40 (b) L40-35 (c) L40-55

Fig. 5. Peak load curves of different slenderness ratios

3.4 Bearing Capacity and Stability Coefficient

The stability coefficient is defined as the ratio of the axial compressive bearing capacity of the long column to the axial compressive bearing capacity of the short column with the same section, the same material and the same reinforcement [18]. From the above analysis, this paper defines the slenderness ratio of the spiral stirrup column when the axial bearing capacity is significantly reduced by the second-order bending moment as the boundary slenderness ratio of the short column and the medium-long column, that is, the stability coefficient of the component is less than one.

Figure 6 shows the curve of the calculated stability coefficient of three groups of spiral stirrup columns with different parameters changing with the slenderness ratio. From Fig. 6, it can be seen that the stability coefficient of the three groups of spiral stirrup columns shows roughly the same trend with the increase of slenderness ratio. When the slenderness ratio of the component reaches the limit value, the axial compression bearing capacity is basically unchanged, and the stability coefficient is always one. At this time, it can be considered that the component is still in the short column range; as the slenderness ratio increases to its limit value, the stability coefficient begins to change, no longer remains one, and shows a decreasing trend. As the slenderness ratio of the component continues to increase, the stability coefficient begins to decrease gradually, and the bearing capacity of the component gradually decreases. When it reaches 9, the stability coefficients of the three groups of components are 0.871, 0.855, and 0.877, respectively. The bearing capacity is reduced by about 13% compared with the corresponding value in the short column range. The second-order effect has a significant effect on its axial compression bearing capacity. It can be considered that the slenderness ratio of the component reaches the range of the medium-long column at this time.

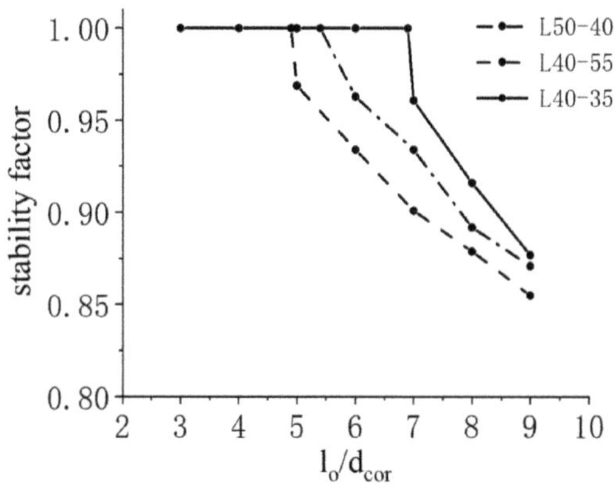

Fig. 6. Stability coefficient curves of different slenderness ratios

4 Parametric Analysis

4.1 Major Effect Parameter

In order to study the influence parameters of the boundary value of the spiral stirrup axial compression short column and the middle and long column, the pitch, concrete strength, stirrup yield strength, longitudinal reinforcement yield strength, stirrup diameter and longitudinal reinforcement ratio were used as control variables to study their influence on the boundary value. The concrete diameter of the core area is 400 mm, and the thickness of the protective layer is 25 mm.

It can be seen from Table 2 that the boundary slenderness ratio of spiral stirrup axial compression short column and medium-long column is mainly affected by concrete strength and pitch, and has no obvious correlation with stirrup yield strength, longitudinal reinforcement yield strength, stirrup diameter, longitudinal reinforcement ratio and other parameters. When the pitch increases from 20 to 60, the limit value gradually decreases from 7.7 to 4.8, and when the pitch continues to increase to 80, the limit value no longer continues to decrease. It can be seen that the boundary slenderness ratio of the short column and the medium-long column under axial compression of spiral stirrups decreases with the increase of pitch, and there is a certain lower limit. For the component numbered B-1, the boundary value shows a gentle downward trend with the increase of the compressive strength of the concrete standard cube, while for the component numbered B-3, the boundary value does not change greatly with the change of the compressive strength of the concrete standard cube. In summary, the limit slenderness ratio of the axially compressed short column and the medium-long column of the spiral stirrup is negatively correlated with the strength of the concrete and the pitch, and there is no obvious correlation with other parameters. Among them, the influence of concrete strength on the limit value will change due to different pitches.

Table 2. Analysis of the influence parameters of the boundary value

	Parameter	Specimen	f_{cu}/MPa	f_y/MPa	f_{sy}/MPa	d/mm	s/mm	$\rho_s/\%$	λ_o
	pitch	A-1	40	335	400	10	20	1.62	7.7
		A-2	40	335	400	10	40	1.62	6.4
		A-3	40	335	400	10	60	1.62	4.4
		A-4	40	335	400	10	80	1.62	4.8
S = 40	concrete strength	B-1-1	30	335	400	10	40	1.62	7.3
		B-1-2	40	335	400	10	40	1.62	6.4
		B-1-3	50	335	400	10	40	1.62	5.4
		B-1-4	60	335	400	10	40	1.62	4.9
	stirrup strength	C-1-1	30	335	300	10	40	1.62	7.3
		C-1-2	30	335	335	10	40	1.62	7.3
		C-1-3	30	335	400	10	40	1.62	7.3
		C-1-4	30	335	500	10	40	1.62	7.3
	stirrup diameter	D-1-1	30	335	400	8	40	1.62	7.3
		D-1-2	30	335	400	10	40	1.62	7.3
		D-1-3	30	335	400	12	40	1.62	7.3
		D-1-4	30	335	400	14	40	1.62	7.3
	longitudinal strength	E-1-1	30	300	400	10	40	1.62	7.3
		E-1-2	30	335	400	10	40	1.62	7.3
		E-1-3	30	400	400	10	40	1.62	7.4
		E-1-4	30	500	400	10	40	1.62	7.4

(continued)

Table 2. (*continued*)

	Parameter	Specimen	f_{cu}/MPa	f_y/MPa	f_{sy}/MPa	d/mm	s/mm	$\rho_s/\%$	λ_o
	volume stirrup ratio	F-1-1	30	335	400	10	40	1.62	7.3
		F-1-2	30	335	400	10	40	2.42	7.3
		F-1-3	30	335	400	10	40	3.38	7.3
S = 60	concrete strength	B-3-1	30	335	400	10	60	1.62	4.9
		B-3-2	40	335	400	10	60	1.62	4.9
		B-3-3	50	335	400	10	60	1.62	4.9
		B-3-4	60	335	400	10	60	1.62	4.8
	stirrup strength	C-3-1	40	335	300	10	60	1.62	4.8
		C-3-2	40	335	335	10	60	1.62	4.8
		C-3-3	40	335	400	10	60	1.62	4.8
		C-3-4	40	335	500	10	60	1.62	4.8
	stirrup diameter	D-3-1	40	335	400	8	60	1.62	4.8
		D-3-2	40	335	400	10	60	1.62	4.9
		D-3-3	40	335	400	12	60	1.62	4.9
		D-3-4	40	335	400	14	60	1.62	4.9
	longitudinal strength	E-3-1	40	300	400	10	60	1.62	5.0
		E-3-2	40	335	400	10	60	1.62	5.0
		E-3-3	40	400	400	10	60	1.62	5.0
		E-3-4	40	500	400	10	60	1.62	5.1
	volume stirrup ratio	F-3-1	40	335	400	10	60	1.62	5.0
		F-3-2	40	335	400	10	60	2.42	5.0
		F-3-3	40	335	400	10	60	3.38	5.0

4.2 Influence Law

In order to further explore the relationship between the influence of concrete strength on the limit value and the pitch, the finite element model of spiral stirrup columns with different pitches was established with concrete strength as the control variable. The specific parameters and limit values are shown in Table 3. In columns with spiral or welded ring stirrups, the spacing of stirrups should not be greater than 80 mm and should not be less than 40 mm. The compressive strength of concrete standard cube ranges from 30 MPa to 60 MPa.

It can be seen from Table 3 that the influence of concrete strength on the limit value is affected by the pitch. With the increase of pitch, the influence of concrete strength on the limit value decreases gradually. When the pitch increases to 60, the limit value no longer changes with the change of concrete strength, and the influence of concrete strength on the limit value is basically zero.

Table 3. Component parameters and boundary values

	Specimen	f_{cu}/MPa	f_y/MPa	f_{sy}/MPa	d/mm	s/mm	$\rho_s/\%$	λ_o
S = 40	B-1-1	30	335	400	10	40	1.62	7.3
	B-1-2	40	335	400	10	40	1.62	6.4
	B-1-3	50	335	400	10	40	1.62	5.4
	B-1-4	60	335	400	10	40	1.62	4.8
	B-1-5	70	335	400	10	40	1.62	4.8
S = 50	B-2-1	30	335	400	10	50	1.62	6.5
	B-2-2	40	335	400	10	50	1.62	5.4
	B-2-3	50	335	400	10	50	1.62	4.8
	B-2-4	60	335	400	10	50	1.62	4.8
	B-2-5	70	335	400	10	50	1.62	4.8
S = 55	B-3-1	30	335	400	10	55	1.62	5.7
	B-3-2	40	335	400	10	55	1.62	4.8
	B-3-3	50	335	400	10	55	1.62	4.8
	B-3-4	60	335	400	10	55	1.62	4.8
	B-3-5	70	335	400	10	55	1.62	4.8
S = 60	B-4-1	30	335	400	10	60	1.62	4.8
	B-4-2	40	335	400	10	60	1.62	4.8
	B-4-3	50	335	400	10	60	1.62	4.8
	B-4-4	60	335	400	10	60	1.62	4.8
	B-4-5	70	335	400	10	60	1.62	4.8
S = 70	B-5-1	30	335	400	10	70	1.62	4.8
	B-5-2	40	335	400	10	70	1.62	4.8
	B-5-3	50	335	400	10	70	1.62	4.8
	B-5-4	60	335	400	10	70	1.62	4.8
	B-5-5	70	335	400	10	70	1.62	4.8

4.3 Formula Fitting

Using Origin software, based on the least square method [19], the nonlinear surface fitting of the relationship between the limit value and the pitch and concrete strength in Table 3 is carried out, and the calculation formula of the limit slenderness ratio of the spiral stirrup axial compression short column and the middle long column under different parameters is obtained:

$$\lambda_o = -0.0901S^2 - 3.7f_{cu}^2 + 0.8233S + 0.03976f_{cu} - 0.00258S \cdot f_{cu} - 8.999 \geq 4.9 \tag{4}$$

It can be concluded from Fig. 7 that the error between the finite element simulation results and the boundary value calculated by Formula 4 is small, and the value of r reaches 0.98 after adjustment. The calculation is in good agreement with the finite element simulation.

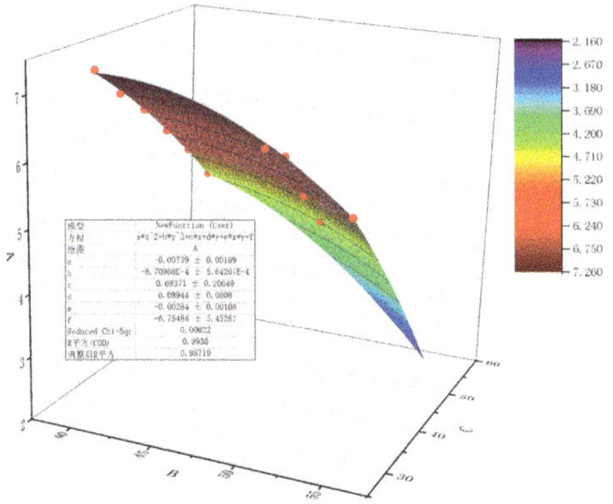

Fig. 7. Fitting surface diagram

5 Conclusion

1) Through the refined elastic-plastic finite element analysis, the variation law of axial compression performance of spiral stirrup columns with different concrete strength, pitch, longitudinal reinforcement ratio, steel yield strength and length-diameter ratio is studied. The results show that the boundary slenderness ratio of spiral stirrup axial compression short column and medium-long column is mainly related to concrete strength and pitch, and negatively correlated with the two. The influence of concrete strength on the limit value is affected by the pitch. As the pitch increases, the influence of concrete strength on the limit value gradually decreases.
2) For the axially compressed short columns with spiral stirrups, the second-order effect is small, and the influence on the bearing capacity can be neglected. At this time, the stability coefficient can be regarded as one. For the medium-long columns with spiral stirrups under axial compression, the influence of the second-order effect on the bearing capacity of the members increases with the increase of the slenderness ratio, and the stability coefficient decreases gradually.
3) Based on the results of finite element analysis, a recommended calculation formula for the slenderness ratio of the axial compression short column and the middle-long column with spiral stirrups is proposed. This formula is applicable to the concrete cube compressive strength standard value between 30 to 70 N/mm^2, spiral stirrup spacing between 40 to 70 between the spiral stirrup axial compression cylinder, when the calculation result is less than 4.9, according to 4.9 value.

References

1. Liu, X., Xie, S.H., Yu, Z., Yang, Y., Qi, M.: Effect of slenderness ratio on axial compression bearing capacity of spiral stirrup columns. J. Build. Struct. **41**(08), 183–190 (2020)
2. Richart, F.E., Brandtzæg, A., Brown, R.L.: Failure of plain and spirally reinforced concrete in compression. University of Illinois. Engineering Experiment Station. Bulletin, no. 190 (1929)
3. Sundara Raja Iyengar, K.T., Desayi, P., Reddy, K.N.: Stress-strain characteristics of concrete confined in steel binders. Mag. Concr. Res. **22**(72), 173–184 (1970)

4. Chan, W.W.L.: The ultimate strength and deformation of plastic hinges in reinforced concrete frameworks. Mag. Concr. Res. **7**(21), 121–132 (1955)
5. Muzhir, S.M., Khan, M.M.A.: Experimental investigation into the deformation potentialities confined concrete in the plastic zone (1982)
6. Sheikh, S.A., Uzumeri, S.M.: Analytical model for concrete confinement in tied columns. J. Struct. Div. **108**(12), 2703–2722 (1982)
7. Guan, P., Wang, Q., Zhao, G.: Experimental study on stress-strain constitutive relation of high strength confined concrete. Ind. Build. (11), 27–30 (1997)
8. Xiao, Y., Wu, Y., Shang, S., Yun, H.W., Esmaeily, A.: Full-scale experimental study and theoretical analysis on seismic behavior of high-strength concrete columns. J. Southeast Univ. (Nat. Sci. Ed.) (05), 746–749 (2002)
9. Li, Q., Zheng, X., Yan, Z., Jiang, W., Pan, S.: Experimental study and theoretical analysis on seismic behavior of high-strength spiral stirrup confined concrete column connections. J. Build. Struct. (08), 90–99 (2013)
10. Zhang, X., Zheng, X., Pan, S., Jiang, W., Li, Q.: Experimental study on seismic behavior of concrete columns confined by high-strength spiral stirrups. J. Archit. Sci. Eng. **30**(2), 7 (2013)
11. Ministry of Construction of the People's Republic of China: Code for Design of Concrete Structures (GB50010-2010). China Building Industry Press, Beijing (2010)
12. Zhang, Y.: High-strength spiral stirrup confined concrete axial compression mechanical properties test and finite element analysis. Xi'an University of Architecture and Technology (2008)
13. Han, L.H.: Concrete-Filled Steel Tubular Structures-Theory and Practice, 3rd edn. Science Press, Beijing (2016)
14. Study on axial compression performance of multi-spiral stirrup confined concrete square columns. Chang'an University (2022)
15. Zhu, P.R.: Research on axial compression mechanical properties of C100 concrete columns confined by 500 MPa spiral stirrups. Shenzhen University (2016)
16. Yang, K., Wang, Y., Liu, R., Sun, S.S., Wang, B.: Experimental study and bearing capacity calculation of concrete columns confined by two-way spiral stirrups under axial compression. J. Archit. Sci. Eng. 1–10
17. Hou, X., Xu, Y.F.: Effect of slenderness ratio on the mechanical behavior of L-shaped steel reinforced concrete special-shaped columns under axial compression. In: Academic Committee of the National Symposium on Modern Structural Engineering. Tianjin University (2009)
18. Ye, L.P.: Design Principle of Concrete Structure, 5th edn. China Construction Industry Press (2012)
19. Huo, X.C., Li, X.P.: Fitting surface equation with least square method. J. Chifeng Univ. (Nat. Sci. Ed.) **25**(06), 11–13 (2009)

Safety Evaluation Model Detection of Jacking Form Work System Based on BP Neural Network

Shilong Jia(✉), Yi Sha, Xiangxu Wang, Hanjie Zhong, Juncheng Ren, and Haojia Sun

Shenyang Jianzhu University, Shenyang 110168, China
cesljia@sjzu.edu.cn

Abstract. Heinrich's law identifies people as the primary cause of accidents. Human factors must be controlled in safety management to prevent production accidents. Therefore, to ensure the safe construction of the top die system, and improve the productivity of the top die system. Based on the theory of human factor engineering, this paper will conduct safety evaluation on the construction of top die system, deeply study the application of human factor engineering in the construction safety evaluation of top die system, and improve the safety management system of top die system construction. A neural network algorithm based on BP is used to test the scientific validity of the model. As a result of learning and training the sample data collected from the actual engineering project, the output value and predicted value of the final BP neural network algorithm are compared in order to determine the model's scientific validity and effectiveness.

Keywords: jacking form work system · BP neural network · Human factor engineering · safety evaluation

1 Introduction

An Artificial Neural Network (ANN) simulates how neurons in the human brain process external information by using mathematics. The human neuron is composed of a cell body, a short dendrite with many branches, and a long axon with a few branches. Neurons receive stimuli through dendrites and process them inside their cells before passing them on to other neurons through their axons. In a neural network, neurons are connected in a specific way to form a complex, non-linear network. As a result of this complex mechanism, the human brain can interact with the world around it. An illustration of a physiological neuron model is shown in Fig. 1.

Back-Propagation was proposed by David Rumelhart and J. M. Calelland in 1985. It adopts nonlinear continuous transformation function to enable hidden layer neurons to have the error backward propagation of learning ability. The typical composition of BP neural network is shown in Fig. 2, and the layers are connected by weights.

I. Tuns et al. (Eds.): CIBv 2023, LNCE 510, pp. 171–189, 2024.
https://doi.org/10.1007/978-3-031-60765-3_15

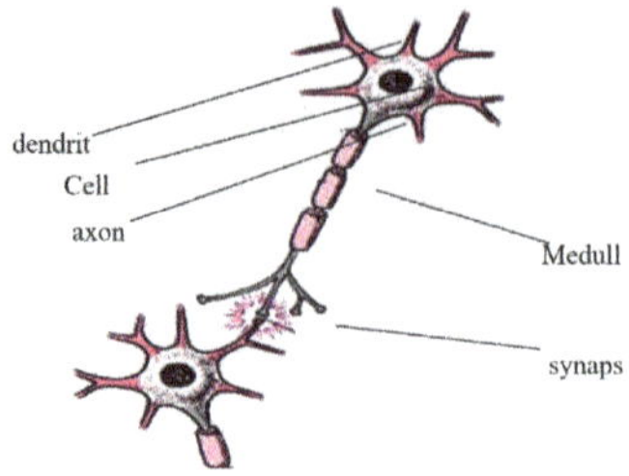

Fig. 1. Neuron physiological model

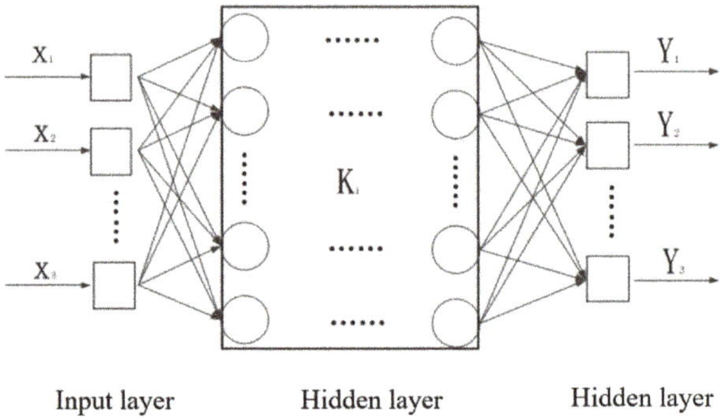

Fig. 2. BP neural network structure diagram

BP neural network is a supervised learning algorithm. After the input of the learning sample, the backpropagation algorithm is used to adjust and train the weight and deviation of the network repeatedly, so that the output result is as close as possible to the actual result. When the flat sum of the error of the output layer of the network is less than the established error, the training is completed and the weight and deviation of the network are saved. The main idea of BP neural network is to divide the training process into two stages:

Forward propagation process: The input information is processed layer by layer through the input layer and the actual output value of each unit is calculated.

Backpropagation process: If the output layer cannot get the desired output value, then the difference between the actual output and the expected output is calculated recursively layer by layer, so that the weight is adjusted according to the difference.

2 Construction of Safety Evaluation System of Jacking Form Work System

By analyzing the construction characteristics in the construction environment of the top die system, this paper sorted out the balance of the relationship between man, machine and environment that affected the top die system, and established the construction safety evaluation system of the top die system based on the human factor engineering theory, as shown in Table 1.

Table 1. Construction Safety Evaluation System of Jacking Form Work System

Target layer	First level	Second level
Construction safety evaluation system of Jacking Form work System O	Human factor E_1	Psychological and physiological state of construction personnel of jacking form work system F_1
		Operation proficiency of jacking form work system construction personnel F_2
		Safety awareness of jacking form work system construction personnel F_3
		Safety management attitude of top mode system managers F_4
		Top mode system management personnel safety management ability F_5
		Special operation personnel of jacking form work system construction are on duty with certificates F_6
	Construction material factor E_2	Construction material quality F_7
		Construction material storage F_8
		Construction material stacking F_9
	Mechanical factor E_3	Mechanical equipment inspection, maintenance and maintenance F_{10}
		Formwork system construction safety F_{11}
		Hanger system construction safety F_{12}
		Support system construction safety F_{13}
		Warning mark condition F_{14}
		Frame system situation F_{15}

(*continued*)

Table 1. (*continued*)

Target layer	First level	Second level
		Power and control system conditions F_{16}
		Monitoring system condition F_{17}
		Integrated equipment status F_{18}
		Accessory system condition F_{19}
	Environmental factor E_4	Operating ambient temperature F_{20}
		Operating environment illumination F_{21}
		Operating environment noise F_{22}
		Site sign setting F_{23}
		Operation area division F_{24}
	Management factor E_5	Safety technology disclosure F_{25}
		Emergency plan measures F_{26}
		Safety education and training F_{27}

2.1 Deterministic Judgment Matrix

The top layer of the figure is the target layer O, which is the construction safety evaluation index system of the jacking form work system. The criterion layer is the first-level index set as C, and the scheme layer is the second-level index represented by P. In order to ensure the accuracy of the results, the analytic hierarchy process is used to determine the index weight.

Table 2. Target Layer Judgment Matrix

O	C_1	C_2	C_3	C_4	C_5
C_1	1	1/5	1	1/3	3
C_2	5	1	5	3	7
C_3	1	1/5	1	1/2	2
C_4	3	1/3	2	2	3
C_5	1/3	1/7	1/3	1/5	1

Table 3. Human Factor Judgment Matrix

C_1	P_1	P_2	P_3	P_4	P_5	P_6
P_1	1	1/2	1	2	2	1/3
P_2	2	1	1	3	3	1/3
P_3	1	1/2	1	2	1	1/3
P_4	1/2	1/2	1/3	1	3	1/3
P_5	1/2	1/3	1	1/2	1	1/3
P_6	3	3	3	3	3	1

Table 4. Judgment Moment of Construction Material Factor

C_2	P_7	P_8	P_9
P_7	1	3	3
P_8	1/3	1	1/2
P_9	1/3	2	1

Table 5. Factor Judgment Matrix of Mechanical Equipment

C_3	P_{10}	P_{11}	P_{12}	P_{13}	P_{14}	P_{15}	P_{16}	P_{17}	P_{18}	P_{19}
P_{10}	1	1/3	1/5	1/7	1/3	1/5	2	3	1/3	1/7
P_{11}	3	1	1/3	1/5	1/5	2	1/3	5	1/2	3
P_{12}	5	2	1	1/5	1/3	1	6	2	1	1/2
P_{13}	7	5	5	1	2	1/2	2	1/3	1/2	3
P_{14}	3	5	3	1/2	1	5	1	3	1	1/4

(*continued*)

Table 5. (*continued*)

C_3	P_{10}	P_{11}	P_{12}	P_{13}	P_{14}	P_{15}	P_{16}	P_{17}	P_{18}	P_{19}
P_{15}	5	1/2	1	2	1/5	1	1/6	1/2	2	1/2
P_{16}	1/2	3	1/6	1/2	1	6	1	3	1/5	1
P_{17}	1/3	1/5	1/2	3	1/3	2	1/3	1	1/2	1/2
P_{18}	3	2	1	2	1	1/2	5	2	1	3
P_{19}	7	1/3	2	1/3	4	2	1	2	1/3	1

Table 6. Environmental Factor Judgment Matrix

C_4	P_{20}	P_{21}	P_{22}	P_{23}	P_{24}
P_{20}	1	1/3	1/3	1/5	1/7
P_{21}	3	1	1/4	1/4	1/6
P_{22}	3	4	1	1/3	1/6
P_{23}	5	4	3	1	1/5
P_{24}	7	6	5	3	1

Table 7. Management Factor Judgment Matrix

C_5	P_{25}	P_{26}	P_{27}
P_{25}	1	1/4	1/5
P_{26}	4	1	1/3
P_{27}	5	3	1

2.2 Required Maximum Eigenvector W

According to Table 2, 3, 4, 5, 6 and 7, the eigenvalue method is adopted to calculate the weight of the values in the judgment matrix, and the required maximum eigenvector W and CR can be obtained as follows:

$$\begin{aligned} W &= \{w_{C1}, w_{C2}, w_{C3}, w_{C4}, w_{C5}\} \\ &= \{0.1049, 0.5115, 0.1049, 0.2320, 0.0468\} \end{aligned} \tag{1}$$

$CR = 0.0113 < 0.1$, there is satisfactory consistency.

$$\begin{aligned} W_1 &= \{w_{P1}, w_{P2}, w_{P3}, w_{P4}, w_{P5}, w_{P6}\} \\ &= \{0.1312, 0.2157, 0.1184, 0.0910, 0.0804, 0.3633\} \end{aligned} \quad (2)$$

$CR = 0.0410 < 0.1$, there is satisfactory consistency.

$$\begin{aligned} W_2 &= \{w_{P7}, w_{P8}, w_{P9}\} \\ &= \{0.5936, 0.1571, 0.2493\} \end{aligned} \quad (3)$$

$CR = 0.0516 < 0.1$, there is satisfactory consistency.

$$\begin{aligned} W_3 &= \{w_{P10}, w_{P11}, w_{P12}, w_{P13}, w_{P14}, w_{P15}, w_{P16}, w_{P17}, w_{P18}, w_{P19}\} \\ &= \{0.0521, 0.1090, 0.2095, 0.2015, 0.1644, 0.0963, 0.0520, 0.0833, 0.0364, 0.0355\} \end{aligned} \quad (4)$$

$CR = 0.0900 < 0.1$, there is satisfactory consistency.

$$\begin{aligned} W_4 &= \{w_{P20}, w_{P21}, w_{P22}, w_{P23}, w_{P24}\} \\ &= \{00436, 0.0706, 0.1496, 0.2383, 0.4979\} \end{aligned} \quad (5)$$

$CR = 0.0900 < 0.1$, there is satisfactory consistency.

$$\begin{aligned} W_5 &= \{w_{P25}, w_{P26}, w_{P27}\} \\ &= \{0.0936, 0.2797, 0.6267\} \end{aligned} \quad (6)$$

$CR = 0.0825 < 0.1$, there is satisfactory consistency.

Table 8. Construction Safety Evaluation System of Jacking Form Work System Based on Human Factors Perspective

First-order index	weight	Secondary index	weight
Human factor C_1	0.1049	Psychological and physiological state of construction personnel of jacking form work system P_1	0.1322
		Operation proficiency of jacking form work system construction personnel P_2	0.2147
		Safety awareness of jacking form work system construction personnel P_3	0.1194
		Safety management attitude of top mode system managers P_4	0.0900
		Top mode system management personnel safety management ability P_5	0.0804

(*continued*)

Table 8. (*continued*)

First-order index	weight	Secondary index	weight
		Special operation personnel of jacking form work system construction are on duty with certificates P_6	0.3633
Construction material factor C_2	0.1049	Construction material quality P_7	0.5836
		Construction material storage P_8	0.1671
		Construction material stacking P_9	0.2393
Mechanical factor C_3	0.5115	Mechanical equipment inspection, maintenance and maintenance P_{10}	0.0521
		Formwork system construction safety P_{11}	0.1090
		Hanger system construction safety P_{12}	0.2095
		Support system construction safety P_{13}	0.2015
		Warning mark condition F_{14}	0.1644
		Frame system situation F_{15}	0.0963
		Power and control system conditions F_{16}	0.0520
		Monitoring system condition F_{17}	0.0833
		Integrated equipment status F_{18}	0.0364
		Accessory system condition F_{19}	0.0355
Environmental factor C_4	0.2320	Operating ambient temperature P_{20}	0.0436
		Operating environment illumination P_{21}	0.0606
		Operating environment noise P_{22}	0.1596
		Site sign setting P_{23}	0.2383
		Operation area division P_{24}	0.4979
Management factor C_5	0.0468	Safety technology disclosure P_{25}	0.1036
		Emergency plan measures P_{26}	0.2797
		Safety education and training P_{27}	0.6367

3 Construction Safety Evaluation Model Detection of Jacking Form Work System Based on BP Neural Network

In terms of image processing, MATLAB can show the data in the form of graphics, so that users can feel the changes of the data more intuitively. Moreover, MATLAB constantly changes and improves, showing excellent processing ability in lighting, chroma processing and thinking data.

3.1 MATLAB Toolbox Training BP Neural Network Implementation Mechanism

Matlab sets a variety of training function toolbox, do not need complex program editing, only need to set certain parameters, you can get the desired results. This paper uses Neural Net Fitting in the Matlab packaging APP to train the samples. The following is part of the program code during the internal training of the toolbox:

```
% X - input data.
% Y - target data.
x = X';
t = Y';
 % Choose a Training Function
trainFcn = 'trainlm';
hiddenLayerSize = 10;
net = fitnet(hiddenLayerSize,trainFcn);
net.input.processFcns = {'removeconstantrows','mapminmax'};
net.output.processFcns = {'removeconstantrows','mapminmax'};
net.divideFcn = 'dividerand'; % Divide data randomly
net.divideMode = 'sample'; % Divide up every sample
net.divideParam.trainRatio = 70/100;
net.divideParam.valRatio = 15/100;
net.performFcn = 'mse';
net.plotFcns = {'plotperform','plottrainstate','ploterrhist', ...
  'plotregression', 'plotfit'};
 % Train the Network
[net,tr] = train(net,x,t);
  % Test the Network
y = net(x);
e = gsubtract(t,y);
......
```

3.2 Construction Safety Evaluation Model Detection Based on BP Neural Network

Determination of Network Structure

The number of neurons in the input layer of BP neural network can be determined to be 21 according to the established construction safety evaluation index of the top die system. The output layer is set according to the evaluation set by specifying the evaluation objective, and the safety is represented by [100, 00, 00] respectively. [0 1 0 0 0] means safer; [0 0 1 0 0] indicates critical; [0 0 0 1 0] indicates danger; [0 0 0 0 1] means dangerous. “Safety” means that the grade of the evaluated factors of the project is safety, and the construction project is in a safe state; “Relatively safe” means that the evaluated factors of the project are relatively safe and there is room for improvement; “Critical” means that the evaluation factor of the project needs to be improved; “Danger” means that there are hidden dangers in the evaluated factors of the project, which need to be rectified in time; “Very dangerous” indicates that the project is not qualified by the evaluation factor management and needs to be redesigned and planned. Figure 3 shows the structure of BP neural network.

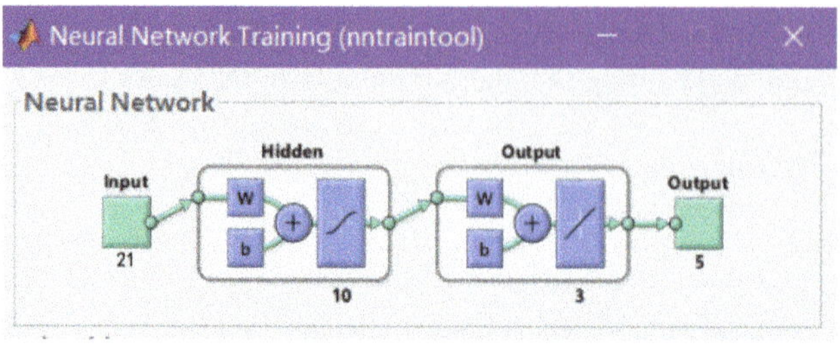

Fig. 3. Neural Network

Sample Selection and Model Training

An Artificial Neural Network (ANN) simulates how neurons in the human brain process external information by using mathematics. The human neuron is composed of a cell body, a short dendrite with many branches, and a long axon with a few branches. Neurons receive stimuli through dendrites and process them inside their cells before passing them on to other neurons through their axons. In a neural network, neurons are connected in a specific way to form a complex, non-linear network. In normal circumstances, the selection principle for the number of samples is that the greater the number of samples, the better. However, due to the constraints of the research topic itself, the number of samples cannot be infinite; too many samples will lengthen the training sample cycle and increase the complexity of the BP neural network structure. Although a small number of samples can easily lead to large errors, there is no reference significance when only a few samples are considered. A sample size of 10 or more is normally considered to be the most economical under normal circumstances. During this evaluation, the data used is based on the expert's rich experience and professional knowledge base combined with specific projects, which are relatively objective and can meet certain consistency rules. To test the data quantity and accuracy of samples, 10 sets of data are used as training samples, of which 8 sets are used as training samples and 2 sets as detection samples. Tables 2, 3, 4, 5, 6, 7, 8, 9, 10, 11 and 12 present the sample data.

Table 9. Training sample 1

1	u_{11}	u_{12}	u_{13}	u_{14}	u_{15}	u_{16}	u_{21}	u_{22}	u_{23}	u_{31}	u_{32}
v_1	1	1	0	3	2	4	1	0	1	1	1
v_2	1	3	4	1	2	1	4	2	1	2	1
v_3	2	1	1	1	1	0	0	2	3	1	2
v_4	1	0	0	0	0	0	0	1	0	1	1
v_5	0	0	0	0	0	0	0	0	0	0	0
	u_{33}	u_{34}	u_{41}	u_{42}	u_{43}	u_{44}	u_{45}	u_{51}	u_{52}	u_{53}	
v_1	1	3	1	0	0	2	0	3	2	0	
v_2	3	1	2	0	1	1	3	1	2	1	
v_3	1	1	2	2	3	2	2	1	1	4	
v_4	0	0	0	1	1	0	0	0	0	0	
v_5	0	0	0	2	0	0	0	0	0	0	

Table 10. Training sample 2

2	u_{11}	u_{12}	u_{13}	u_{14}	u_{15}	u_{16}	u_{21}	u_{22}	u_{23}	u_{31}	u_{32}
v_1	2	1	0	1	2	3	1	1	1	1	2
v_2	1	1	1	2	1	1	3	1	2	2	2
v_3	2	3	2	2	2	1	1	3	2	1	1
v_4	0	0	1	0	0	0	0	0	0	1	0
v_5	0	0	0	0	0	0	0	0	0	0	0
	u_{33}	u_{34}	u_{41}	u_{42}	u_{43}	u_{44}	u_{45}	u_{51}	u_{52}	u_{53}	
v_1	2	1	0	0	1	3	1	2	1	0	
v_2	1	2	1	1	2	2	2	2	3	2	
v_3	2	2	3	2	2	0	2	1	1	3	
v_4	0	0	1	2	0	0	0	0	0	0	
v_5	0	0	0	0	0	0	0	0	0	0	

Table 11. Training sample 3

3	u_{11}	u_{12}	u_{13}	u_{14}	u_{15}	u_{16}	u_{21}	u_{22}	u_{23}	u_{31}	u_{32}
v_1	2	1	1	1	3	3	0	1	1	2	1
v_2	2	2	3	2	1	2	2	1	2	1	4
v_3	1	2	0	2	1	0	3	3	2	2	0
v_4	0	0	1	0	0	0	0	0	0	0	0
v_5	0	0	0	0	0	0	0	0	0	0	0
	u_{33}	u_{34}	u_{41}	u_{42}	u_{43}	u_{44}	u_{45}	u_{51}	u_{52}	u_{53}	
v_1	2	1	2	1	0	2	1	1	2	2	
v_2	3	3	2	1	1	2	3	3	2	1	
v_3	0	1	1	3	1	1	1	1	1	2	
v_4	0	0	0	0	3	0	0	0	0	0	
v_5	0	0	0	0	0	0	0	0	0	0	

Table 12. Training sample 4

4	u_{11}	u_{12}	u_{13}	u_{14}	u_{15}	u_{16}	u_{21}	u_{22}	u_{23}	u_{31}	u_{32}
v_1	2	1	1	1	2	1	3	0	2	3	3
v_2	1	1	1	3	2	3	2	0	1	1	1
v_3	1	3	2	1	1	1	0	2	2	1	1
v_4	1	0	1	0	0	0	0	3	0	0	0
v_5	0	0	0	0	0	0	0	0	0	0	0
	u_{33}	u_{34}	u_{41}	u_{42}	u_{43}	u_{44}	u_{45}	u_{51}	u_{52}	u_{53}	
v_1	2	2	0	0	0	1	3	2	2	0	
v_2	0	2	3	1	0	1	1	1	3	2	
v_3	3	1	2	3	1	3	1	1	0	2	
v_4	0	0	0	1	4	0	0	1	0	1	
v_5	0	0	0	0	0	0	0	0	0	0	

Table 13. Training sample 5

5	u_{11}	u_{12}	u_{13}	u_{14}	u_{15}	u_{16}	u_{21}	u_{22}	u_{23}	u_{31}	u_{32}
v_1	1	0	1	2	3	2	2	1	0	2	2
v_2	2	3	1	2	1	2	2	1	3	1	2
v_3	2	2	3	1	1	1	1	3	2	2	1
v_4	0	0	0	0	0	0	0	0	0	0	0
v_5	0	0	0	0	0	0	0	0	0	0	0
	u_{33}	u_{34}	u_{41}	u_{42}	u_{43}	u_{44}	u_{45}	u_{51}	u_{52}	u_{53}	
v_1	3	3	0	0	0	1	2	1	1	1	
v_2	1	2	3	2	2	3	2	3	2	2	
v_3	1	0	1	3	3	1	1	1	2	2	
v_4	0	0	1	0	0	0	0	0	0	0	
v_5	0	0	0	0	0	0	0	0	0	0	

Table 14. Training sample 6

6	u_{11}	u_{12}	u_{13}	u_{14}	u_{15}	u_{16}	u_{21}	u_{22}	u_{23}	u_{31}	u_{32}
v_1	1	1	0	1	3	2	3	2	1	2	2
v_2	3	2	2	3	1	1	1	0	2	1	2
v_3	1	2	3	1	1	2	1	3	1	2	1
v_4	0	0	0	0	0	0	0	0	1	0	0
v_5	0	0	0	0	0	0	0	0	0	0	0
	u_{33}	u_{34}	u_{41}	u_{42}	u_{43}	u_{44}	u_{45}	u_{51}	u_{52}	u_{53}	
v_1	3	3	1	1	0	2	3	2	2	1	
v_2	1	2	1	2	1	2	1	2	1	3	
v_3	1	0	3	1	2	1	1	1	2	1	
v_4	0	0	0	1	2	0	0	0	0	0	
v_5	0	0	0	0	0	0	0	0	0	0	

Table 15. Training sample 7

7	u_{11}	u_{12}	u_{13}	u_{14}	u_{15}	u_{16}	u_{21}	u_{22}	u_{23}	u_{31}	u_{32}
v_1	2	1	1	3	1	2	3	1	0	2	3
v_2	3	1	2	1	3	2	2	2	1	2	1
v_3	0	3	2	1	1	1	0	2	4	1	1
v_4	0	0	0	0	0	0	0	0	0	0	0
v_5	0	0	0	0	0	0	0	0	0	0	0
	u_{33}	u_{34}	u_{41}	u_{42}	u_{43}	u_{44}	u_{45}	u_{51}	u_{52}	u_{53}	
v_1	2	3	0	1	1	3	2	2	3	2	
v_2	1	2	2	2	2	1	3	1	2	2	
v_3	2	0	3	2	1	1	0	2	0	1	
v_4	0	0	0	0	1	0	0	0	0	0	
v_5	0	0	0	0	0	0	0	0	0	0	

Table 16. Training sample 8

8	u_{11}	u_{12}	u_{13}	u_{14}	u_{15}	u_{16}	u_{21}	u_{22}	u_{23}	u_{31}	u_{32}
v_1	1	0	1	3	3	2	3	2	1	1	2
v_2	2	1	1	1	2	1	2	2	1	2	2
v_3	1	4	3	1	0	2	0	1	3	2	1
v_4	1	0	0	0	0	0	0	0	0	0	0
v_5	0	0	0	0	0	0	0	0	0	0	0
	u_{33}	u_{34}	u_{41}	u_{42}	u_{43}	u_{44}	u_{45}	u_{51}	u_{52}	u_{53}	
v_1	3	3	2	2	1	2	2	3	2	0	
v_2	1	1	1	2	1	2	2	2	1	3	
v_3	1	1	2	1	3	1	1	0	2	2	
v_4	0	0	0	0	0	1	0	0	0	0	
v_5	0	0	0	0	0	0	0	0	0	0	

Table 17. Tested sample 9

9	u_{11}	u_{12}	u_{13}	u_{14}	u_{15}	u_{16}	u_{21}	u_{22}	u_{23}	u_{31}	u_{32}
v_1	1	1	0	3	2	1	2	3	0	2	3
v_2	3	2	2	1	3	3	2	1	3	1	2
v_3	1	2	3	1	0	1	1	1	2	2	0
v_4	0	0	0	0	0	0	0	0	0	0	0
v_5	0	0	0	0	0	0	0	0	0	0	0
	u_{33}	u_{34}	u_{41}	u_{42}	u_{43}	u_{44}	u_{45}	u_{51}	u_{52}	u_{53}	
v_1	2	4	1	2	1	3	1	3	3	2	
v_2	3	1	1	2	2	2	3	2	1	2	
v_3	0	0	3	1	2	0	1	0	1	1	
v_4	0	0	0	0	0	0	0	0	0	0	
v_5	0	0	0	0	0	0	0	0	0	0	

Table 18. Tested sample 10

10	u_{11}	u_{12}	u_{13}	u_{14}	u_{15}	u_{16}	u_{21}	u_{22}	u_{23}	u_{31}	u_{32}
v_1	2	1	1	2	3	2	3	2	1	3	2
v_2	2	3	3	2	1	2	1	2	2	2	2
v_3	1	1	1	1	1	1	1	1	2	0	1
v_4	0	0	0	0	0	0	0	0	0	0	0
v_5	0	0	0	0	0	0	0	0	0	0	0
	u_{33}	u_{34}	u_{41}	u_{42}	u_{43}	u_{44}	u_{45}	u_{51}	u_{52}	u_{53}	
v_1	4	3	2	3	2	4	3	2	3	3	
v_2	1	2	2	1	1	1	0	3	2	2	
v_3	0	0	1	1	2	0	2	0	0	0	
v_4	0	0	0	0	0	0	0	0	0	0	
v_5	0	0	0	0	0	0	0	0	0	0	

Based on the above data, the input sample of neural network evaluation model can be obtained through normalization processing:

Table 19. Sample data table

Num	1	2	3	4	5	6	7	8	9	10
u_{11}	0.52	0.40	0.36	0.44	0.44	0.40	0.32	0.48	0.40	0.36
u_{12}	0.40	0.48	0.48	0.48	0.48	0.44	0.48	0.56	0.44	0.40
u_{13}	0.32	0.44	0.44	0.40	0.36	0.40	0.32	0.32	0.32	0.36
u_{14}	0.36	0.40	0.32	0.36	0.32	0.32	0.40	0.28	0.32	0.32
u_{15}	0.44	0.48	0.48	0.48	0.48	0.44	0.48	0.56	0.52	0.40
u_{16}	0.24	0.32	0.40	0.40	0.36	0.40	0.36	0.40	0.40	0.36
u_{21}	0.36	0.40	0.28	0.28	0.36	0.32	0.28	0.28	0.36	0.32
u_{22}	0.58	0.48	0.72	0.72	0.48	0.44	0.44	0.36	0.32	0.36
u_{23}	0.48	0.44	0.40	0.40	0.48	0.48	0.56	0.48	0.48	0.44
u_{31}	0.48	0.48	0.32	0.32	0.40	0.32	0.36	0.44	0.40	0.28
u_{32}	0.52	0.36	0.32	0.32	0.36	0.32	0.32	0.36	0.28	0.36
u_{33}	0.40	0.40	0.44	0.44	0.32	0.28	0.40	0.32	0.32	0.24
u_{34}	0.32	0.44	0.36	0.36	0.28	0.24	0.28	0.32	0.24	0.28
u_{35}	0.40	0.40	0.44	0.44	0.32	0.28	0.40	0.32	0.32	0.24
u_{36}	0.32	0.44	0.36	0.36	0.28	0.24	0.28	0.32	0.24	0.28
u_{37}	0.48	0.44	0.32	0.32	0.36	0.32	0.32	0.36	0.40	0.36
u_{38}	0.32	0.36	0.44	0.44	0.40	0.36	0.40	0.28	0.28	0.32
u_{39}	0.36	0.40	0.32	0.32	0.44	0.40	0.28	0.40	0.32	0.28
u_{310}	0.56	0.52	0.56	0.56	0.44	0.40	0.36	0.48	0.36	0.28
u_{41}	0.48	0.48	0.32	0.32	0.40	0.32	0.36	0.44	0.40	0.28
u_{42}	0.52	0.36	0.32	0.32	0.36	0.32	0.32	0.36	0.28	0.36
u_{43}	0.40	0.40	0.44	0.44	0.32	0.28	0.40	0.32	0.32	0.24
u_{44}	0.32	0.44	0.36	0.36	0.28	0.24	0.28	0.32	0.24	0.28
u_{45}	0.48	0.44	0.32	0.32	0.36	0.32	0.32	0.36	0.40	0.36
u_{51}	0.32	0.36	0.44	0.44	0.40	0.36	0.40	0.28	0.28	0.32
u_{52}	0.36	0.40	0.32	0.32	0.44	0.40	0.28	0.40	0.32	0.28
u_{53}	0.56	0.52	0.56	0.56	0.44	0.40	0.36	0.48	0.36	0.28

Table 20. Sample data table

Num	11	12	13	14	15	16	17	18	19	20
u_{11}	0.48	0.44	0.40	0.40	0.48	0.48	0.56	0.48	0.48	0.44
u_{12}	0.48	0.48	0.32	0.32	0.40	0.32	0.36	0.44	0.40	0.28
u_{13}	0.52	0.36	0.32	0.32	0.36	0.32	0.32	0.36	0.28	0.36
u_{14}	0.40	0.40	0.44	0.44	0.32	0.28	0.40	0.32	0.32	0.24
u_{15}	0.32	0.44	0.36	0.36	0.28	0.24	0.28	0.32	0.24	0.28
u_{16}	0.24	0.32	0.40	0.40	0.36	0.40	0.36	0.40	0.40	0.36
u_{21}	0.36	0.40	0.28	0.28	0.36	0.32	0.28	0.28	0.36	0.32
u_{22}	0.58	0.48	0.72	0.72	0.48	0.44	0.44	0.36	0.32	0.36
u_{23}	0.48	0.44	0.40	0.40	0.48	0.48	0.56	0.48	0.48	0.44
u_{31}	0.48	0.48	0.32	0.32	0.40	0.32	0.36	0.44	0.40	0.28
u_{32}	0.52	0.36	0.32	0.32	0.36	0.32	0.32	0.36	0.28	0.36
u_{33}	0.40	0.40	0.44	0.44	0.32	0.28	0.40	0.32	0.32	0.24
u_{34}	0.32	0.44	0.36	0.36	0.28	0.24	0.28	0.32	0.24	0.28
u_{35}	0.36	0.40	0.28	0.28	0.36	0.32	0.28	0.28	0.36	0.32
u_{36}	0.40	0.40	0.44	0.44	0.32	0.28	0.40	0.32	0.32	0.24
u_{37}	0.32	0.44	0.36	0.36	0.28	0.24	0.28	0.32	0.24	0.28
u_{38}	0.48	0.44	0.32	0.32	0.36	0.32	0.32	0.36	0.40	0.36
u_{39}	0.58	0.48	0.72	0.72	0.48	0.44	0.44	0.36	0.32	0.36
u_{310}	0.48	0.44	0.40	0.40	0.48	0.48	0.56	0.48	0.48	0.44
u_{41}	0.48	0.48	0.32	0.32	0.40	0.32	0.36	0.44	0.40	0.28
u_{42}	0.52	0.36	0.32	0.32	0.36	0.32	0.32	0.36	0.28	0.36
u_{43}	0.40	0.40	0.44	0.44	0.32	0.28	0.40	0.32	0.32	0.24
u_{44}	0.32	0.44	0.36	0.36	0.28	0.24	0.28	0.32	0.24	0.28
u_{45}	0.48	0.44	0.32	0.32	0.36	0.32	0.32	0.36	0.40	0.36
u_{51}	0.32	0.36	0.44	0.44	0.40	0.36	0.4	0.28	0.28	0.32
u_{52}	0.36	0.40	0.32	0.32	0.44	0.40	0.28	0.40	0.32	0.28
u_{53}	0.56	0.52	0.56	0.56	0.44	0.40	0.36	0.48	0.36	0.28

After obtaining the sample, the network is trained by Neural Net Fitting in Matlab toolbox. The trained sample data reached the expected result at epoch 8, as shown in Fig. 4.

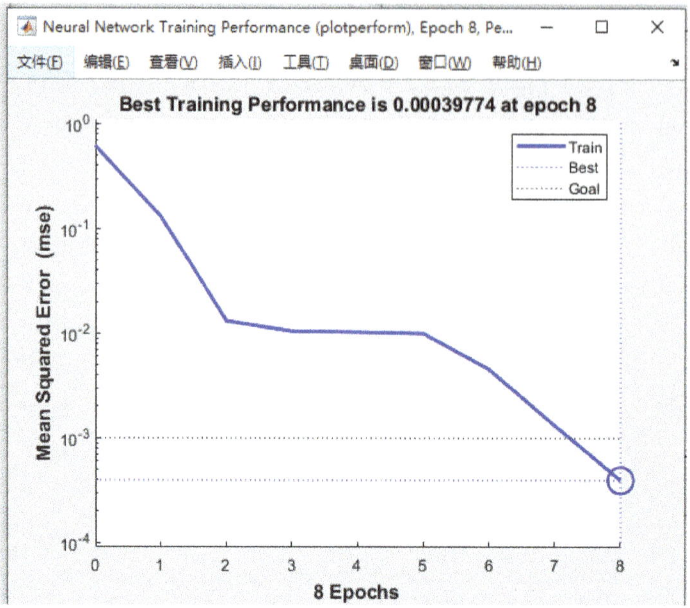

Fig. 4. Matlab training result

The test results (Table 13) indicate that the expected value is essentially consistent with the output value, indicating that the BP neural network developed in this paper has a reasonable structure. As long as ten groups of training samples are used, the accuracy requirements of the model can be met after a certain number of iterations, and the training results are generally consistent with the fuzzy comprehensive evaluation method (Tables 14, 15, 16, 17, 18, 19, 20 and 21).

Table 21. Testing result

Expected output	Actual output	Safety level
[01000]	[0.7462 0.7560 0.0285 0.0000 0.0000]	safer
[00100]	[0.2913 0.2845 0.3308 0.0000 0.0000]	critical

To sum up, the construction safety evaluation result of core tube top molding system of a super high-rise building in Shenyang is relatively safe. The overall safety situation is good, but there is still room for progress, and it is necessary to pay attention to prevention and rectification at all times.

4 Conclusion

This paper describes the application example of the construction safety evaluation model for the top die system. This paper analyzes in detail the top form system engineering of a super high-rise building construction project in Shenyang. The first step in the process is to provide an overview of the project's basic situation. A second step involves evaluating tthe top die system construction ision using a fuzzy comprehensive evaluation method, as well as comparing and verifying the BP neural network algorithm, which demonstrates its practical application.

References

1. Ministry of Housing and Urban-Rural Development, PRC. Provisions on the safety management of parts and items with greater risk. Architectural Des. Manag. **35**(05), 8–9
2. Wenzhou, Y., Fanyan, Y., Botao, Y.: SEM study on the effect of safety atmosphere on human factor risk transmission of subway construction crew. J. Civ. Eng. Manag. **35**(02), 39–44 (2018)
3. Hui, L., Chao, Z.: Study on comprehensive safety evaluation of human-machine-environment system construction site. J. Chongqing Jianzhu Univ. (05), 107–111 (2007)
4. Yicheng, H.: Research on Effectiveness Evaluation of site safety signs of engineering projects based on human factors. Anhui University of Technology, Wuhu (2014)
5. Tingjun, J.: Using human factor engineering theory to reduce the probability of man-made accidents. Theor. Study Explor. (06), 82–83 (2017)
6. Hyla, P.: Overhead travelling crane construction deflection measurements with telematic approach. J. KONES **26**(3), 53–60 (2019)
7. Faccio, M., Ferrari, E., Gamberi, M., Pilati, F.: Human factor analyser for work measurement of manual manufacturing and assembly processes. Int. J. Adv. Manuf. Technol. **103**(1–4), 861–877 (2019)
8. Evgeniy, G.: The influence of the human factor on the safety of operation of the industrial buildings and structures. In: IOP Conference Series: Materials Science and Engineering, vol. 1030, no. 1, p. 012031 (2021)
9. Chua, D.K.H., Goh, Y.M.: Application of case based reasoning in construction safety planning. Inf. Technol. (6), 298–307 (2004)
10. Xuyang, J.: Safety evaluation of high-rise building construction based on BP neural network. Dalian University of Technology, Dalian (2020)

Compression Stress-Strain Curve Model of High-Strength Spiral Stirrup Confined Concrete

Xin Chen, Maosheng Li(✉), and Yunlong Bai

Shenyang Jianzhu University, No. 25 Hunnan Middle Road, Hunnan District, Shenyang, Liaoning, China
17360075867@163.com

Abstract. This paper aims to establish a stress-strain curve model for high-strength spiral hoop-restrained concrete that can be applied to different cross-sectional forms and different constraint conditions. Based on mathematical statistics methods, collect and process the axial compression test data of high-strength spiral reinforced concrete columns with 5 types of hoop reinforcement constraint forms under 2 cross-sectional forms from existing literature are processed. The material parameters of the collected test specimens are concrete strength of 35.2–83.07 MPa, hoop yield strength of 450–1318 MPa, volumetric reinforcement ratio of 0.48%–4.43%, and longitudinal reinforcement ratio of 0.45%–1.79%. Subsequently, regression analysis is conducted on the five characteristic parameters of the stress-strain curve to establish a model for the stress-strain curve of high-strength spiral confined concrete. Comparing the curve model in this paper with the experimentally measured stress-strain curves of the test components, it is concluded that the curve model in this paper has the ability to predict the behavior of high-strength spiral reinforced concrete circular and rectangular components under different constraint conditions. The results show that the proposed formulas for calculating the stress of hoop reinforcement, peak compressive stress, peak compressive strain, and ultimate compressive strain under peak compressive stress are more accurate. The established model of stress-strain curve for high-strength spiral hoop-reinforced confined concrete has a wide range of applicability, accurately predicts the behavior of structural tests, and is convenient for finite element analysis and engineering calculations.

Keywords: High-strength spiral stirrup constraint · stirrup stress · compressive stress-strain relationship · constitutive equation

1 Introduction

With the rapid development of the construction industry, high-strength stirrups and high-strength concrete have been widely promoted. The utilization of high-strength concrete can significantly enhance project performance; however, as strength increases, the brittleness of concrete becomes more pronounced. To mitigate this brittleness and

I. Tuns et al. (Eds.): CIBv 2023, LNCE 510, pp. 190–202, 2024.
https://doi.org/10.1007/978-3-031-60765-3_16

improve its load-bearing capacity, we employ the most effective confinement method high-strength spiral stirrup confinement. The application of high-strength spiral stirrups can enhance both the load-bearing capacity and deformation capacity of concrete. Nevertheless, numerous experimental studies have revealed that when concrete reaches its peak stress, stirrups may not yield, leading to an overestimation of confined concrete's load-bearing capacity. The existing constitutive model of confined concrete presents several challenges, including a multitude of parameters, limited application scope, and complex engineering calculations. To address these issues effectively, it is imperative to establish a unified stress-strain curve model for high-strength spiral stirrup confined concrete by collating available experimental data.

2 Collection of Experimental Data of High-Strength Spiral Stirrup Confined Concrete

2.1 Unified Processing Method of Test Data

The collected experimental data are unified to eliminate the influence of concrete cover and longitudinal reinforcement stress. Assuming that the longitudinal reinforcement strain is the same as the concrete strain [1], the stress of the protective layer is calculated by using the standard concrete constitutive model [2]. The influence of the concrete protective layer can be eliminated by substituting the stress corresponding to the longitudinal reinforcement strain and the stress when the protective layer falls into the bearing capacity calculation Eq. (1) of the core concrete.

$$F_0(\varepsilon) = F(\varepsilon) - A_{\text{cov}} f_{\text{cov}}(\varepsilon) - A_s f_s(\varepsilon) \tag{1}$$

In the Eq. (1), $F_0(\varepsilon)$ is the bearing capacity of the core concrete of the specimen when the strain is ε; $F(\varepsilon)$ is the bearing capacity of the whole section of the specimen when the strain is ε; A_{cov} is the cross-sectional area of the concrete protective layer; A_s is the total cross-sectional area of the longitudinal reinforcement; $f_{\text{cov}}(\varepsilon)$ is the corresponding concrete stress of the protective layer when the strain is ε; $f_s(\varepsilon)$ is the longitudinal reinforcement stress when the strain is ε.

Assuming that there is no relative slip between the longitudinal reinforcement and the core concrete, the peak strain of the core concrete is the longitudinal reinforcement strain. The corresponding stress is calculated by using the two-stage model of the uniaxial compression constitutive model of the steel bar. The longitudinal reinforcement stress is deducted from the experimental data, and the peak compressive stress of the confined concrete can be obtained accurately.

For the concrete used in all tests, the peak strain of unconstrained concrete adopts a unified calculation method [3], as shown in Eq. (2).

$$\varepsilon_{co} = 0.0028 - 0.0008(40/f_{co}) \tag{2}$$

In the Eq. (2), ε_{co} is the peak compressive strain of unconstrained concrete; f_{co} is the peak compressive stress of unconstrained concrete.

2.2 Test Data Collection

The test data collection criteria are as follows: the specimens are all high-strength stirrup confined concrete columns, the height-width ratio of the specimens is 2–6, the stirrup form is a single spiral stirrup and a composite spiral stirrup, and the test data are obtained under axial compression. In this paper, 203 axial compression test data of high strength spiral stirrup confined concrete at home and abroad are collected and sorted out [4–11], including 103 circular sections and 100 rectangular sections. The axial compression stress-strain relationship curves of 203 high-strength stirrup confined concrete columns are shown in Fig. 1. The collected data are unified, and the data are summarized in Table 1.

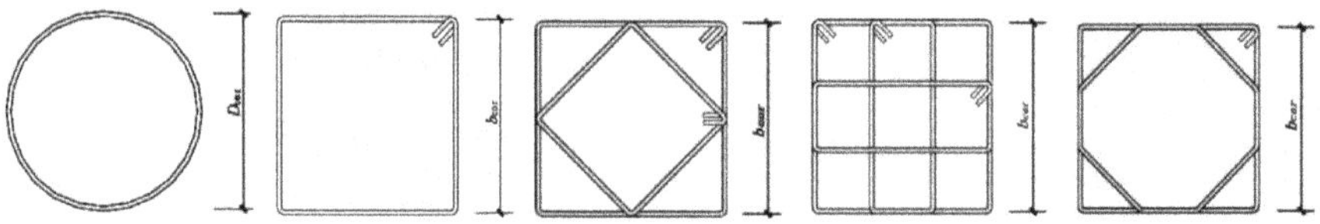

Fig. 1. Stirrup form of component

Table 1. Axial compression test data of high strength spiral stirrup confined concrete

Researchers	Dimension	f_c'(MPa)	ρ_s%	ρ_v%	f_{yv}(MPa)
Zheng [4]	280	40–70	0.78–0.79	1–1.6	480–1219
Hou CC [5]	265	43.2–74.74	0.71–0.72	0.9–2	681–863
Li Bing [6]	240	41–97	1.5	0.82–2.94	445–1318
Wang [7]	300	68.16–83.07	0.45	0.48–1.41	515
Yang K [8]	150	50.2–56.6	0.89–1.79	0.98–2.45	411–1120
Wang N [9]	200	39.3–44.5	1.4–3.06	1.5–2.8	486–1143
Ding HY [10]	250	69.07–71.7	1.45	0.83–1.41	840–872
Li Bing [6]	240	35.2–72.3	0.78–0.79	0.8–5	445–1318
He X [11]	250	55.36–69.43	0.89–1.79	1.3–4.43	450–1100

In the Table 1, D is the section size of the specimen; f_c' is the compressive strength of unconstrained concrete; ρ_s is longitudinal reinforcement ratio; ρ_v is the volume stirrup ratio; f_{yv} is the yield strength of stirrups.

3 Stress and Strain Calculation of High-Strength Spiral Stirrup Confined Concrete

Revised sentence: When establishing the compressive stress-strain curve of high-strength spiral stirrup confined concrete, it is essential to accurately determine the two parameter values of effective lateral restraint stress and stirrup stress at the peak point. Based on this, three characteristic point parameters peak point stress, strain, and ultimate compressive strain can be determined to establish five parameter calculation formulas.

3.1 Effective Lateral Restraint Stress

The specimen is deformed by the load, and the expansion of the core concrete causes the stirrup to be pulled to produce tensile stress. The stirrup reacts on the concrete and forms a passive constraint on the concrete. The constraint stress provided by the circular stirrup is relatively uniform, while the constraint stress of the rectangular stirrup is unevenly distributed along the circumference due to the arch action. Therefore, the effective constraint coefficient proposed by Mander is used to calculate the effective constraint stress of stirrups [12]. The effective lateral stress (f_{le}) is calculated according to the force balance of the section. The effective constraint coefficient (k_e) is to consider the influence of arch effect, which reflects the influence of stirrup form, longitudinal reinforcement distribution, stirrup spacing and other factors on the constraint effect.

For rectangular sections, as in Eqs. (3) and (4):

$$k_e = \frac{(1 - \frac{\sum_i^n \omega_i^2}{6c_x c_y})(1 - \frac{s'}{2c_x})(1 - \frac{s'}{2c_y})}{(1 - \rho_c)} \tag{3}$$

$$f_{le} = \frac{k_e f_s}{s}(\frac{A_{shx} + A_{shy}}{c_x + c_y}) \tag{4}$$

In the formulas: c_x and c_y are the size of the concrete in the x and y directions in the core area of the column; ω_i is the net spacing of adjacent longitudinal bars; s' is the net spacing of stirrups; ρ_c is the reinforcement ratio of longitudinal reinforcement; f_s is the stress of the stirrup under the peak compressive stress; A_{shx} and A_{shy} are the total cross-sectional area of stirrups in the x and y directions, respectively [13].

For circular cross section, as in Eqs. (5) and (6):

$$k_e = \frac{(1 - \frac{s'}{2d_s})^n}{1 - \rho_{cc}} \tag{5}$$

$$f_{le} = 0.5 k_e \rho_s f_s \tag{6}$$

In the Eq. (5) and (6): $n = 1$, is a circular spiral stirrup; $n = 2$, is a circular stirrup. s' is the net spacing of stirrups; d_s is the outer diameter of stirrups; ρ_{cc} is the longitudinal reinforcement ratio; ρ_v is the stirrup volume stirrup ratio; the stirrup stress at the peak stress of f_s.

3.2 Steel Yield Stress

The strain gauge is pre-fixed on the stirrup in the axial compression test to measure the strain, enabling assessment of whether the stirrup yields at peak concrete stress. Before concrete pouring, strain gauges were pasted on the two rings of spiral stirrups in the middle of the test section, and the strain gauge reading was the tensile strain of the stirrup. Extensive research by numerous scholars has revealed that ordinary confined concrete with stirrups can yield when subjected to peak compressive stress, leading many proposed models for confined concrete to calculate lateral restraint stress based on stirrup yield strength. However, high-strength spiral stirrups typically do not yield when confined concrete reaches its peak compressive stress. Cusson [13] conducted axial compression tests on four kinds of high-strength stirrup confined concrete specimens, and found that high-strength stirrups basically could not reach the yield strength. Therefore, it is necessary to utilize experimental data from the peak point to accurately determine the effective lateral restraint stress of concrete.

Based on the statistical regression analysis of 115 experimental data collected, the calculation model of high-strength spiral stirrup stress of confined concrete at the peak point is obtained, as shown Eq. (7) and Fig. 2. The comparison between the stress test values and the calculated values of the five types of stirrup specimens is shown in Fig. 3. It can be seen from the figure that most of the calculated values of stirrup stress are in good agreement with the experimental values. The average value of the ratio of the calculated value of stirrup stress to the experimental value at the peak point of confined concrete is 1.08, and the standard deviation is 0.23.

$$f_s = 0.094 E_s k_e^2 \sqrt{\frac{\rho_v}{f_{co}}} + 350 \leq f_{yv} \tag{7}$$

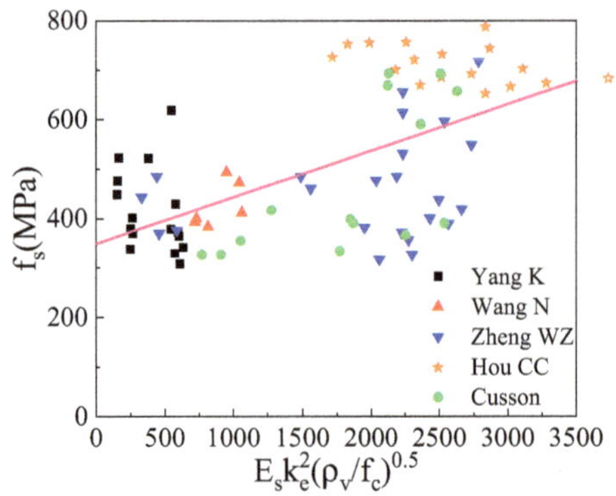

Fig. 2. Stirrup stress fitting curve

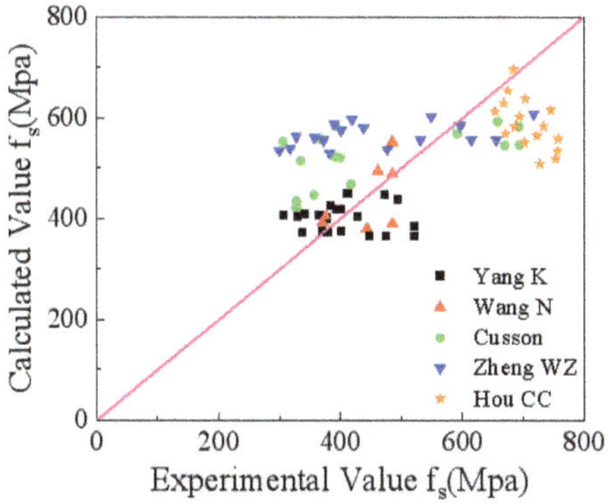

Fig. 3. Comparison diagram of test value and calculated value

3.3 Peak Stress

Under the constraint of stirrups, the concrete is in a three-dimensional compression state, and the stress of the confined concrete at the peak point will be improved. According to the theoretical calculation formula of peak stress of ordinary stirrup confined concrete established by Mander [12] in 1988 using William-Warnke five-parameter multi-axial failure criterion [14], the calculation formula of peak stress of high-strength spiral stirrup confined concrete is established. To simplify the calculation, the formula is as follows.

$$\frac{\tau_{oct}}{f_{co}} = b_0 + b_1 \frac{\tau_{oct}}{f_{co}} + b_2 (\frac{\sigma_{oct}}{f_{co}})^2 \tag{8}$$

$$f_{cc} = f_{co}\left[\frac{3(b_1+\sqrt{2})}{2b_2} + \sqrt{\left(\frac{3(b_1+\sqrt{2})}{2b_2}\right)^2 - \frac{9b_0}{b_2} - \frac{9\sqrt{2}f_{le}}{b_2 f_{co}}} - 2\frac{f_{le}}{f_{co}}\right] \tag{9}$$

In the Eq. (8) and (9): σ_{oct} is the octahedral normal stress and τ_{oct} is the octahedral shear stress; b_0, b_1, b_2 are parameters.

Based on the collected 160 experimental data, the fitted compression meridian is shown in Fig. 4. By bringing $b_0 = 0.107, b_1 = -1.19, b_2 = -0.339$ into the equation, the formula for calculating the peak stress of high-strength spiral stirrup confined concrete can be obtained, as shown in Eq. (11). Comparing the calculated values with the actual results of the test, the prediction effect of the formula is very good, as shown in Fig. 5. The average value of the ratio of the calculated value to the experimental value is 1.01, and the standard deviation is 0.11, which can effectively predict the peak point stress of high strength rectangular and circular stirrup confined concrete.

$$\frac{\tau_{oct}}{f_{co}} = 0.107 - 1.19\frac{\sigma_{oct}}{f_{co}} - 0.339(\frac{\sigma_{oct}}{f_{co}})^2 \tag{10}$$

$$f_{cc} = f_{co}\left(-0.992 + 1.956\sqrt{1 + 9.816\frac{f_{le}}{f_{co}}} - 2\frac{f_{le}}{f_{co}}\right) \tag{11}$$

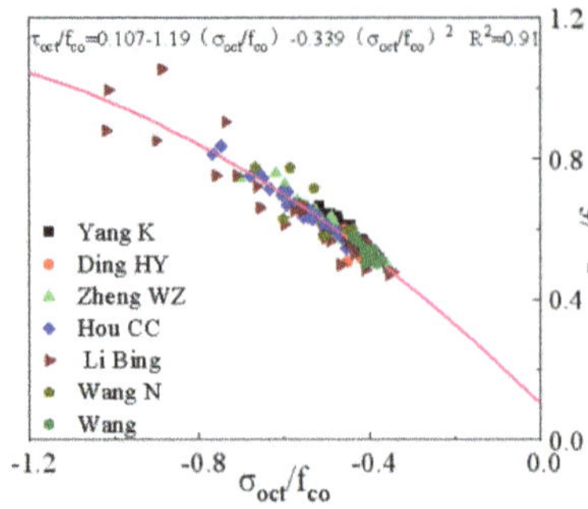

Fig. 4. Octahedral compression meridian

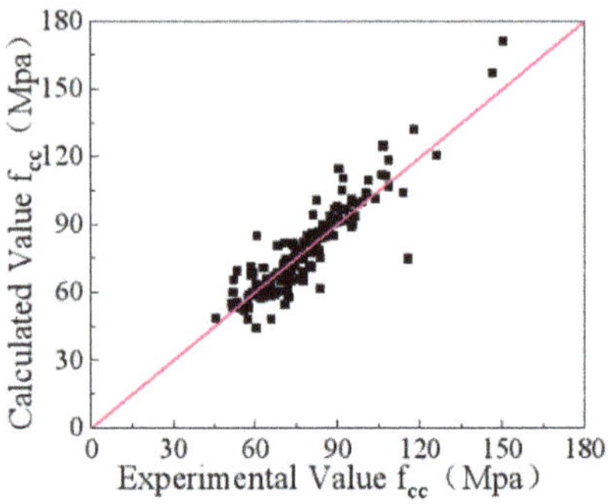

Fig. 5. Comparison diagram of test value and calculated value

3.4 Peak Strain

The confinement of stirrups can better improve the deformation ability of concrete, so that the improvement effect of peak strain is more obvious than that of peak stress, and the improvement coefficient can reach more than ten. In this paper, based on the collected 110 experimental data, the peak strain calculation formula is obtained by regression as Eq. (12), and the fitting result is shown in Fig. 6. Comparing the peak point strain calculated by Eq. (12) with the test results, as shown in Fig. 7, the calculation results are consistent with the test values. The average value of the ratio of the test value to the calculated value is 1.18, and the standard deviation is 0.31, indicating that it can effectively predict the peak point strain of high strength rectangular and circular stirrup confined concrete..

$$\varepsilon_{cc} = \varepsilon_{cc}\left[1 + 68.6(\frac{f_{le}}{f_{co}})^{1.6}\right] \tag{12}$$

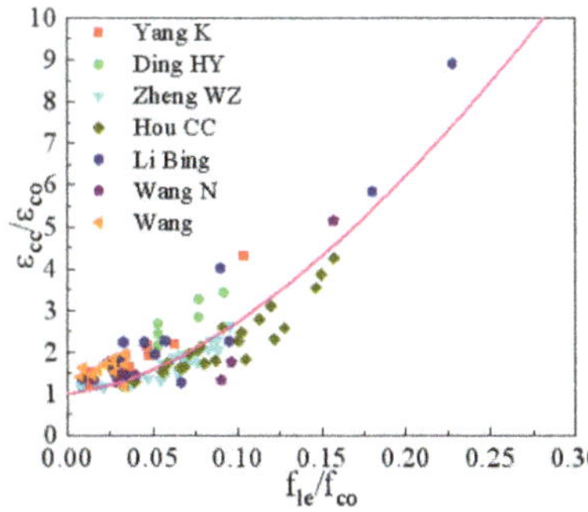

Fig. 6. Relationship between peak strain of confined concrete and effective confinement efficiency of stirrups

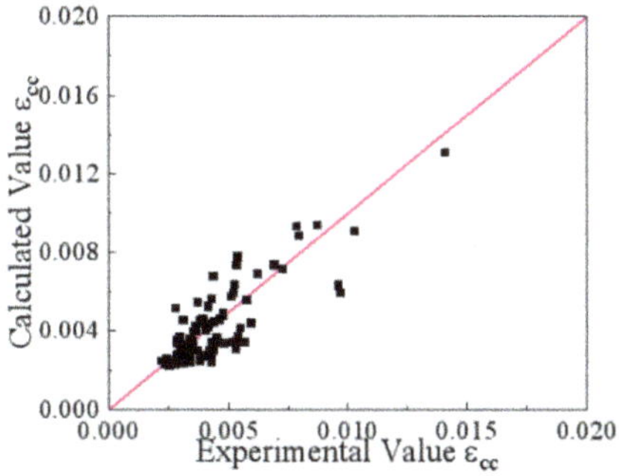

Fig. 7. Comparison diagram of test value and calculated value

3.5 Ultimate Compressive Strain

When the confined concrete reaches the ultimate compressive strain, both ordinary stirrups and high-strength stirrups yield. The stirrup yield strength can be used to calculate the effective lateral confinement stress (f_{le}) of concrete. In this paper, based on the collected 80 experimental data, the formula for calculating the ultimate compressive strain is obtained by regression analysis as Eq. (13), and the fitting result is shown in Fig. 8. When the effective lateral constraint stress of stirrup is 0, the ultimate compressive strain calculated by the fitting formula is 2 ε_{co}, which is similar to the ultimate compressive strain of unconstrained concrete, indicating that the fitting formula can converge at the lower boundary end. Comparing the ultimate compressive strain calculated by Eq. (13) with the experimental value, as shown in Fig. 9, the calculation results are consistent with the experimental values. The average value of the ratio of the experimental value to the calculated value is 1.12, and the standard deviation is 0.26, indicating that the ultimate compressive strain of rectangular and circular stirrup confined concrete can be predicted to a certain extent.

$$\varepsilon_{ccu} = \varepsilon_{co}\left[2 + 88.58\left(\frac{f_{le}}{f_{co}}\right)^{1.13}\right] \tag{13}$$

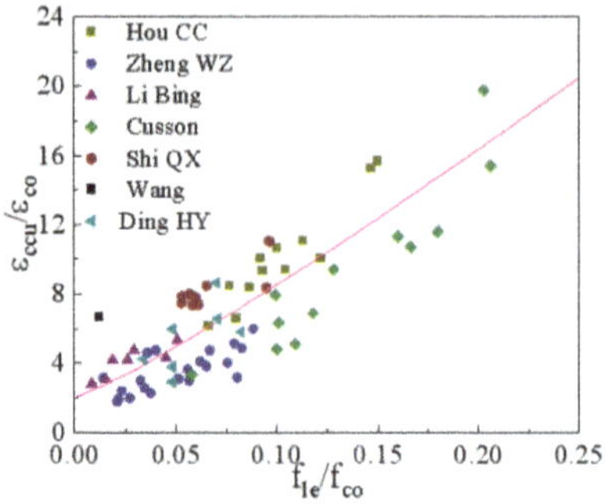

Fig. 8. Relationship between ultimate compressive strain and effective confinement efficiency of stirrups

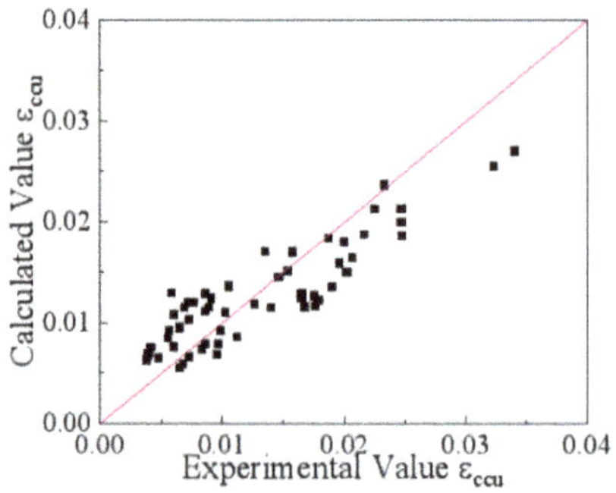

Fig. 9. Comparison diagram of test value and calculated value

Equations (7)–(13) using the equivalent column theory of Legeron [17], the constraint of rectangular stirrup on the specimen is equivalent to the simplified calculation of the cylinder. Through the comparison between the calculation model and the actual specimen data, the calculation formula established in this paper is suitable for five kinds of stirrup forms.

4 Constitutive Model of High-Strength Spiral Stirrup Confined Concrete

For the study of confined concrete, scholars at home and abroad have proposed many stress-strain relationship curve models of confined concrete. The ways to establish the models are rich and varied, including pure theoretical derivation, numerical calculation, semi-theoretical and semi-empirical methods. Among them, the more typical confined concrete models are: Mander model [12], Guo Zhenhai-Zhang Xiuqin model [15], Saatcioglu-Razvi model [16], Cusson-Paultre model [13], Li Bing model [6], Legeron model [17]; these models reflect the influence of various factors on the confinement of stirrups to a certain extent. Based on the previous models, this paper analyzes the collected test data and proposes a stress-strain curve model suitable for high-strength spiral stirrup confined concrete.

4.1 Establishment of Constitutive Model

According to the existing constitutive model of confined concrete, although the equations used in the rising section of each model are different, the results of the rising section of the curve are not much different, and the mechanical properties of confined concrete before reaching the peak strength can be well simulated. This paper proposes to use the ascending curve equation proposed by Popovics et al. [18], such as Eq. (14), which is simple in form and clear in parameter meaning.

$$f_c = f_{cc}\left[\frac{k\frac{\varepsilon_c}{\varepsilon_{cc}}}{k-1+(\frac{\varepsilon_c}{\varepsilon_{cc}})^2}\right], \varepsilon_c \leq \varepsilon_{cc} \tag{14}$$

In the Eq. (14), where k reflects the cracking degree of confined concrete at peak stress, and $k = E_C/(E_C - E_{\text{sec}})$, where $E_{\text{sec}} = f_{cc}/\varepsilon_{cc}$.

In this paper, the descending section of the curve model is linear, that is, the descending section of the Li-Bing model [6], as shown in Eq. (15). It is considered that the stirrup constraint form, volume stirrup ratio, stirrup yield strength, concrete strength, section size and longitudinal reinforcement distribution have different degrees of influence on the descending section of the stress-strain curve of confined concrete. By introducing the parameter β, the influence of stirrup form, volume stirrup ratio, stirrup yield strength and concrete strength on the descending section of the curve is comprehensively considered. The parameter β is the ratio of the slope of the descending section of the model curve to the secant modulus of the peak point.

In this paper, based on the collected 81 experimental data, the calculation formula of the control parameter β in the descending section is obtained by regression analysis, such as Eq. (16), and the fitting result is shown in Fig. 10.

$$f_c = f_{cc} - \beta\frac{f_{cc}}{\varepsilon_{cc}}(\varepsilon_c - \varepsilon_{cc}) \leq 0.4f_{cc}, \varepsilon_c > \varepsilon_{cc} \tag{15}$$

$$\beta = 0.883 - 2.01(\frac{f_{le}}{f_{co}})^{0.51} \tag{16}$$

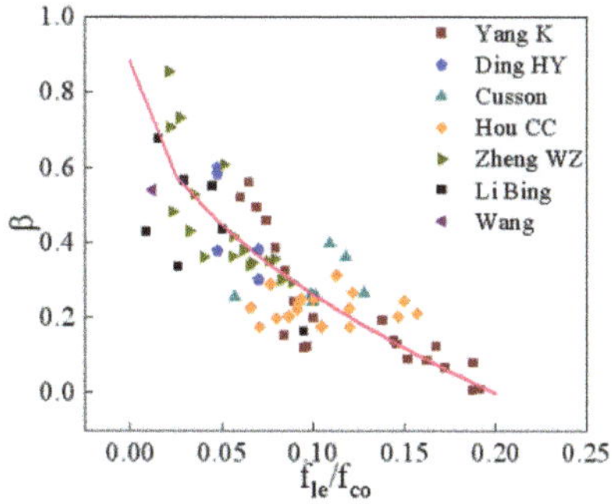

Fig. 10. The relationship between β and f_{le}/ε_{co}

According to the characteristic parameter calculation formula and curve equation of the curve, the constitutive relation curve equation of high strength spiral stirrup confined concrete can be established, as shown in Eq. (14) and (15).

4.2 Verification of Constitutive Model

To further verify the accuracy of the unified equation in this paper, the collected test results are calculated. The comparison between the calculation results and the test results of some specimens is shown in Fig. 11. It can be seen from Fig. 11 that the curve model in this paper is in good agreement with the experimental curve, and it is more convenient in calculation and convenient for practical use in engineering.

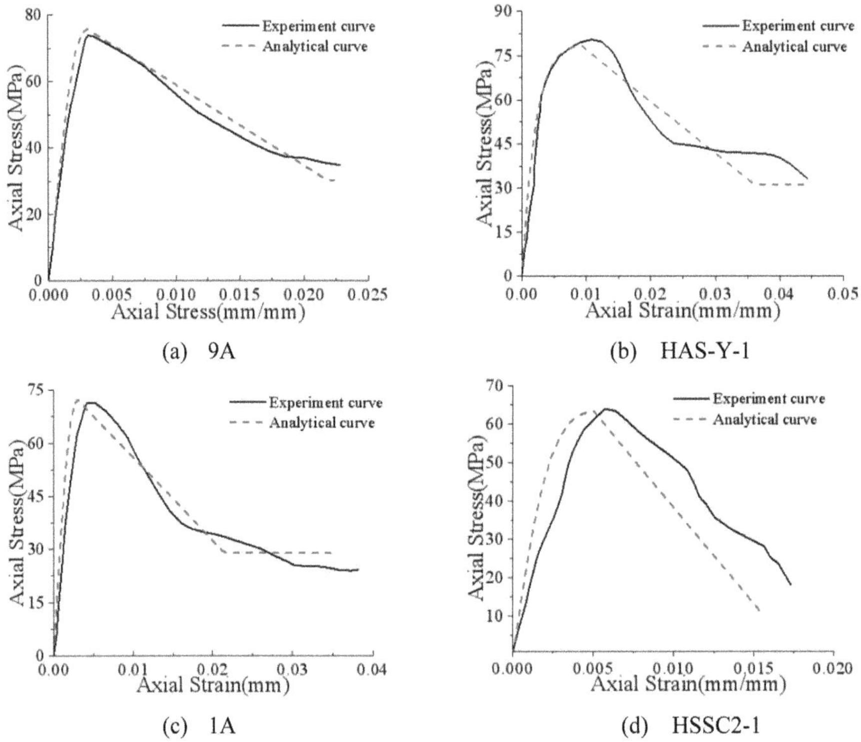

Fig. 11. Comparison between analytical and experimental curves of samples

5 Conclusion

The present study conducts a statistical analysis on a total of 203 experimental data sets obtained from axial compression tests conducted on high-strength spiral stirrup confined concrete.

1. According to the experimental results of scholars and the calculation model of each feature point, the constraint effect of circular spiral stirrups is better than that of rectangular spiral stirrups. Among the rectangular spiral stirrups, the well-shaped spiral stirrups have the best constraint effect, and the diamond-shaped spiral stirrups are optimal considering the convenience of practical engineering.
2. The proposed calculation formula accurately predicts the stirrup stress when the confined concrete reaches its peak compressive stress. By comparing with experimental data, this formula can effectively determine whether the stirrup is yielding and predict the precise value of the stirrup stress. As a result, it enhances the accuracy of calculating the characteristic parameters of the constitutive curve.
3. The stress-strain curve model of high-strength spiral stirrup confined concrete is established, which is suitable for five kinds of stirrup forms. In comparison to existing constitutive models, this approach offers simplified calculations and a broader application range, facilitating practical engineering utilization as well as finite element analysis and computation.

References

1. Zhang, W.J., Zhang, B., Jiang, X.P.: Experimental study on axial compressive behavior of full-scale reinforced concrete columns with close confinement. J. Earthq. Eng. Eng. Vibr. **33**(1), 106 (2013)
2. Guo, Z.H., Zhang, X.Q., Zhang, D.C.: Experimental investigation of the complete stress-strain curve of concrete. J. Build. Struct. (1), 1 (1982)
3. Legeron, F., Paultre, P., Asce, M.: Uniaxial confinement model for normal and high-strength concrete columns. J. Struct. Eng. (2), 129 (2003)
4. Zheng, W.Z., Zhang, J., Wang, G.: Experimental study on axial compression performance of high strength concrete columns confined by spiral stirrups. Chin. J. Highway **35**(06), 22–35 (2022)
5. Zheng, W.Z., Hou, C.C., Chang, W.: Experimental study on the mechanical behavior of circular concrete columns with high-strength spiral stirrups. J. Build. Struct. **39**(6), 21–31 (2018)
6. Li, B., Park, R., Tanaka, H.: Stress-strain behavior of high-strength concrete confined by ultra-high- and normal-strength transverse reinforcements. ACI Struct. J. **98**(3), 395–406 (2001)
7. Weilun, W., Mingyang, Z., Yun, T.: Behaviour of high-strength concrete columns confined by spiral reinforcement under uniaxial compression (2023)
8. Yang, K., Shi, Q.X., Zhao, J.H., Jiang, W.S., Meng, H.: Study on constitutive model of high strength stirrup confined high strength concrete. J. Civil Eng. **46**(01), 34–41 (2013)
9. Shi, Q.X., Wang, N., Tian, Y., Wang, P., Li, K.: Study on the complete stress-strain curve of high-strength concrete confined by high-strength stirrups under axial compression. J. Build. Struct. **34**(04), 144–151 (2013)
10. Ding, H.Y., Liu, Y., Qiu, S.: Experimental study on axial compression of high strength concrete confined by high strength stirrups. Build. Struct. **45**(12), 7–12 (2015)
11. He, X.: Experimental study on mechanical properties of high strength concrete confined by high strength spiral stirrups. Xi 'an University of Architecture and Technology (2008)
12. Mander, J.B.: Theoretical stress-strain model or for confined concrete. J. Struct. Div. **114** (1988)

13. Cusson, D., Paultre, P.: High-strength concrete columns confined by rectangular ties. J. Struct. Eng. **120**(3), 783–804 (1994)
14. William, K.L., Warnke, E.P.: Constitutive model for the triaxial behavior of concrete. In: International Association for Bridge and Structural Engineering, Proceedings, vol. 19 (1975)
15. Zhang, X.Q., Guo, Z.H., Wang, C.Z.: Stress-strain curve equation of stirrup confined concrete under cyclic loading. J. Build. Struct. **12**, 16–20 (1985)
16. Saatcioglu, M., Razvi, S.R.: Strength and ductility of confined concrete. J. Struct. Eng. **118**(6), 1590–1607 (1999)
17. Legeron, F., Paultre, P.: Uniaxial confinement model for normal- and high-strength concrete columns. J. Struct. Eng. **129**, 241–252 (2003)
18. Popovics, S.: A numerical approach to the complete stress-strain curve of concrete. Cem. Concr. Res. **3**(5), 583–599 (1973)

Research on Leakage Characteristics and Accurate Discharge Formula of Water Supply Pipeline

Baoxin Qi[1](✉), Haolin Han[1], and Haoxiang He[2]

[1] School of Civil Engineering, Shenyang Jianzhu University, Shenyang 110168, Liaoning, China
bxqi@sjzu.edu.cn

[2] The Key Laboratory of Urban Security and Disaster Engineering of Ministry of Education, Beijing University of Technology, Beijing 100124, China

Abstract. In this paper, based on the computational fluid dynamics method, the k-ε turbulence model is used to establish a steady state model to investigate the leakage law of the pipeline with a single leakage point. Starting from the three basic information of leakage port shape, leakage port area, and pipeline pressure, when the variables are changed, we investigate the change of the pressure flow field inside the leakage port and the influence on the leakage volumetric flow rate of the pipeline. Numerical simulation results show that: The shape of the leakage port, the internal pressure of the pipe, and the area of the leakage port all have influence on the pressure flow field inside the leakage port; the leakage coefficient is directly proportional to the area of the leakage port, which in turn introduces the leakage influence coefficient, and accurately optimizes the discharge formula to obtain the leakage loss formula of the pipeline under different leakage port shapes.

Keywords: Pipeline Leakage · Numerical Stimulation · Pressure Flow Field

1 Introduction

Water supply pipeline is the infrastructure on which the city relies for survival and development, while leakage phenomenon occurs from time to time [1], how to reduce the water waste generated by pipeline leakage has become the focus of research. Nowadays, pipeline leakage detection methods are mainly divided into two categories, hardware detection and data processing. Hardware-based methods mainly include leakage magnetic flaw detection [2], tracer detection [3], pipeline robotic detection [4], etc.; data-processing-based methods mainly include flow balance method [5], negative pressure wave method [6], pressure gradient method [7], acoustic wave method [8, 9] and so on.

I. Tuns et al. (Eds.): CIBv 2023, LNCE 510, pp. 203–212, 2024.
https://doi.org/10.1007/978-3-031-60765-3_17

2 Fundamental Equations of Numerical Simulation

2.1 Fundamental Control Equation

Fluid motion is one of the most complex physical behaviors and of course follows the basic conservation laws. The energy equation is not turned on in this numerical simulation because the variation of the temperature field has a small effect on the pipe leakage problem.

Continuity equation

$$\frac{\partial \rho}{\partial t} + \nabla \cdot \left(\rho \vec{u}\right) = 0 \tag{2.1}$$

where ρ is the density, t is the time, and $\vec{u}$ is the velocity vector.

Momentum equation

$$\rho \frac{d\vec{u}}{dt} = \rho \vec{g} - \nabla p + \mu \Delta \vec{u} \tag{2.2}$$

where ρ is the density, $\vec{u}$ is the velocity vector, t is the time, $\rho \vec{g}$ is the gravitational body force, p is the pressure, μ is the dynamic viscosity.

2.2 Conservation for the Turbulence Model

Turbulent energy equation k

$$\frac{\partial}{\partial t}(\rho k) + \frac{\partial}{\partial x_i}(\rho k u_i) = \frac{\partial}{\partial x_i}\left[\left(\mu + \frac{\mu_t}{\sigma_k}\right)\right] + G_k + G_b - \rho\varepsilon - Y_m + S_k \tag{2.3}$$

where ρ is the density, t is the time, k is the turbulent kinetic energy, x_i and x_j are the corresponding locations, u_i is the velocity at the direction of x_i, μ is the viscosity, μ_t is the eddy viscosity, σ_k is the constant which equals to 1, G_k is the turbulent flow energy produced by laminar velocity gradient, G_b is the turbulent flow energy generated by buoyancy, Y_m is the fluctuations caused by excessive diffusion in turbulence, ε is the rate of dissipation of the turbulent kinetic energy, and S_k is the source term for turbulent kinetic energy.

Dissipation equation ε

$$\frac{\partial}{\partial t}(\rho\varepsilon) + \frac{\partial}{\partial x_i}(\rho\varepsilon u_i) = \frac{\partial}{\partial x_j}\left[\left(\mu + \frac{\mu_t}{\sigma_\varepsilon}\right)\frac{\partial \varepsilon}{\partial x_j}\right] + C_{1\varepsilon}\frac{\varepsilon}{k}(G_k + C_{3\varepsilon}G_b) - C_{2\varepsilon}\rho\frac{\varepsilon^2}{k} + S_\varepsilon \tag{2.4}$$

where $C_{1\varepsilon}$, $C_{2\varepsilon}$, $C_{3\varepsilon}$ are the constant parameters, S_ε is the source term for the dissipation rate, σ_ε is the constant.

3 Physical Modeling and Meshing

The shape of the leakage port when the pipeline leaks is mainly divided into irregular round holes, axial cracks, circumferential cracks, etc. In order to improve the authenticity and accuracy of the water supply pipeline leakage model and better complete the numerical simulation, the research objects of this paper are round hole leakage, circumferential rectangular leakage port and Axial rectangular leakage port. Pipe leakage models with different leak shapes are shown in Fig. 1. Its nominal outer diameter is 50 mm, the length of the pipe is 1000 mm, and the leakage is located in the center of the upper wall of the pipeline.

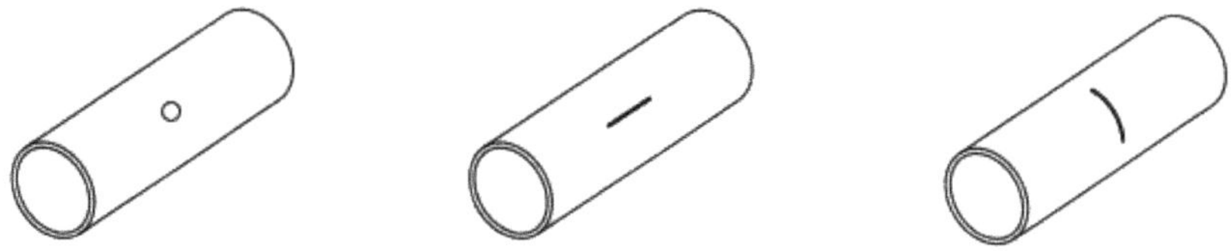

a)Round hole leakage port b)Circumferential rectangular leakage port c)Axial rectangular leakage port

Fig. 1. The pipeline model

3.1 Meshing

The mesh division of the pipeline leakage physical model adopts the combination of body mesh and sizing mesh, as shown in Fig. 2. The tetrahedral mesh is selected for the main part of the pipeline fluid, and the sizing mesh is used for local mesh refinement for the pipeline leakage port. After the verification of grid independence, the total number of grids is selected to be 1.13 million, the average mesh metric is 0.85, the average aspect ratio is 8.66, the average skewness is 0.20, and the maximum skewness is 0.84. To sum up, the values used to measure the quality of the grids are in a reasonable range, and the quality of the grids meets the requirements of the simulation.

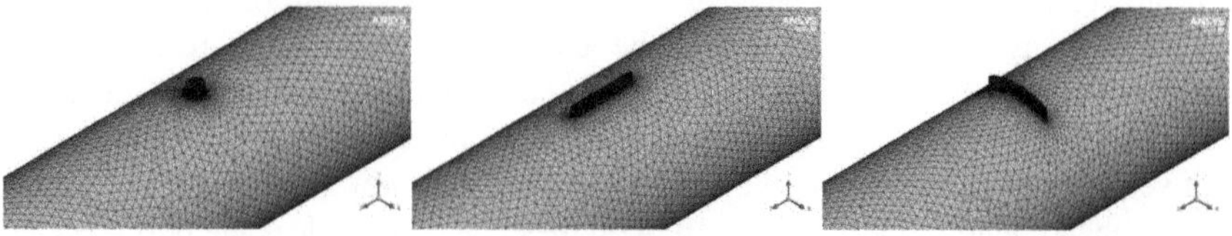

a)Round hole leakage port b)Circumferential rectangular leakage port c)Axial rectangular leakage port

Fig. 2. Meshing of the fluid domain

4 Analysis of Numerical Simulation Results

4.1 Effect of Leakage Port Area on the Pressure Flow Field at the Leakage Port

In order to visually observe the pressure field, the axial center line of the leakage port is selected as the position carrier to analyze the pressure change, and the pressure change curve along the axial direction of the leakage port is shown in the figure.

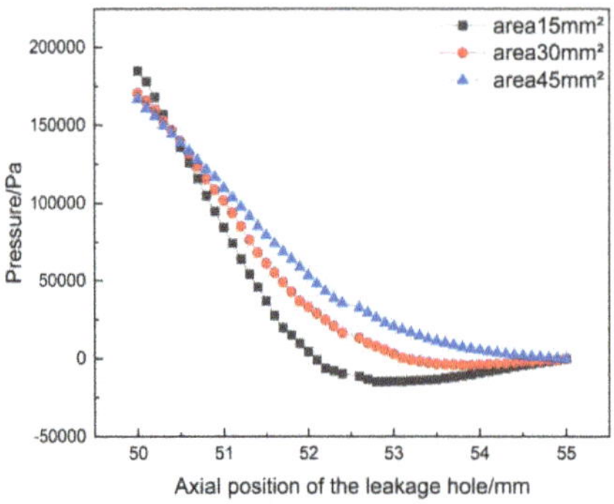

a)Round hole leakage port

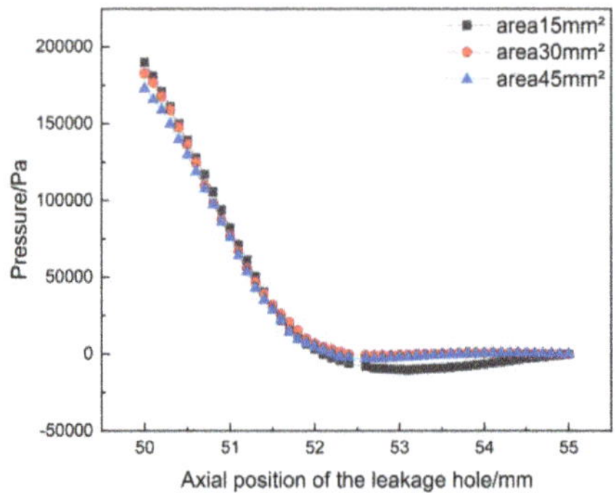

b)Circumferential rectangular leakage port

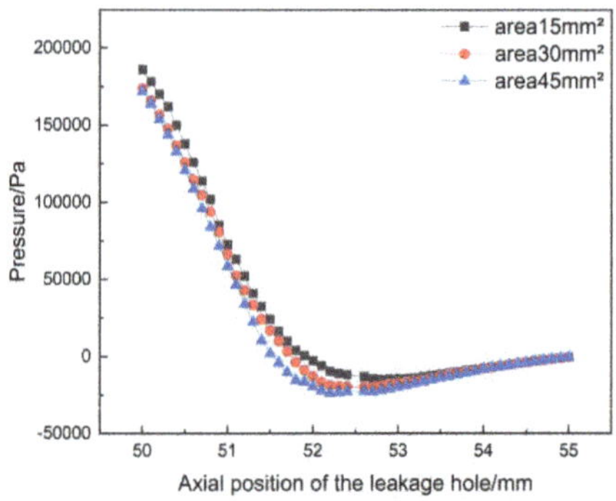

c)Axial rectangular leakage port

Fig. 3. Variation of pressure flow field at the leakage port for different leakage port areas

From Fig. 3(a), it can be seen that as the area of the leakage port increases, the change in pressure along the axial direction of the leakage port becomes more gentle, and at about 52.5 mm in the axial direction of the leakage port, the area of the leakage port of 15 mm^2 has an obvious negative pressure, while the area of 30 mm^2, only a small negative pressure phenomenon, the area of 45 mm^2 does not have an obvious negative pressure phenomenon. For the Fig. 3(b), with the change of the leakage port area, along the pipeline leakage port axial direction of the pressure change does not appear obvious difference; for the axial rectangular leakage port, the difference is not significant enough, but can be seen from the Fig. 3(c) with the increase of the leakage port area, the maximum negative pressure at the pipeline increases, which is the opposite of the law of the round hole leakage port.

4.2 Effect of Leakage Port Shape on the Pressure Flow Field at the Leakage Port

When the area of the leakage port is constant at 45 mm^2, it can be seen Fig. 4(a) that the pressure change along the axial direction of the round hole leakage port is relatively flat, with the smallest slope; the axial rectangular leakage port has the steepest change in pressure, with the largest slope; the slope of the pressure change curve of the circumferential rectangular leakage port is in the axial rectangular leakage port and the round hole leakage port between the slopes. When the area of the leakage port is constant at 30 mm^2, it can be seen Fig. 4(b) that the law is the same as that of the area of 45 mm^2, but the distance between the different pressure change curves is gradually shortened; and when the area of the leakage port is constant at 15 mm^2, the pressure changes along the axial direction of the leakage port are similar in all the three types of leakage ports, with no significant difference. This shows that when the leakage port area is small, the shape of the leakage port does not play a significant role in the pressure flow field at the leakage port, while the shape of the leakage port plays a more and more significant role when the area of the leakage port gradually increases.

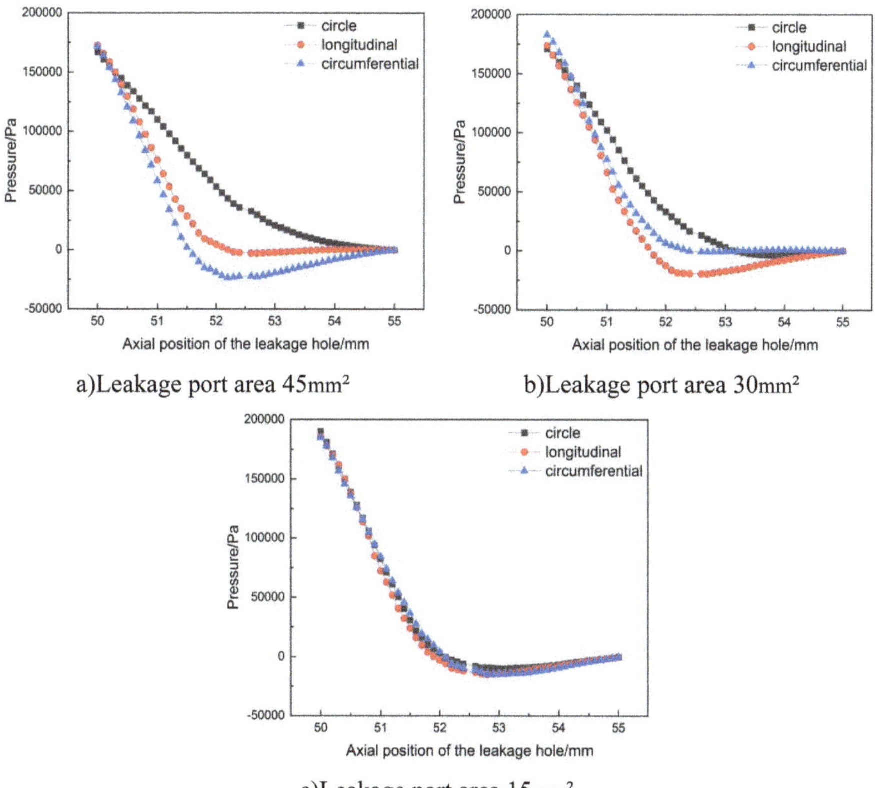

Fig. 4. Variation of pressure flow field at the leakage port for different leakage port shapes

4.3 Effect of Pipeline Pressure on the Pressure Flow Field at the Leakage Port

From the Fig. 5, it can be seen that the pressure change curves under different leakage port shapes are similar, which proves that the shape of the leakage port does not have a significant effect on the pressure flow field at the leakage port under different pipe pressures. And the greater the pressure inside the pipe, the greater the pressure at the starting point, the pressure change curve of the leakage port under different pressure intersects at 52 mm, the greater the pipe pressure, the greater the negative pressure at the leakage port, and this law applies to different leakage port shapes.

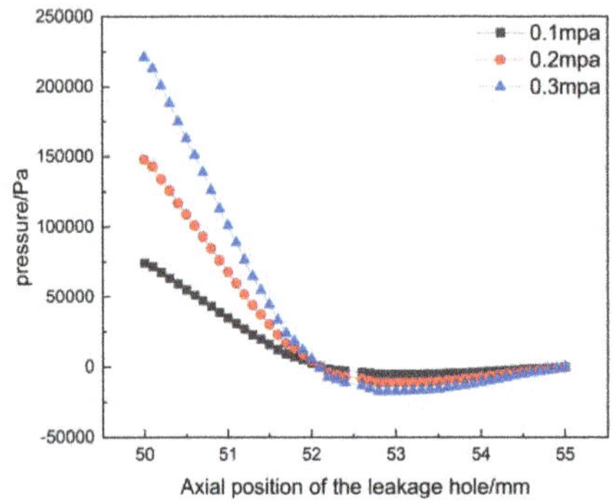

a)Round hole leakage port

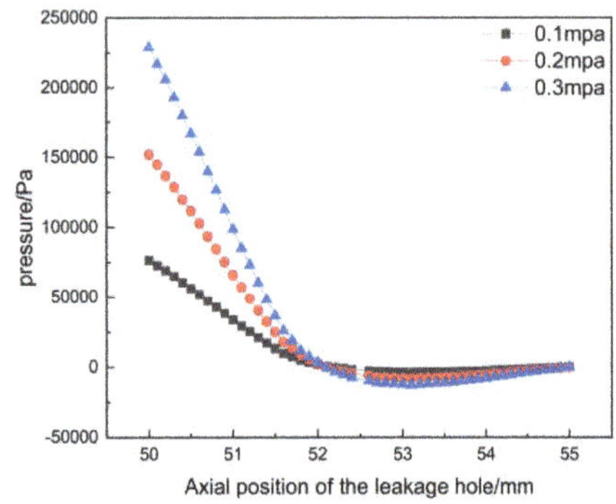

b)Circumferential rectangular leakage port

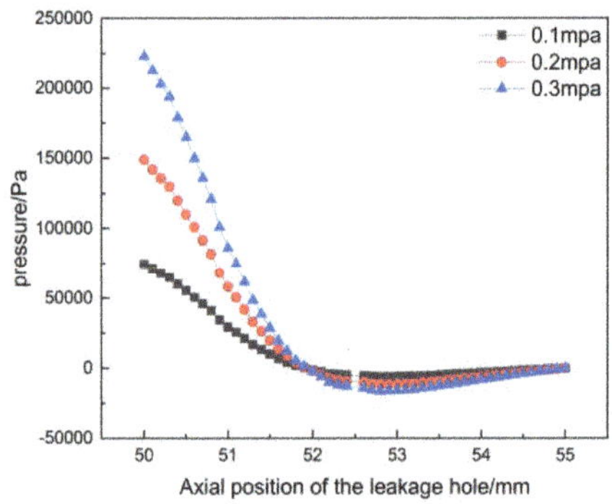

c)Axial rectangular leakage port

Fig. 5. Variation of pressure flow field at the leakage port for different pipe pressures

4.4 Precise Discharge Formula

Orifice discharge formula [10].

Initially, in order to describe the law of physical leakage flow at a single point, scholars who study water supply pipeline leakage derive the orifice discharge formula.

$$q = k \cdot H^{\beta} \tag{4.1}$$

k is the leakage coefficient and β is the leakage index. The study in this paper ignores the plastic deformation of the pipe, so theoretically the leakage index β should be close to 0.5. By fitting the relationship between the volume flow rate at the leakage port and the pressure inside the pipe, as shown in the Fig. 6, the leakage index and the leakage coefficient can be obtained, and the leakage index is about 0.5.

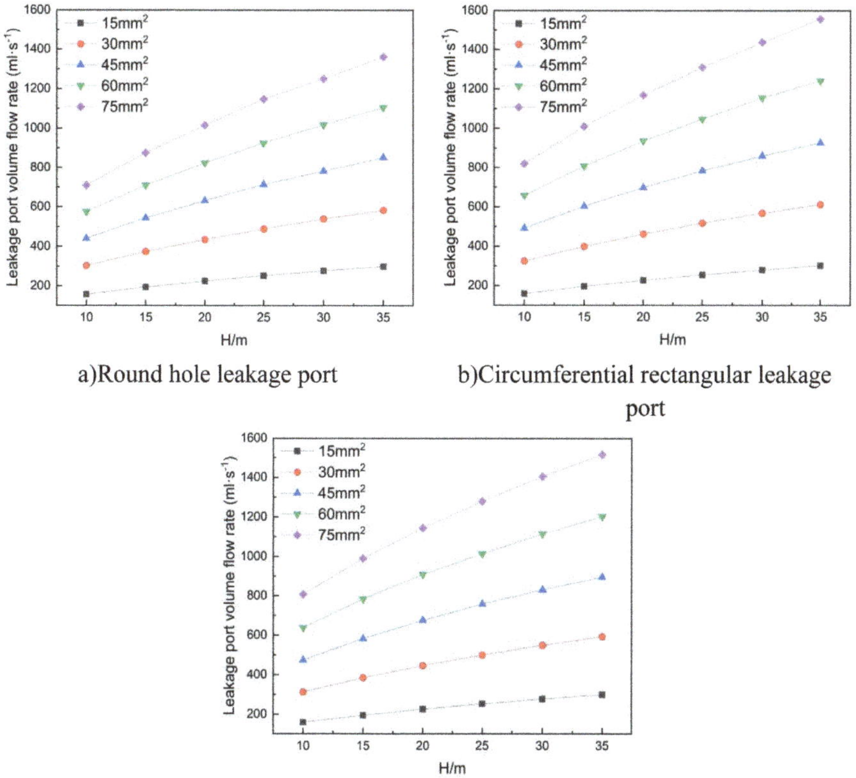

a)Round hole leakage port

b)Circumferential rectangular leakage port

c)Axial rectangular leakage port

Fig. 6. Leakage port volume flow rate-pipe pressure relationship for different leakage port areas

The leakage coefficient is related to the area of the leakage port, and the fitting results in Fig. 7. It can be seen that the leakage coefficient is proportional to the area of the leakage port, and we define the proportionality coefficient as α. The α of the round hole leakage port is 2.72, the α of the axial rectangular leakage port is 3.39, and the α of the circumferential rectangular leakage port is 3.41. The α of the circumferential rectangular leakage port is slightly larger than the axial rectangular leakage port, and both of them are obviously larger than the round hole leakage port.

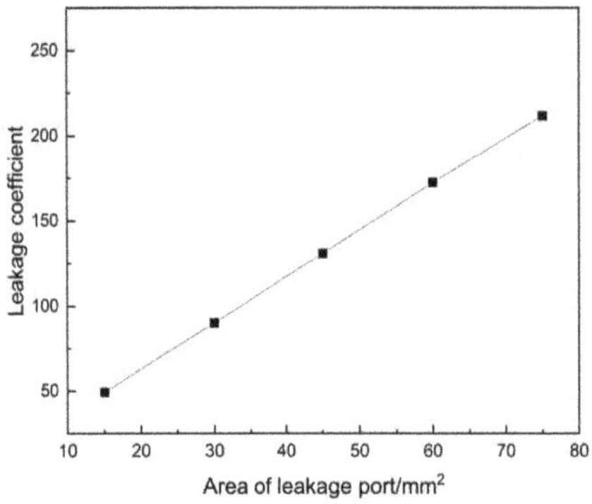

a)Round hole leakage port

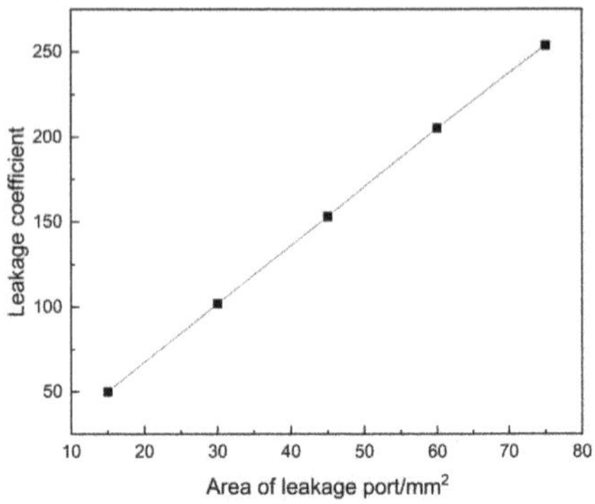

b)Circumferential rectangular leakage port

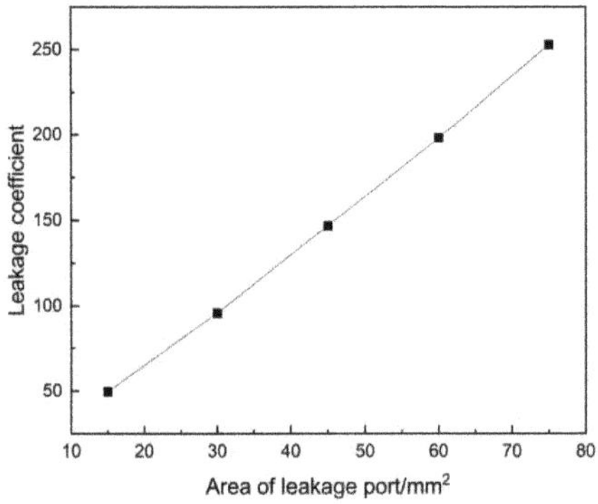

c)Axial rectangular leakage port

Fig. 7. Leakage influence factor - leakage port area relationship

From the discharge formula can be seen that the leakage port volume flow rate in addition to the internal pressure of the pipeline, but also with the leakage coefficient, and the leakage coefficient and the leakage port area is approximately proportional to the leakage influence coefficient can be more accurately respond to the leakage port area and the relationship between the leakage port volume flow rate. Through the leakage influence coefficient, it can be seen that the circumferential rectangular leakage port volume flow rate slightly > axial leakage port volume flow rate > round hole leakage port volume flow rate. The introduction of the leakage influence coefficient α can be obtained after the introduction of accurate discharge formula:

$$q = a \cdot \alpha \cdot H^{\beta} \tag{4.2}$$

The accurate discharge Eq. (4.2) is deformed to Eq. (4.3):

$$a = \frac{q}{\alpha \cdot H^{\beta}} \tag{4.3}$$

The volume flow rate of the leakage port can be obtained by subtracting the volume flow rate of the outlet from the volume flow rate of the inlet; the pressure inside the pipeline is artificially controlled, which is also known; for the shape of the leakage port, since the leakage influence coefficient of the axial rectangular leakage port and the circumferential rectangular leakage port are similar, it is only necessary to roughly determine the shape of the leakage port whether it is approximated to be a round hole or approximated to be a rectangle, which can be determined by the ultrasonic detection. When the above three values are determined, the area of the leakage port can be estimated by the formula, which has certain reference value when dealing with pipeline leakage problems in practice.

4.5 Conclusions

When the shape of the leakage port is a round hole, the negative pressure at the leakage port gradually increases with the increase of the leakage port area, while the shape of the leakage port is an axial rectangular leakage port and a circumferential rectangular leakage port, the effect of the leakage port area on the pressure flow field at the leakage port is small.

When the leakage port area is certain and the area is large, the negative pressure of axial rectangular leakage port > negative pressure of circumferential rectangular leakage port > negative pressure of round hole leakage port, and when the area of leakage port is small, the change of pressure at different leakage ports does not have a significant difference.

The greater the pressure inside the pipe, the greater the pressure at the starting point of the leakage port axis, the greater the negative pressure at the leakage port, and different leakage port shapes are in line with this law.

Different leakage coefficients of different leakage port shapes are different, the leakage coefficient of round hole leakage port is the smallest, and the leakage coefficient of circumferential rectangular leakage port is similar to that of axial leakage port, and the precise formula of discharge and the deduced formula of leakage port area can be used as reference for the actual project.

References

1. Wu, Z.Y., Farley, M., Turtle, D., et al.: Water loss reduction. Am. Water Works Assoc. J. **103**(9), 124–126 (2011)
2. Liu, B., Cao, Y., Zhang, Y., et al.: Weak magnetic flux leakage: a possible method for studying pipeline defects located either inside or outside the structures. NDT E Int. **74**, 81–86 (2015)
3. Hiroki, S., Tanzawa, S., Arai, T., et al.: Development of water leak detection method in fusion reactors using water-soluble gas. Fusion Eng. Des. **83**(1), 72–78 (2008)
4. Fletcher, R.S.: A new approach in pipeline leak detection. In: Proceedings of the International Pipeline Conference, IPC 2008 (2008)
5. Zhang, C., Lambert, M.F., Gong, J., et al.: Bayesian inverse transient analysis for pipeline condition assessment: parameter estimation and uncertainty quantification. Water Resour. Manag. **34**(9), 2807–2820 (2020)
6. Li, J., Zheng, Q., Qian, Z.H., et al.: A novel location algorithm for pipeline leakage based on the attenuation of negative pressure wave. Process. Saf. Environ. Prot. **123**(B), 309–316 (2019)

7. Lu, W., Liang, W., Zhang, L., et al.: A novel noise reduction method applied in negative pressure wave for pipeline leakage localization. Process. Saf. Environ. Prot. **104**(A), 142–149 (2016)
8. Kafle, M.D., Narasimhan, S.: Active acoustic leak detection in a pressurized PVC pipe. Urban Water J. **17**(4), 315–324 (2020)
9. Xu, C., Du, S., Gong, P., et al.: An improved method for pipeline leakage localization with a single sensor based on modal acoustic emission and empirical mode decomposition with Hilbert transform. IEEE Sens. J. **20**(10), 5480–5491 (2020)
10. Khadam, M.A.: Water losses from municipal utilities and their impacts. Water Int. **16**, 254–261 (1991)

Experimental Study on Leakage Damage Monitoring of Liquid Filled Reinforced Concrete Pipeline

Baoxin Qi[1(✉)], Jingyu Wei[1], Yi Wu[1], and Haoxiang He[2]

[1] School of Civil Engineering, Shenyang Jianzhu University, Shenyang 110168, Liaoning, China
bxqi@sjzu.edu.cn

[2] The Key Laboratory of Urban Security and Disaster Engineering of Ministry of Education, Beijing University of Technology, Beijing 100124, China

Abstract. Nowadays, with the continuous development of industrialization, liquid-filled pipeline is the most representative pipeline transportation system. The damage forms of pipes with cracks will affect the mechanical properties of reinforced concrete pipes to varying degrees and make their properties decline. We investigated the undamaged, damaged and liquid-filled reinforced concrete pipelines, and analyzed the propagation mechanism of piezoelectric sensing signals in reinforced concrete pipelines. When the ultrasonic guided wave touches the damaged position, part of the ultrasonic guided wave will reflect, while the other part of the guided wave will continue to propagate through the damaged position, which is the reflection and transmission phenomenon of the ultrasonic guided wave. The damage assessment of pipeline structure can be realized by using the reflection and transmission phenomenon of ultrasonic guided wave. The signals of undamaged hollow pipe and undamaged liquid-filled pipe were compared and compared undamaged signal and damaged signal in liquid filled pipeline. Based on pulse-echo method, the influence of damage on signal propagation in liquid-filled pipeline is obtained.

Keywords: liquid-filled reinforced concrete pipelines · damaged position · damage assessment · reflection and transmission · pulse-echo method

1 Introduction

In recent years, with the continuous development of industrial and agricultural modernization, pipelines, as transport carriers of natural gas, water and oil, have played an increasingly important role in urban construction [1]. Compared with the traditional transportation mode, pipeline transportation has the advantages of fast, safe and low cost. However, after the pipeline works for a long time, due to adverse factors, such as aging of pipeline materials, environmental corrosion, etc., the pipeline structure will often be damaged to different degrees, and even major accidents such as explosion or collapse will occur [2, 3]. Therefore, before a major accident occurs, the health monitoring of the pipeline structure should be carried out in advance to ensure people's safety and social security [4].

I. Tuns et al. (Eds.): CIBv 2023, LNCE 510, pp. 213–225, 2024.
https://doi.org/10.1007/978-3-031-60765-3_18

The research of pipeline structure inspection has been carried out for many years. The development of theoretical research began as early as 1917, Rayleigh [5] and Lamb [6] derived the Rayleigh-Lamb equation in plates, and further understood the propagation mode of elastic waves in plates. In 1923, Ghosh [7] studied the propagation characteristics of guided waves in tubes and obtained the linear elastic solution of propagation. Love [8], Cooper and Naghdi [9, 10] further studied the propagation law in hollow cylinders. Lin et al. obtained the relation between frequency and wave number of guided wave in hollow cylinder. In 1969, Aristegui et al. [11] expanded the research field and studied the pipeline propagation characteristics of ultrasonic guided waves under liquid-filled conditions. Cawley [12] and Lowe et al. [13–15] used finite element simulation to study the propagation of lamb in plates. In 1963, A. H. Foch [16] obtained the dispersion curve of guided waves in tubes through experiments. M. G. Seilk et al. [17] successfully stimulated initial signals of different modes in the ultrasonic guided wave test, which verified the feasibility of using ultrasonic guided wave to monitor structural health. M. V. Book et al.[18] excited guided waves at one end of the pipe and successfully received signals at the other end, which verified the feasibility of using guided waves to monitor pipelines. Alleyna et al. [19–21] used the sensitivity of ultrasonic guided waves to defects to carry out health monitoring of defective pipelines. The damage location is determined by calculating the time of receiving the signal. Then the health monitoring system of ultrasonic guided wave structure is applied in engineering practice, and the piezoelectric sensor for sending excitation and receiving signal is developed. Tay [22] used signals of different frequencies to monitor the structural health of liquid-filled pipelines. HarumichiSATo [23] studied the modes of ultrasonic guided waves in liquid-filled pipelines through finite element software and test verification. C. Ristegui et al. [24] replaced the medium with different fluids and introduced different working conditions in the ultrasonic guided wave liquid-filled pipeline monitoring test, so as to explore the influence of different fluid medium conditions on ultrasonic guided waves.

Based on the summary of current research, ultrasonic guided wave is used to monitor non-destructive hollow pipe, non-destructive liquid filled pipe and damaged liquid filled pipe. When the ultrasonic guided wave touches the damaged position, part of the ultrasonic guided wave will reflect, while the other part of the guided wave will continue to propagate through the damaged position, which is the reflection and transmission phenomenon of the ultrasonic guided wave. The damage assessment of pipeline structure can be realized by using the reflection and transmission phenomenon of ultrasonic guided wave. The pulse-echo method is a method that utilizes the reflection and transmission phenomenon of ultrasonic guided wave to extract the ultrasonic guided wave echo signal, and analyzes the damage echo in the echo signal, so as to evaluate the damage. Based on the pulse echo method to monitor the structural damage, the piezoelectric ceramic plate is usually arranged in the same end of excitation, the same end of receiving. In this paper, the propagation characteristics of ultrasonic guided waves in hollow and liquid-filled reinforced concrete pipelines are monitored by piezoelectric ceramic plates. Based on pulse echo method, the monitoring test of reinforced concrete pipe structure with hollow liquid filling is established, and the influence of hollow liquid filling state on signal propagation is obtained.

2 Monitoring Test of Hollow Reinforced Concrete Pipeline

2.1 Experimental Equipment

The signal generator model is DG4062 of Puyuan Precision Electronics Technology, which adopts DDS direct digital synthesis technology, which can generate more stable, more accurate and more pure output signals. The upper frequency of the signal reaches 60 MHz, the sampling rate is 500 MSa/s, and the instrument has rich digital modulation functions. It also supports signals imported from external storage devices. The purpose of using signal amplifiers is mainly because the signal will have a certain attenuation during the propagation process, and the signal amplitude will be reduced. In order to make the signal long-term, stable and low noise output, the signal needs to be amplified. The AigtekATA-2041 power amplifier is used, which can output a maximum of 400 Vp high voltage and a maximum output current of 100 mAp, which can drive a high-voltage load and achieve high-magnification of the excitation signal generated by the signal generator. The digital oscilloscope uses the DS2302A of pure electronics Technology, its functions include signal triggering, acquisition, operation, support 300 MHz bandwidth, sampling rate of 2GSa/s, can be more convenient for the digital signal logic analysis, to meet the test requirements. Experimental apparatus are shown in Fig. 1.

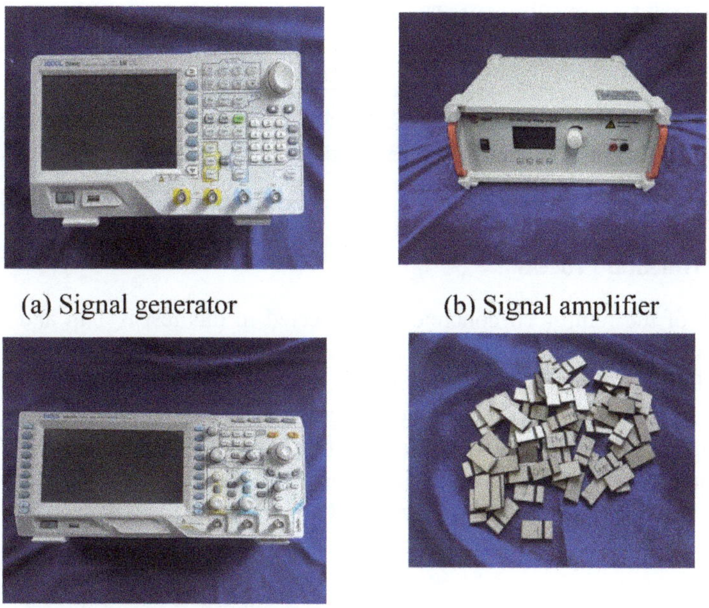

(a) Signal generator (b) Signal amplifier

(c) Signal oscilloscope (d) Piezoelectric ceramic plates

Fig. 1. Experimental apparatus

The test process designed in this paper is that (1) the self-made signal is connected to the USB port through the U disk and imported into the signal generator. (2) Connect the power amplifier to the signal generator, and connect the power amplifier to the positive and negative poles of the excitation end PZT through the wire. (3) The positive and negative poles of the receiving end PZT plate are connected to the digital oscilloscope through the wire. (4) Receive piezoelectric signals in the digital oscilloscope.

The propagation process of ultrasonic guided wave in the whole pipeline is: (1) the signal generator excites the excitation signal, and through the power amplifier, acts on the positive and negative poles of the excitation end PZT. (2) The excitation end PZT will receive the signal through the inverse piezoelectric effect, generate vibration inside the PZT, and propagate forward from the excitation end PZT to the receiving end PZT. (3) The receiving end PZT is subjected to vibration, and through the positive piezoelectric effect, the internal charge begins to move on the upper and lower surfaces, resulting in polarization. Polarization causes the receiving end PZT to generate an electric charge. (4) The piezoelectric signal of the receiving end PZT is displayed on the digital oscilloscope. Damage to the structure can be assessed by analyzing the piezoelectric signal.

2.2 Selection of Excitation Signal

The excitation signal with center frequency of 15 kHz was used in this experiment. In this way, the energy of the guided wave is more concentrated around the center frequency, the propagation distance in the pipeline is longer, the energy dissipation is less, and the damage recognition effect is good. The signal expression is shown in (1)

$$x(t) = [\frac{1}{2}(1 - \cos\frac{2\pi f_c t}{n})]\sin(2\pi f_c t) \tag{1}$$

where, f_c is the center frequency of the excitation signal, and n is the number of single audios.

2.3 Experimental Process

The length of the reinforced concrete pipe selected in this test is 1 m and the outer diameter is 125 mm. In order to suppress the generation of guided waves in the bending mode, 16 PZT-5 piezoelectric ceramic plates with dimensions of 12 mm × 6 mm × 1 mm were uniformly pasted along the circumference of the pipeline as actuators to excite signals in the test. At the receiving end of the same end of the pipeline, PZT plates of the same type and size are arranged to receive the pipeline echo signal. PZT sheet layout and test instruments are shown in Fig. 2 and Fig. 3.

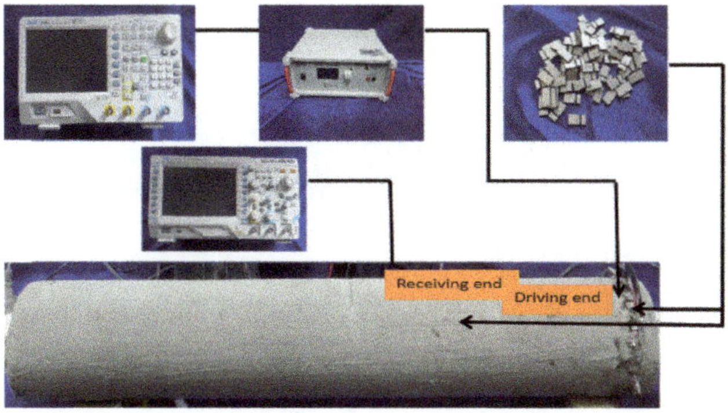

Fig. 2. Arrangement of experimental instruments

The wire connected to the PZT is fixed on the positive and negative stages of the PZT by solder, and then the PZT is fixed on the surface of the pipeline structure under the joint action of glue and epoxy resin.

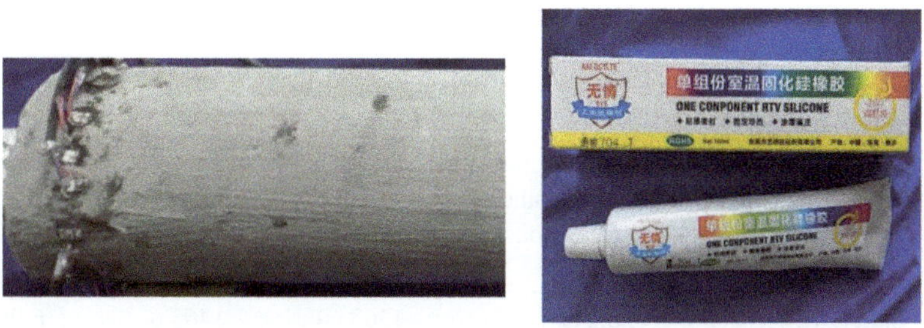

(a) PZT pipe layout (b) Epoxy resin adhesive

Fig. 3. PZT pipe layout and bonding tools

2.4 Experimental Results and Analysis

Using the structural health monitoring system, the ultrasonic guided wave is obtained in the signal receiver and the response signal in the hollow reinforced concrete pipe is shown in Fig. 4.

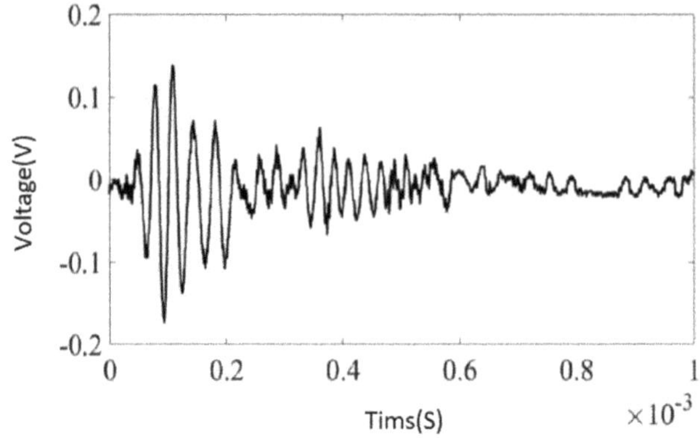

Fig. 4. Response signal of hollow reinforced concrete pipe

The propagation velocity of ultrasonic guided wave in hollow reinforced concrete pipe can be calculated by time-of-flight damage location method. The propagation velocity obtained is compared with that of the theoretical dispersion curve, As shown in Table 1.

Table 1. Guide propagation velocity error table in hollow pipe

Data type	Propagation distance/mm	Occurrence time of boundary reflection wave 1/ms	Propagation speed/m/s	Theoretical velocity/m/s	error/%
Hollow state	30 + 300 = 330	0.242	1363.64	1465.6	6.96

Where, the propagation velocity error formula of experimental and theoretical dispersion is

$$\text{error} = \left| \frac{\text{Experimental value} - \text{Theoretical value}}{\text{Theoretical value}} \right| \times 100\% \tag{2}$$

The error between the experimental propagation velocity value and the theoretical dispersion velocity value is 6.96%, and the error is small. It shows that the calculated dispersion curve of reinforced concrete pipeline can reflect the modes of the actual ultrasonic guided wave excitation in reinforced concrete pipeline and the corresponding group velocity and phase velocity of the modes.

3 Non-destructive Monitoring Test of Liquid-Filled Reinforced Concrete Pipeline

In practical life, pipeline is often used to transport liquid, so the monitoring test research of liquid-filled reinforced concrete pipeline becomes particularly important. In this paper, water is used as the filling liquid, that is, the water-filled pipe is studied. Based on the monitoring test of hollow reinforced concrete pipeline, the pipeline is filled with water to monitor the propagation characteristics of ultrasonic guided wave in the liquid-filled reinforced concrete pipeline under non-destructive and lossy conditions.

3.1 Selection of Excitation Signal

According to the calculated dispersion curve, the excitation signal with center frequency of 10 kHz is used in this experiment. In this way, the energy of the guided wave is more concentrated around the center frequency, the propagation distance in the pipeline is longer, the energy dissipation is less, and the damage recognition effect is good. The signal expression is

$$x(t) = [\frac{1}{2}(1 - \cos\frac{2\pi f_c t}{n})]\sin(2\pi f_c t) \tag{3}$$

3.2 Experimental Equipment

In order to further study the characteristics of guided wave propagation in liquid-filled reinforced concrete pipes, it is necessary to carry out liquid-filled treatment based on the monitoring test of hollow reinforced concrete pipeline, such as Fig. 5.

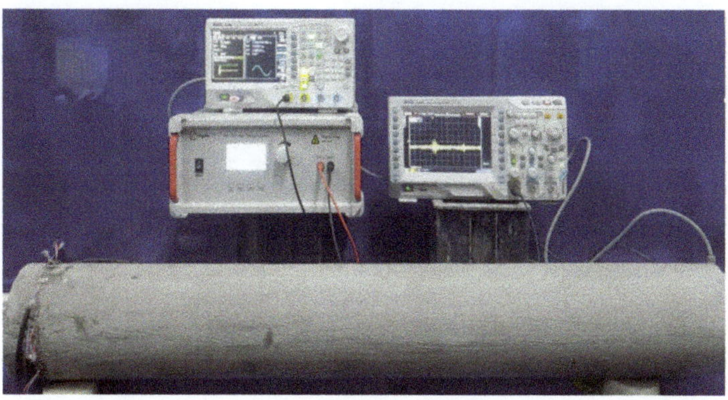

Fig. 5. Health monitoring system for liquid-filled reinforced concrete pipeline structure

When filling a liquid pipeline, tilt one end of the pipe to a certain height. After the liquid is filled, plug the two ends of the pipe to prevent liquid overflow. As shown in Fig. 6.

Fig. 6. A rubber plug used to plug pipes

3.3 Experimental Results and Analysis

The signal received on the oscilloscope, As shown in Fig. 7.

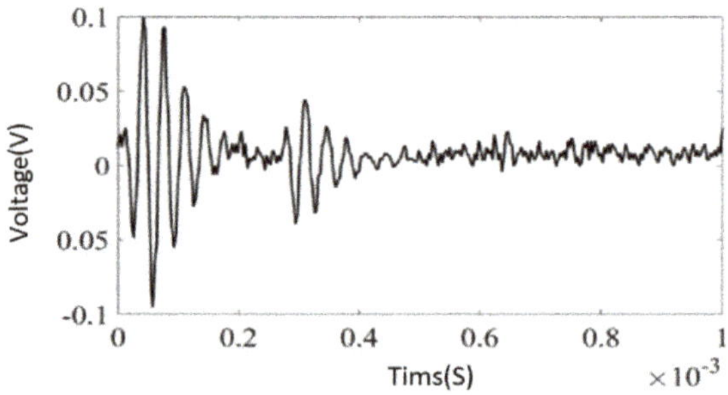

Fig. 7. Response signal of liquid-filled reinforced concrete pipeline

Comparing the signals of the hollow and liquid-filled pipes obtained by the test, As shown in Fig. 8.

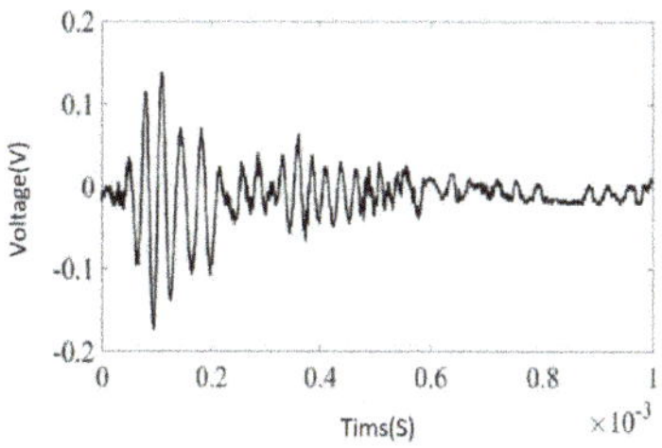

(a) Response signal of hollow pipe

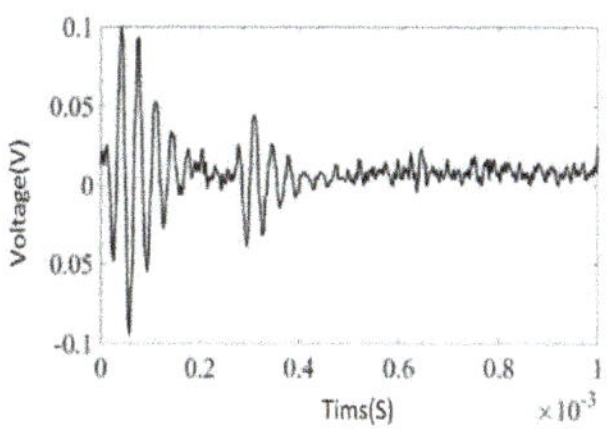

(b) Response signal of liquid filling test

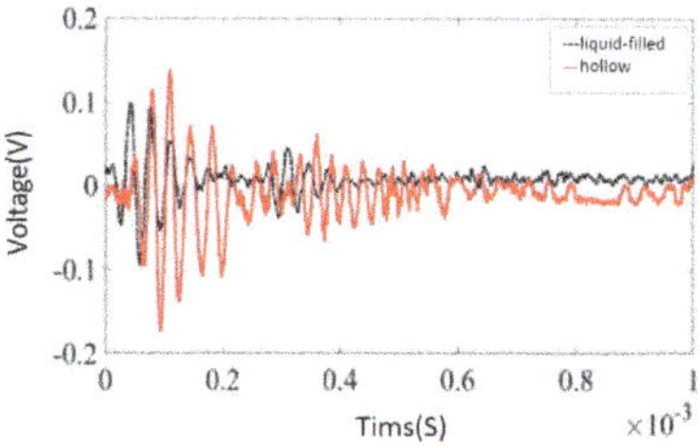

Fig. 8. Signal comparison of hollow and liquid filled pipes

As can be seen from Fig. 9, the strength of the guided wave signal in the liquid-filled state is significantly lower than that in the hollow state, and the signal attenuation is faster and faster with the increase of the propagation distance. It shows that the fluid has a certain signal absorption effect, which reduces the overall strength of the signal to a large extent, and also increases the difficulty of analyzing the signal of the liquid-filled pipeline.

The propagation velocity of the ultrasonic guided wave in the hollow reinforced concrete pipeline in the test can be calculated by time-of-flight damage location method, and the propagation velocity obtained is compared with the propagation velocity of the theoretical dispersion curve, as shown in Table 2.

Table 2. Guide propagation velocity error table in liquid-filled pipeline

Data type	Propagation distance/mm	Occurrence time of boundary reflection wave 1/ms	Propagation speed/m/s	Theoretical velocity/m/s	error/%
liquid-filled	30 + 300 = 330	0.281	1166.08	1263	7.77

The error between the experimental propagation velocity value and the theoretical dispersion velocity value is 7.77%, which is small, indicating that the guided wave dispersion curve of the reinforced concrete pipeline under the liquid-filled state can better reflect the various modes excited by the ultrasonic guided wave in the actual reinforced concrete pipeline under the liquid-filled state, as well as the corresponding group velocity and phase velocity of the modes.

4 Damage Monitoring Test of Liquid-Filled Reinforced Concrete Pipeline

4.1 Establishment of Damage

The damage condition of the liquid-filled reinforced concrete pipeline was established, and the circumferential damage with a circumferential Angle of 45° and a longitudinal depth of 1 cm from the center of the circle was established at 0.5 m away from the pipeline. As shown in Fig. 9.

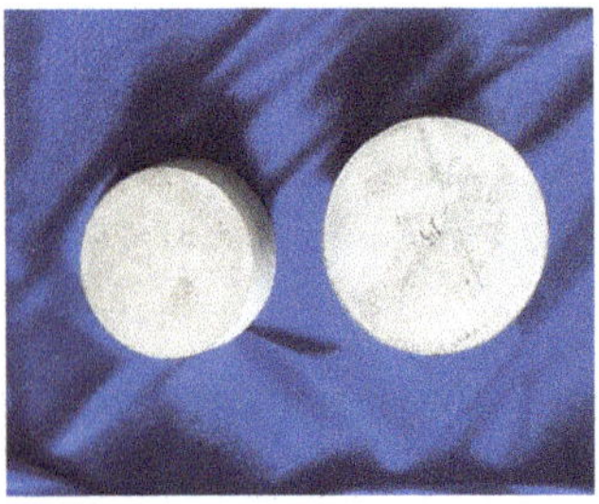

(a) Rubber plugs

(b) Establishment of pipeline damage

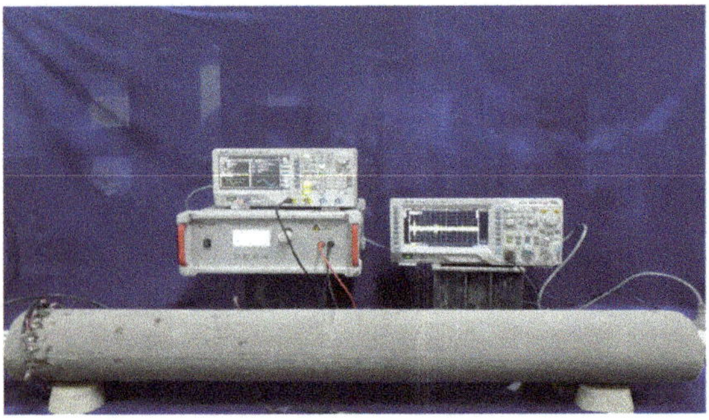

Fig. 9. Diagram of pipeline damage monitoring test

4.2 Result Analysis

The signal received on the oscilloscope is shown in Fig. 10.

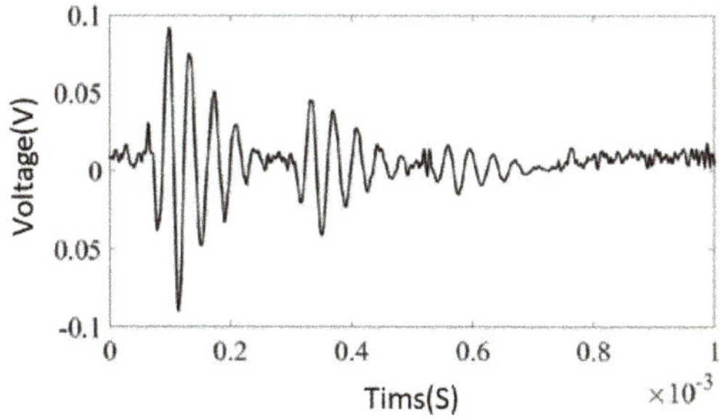

Fig. 10. Signal of liquid-filled reinforced concrete pipeline under damage condition

The damage signal in the liquid-filled state obtained by the test was compared with the non-destructive signal, as shown in Fig. 11.

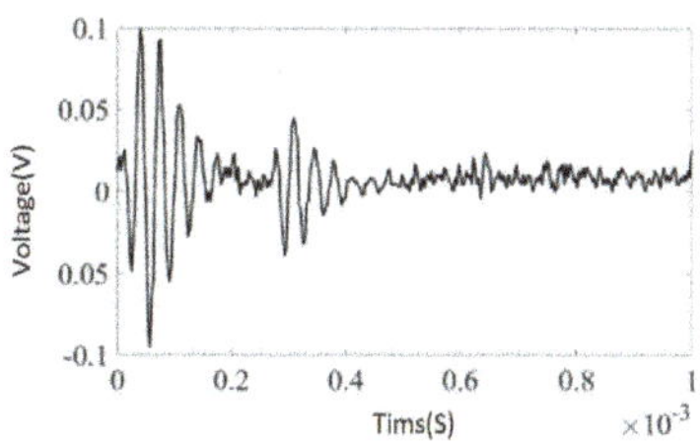

(a) Non-destructive test signals

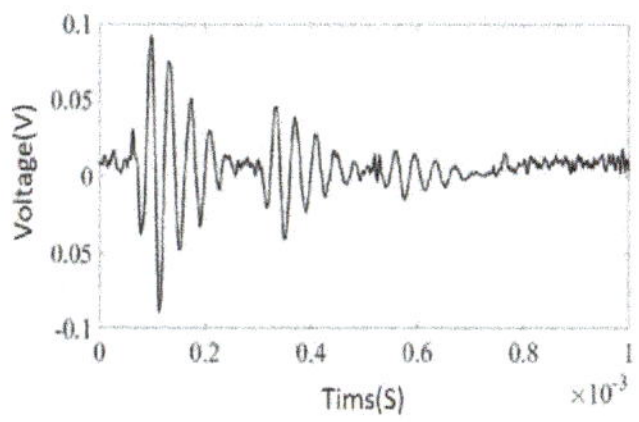

(b) Lossy test

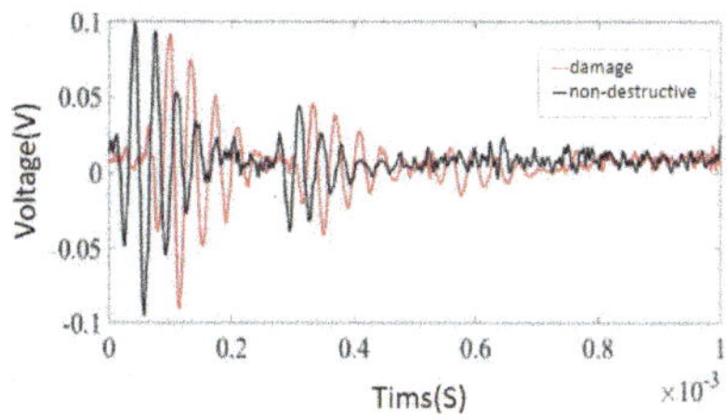

Fig. 11. Comparison of lossless and lossy test signals in liquid state

As can be seen from Fig. 12, by comparing the two signals in the non-destructive and damage state in the liquid-filled pipeline monitoring test, it can be seen that the guided wave in the pipeline has obvious changes in the two states: In the damaged state, the overall amplitude of the guided wave is lower than that in the non-destructive state. With the continuous propagation of the guided wave, the attenuation rate of the signal is faster and faster. Although the overall strength of the signal is low, the echoes at the damaged location are more obvious in the low-amplitude echoes around. The damage echo is generated when the guided wave propagates to the damage position, and the damage echo is received at 0.534 ms, which is consistent with the propagation rule. Low amplitude echoes can be seen from the overall signal, which may be due to the fact that the actual PZT plates are not perfectly aligned or the anisotropy of the pipeline material generates guided waves of other modes in the pipeline.

5 Conclusion

In this paper, piezoelectric ceramic plates are used to monitor the propagation characteristics of ultrasonic guided waves in hollow and liquid-filled reinforced concrete pipelines, and the following conclusions are obtained: The piezoelectric effect of piezoelectric ceramics can be used to generate excitation signals in different pipeline structures. The calculated dispersion curves of reinforced concrete pipelines can clearly reflect the actual dispersion characteristics in both hollow state and liquid-filled state. We have established a health monitoring system for reinforced concrete pipeline structures using ultrasonic guided waves in a hollow and liquid filled state, and compare the obtained experimental propagation speed with the theoretical dispersion curve propagation speed, and obtain the experimental error, which is relatively small, The received signals of ultrasonic guided waves in hollow and liquid filled reinforced concrete pipelines in comparative experiments, it can be seen that the amplitude signal of the voltage in the liquid filled reinforced concrete pipeline is much smaller than that in the hollow state of the reinforced concrete pipeline due to the absorption of ultrasonic guided wave signals by the liquid layer; The correctness of the dispersion curve provides theoretical guidance for the selection of excitation signals and modes. This paper only conducted experimental research on water-filled pipes, so the experimental research on different types of liquid-filled pipes will be the further research object. In addition, this paper draws the dispersion curve of reinforced concrete pipeline through data processing software. Therefore, the influence of pipeline material and pipeline thickness on the dispersion curve still needs to be discussed.

Acknowledgments. This research was performed at the Shenyang Jianzhu University. The research was funded by the research was funded by the National Natural Science Fund (52278196), the Key Laboratory of Urban Security and Disaster Engineering of Ministry of Education, Beijing University of Technology (2021B06) and Natural Science Foundation of Liaoning Province (20181023).

References

1. Qi, X.: Experimental Research on Damage Detection of liquid-filled Pipeline based on ultrasonic guided Wave Technology. Jinan University (2012)
2. Li, Y.: Weak Signal Stochastic Resonance Method for detecting pipeline deformation with ultrasonic guided wave. Jinan University (2018)
3. Zhang, S.: Research on damage Identification and Location of pipeline structure based on guided wave. Shanghai Jiao Tong University (2017)
4. Young, J.R., Alfrcdo, H., Ang, S.: Seismic damage analysis of reinforced concrete buildings. J. Struct. Eng. ASCE **111**(4), 740–756 (1985)
5. Rayleigh, J.: The theory of Sound. Dover Publications (2), pp. 100–150 (1945)
6. Lamb, H.: On waves in an elastic plate. Proceedings of Royal Society (A93), 114 (1917)

7. Ghosh, J.: Longitudinal vibrations of a hollow cylinder. Bull. Caleutta Math. Soc. **24**(14), 31–40 (1923)
8. Love, A.: A treatise on the mathematical theory of elasticity. Dover Publications (1944)
9. Cooper, R.M., Naghdi, P.M.: Propagation of nonaxially symmetric waves in elastic cylindrical shells. J. Acoust. Soc. Am. **29**, 1365 (1957)
10. Naghdi, P.M., Cooper, R.M.: Propagation of elastic waves in cylindrical shells, including the effects oftransverse shear and rotatory inertia. J. Acoust. Soc. Am. **28**(1), 56–63 (1956)
11. Armenakas, N.A., Gazis, D.C., Hemnann, G.: Free vibrations of circular cylindrical shells. Oxford Pergamon Press (1969)
12. Alleyne, D.N., Cawley, P.: The interaction of Lamb wave with defects. IEEE Trans. Ultrason. Ferroelectr. Freq. Control **39**(3), 381–397 (1992)
13. Lowe, M.J.S., Alleyne, D.N., Cawley, P.: The mode conversion of a guided wave by a part-circurferential noteh in a pipe. J. Appl. Mech. **65**, 649–656 (1998)
14. Lowe, M.J.S., Alleyne, D.N., Cawley, P.: Mod conversion of guided waves by defects in pipes. Rev. Progress Quant. Nondestructlve Eval. **16**, 1261–1268 (1997)
15. Wilecx, P., Lowe, M.J.S., Cawley, R.: The effect of dispersion on long- range in section using ultrasonic guided waves. NDT&E Int. **34**, 1–9 (2001)
16. Fitch, A.H.: Observation of elastic pulse propagation in axially symmetric and nonaxially symmetriclongitudinal modes of hollow cylinders. J. Acoust. Soc. Am. **35**, 706–707 (1963)
17. Silk, M.G., Bainton, K.F.: The propagation in metal tubing of ultrasonic wave modes equivalent to, waves. Ultrasonics **17**(1), 11–19 (1979)
18. Brook, M.V., Ngoc, T.D.K., Eder, J.E.: Ultrasonics inspection of steam generator tubing by cylindricalguided waves. Rev. Progress Quant. Nondestructive Eval. **9**, 243–249 (1980)
19. Alleyne, D.N., Cawley, P.: The excitation of Lamb waves in pipes using dry-coupled piezoelectrictransduce. J. Nondestructive Eval. **15**(15), 11–20 (1996)
20. Cawley, P., Alleyne, D.: The use of Lamb waves for the long range inspection of large structures. Ultrasonies **34**, 287–290 (1996)
21. Alleyne, D.N., Cawley, P.: Long range propagation of Lamb waves in chemical plant pipe work. Mater. Eval. **53**(4), 504–508 (1997)
22. Hay, T., Rose, J.L.: Flexible PVDF comb transducers for excitation of axisymmetricguided waves in pipe. Sens. Actuators **100**, 18–23 (2002)
23. Sato, H., Lebedev, M., Akedo, J.: Theoretical and experimental investigation of propagation of guide waves in cylindrical pipe filled with fluid. Jpn. J. Appl. Phys. Part 2 (Letters) **45**(5B), 4573–4576 (2006)
24. Aristegui, C., Lowe, M.J.S., Cawley, P.: Guided waves in fluid-filled pipes surrounded by different fluids. Ultrasonics **39**(5), 367–375 (2001)

Research on Dynamic Response of Buried Pipeline Under Explosion Load Based on SPH-FEM Method

Baoxin Qi[1(✉)], Fan Yang[1], and Haoxiang He[2]

[1] School of Civil Engineering, Shenyang Jianzhu University, Liaoning 110168, Shenyang, China
bxqi@sjzu.edu.cn

[2] The Key Laboratory of Urban Security and Disaster Engineering of Ministry of Education, Beijing University of Technology, Beijing 100124, China

Abstract. Numerical simulation research on pipelines subjected to explosion impact basically turns to ALE. Yet it suffers from unclear interfaces between flowing materials and fails to value the impact of soil compression state and pipe-soil sliding on pipeline stress. To streamline calculation and better demonstrate the damage process of soil under explosion load, this study applies SPH-FEM coupling to study the dynamic response of buried pipelines under explosion load, and compares results of this study with experimental results of Ambrosini et al., and good consistency was observed. This study dissects the dynamic response of the buried X65 pipeline under explosion load, considers the influence of different diameter-thickness ratios of the pipeline and the distance from the explosion, and compares the simulation results in virtue of numerical analysis. SPH-FEM coupling intuitively simulates the entire development process of explosion craters under the action of explosion load, and handles the SPH boundary problem well, found study outcomes. As the diameter-thickness ratio of the pipeline decreases and the explosion distance increases, the deformation and maximum equivalent strain of the pipeline gradually decreases, and the impact of the explosion distance is much greater than the diameter-thickness ratio. This study offers reference for upgrading the protection design of steel pipes.

Keywords: SPH-FEM coupling method · Pipeline · Diameter-to-thickness ratio · Explosion distance

1 Introduction

Buried pipelines are widely used to transport residents' domestic water and gas. Buried pipelines are directly buried in the soil layer with shallow burial depth and reduced protective measures, and are susceptible to various external loads, resulting in deformation or damage to different degrees of buried pipelines, which greatly affects residents' daily supply. In recent years, various accidental explosion accidents have occurred frequently, and explosion load has become an important factor affecting the safety of buried pipelines [1–4]. Based on this, the dynamic response of buried pipeline under explosion load is

I. Tuns et al. (Eds.): CIBv 2023, LNCE 510, pp. 226–237, 2024.
https://doi.org/10.1007/978-3-031-60765-3_19

studied in this paper. At present, in the experimental research on the dynamic response of buried pipelines under the action of blasting vibration, although the monitoring data in the field blasting experiment can truly reflect the pipeline response, the blasting experiment research has high cost, long time consuming, few working conditions covered and limited types of pipelines involved in the experiment. Numerical simulation can effectively overcome the problem, and on the premise of meeting the solution accuracy. The dynamic response of buried pipeline under explosion under different working conditions is calculated [5–7].

Smooth particle hydrodynamics SPH method [8, 9] (Smoothed Particle Hydrodynamics), is a collection of numerous grid method, the Lagrange method and the particles method of the numerical method, compared with the traditional FEM numerical algorithms, such as though laser, Euler and the ALE algorithm, The SPH method avoids the grid distortion problem of the conventional Lagrange method in large deformation. Currently, ALE method is mostly used in numerical simulation studies of pipelines subjected to explosion impact at home and abroad. However, ALE method has the problem of blurred interface between flowing materials [10], which fails to pay attention to the influence of soil compression state and soil sliding on the force of pipelines. The SPH method can avoid this problem and avoid the precision damage caused by mesh distortion, and its particle properties can better show the damage phenomena such as crater formation and material spatter caused by explosion impact. In recent years, SPH method has been applied by many researchers to the simulation study of blasting process [11, 12]. However, due to the difference of algorithms, SPH method requires strict calculation conditions and longer time than traditional finite element method, and has obvious defects in solving boundary problems. Therefore, in order to make up for this defect, many scholars put forward the SPH-FEM coupling method [13–16], which adopts SPH method for modeling in the near area of explosion and FEM method for finite element method for modeling in the far area, which not only solves the problem of mesh distortion, but also ensures the stability and efficiency of calculation [17].

2 Simulation Method

2.1 Basic Principles of SPH Algorithm

The essence of SPH algorithm is a Lagrange method, so its governing equation adopts the Lagrange form of Navier-Stockes partial differential equation, and the core of SPH method is actually an interpolation. In SPH, any macroscopic physical variable (such as density, pressure, temperature, etc.) $f(x)$ can be calculated by means of integral interpolation represented by values on a set of disordered points, the form is:

$$f(x_i) = \sum_{j=1}^{N} \frac{m_j}{\rho_j} f(x_j) W(|x_i - x_j|, h) \tag{1}$$

where m_j, ρ_j is the mass and density of the particle respectively; W(·) is the difference kernel. Based on the concepts of classical solid mechanics and particle approximation, several main governing equations of solid mechanics are given as follows:

Mass conservation equation:

$$\frac{d\rho}{dt}(x_i) = \sum\nolimits_{j=1}^{N} m_j\left[v(x_j) - v(x_i)\right]A_{ij} \tag{2}$$

Momentum conservation equation:

$$\frac{dv^\alpha}{dt}(x_i) = \sum\nolimits_{j=1}^{N} m_j\left[\frac{\sigma^{\alpha,\beta}(x_j)}{\rho_i^2}A_{ij} - \frac{\sigma^{\alpha,\beta}(x_i)}{\rho_j^2}A_{ji}\right] \tag{3}$$

Energy conservation equation:

$$\frac{dE}{dt}(x_i) = -\frac{P_i}{\rho_i^2}\sum\nolimits_{j=1}^{N} m_j\left[v(x_j) - v(x_i)\right]A_{ij} \tag{4}$$

where v(·) is the particle velocity and E is the particle energy.

2.2 SPH-FEM Coupling Mode

The most critical point of SPH and FEM coupling is how a row of SPH particles closest to the finite element mesh can smoothly transfer physical information such as stress and strain to the finite element element. One of the commonly used methods is to define the direct solid contact between nodes and surfaces to transfer information such as the equation of motion. In this paper, the SPH particles and the finite element mesh are consolidated together, and the coordination of their displacement is ensured by point-plane consolidation. At the same time, the physical information such as energy and momentum is transferred to the finite element by the particles close to the finite element. This paper defines this contact by adding the keyword CONTACT_TIED_NODES_TO_SURFACE to the K file in LS-PREPOST, defines the SPH node as the slave node, and defines the Lagrange unit on the interface that contacts the SPH node as the primary surface (Fig. 1).

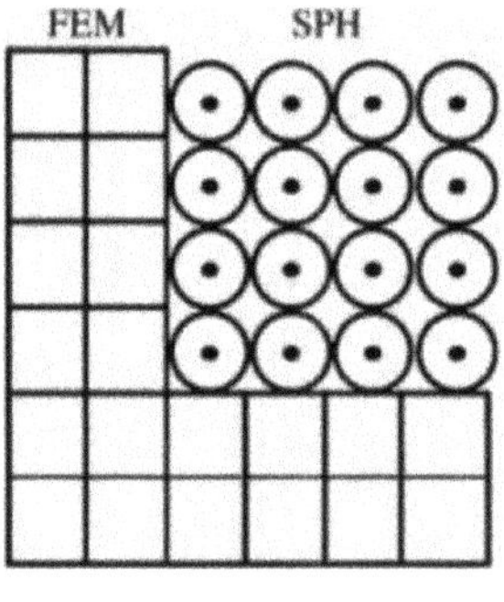

Fig. 1. SPH-FEM coupling

3 Constitutive Model

3.1 Material Model of X65 Steel

Due to the high strain rate on the pipe wall caused by the explosion load, the material used to simulate the behavior of steel pipe material needs to consider the strain rate of the material. Therefore, the simplified JOHNSON_COOK material model is adopted in the steel pipe model to define the behavior of the pipe material. The effect of strain hardening and strain rate on material flow stress (σ_y) is considered in the model, which improves the simulation accuracy. The simplified JOHNSON_COOK material model is defined as follows [18–20]

$$\sigma_y = \left(A + B\overline{\varepsilon}^{Pn}\right)\left(1 + C\ln\dot{\varepsilon}^*\right) \tag{5}$$

where A, B, C and n are material constants; $\overline{\varepsilon}^{Pn}$ is the equivalent plastic strain; $\dot{\varepsilon}^* = \overline{\varepsilon}^{P}/\dot{\varepsilon}_0$ is the dimensionless plastic strain rate. $\dot{\varepsilon}_0$ is the reference strain rate. The material constants of the simplified Johnson-Cook material model for X65 steel are given in Table 1.

Table 1. Material constants for the simplified Johnson-Cook model

E(GPa)	A(MPa)	B(MPa)	n	C	$\dot{\varepsilon}_0$ (S^{-1})
210	500	857	0.34257	0.032604	1

3.2 Soil Material Model

Soil mass adopts MAT_SOIL_AND_FORM material model, whose ideal plastic yield function is [21]

$$\varphi = J_2 - (\alpha_0 + \alpha_1\sigma + \alpha_2\sigma^2) \tag{6}$$

where $J_2 = \sigma_{ij}\sigma_{ij}/2$, σ_{ij} is the component of deviatoric stress; σ is the average stress; The constants α_0, α_1 and α_2 are the constants of the quadratic fitting curve of the dimensionless curve $J_2 - \sigma$.

The main calculation parameters of soil mass are as follows: density is 2.20 g /cm^3, shear modulus is 240 MPa, volume unloading modulus is 1.26 GPa, shear yield surface parameters $\alpha_0 = 2.4 \times 10^6$ Pa^2, $\alpha_1 = 1.360 \times 10^4$Pa and $\alpha_0 = 2.4 \times 10^6$ Pa^2, $\alpha_1 = 1.360 \times 10^4$Pa.

3.3 Model for TNT Explosive

The material model of MAT_HIGH_EXPLOSIVE_BURN and the JWL equation of state [22] were used to accurately describe the volume, pressure and energy characteristics of the explosive products during the explosion process. The expression is as follows [23]

$$P = C_1\left(1 - \frac{\omega}{R_1 v}\right)\exp(-R_1 v) + C_2\left(1 - \frac{\omega}{R_2 v}\right)\exp(-R_2 v) + \frac{\omega}{v}e \tag{7}$$

where P is the detonation pressure determined by JWL equation of state; V is the relative volume; e is the initial specific internal energy; C_1, C_2, R_1, R_2 and ω are all independent constants describing the JWL equation. For the values of the above parameters, refer to the LS-DYNA software manual. The relevant explosive parameters are shown in Table 2.

Table 2. Parameters of explosive

C_d (m/s)	C_1 (GPa)	C_2 (GPa)	R_1	R_2
6930	373.8	3.747	4.15	0.9
ω	ρ (Kg/m^3)		e_0 (KJ/m^3)	
0.35	1630		6.0e6	

4 Model Establishment

This model consists of three parts: steel pipe, soil and explosive. Due to the symmetry of the model, in order to save calculation time, a quarter model is adopted, SPH particle modeling is adopted for the explosive and the soil near the explosive, and Lagrange modeling is adopted for the soil and steel pipe at the far end, as shown in the Fig. 2. The dimensions are developed based on a parametric analysis carried out in the current study so that larger dimensions do not influence the results. The soil model is a 600 cm cube. The SPH part of the soil mass is 200 cm × 200 cm × 160 cm, and the pipeline is 250 cm below the soil mass. In order to explore the influence of different diameter-to-thickness ratio of the pipeline on the dynamic response of the pipeline after explosion, the pipeline model with different diameter to thickness ratio is established for calculation. In particular, the typical size of oil and gas transmission pipeline is adopted. The outer diameter of the pipeline is 914.4 mm, the thickness of the pipeline is 6.35 mm, 9.53 mm, 12.7 mm and 15.88 mm respectively, and the corresponding diameter to thickness ratio (D/tp) is 144, 96, 72 and 57.6 respectively. According to the field test conducted by Ambrosini et al., the mass of TNT is 10 kg and the burial depth is 98 cm. The size is 15 cm × 10 cm × 20 cm, and the mesh is divided by mapping. In order to facilitate calculation under the premise of ensuring accuracy, the mesh size of the model near the explosive is smaller, while the mesh size of the part away from the explosive is larger. In order to explore the effect of explosion distance on the dynamic response of the pipeline and compare the effect with that of the diameter to thickness ratio after explosion, based on the original model, a pipeline with a diameter to thickness ratio of 144 was adopted, and only the position of the pipeline was changed, which was 250 cm, 300 cm, 350 cm and 400 cm below the soil mass respectively. Other conditions remained unchanged (Table 3).

Table 3. Data of pipe diameter thickness ratio

Working condition	Outside diameter (mm)	thickness (mm)	diameter-to-thickness (D/t_p)
ST1	914.4	6.35	144
ST2	914.4	9.53	96
ST3	914.4	12.7	72
ST4	914.4	15.88	57.6

The model was mainly established in LS-DYNA, and was improved by modifying keywords. After FEM model was established in LS-DYNA, SPH model was established by soild-center and soild-nodes in LS-PREPOST, and finally solved by LS-DYNA. Post-processing was carried out through LS-PREPOST. In boundary processing, free boundary was set on the upper surface of the model to describe the free surface of soil and air contact. Meanwhile, non-reflective boundary was applied to the outer and bottom surfaces of the soil to describe the infinite soil space. SPH symmetric boundary and FEM symmetric boundary are set at the X = 0 and Z = 0 planes of symmetry. The SPH symmetric surface constraint is imposed by BOUNDARY_SPH_SYMMETRY_PLANE, and the symmetric surface constraint of finite element is imposed by CONSTRAINED_GLOBLE to constrain the normal constraint of elements and particles on the symmetric surface. A solid contact is set between the particle near the FEM part and the FEM, which is defined by the keyword CONTACT_TIED_NODES_TO_SURFACE.

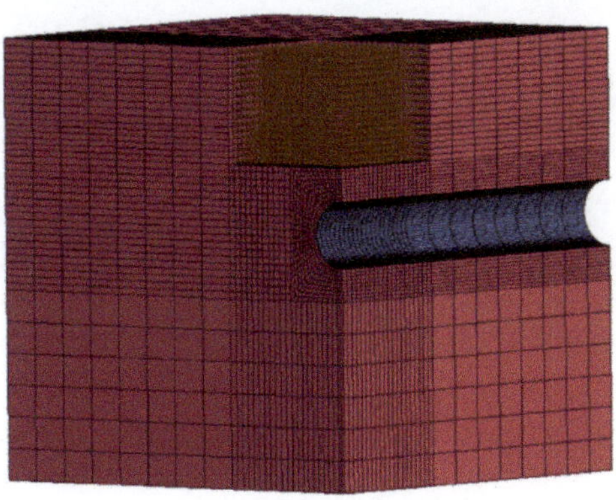

Fig. 2. Finite element model (quarter)

5 Result Analysis

5.1 Crater Size and Soil Failure State

The Fig. 4 shows the formation process of soil pit when 10 kg TNT is buried 98 cm deep and the state of soil spatter under the action of explosion. As shown in the Fig. 4, in the early stage of explosion, with the release of explosive energy, the soil around the explosive is rapidly broken by the high-pressure gas generated by the explosion because the pressure of the shock wave is much larger than the dynamic compressive strength of the soil. The diameter of the explosive pit is about 4 m, and the peak pressure decreases with the increase of the distance from the explosive center. The deformation and throwing process of soil mass described in the figure accords with the characteristics of conventional buried explosive blasting in practice, indicating that SPH-FEM coupling analysis can be used to simulate the deformation and movement of soil mass in buried explosive blasting.

The Fig. 5 shows the soil explosion results [24] 49 ms after the explosion obtained by Ambrosini et al., as well as the simulation effect at 49 ms in this simulation, as shown in the Fig. 5, the numerical effect is in good agreement with the test effect. Moreover, in Ambrosini's test, the results show that the average diameter of the explosion crater is 3.93 m. The numerical simulation results show that the average diameter of the crater is 4 m, which is almost consistent with the experimental results of Ambrosini et al. (Fig. 3).

Fig. 3. Test underground explosion

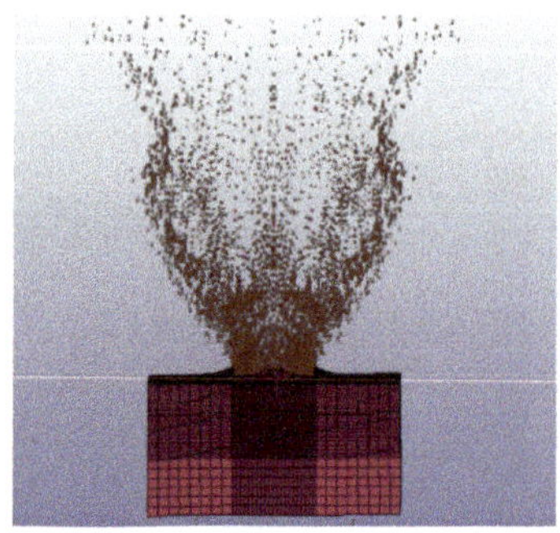

Fig. 4. Simulated underground explosion

5.2 Distribution of Equivalent Stress and Deformation of Pipeline

In this section, the influence of the change of the diameter to thickness ratio on the deformation and stress of X65 steel pipe is studied. The Fig. 6 shows the variation of the equivalent stress and deformation distribution in the wall of a pipe with a diameter to thickness ratio of 144 during detonation. It can be seen from the figure that the maximum equivalent stress of the pipeline appears at the top of the pipeline, and at the beginning of the deformation of the pipeline, the maximum equal effect change appears right below the explosive. As the explosion time goes on, the deformation of the pipeline becomes larger and larger. The cross section of the pipeline first changes in an oval shape, and then the center of the upper part of the pipeline appears obvious depression, and finally

Fig. 5. Comparison of test and simulation effects

the pipeline is severely flattened and the depression is larger. At the same time, the reduction of the diameter to thickness ratio also leads to the reduction of the length of the deformation belt and the deformation of the cross section of the pipeline.

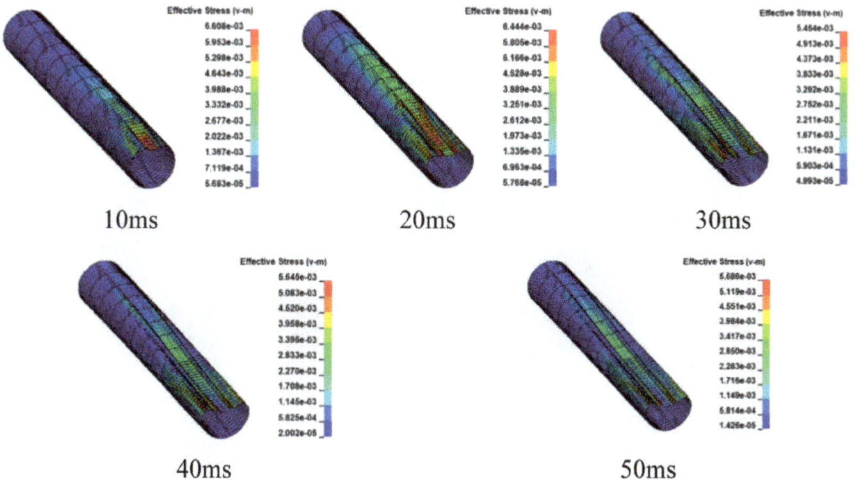

Fig. 6. Changes of equivalent stress and deformation with time

Figure 7 shows the influence of the ratio of diameter-to-thickness on the deformation of the central cross section of the pipeline at the moment of the maximum sag of the pipeline. It can be seen that with the increase of the ratio of diameter to thickness of the pipeline, the deformation of the pipeline cross section becomes more serious.

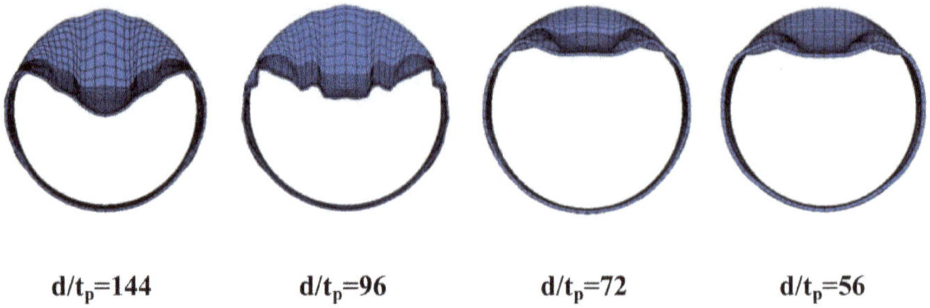

Fig. 7. Cross section deformation of pipeline

5.3 Comparison of the Effects of Diameter to Thickness Ratio and Explosion Distance on Pipelines

In the result analysis, the deformation of the pipeline under explosion load is compared by comparing the diameter to thickness ratio of the pipeline and the explosion distance, so as to compare the two factors that have a greater impact on the pipeline under explosion load. In the specific analysis, as shown in Fig. 8, the center point of the pipeline facing the explosion surface is taken as the measuring point. The displacement time history curve in Y direction at the same explosion distance and different diameter to thickness ratio, and the displacement in Y direction at the same diameter to thickness ratio and different explosion distance are extracted.

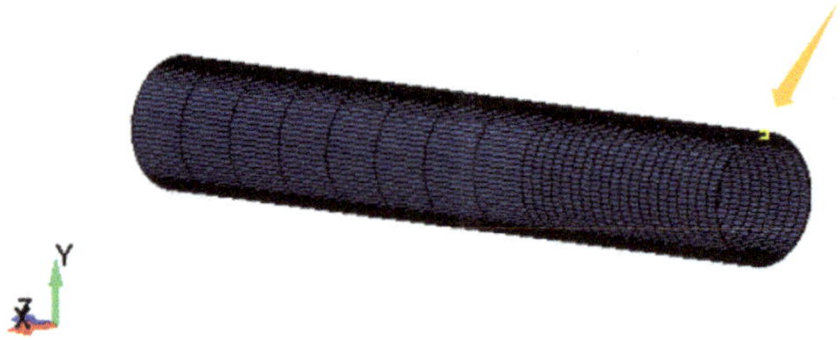

Fig. 8. Selection of measuring points

The results are shown in Fig. 9 and Fig. 10. As can be seen from the figure, after a period of time after detonation of explosives, the displacement of the measuring point increases rapidly after reaching the measuring point, and becomes gentle at 40 ms. When only the diameter to thickness ratio is changed, the displacement of the measuring point in the Y direction differs by about 14 cm at most. The displacement difference of the measuring point in the Y direction is about 25 cm at most, and it can be found by observing the curve that when the explosion distance reaches more than 400 cm, the deformation of the pipeline has been very small and no serious damage has occurred. By comparing the two cases, it can be concluded that the influence of diameter to thickness ratio on pipeline damage is less than that of explosion distance.

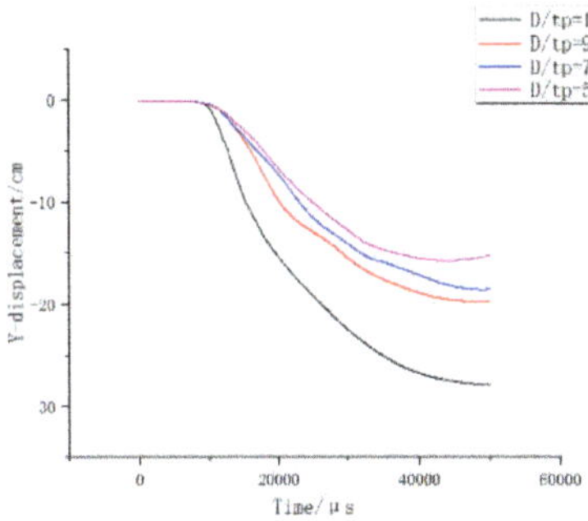

Fig. 9. Time history curve of displacement in Y direction (different diameter to thickness ratio)

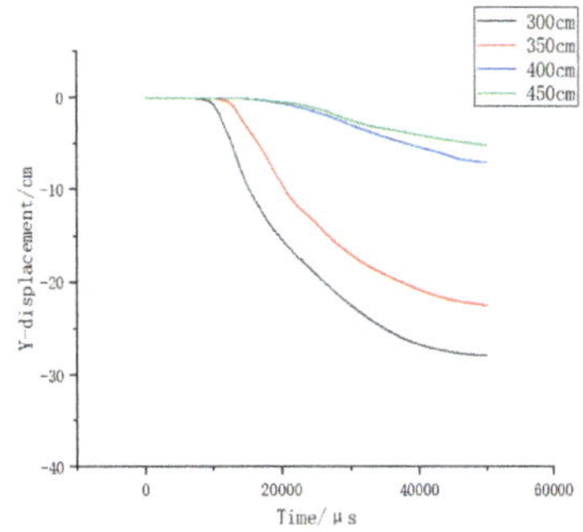

Fig. 10. Y displacement time-history curve (different explosion distances)

6 Future Research Prospect

At present, most studies on the dynamic response of buried pipelines under explosion load focus on single point explosion. However, in actual situations, the explosion of one pipeline will cause the explosion of adjacent pipelines, thus triggering multi-point explosion. The current research lacks the study on the dynamic response of pipelines under multi-point continuous explosion. Therefore, in the future research, the dynamic response of the pipeline under multi-point explosion will be studied.

7 Conclusions

(1) The SPH-FEM coupling method is used to simulate explosive explosion in soil, which can intuitively show the process of soil spatter and crater formation, and the simulation results are almost consistent with the test results of Ambrosini et al. Meanwhile, the calculation efficiency of SPH-FEM coupling method is also high.

(2) By using SPH-FEM coupling method, the stress wave generated by the explosion can be well transmitted to the pipeline through the soil mass. Meanwhile, by comparing the cross section deformation of the pipeline with different diameter to thickness ratios, it is found that the deformation of the pipeline becomes more serious with the increase of the diameter to thickness ratio. When the diameter to thickness ratio is less than 96, the diameter to thickness ratio has little influence on the deformation of the pipeline. Therefore, in practical applications, in order to save costs, reduce the use of steel pipes with diameter to thickness ratios far less than 96.

(3) Simulation analysis was carried out for different pipe diameter thickness ratio and different blasting distance to study their influence on the dynamic response of the pipeline. It is found that the displacement of measuring point increases rapidly when the stress is transferred to the blasting surface of the pipeline, and the effect of explosion distance is much greater than that of diameter thickness ratio. At the same time, it is recommended that the minimum depth of pipeline burial in actual engineering be 4 m to avoid serious damage to the pipeline.

Acknowledgments. This research was performed at the Shenyang Jianzhu University. The research was funded by the research was funded by the National Natural Science Fund (52278196), the Key Laboratory of Urban Security and Disaster Engineering of Ministry of Education, Beijing University of Technology (2021B06) and Natural Science Foundation of Liaoning Province (20181023).

References

1. Zhang, F., Jing, Z., Xu, G.F.: Overview of dynamic response of buried pipelines under explosion impact. Shanxi Archit. **43**(14), 116–117 (2017)
2. Pape, R., Mniszewski, K.R., Longinow, A.: Explosion phenomena and effects of explosions on structures. I: phenomena and effects. Pract. Periodical Struct. Des. Constr. **135**(15), 135–140 (2010)
3. Feldgun, V.R., Kochetkov, A.V., Karinski, A.V., et al.: Internal blast loading in a buried lined tunnel. Int. J. Impact Eng. **35**(3), 172–183 (2008)
4. Ding, Y.Q., Zhou, H.Y., Dai, Z.W., et al.: Calculation of near field explosion load and pipeline response of buried pipelines. Pressure Vessel **35**(12), 18–25 (2018)
5. Zhang, L., Liang, Z., Zhang, J.: Mechanical response of a buried pipeline to explosion loading. J. Fail. Anal. Prev. **16**(4), 576–582 (2016)
6. Kouretzis, G.P., et al.: Analytical calculation of blast-induced strains to buried pipelines. Int. J. Impact Eng. **34**, 1683–1704 (2006). https://doi.org/10.1016/j.ijimpeng.2006.08.008
7. Zhang, Z., Zhou, C.B., Lu, S.W., et al.: Dynamic response characteristics of adjacent buried concrete pipelines under blasting vibration. J. Harbin Inst. Technol. **49**(9), 79–84 (2017)
8. Han, X., Yang, G., Qiang, H.F.: Smooth Particle Hydrodynamics, A Meshless Particle Method. Hunan University Press, Changsha (2005)
9. Monaghan, J.J.: An introduction to SPH. Comput. Phys. Commun. **48**(1), 89–96 (1988)
10. Liu, G.R., Liu, M.B.: Smoothed particle hydrodynamics: a meshfree particle method. World Sci. 92–93 (2003)
11. Huang, Y., Hao, L., et al.: The state of the art of SPH method applied in geotechnical engineering. Chin. J. Geotech. Eng. **30**(2), 256–262 (2008)
12. Liu, M.B., Liu, G.R., Zong, Z., et al.: Computer simulation of high explosive explosion using smoothed particle hydrodynamics methodology. Comput. Fluids **32**(3), 305–322 (2003)
13. Hu, Y.G., Lu, W.B., Chen, M., et al.: Implementation and verification of SPH-EM coupling blasting damage analytical method. Chin. J. Rock Mech. Eng. **34**(suppl 1), 2740–2748 (2015)
14. Wang, W.G., Liu, H.L., Chen, Y.M., et al.: Coupled SPH-FEM method for analyzing touchdown explosion in sand foundation. J. PLA Univ. Sci. Technol. **14**(3), 271–276 (2013)
15. Tian, X.J., Yao, A.L., Xu, T.L., et al.: Study on explosion shock response of defective gas pipeline based on SPH-FEM method. China Saf. Prod. Sci. Technol. **9**, 55–62 (2018)
16. Wanc, W.G., Chen, Y.M., Liu, H.L., et al.: Numerical simulation of explosion in soil based on a coupled SPH-FEM algorithm. Rock Soil Mech. **34**(7), 2104–2110 (2013)
17. Xiao, Y.H., Hu, D., Han, X., et al.: An adaptive axisymmetric FEM-SPH fusion algorithm and its application in high-speed impact simulation. Explosion Shock **32**(4), 384–392 (2012)
18. El-Danaf, E., Baig, M., Almajid, A., Alshalfan, W., Al-Mojil, M., Al-Shahrani, S.: Mechanical, microstructure and texture characterization of API X65 steel. Mater. Des. **47**, 529–538 (2013)
19. William, K., Johns, S.E.: A revised form for the Johnson-Cook strength model. Int. J. Impact Eng. **21**(8), 609–624 (1998)
20. Johnson, G.R., Cook, W.H.: Fracture characteristics of three metals subjected to various strains, strain rates, temperatures and pressures. Eng. Fract. Mech. **21**, 31–48 (1985)

21. LS-DYNA keyword user' manual. Version 971, Livermore Software Technology Corporation (2007)
22. Lee, E.L., Hornig, H.C., Kury, J.: Adiabatic Expansion of High Explosive Detonation Products, UCRL-50422, Lawrence Radiation Laboratory, University of California, USA (1968)
23. Wang, Z., Lu, Y., Hao, H., Chong, K.: A full coupled numerical analysis approach for buried structures subjected to subsurface blast. Comput. Struct. **83**, 339–356 (2005)
24. Ambrosini, R.D., Luccioni, B.M., Danesi, R.F., Riera, J.D., Rocha, M.M.: Size of craters produced by explosive charges on or above the ground surface. Shock Waves **12**, 69–78 (2002)

Analysis on Dynamic Behavior of Simply Supported GFRP Tube Concrete-Encased Steel Members Under Lateral Impact Loads

Haixia Zhang(✉), Huan Chen, Yuhang Zhou, and Fan Yang

Shenyang Jianzhu University, Shenyang 110168, China
iriszhx@163.com

Abstract. In order to investigate the lateral impact resistance of GFRP concrete-encased steel members, a total of 26 members were established using ABAQUS. On the basis of correct models, the failure mode has been identified, and the parameters have been analyzed. The results indicate that the failure mode of the members is flexural failure under lateral impact. The increase of the impact energy and impulse caused further failure of the members. Conversely, the impact resistance of the members was improved with the increase of the concrete strength, yield strength of the steel, thickness of the GFRP tube, cross-section diameter, and steel ratio. It is recommended that the yield strength of the steel not be more than 390 MPa, while the optimum thickness of the GFRP tube is 6 mm.

Keywords: GFRP Tube Concrete-encased Steel · Finite Element · Failure Mode · Parameters Analysis

1 Introduction

With the continuous process of urbanization, there is a growing demand from individuals for higher standards of building safety. However, existing diverse uncertain outside loads like impacts and explosions, jeopardize the stability of architectural structures [1, 2]. In this context, the performance of structural materials and members is crucial, particularly the dynamic response of the members under impact loads.

In recent years, the glass fiber reinforced polymer (GFRP) tube concrete-encased steel composite members have received wide attention as a novel type of composite structural framework. This is attributed to the fact that GFRP tube not only prevents steel corrosion but also provides confinement for the core concrete. Due to the compression of core concrete in three directions, the overall load-bearing capacity is improved. The existence of encased steel enhances the bending and shear resistance of the member. The member can be applied to large-span bridges and ocean engineering. The researchers found that the concrete strength, steel strength, GFRP tube, and the steel ratio have a positive effect on the GFRP tube-encased steel concrete member under axial compression, while the slenderness ratio has a negative effect. The researcher proposed a formula for calculating the bearing capacity of the members [3–10]. The static mechanical properties

I. Tuns et al. (Eds.): CIBv 2023, LNCE 510, pp. 238–249, 2024.
https://doi.org/10.1007/978-3-031-60765-3_20

of the particular members have been conducted in numerous investigations. Nevertheless, there is a lack of sufficient investigation into its behavior under impact loads.

In this study, the performance of GFRP tube concrete-encased steel composite members is investigated under lateral impact through numerical simulation. Specifically, the failure mode of a typical member under lateral impact has been analyzed. Meanwhile, the main parameters of the mass and velocity of the impactor, the thickness of the GFRP tube, and the material strength have been discussed. The axial force is not considered in the numerical model due to there are simple supports at both ends. The effect of the axial force will be considered in subsequent studies. It is hoped to provide a scientific basis for the design and application of GFRP tube concrete-encased steel composite members under lateral impact.

2 Numerical Model

2.1 Establish Model

The numerical model of GFRP tube concrete-encased steel composite members was established using ABAQUS. The model consists of five parts: the impactor, support, end plate, GFRP tube, concrete, and encased steel. The impactor, support, and end plates were modeled as the rigid body, the GFRP tube as the shell unit, and the concrete and encased steel as 3D solid units.

For the contact settings, the hard contact is used to simulate the normal behavior between the GFRP tube and concrete, concrete and encased steel. Moreover, the friction contact is used to model tangential behavior with a coefficient of 0.6 [5, 11]. "Tie" is used to simulate contact behavior between the end plate and other parts. Additionally, the normal behavior between the other parts is hard contact, while the tangential behavior is penalty contact with a coefficient of 0 [11].

The boundary conditions are defined such that the support is completely fixed while allowing the impactor to solely move in the direction of impact. Additionally, a predefined field is used to assign a velocity to the impactor. To ensure the accuracy of the simulation results, the mesh size was 1/200th of the length of the member. It is worth noting that the mesh of the impact area was conducted to encrypt, with a size of 1/400th of the length of the member. The established model is shown in Fig. 1.

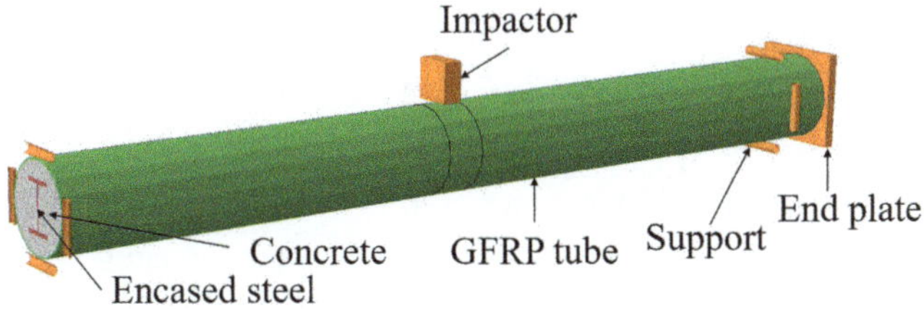

Fig. 1. Schematic of the model.

2.2 Material Constitutive Model

The constitutive model of the GFRP tube adopts the linear elastic model. The compressive behavior of concrete follows Professor Teng's model [12], while the tensile behavior of concrete uses the energy fracture standard [13]. The encased steel opts for the five-stage stress-strain model [14]. The strain rate effect of the GFRP tube is ignored in this paper, and the strain rate effect of concrete is considered by CEB-FIP (2010) [15]. The compression of concrete is displayed in Eq. 1, and the tension of concrete is described in Eq. 2. The strain rate effect of encased steel is considered by Cowper-Symonds, as demonstrated by Eq. 3.

$$\frac{f_d}{f_s} = \begin{cases} (\dot{\varepsilon}_d/\dot{\varepsilon}_s)^{0.014} & \left(\dot{\varepsilon}_d \le 30\,s^{-1}\right) \\ 0.012(\dot{\varepsilon}_d/\dot{\varepsilon}_s)^{1/3} & \left(\dot{\varepsilon}_d > 30\,s^{-1}\right) \end{cases} \tag{1}$$

$$\frac{f_{td}}{f_{ts}} = \begin{cases} (\dot{\varepsilon}_t/\dot{\varepsilon}_{t0})^{0.014} & \dot{\varepsilon}_t \le 10\,s^{-1} \\ 0.0062(\dot{\varepsilon}_t/\dot{\varepsilon}_{t0})^{1/3} & \left(\dot{\varepsilon}_t > 10\,s^{-1}\right) \end{cases} \tag{2}$$

where f_d and f_s are the compressive strength of concrete under static load and dynamic load, respectively. $\dot{\varepsilon}_d$ and $\dot{\varepsilon}_s$ are compressive strain rates under static load and dynamic load, respectively. F_{td} and f_{ts} are the tensive strength of concrete under static load and dynamic load, respectively. $\dot{\varepsilon}_t$ and $\dot{\varepsilon}_{t0}$ are tensive strain rates under static load and dynamic load, respectively. $\dot{\varepsilon}_t$ has a range of values from 10^{-6} to 300 s^{-1}.

$$\frac{\sigma_d}{\sigma_y} = 1 + \left(\frac{\dot{\varepsilon}}{D}\right)^{1/P} \tag{3}$$

where $\dot{\varepsilon}$ = strain rate of the steel; σ_d = stress in the steel at strain rate $\dot{\varepsilon}$; σ_y = initial yield stress of the steel; $D = 6844s^{-1}$; $P = 3.91$.

Table 1. Parameters for low velocity impact model.

Label	f_c /MPa	f_y /MPa	t /mm	D_1 /mm	α /%	v /m·s^{-1}	m /kg	F_{max} /kN	F_{plat} /kN	Δ_{max} /mm
JC	50	345	4	180	6	8	1000	418.8	128.1	279.2
V6	50	345	4	180	6	6	1000	314.7	96.7	203.2
V10	50	345	4	180	6	10	1000	530.4	205.1	340.1
V12	50	345	4	180	6	12	1000	519.7	170.2	407.8
I6	50	345	4	180	6	6	1780	320.0	134.8	303.5
I10	50	345	4	180	6	10	640	519.8	123.6	261.3
I12	50	345	4	180	6	12	444	547.1	113.5	250.2
C30	30	345	4	180	6	8	1000	386.0	125.9	295.9

(continued)

Table 1. (*continued*)

Label	f_c /MPa	f_y /MPa	t /mm	D_1 /mm	α /%	v /m·s^{-1}	m /kg	F_{max} /kN	F_{plat} /kN	Δ_{max} /mm
C70	70	345	4	180	6	8	1000	488.1	132.8	267.5
C90	90	345	4	180	6	8	1000	537.6	135.7	258.7
C110	110	345	4	180	6	8	1000	585.6	140.2	250.9
Q235	50	235	4	180	6	8	1000	354.7	120.8	306.2
Q275	50	275	4	180	6	8	1000	365.8	124.1	294.9
Q390	50	390	4	180	6	8	1000	423.6	130.7	270.4
Q420	50	420	4	180	6	8	1000	422.8	134.3	261.4
Q460	50	460	4	180	6	8	1000	424.8	134.2	257.0
T2	50	345	2	176	6	8	1000	358.2	118.7	279.2
T6	50	345	6	184	6	8	1000	452.9	136.3	256.8
T8	50	345	8	188	6	8	1000	467.2	133.7	246.7
T10	50	345	10	192	6	8	1000	494.4	131.0	232.8
T12	50	345	12	196	6	8	1000	520.4	130.1	220.0
S150	50	345	4	150	6	8	1000	364.8	121.7	316.8
S210	50	345	4	210	6	8	1000	479.5	140.8	231.2
S240	50	345	4	240	6	8	1000	551.6	151.8	190.0
P7	50	345	4	180	7	8	1000	533.6	137.0	238.2
P9	50	345	4	180	9	8	1000	659.4	156.4	200.6

2.3 Verification

As mentioned above, the concrete-filled GFRP-steel double skin tubular column D7-III [16] model was established to verify the effectiveness of the model setting. The impact mass and velocity are 1580 kg and 7 m/s, respectively. From Fig. 2, the errors between the experiment and numerical modeling are within a suitable range and acceptable. It also can be shown in the failure modes that the GFRP tube damage situation of simulation and experiment both appear at the impact and near the concrete base. It has been confirmed that the numerical simulation is feasible for the performance prediction of the GFRP tube concrete-steel member under impact.

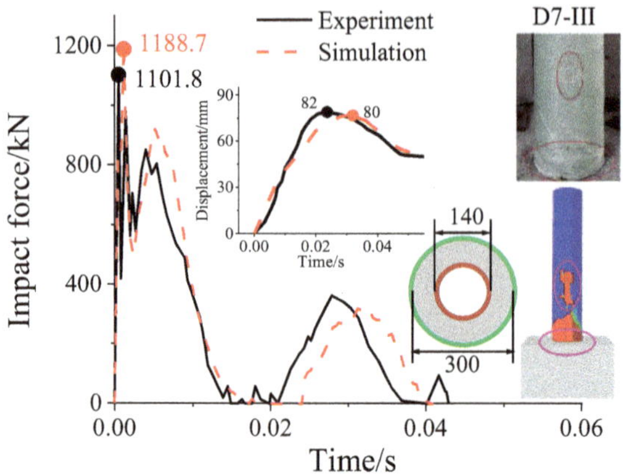

Fig. 2. Comparison of numerical and experimental results [16].

2.4 Design Model

All members have a length of 2000 mm. The GFRP tube all have a winding angle of 55° and material parameters are selected following Karimi [3]. The steel ratio α, which according to "Code for design of composite structures" (JGJ 138-2016) [17], is selected in the range of $4\% \leq \alpha \leq 15\%$. The selection of impactor velocity and mass by Sharma [18] divides the low-velocity impact range. A total of 26 members were designed, and the specific parameters of model in Table 1. The symbols f_c, f_y, t, D_1, α, v, and m are concrete strength, steel yield strength, GFRP tube thickness, cross-section diameter (outer diameter), steel ratio, impact velocity, and impactor mass, respectively.

3 Failure Mode

Figure 3 shows the failure mode of a typical member JC under impact load. From Fig. 3(a), matrix compression damage occurred at the impact area of the GFRP tube, whereas matrix tensile damage was exhibited in a small portion of this area. Most matrix tensile damage occurred on the opposite side of the impact, which was induced by the flexural deformation of the GFRP tube. The matrix compression damage and tension damage also occurred near the support.

From Fig. 3(b), small inclined cracks occurred at the impact area of the concrete, while significant vertical cracks occurred on the opposite side of the impact. This shows that the concrete experienced flexural deformation. From Fig. 3(c), it can be observed that the upper flange of the encased steel in the impact area was compressed, while the web and lower flange were subjected to tension. This confirms the occurrence of flexural deformation in the encased steel. GFRP tube concrete-encased steel members exhibit flexural failure characteristics when subjected to lateral impact loads, as determined by analyzing the interior of the member.

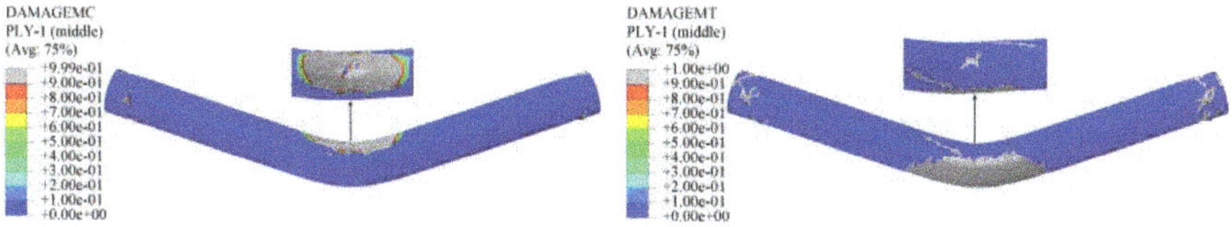

(a) GFRP tube

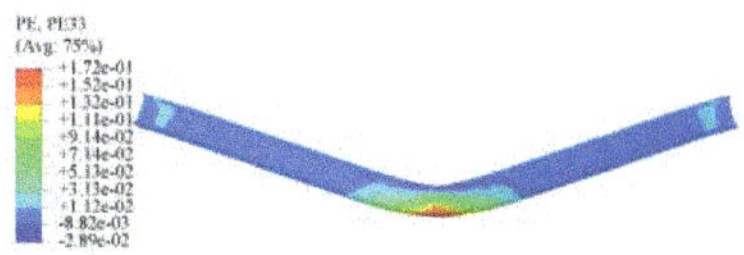

(b) Concrete (c) Encased steel

Fig. 3. Failure cloud diagram.

4 Analysis of Parameters

4.1 Impact Energy

The mass of the impactor remains constant, adjusting the velocity of the impactor to affect the impact energy. Comparative studies are conducted on members with impact velocities of 6 m/s, 8 m/s, 10 m/s, and 12 m/s, which correspond to impact energy of 18 kJ, 32 kJ, 50 kJ, and 72 kJ, respectively. Figure 4 displays the impact force versus mid-span displacement curve and impact durations for different impact energies. Peak impact force $F_{\max}$, plateau impact force F_{plat}, and maximum mid-span displacement $\Delta_{\max}$ are presented in the Table 1. The impact energy increases from 18 kJ to 72 kJ, the peak impact force, plateau impact force, and maximum mid-span displacement are increased by 69%, 53%, and 101%, respectively. Increasing impact velocity leads to a higher material strain rate and the increasing of material strength results in a more advanced plateau impact force. The impact energy increases from 50 kJ to 72 kJ, while the peak impact force increases slowly. This confirms that the member may have failed. The plateau impact force increases with the impact energy, as well as with increased deformation of the member.

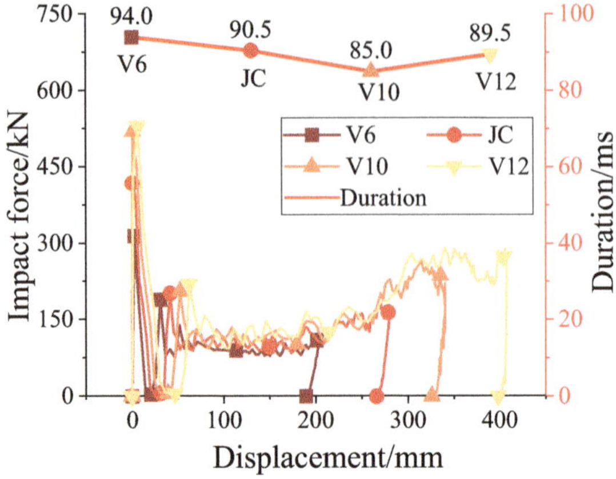

Fig. 4. The impact force versus mid-span displacement curve and impact durations (impact energy).

4.2 Impulse

The impact energy remains constant, and the members are studied for impulses of 5333N·s, 6400 N·s, 8000 N·s, and 10667 N·s. Figure 5 exhibits the impact force versus mid-span displacement curve and impact durations for different impulses with the same impact energy. It is found that the impulse increases from 5333 N·s to 10667 N·s, the plateau impact force and mid-span displacement are increased by 19% and 21%, respectively. Meanwhile, the peak impact force of the member is reduced by 42%. This is because the impact energy of the four members is equal, and the magnitude of the impulse is positively correlated with the mass of the impactor, so the inertia of the impactor also increases. Consequently, the time and deformation for resistance of the member are increased with the inertia of the impactor. When the impact energy is

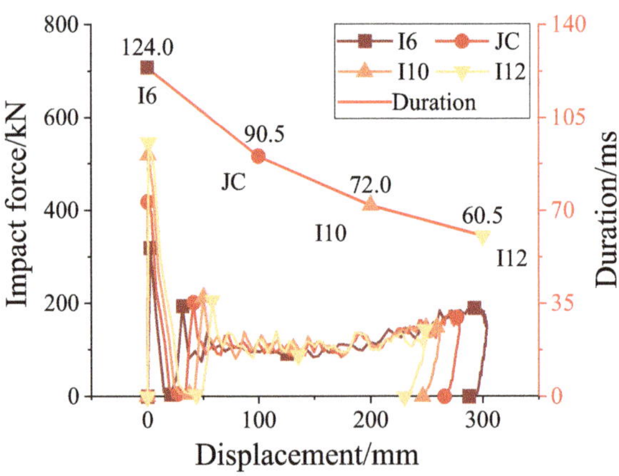

Fig. 5. The impact force versus mid-span displacement curve and impact durations (impulse).

constant, the impactor with a large mass and low velocity causes more damage to the member.

4.3 Concrete Strength

Figure 6 displays the impact force versus mid-span displacement curve and impact durations for different concrete strengths. The concrete strength is improved from 30 MPa to 110 MPa, as a result, the peak impact force and the plateau impact force are increased by 52% and 11%, respectively. Meanwhile, the mid-span displacement is decreased by 15%. Improving concrete strength significantly enhances the overall stiffness of the member, while reducing the deformation, shortening the impact duration, and improving the impact resistance.

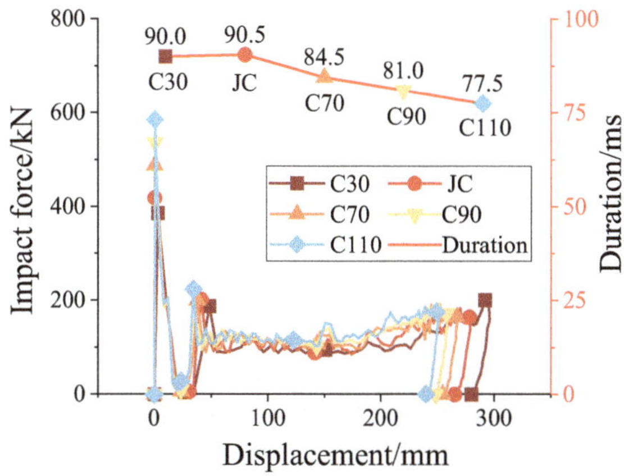

Fig. 6. The impact force versus mid-span displacement curve and impact durations (concrete strength).

4.4 Yield Strength of the Steel

Figure 7 shows the impact force versus mid-span displacement curve and impact durations for the members with steel under different yield strengths. The yield strength of the steel is improved from 235 MPa to 460 MPa, while the peak impact force and plateau impact force are increased by 20% and 11%, respectively. Whereas the mid-span displacement is decreased by 16%. The overall stiffness of the member is enhanced as the yield strength of the steel improves, resulting in a reduction of deformation. However, it has a minor effect on the impact duration. It is found that the yield strength of the steel is improved from 390 MPa to 460 MPa. The peak impact force and plateau impact force are increased by 1% and 3%, respectively. While the mid-span displacement is decreased by 5%. Indicating that the impact resistance of the member shows a slight improvement when the yield strength of the steel exceeds 390 MPa.

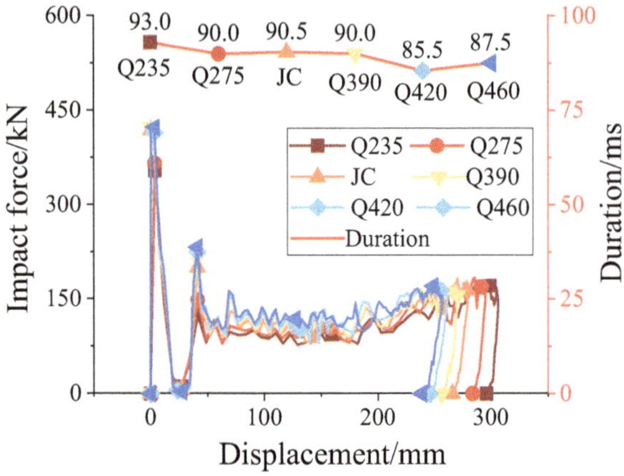

Fig. 7. The impact force versus mid-span displacement curve and impact durations (yield strength of the steel).

4.5 Thickness of the GFRP Tube

Figure 8 displays the impact force versus mid-span displacement curve and impact durations for the member with GFRP tube under different thicknesses. The thickness of the GFRP tube is increased from 2 mm to 12 mm, and the peak impact force is improved by 45%, whereas the mid-span displacement is reduced by 21%. As the thickness of the GFRP tube is increased, the impact resistance of the member is increased as well as the deformation is reduced. The plateau impact force reflects the overall impact resistance and flexural resistance of lateral impact [19]. The thickness of the GFRP tube is increased from 2 mm to 6 mm, and the plateau impact force is increased by 15%, whereas the plateau impact force is decreased by 5% from 6 mm to 12 mm of the GFRP tube. This shows that when the thickness of the GFRP tube exceeds 6 mm, improvement of the thickness of the GFRP tube has a little effect on the impact resistance of the member. Consequently, it is suggested that the optimum thickness of the GFRP tube of this member is chosen as 6 mm under impact load, considering both economic and performance factors.

4.6 Cross-Section Diameter

Figure 9 shows the impact force versus mid-span displacement curve and impact durations for the members under different cross-section diameters. The cross-section diameter is increased from 150 mm to 240 mm, and the peak impact force and plateau impact force significantly are increased by 51% and 25%, respectively. Whereas the mid-span displacement is decreased by 40%. As the cross-section diameter is increased, the overall stiffness of the member is significantly enhanced. This leads to improve impact resistance, reduce deformation, and shorten impact duration.

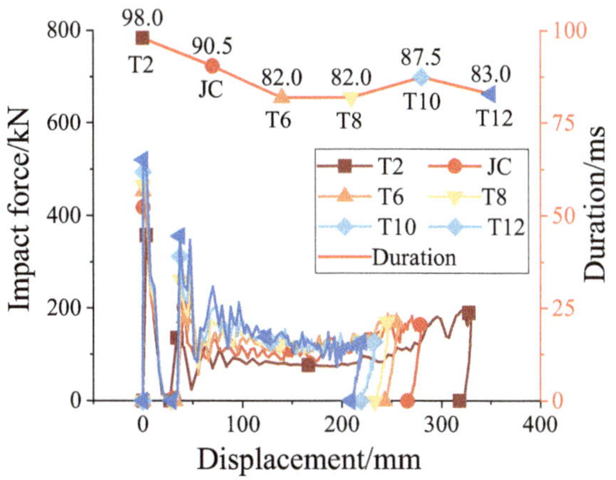

Fig. 8. The impact force versus mid-span displacement curve and impact durations (thickness of the GFRP tube).

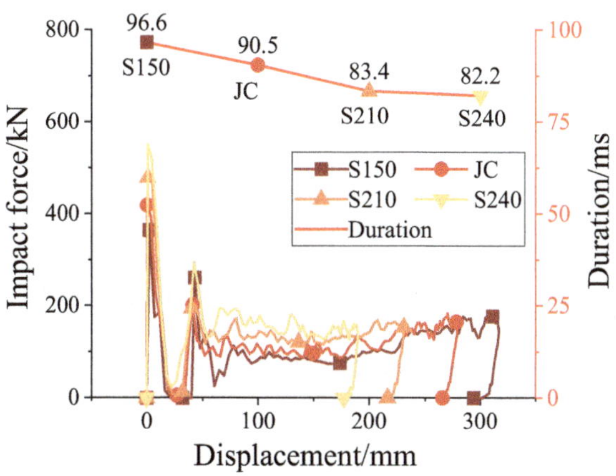

Fig. 9. The impact force versus mid-span displacement curve and impact durations (cross-section diameter).

4.7 Steel Ratio

Figure 10 shows the impact force versus mid-span displacement curve and impact durations for the members under different steel ratios. The steel ratio is increased from 6% to 9%, resulting in a substantial 57% rise in the peak impact force and a notable 22% increase in the plateau impact force. The mid-span displacement exhibits a decrease of 28%. As the steel ratio is increased, the overall stiffness of the member is increased significantly. This phenomenon results in enhancing impact resistance, as well as decreasing deformation and impact duration.

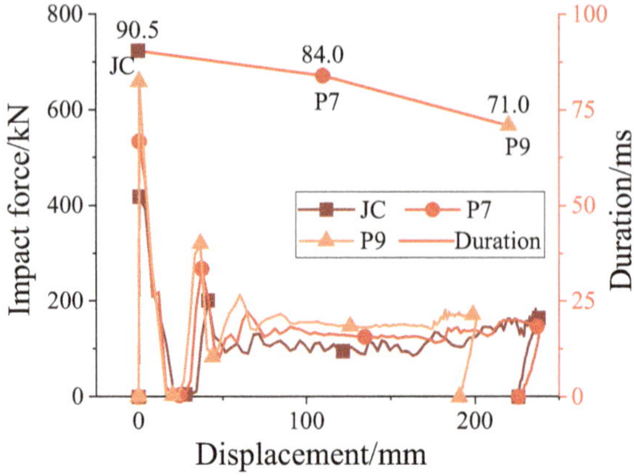

Fig. 10. The impact force versus mid-span displacement curve and impact durations (steel ratio).

5 Conclusion

The structure responses of GFRP tube concrete-encased steel composite members under lateral impact were investigated. The results from this study are as follows:

1. The flexural modes of the GFRP tube concrete-encased steel members were observed under impact load. For the members that fail in flexural, the GFRP tube impact area appeared matrix damage with the concrete existing crack.
2. The failure of the members further increases with the increase of the impact energy. When subjected to the impact energy of 50 kJ, the member experiences complete failure. If the impact energy remains constant, the failure of the member is more severe with high impulses.
3. The impact resistance is improved with the increase of the concrete strength, yield strength of the steel, thickness of the GFRP tube, cross-section diameter, and steel ratio. Considering economic and performance factors, it is not beneficial to increase the yield strength of the steel beyond 390 MPa or the thickness of the GFRP tube above 6 mm.

It should be noted that the influence of axial force and boundary conditions on the impact resistance of the GFRP tube concrete-encased steel member is not consider, as well as how to determine actual practical values for the impact loads in this study. Therefore, a more in-depth analysis in the following study is necessary.

References

1. Sun, J.M., Yi, W.J., Chen, H., et al.: Dynamic responses of RC columns under axial load and lateral impact. J. Struct. Eng. **149**(1), 04022210 (2023)
2. Liu, Y., Dong, A., Zhao, S., et al.: The effect of CFRP-shear strengthening on existing circular RC columns under impact loads. Constr. Build. Mater. **302**, 124185 (2021)
3. Karimi, K., Tait, M.J., El-Dakhakhni, W.W.: Analytical modeling and axial load design of a novel FRP-encased steel–concrete composite column for various slenderness ratios. Eng. Struct. **46**, 526–534 (2013)

4. Zhang, H.X., Ju, S.L., Wang, S.Q.: Experimental study on seismic performance of steel-encased concrete filled GFRP tubes pure bending member. J. Build. Struct. **42**(S2), 348–357 (2021)
5. Zhang, H.X., He, L.Y., Cao, P.P.: Fiber winding angle influence analysis on mechanical behavior of steel-encased concrete filled FRP tubular column subjected to eccentric compression load. J. Build. Struct. **34**(S1), 345–352 (2013)
6. Ren, F.M., Liang, Y.W., Ho, J.C.M., et al.: Behaviour of FRP tube-concrete-encased steel composite columns. Compos. Struct. **241**, 112139 (2020)
7. Cao, Q., Li, X., Lin, Z., et al.: Compression behavior of expansive concrete-encased-steel filled square CFRP tubes. Compos. Struct. **225**, 111106 (2019)
8. Yu, T., Lin, G., Zhang, S.S.: Compressive behavior of FRP-confined concrete-encased steel columns. Compos. Struct. **154**, 493–506 (2016)
9. Huang, L., Yu, T., Zhang, S.S.: FRP-confined concrete-encased cross-shaped steel columns: effects of key parameters. Compos. Struct. **272**, 114252 (2021)
10. Xie, W., Chen, Y., Han, S., et al.: Research on I steel reinforced concrete-filled GFRP tubular short columns. Thin-Walled Struct. **120**, 282–296 (2017)
11. Kang, M., Zhu, X., Wang, R., et al.: Study on impact resistance of circular steel tube columns filled with steel-reinforced concrete under lateral impact. J. Build. Struct. **41**(S1), 128–135 (2020)
12. Teng, J.G., Jiang, T., Lam, L., et al.: Refinement of a design-oriented stress-strain model for FRP-confined concrete. J. Compos. Constr. **13**(4), 269–278 (2009)
13. Tao, Z., Wang, Z.B., Yu, Q.: Finite element modelling of concrete-filled steel stub columns under axial compression. J. Constr. Steel Res. **89**, 121–131 (2013)
14. Han, L.H.: Concrete Filled Steel Tubular Structures—Theory and Practice. Science Press, Beijing (2017)
15. Comite Euro-International du Beton. CEB-FIP (2010) Fib model code for concrete structures. Wilhelm Ernst & Sohn, Berlin (2013)
16. Chen, Z., Wang, J., Chen, J., et al.: Responses of concrete-filled FRP tubular and concrete-filled FRP-steel double skin tubular columns under horizontal impact. Thin-Walled Struct. **155**, 106941 (2020)
17. JGJ 138-2016 Code for design of composite structures JGJ138-2016. China Architecture & Building Press, Beijing (2016)
18. Sharma, H., Hurlebaus, S., Gardoni, P.: Performance-based response evaluation of reinforced concrete columns subject to vehicle impact. Int. J. Impact Eng **43**, 52–62 (2012)
19. Yang, X., Yang, H., Zhang, S.: Transverse impact behavior of high-strength concrete filled normal-/high-strength square steel tube columns. Int. J. Impact Eng **139**, 103512 (2020)

Analytical Solutions for Torsional Response of Single-Story Frame Structure Subjected to Seismic Wave Passage Excitation

Bowen Gu(✉) and Tielin Liu

School of Civil Engineering, Shenyang Jianzhu University, Shenyang, China
g463816377@163.com

Abstract. The torsional response of single-story symmetric frame structure subjected to seismic wave passage excitation is studied by analytical method. The dynamic equilibrium equation is established for the single-story frame structure subjected to wave passage excitation. The analytical solutions of translational and angular displacements of the roof centroid, torque and column shear forces of the structure with wave passage excitation are derived. Four sets of column shear forces of two single-story frame structures with two wave passage excitations are calculated and compared with the corresponding results of the uniform excitation. The calculating results show that the column shear forces in the exciting direction with wave passage excitation are related to the distance between the column and the roof centroid in the propagation direction. Compared with uniform excitation, the peak shear forces of columns near the centroid reduce respectively by 79.5%, 81.5%, 23.6% and 31.2%, while the peak shear forces of columns away from the centroid increase respectively by 17.2%, 73.1%, 83.6% and 5.6% with wave passage excitation. The peak shear forces of the distal columns in the direction of wave passage propagation are bigger than those of the proximal columns (facing the wave input). The peak shear forces of the distal columns are 22.9%, 4.4%, 50.8% and 11.3% higher than those of the proximal columns respectively.

Keywords: Torsional Response · Seismic Wave Passage · Analytical Solution

1 Introduction

Destruction caused by building torsion occurs from time to time [1]. The main reasons for the torsional response of the structure are physical eccentricity, accidental eccentricity, foundation torsion and so on.

Some scholars have made some researches on the torsional response of symmetric structures. Newmark [2] first proposed the influence of torsional input of foundation on symmetric structure. It was concluded that the ratio of the accidental eccentricity e to the long plan dimension a varies almost directly with the fundamental frequency of vibration of the building and with the transit time of the earthquake wave motion. De La Llera [3] suggested that a building with nominally symmetric plan is actually asymmetric to some unknown degree and will undergo torsional vibration when subjected to

I. Tuns et al. (Eds.): CIBv 2023, LNCE 510, pp. 250–262, 2024.
https://doi.org/10.1007/978-3-031-60765-3_21

purely translational ground motion. The response of nominally symmetric-plan systems increases by a greater amount relative to asymmetric-plan systems. Yenan Cao [4] found that the torsional response is most sensitive to the pulse period and the wave passage effects are controlled by the wave apparent velocity. In Lila Kalani Sarokolayi's study [5], the rotational components showed a direct relation with a higher order derivation of translational components. The relationship between the peak acceleration values of the torsion component and the translation component is linear. W. K. Tso [6] found the system is particularly susceptible to induced torsional oscillations when the torsional frequency is close to the lateral frequency. Hong Hao [7] found that if the structure is rigid or its size is large, only considering the dynamic response caused by uniform excitation will greatly underestimate the column shear force.

Some scholars have made some researches on the torsional response of asymmetric structures. H. Hao [8] found that torque is closely related to the uncoupled torsional frequency ratio $\Omega = \omega_\theta/\omega_y$. For torsional stiff structures, the torque is mainly caused by the asymmetry of the building, and in asymmetric structures with greater eccentricity, multiple excitation actually reduces the torque. For torsional flexible structures, the influence of multiple excitation is the main factor of generating torque, while the influence of structural asymmetry is less. W. K. Tso [9] found that when the eccentricity of the building was small and when the decoupling frequency ratio of the building was close to the unit, four of the five types of seismic code are underestimated of the building's torque and edge displacement.

Some scholars have studied the characteristics of symmetric structure and asymmetric structure in the reverse input of the foundation. De La Llera [10] found that the accidental torsion effects, when significant, are larger for symmetric buildings compared with asymmetric buildings. The response of short period (less than half a second) and torsional flexible systems may increase significantly due to accidental torsion. H. Shakib [11] found that the seismic rotation component has a greater impact on low period buildings. The influence of seismic rotation component on the structure decreases with the increase of Ω. The effect of earthquake torsional component is much higher in symmetric buildings than in asymmetric buildings. Hong Hao [12] found that both structural asymmetry and multiple excitation reduce base shear and induce torque. The multiple excitations produce more torsional responses than structural eccentricity if the structure is symmetric or asymmetric with small eccentricities.

The purpose of this paper is to study the response characteristics of single-story symmetrical frame structure in elastic stage with seismic wave passage ex-citation in an analytical way. By establishing and solving the dynamic equilibrium equation of single-story frame structure with wave passage excitation, the analytical solutions of displacement and column shear force of single-layer frame structure are given. By comparing with the uniform excitation response, the necessity of considering traveling wave excitation in the design of symmetrical structure is explained, The analytical solutions and response characteristics obtained can provide reference for the aseismic design of single-layer symmetric frame structures.

2 The Dynamic Equilibrium Equation

Figure 1 is a diagram of a single-story frame structure, where D and A are the total length in both directions respectively, and h is the height of the structure. The mass of the rigid roof is m, and the moment of inertia around the vertical axis of the centroid is J_c. The lateral stiffness of column number n in the x and y directions is k_{nx} and k_{ny}, respectively, and the distance from the centroid is r_{nx} and r_{ny}, respectively. The seismic wave propagates in the x direction and excites in the y direction. The leftmost columns are the first to be excited, and the rightmost columns are the last to be excited. The wave apparent speed is v.

Let the absolute acceleration of the roof centroid in the y direction be $\ddot{U}_{cy}$, and the absolute rotational acceleration of the roof centroid be $\ddot{\phi}_c$.

Let the absolute acceleration of the foundation in the y direction be $\ddot{U}^g_{cy}$, the absolute rotational acceleration of the foundation be $\ddot{\phi}^g_c$, relative acceleration be $\ddot{U}^r_{cy}$ and the relative rotational acceleration be $\ddot{\phi}^r_c$.

For the nth column, the absolute displacement of the column top in the x direction is U_{nx}, and the absolute displacement in the y direction is U_{ny}; The absolute displacement of the column bottom in the x direction is U^g_{nx}, and the absolute displacement in the y direction is U^g_{ny}.

$$\begin{aligned} U_{nx} &= U_{cx} - \phi_c r_{ny}, \quad U_{ny} = U_{cy} + \phi_c r_{nx} \\ U^g_{nx} &= U^g_{cx} - \phi^g_c r_{ny}, \quad U^g_{ny} = U^g_{ny} + \phi^g_c r_{nx} \end{aligned} \tag{1}$$

For the roof centroid, the translational equation in the y direction and the torsional equation in the z direction are established as follows:

$$m\ddot{U}_{cy} = m\left(\ddot{U}^r_{cy} + \ddot{U}^g_{cy}\right) = \sum_{n=1}^{N} -k_{ny}\left(U_{ny} - U^g_{ny}\right) \tag{2}$$

$$J_c\ddot{\phi}_c = J_c\left(\ddot{\phi}^r_c + \ddot{\phi}^g_c\right) = \sum_{n=1}^{N} -k_{ny}\left(U_{ny} - U^g_{ny}\right)r_{nx} + \sum_{n=1}^{N} k_{nx}\left(U_{nx} - U^g_{nx}\right)r_{ny} \tag{3}$$

Substitute Eq. (1) into Eqs. (2) and (3)

$$m\left(\ddot{U}^r_{cy} + \ddot{U}^g_{cy}\right) = -\sum_{n=1}^{N} k_{ny} U^r_{cy} - \sum_{n=1}^{N} k_{ny} r_{nx} \phi^r_c \tag{4}$$

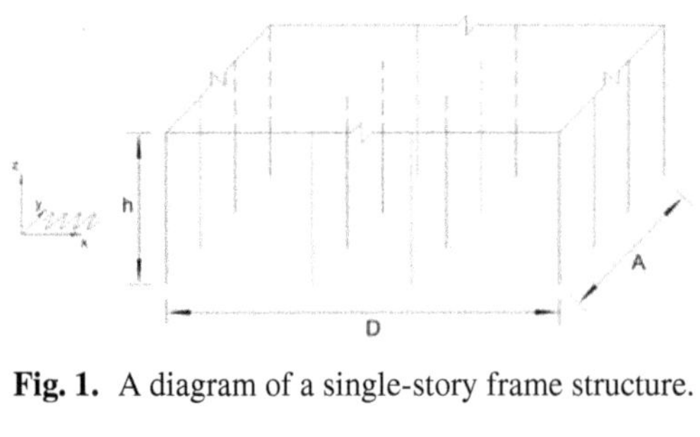

Fig. 1. A diagram of a single-story frame structure.

$$J_c(\ddot{\phi}_c^r + \ddot{\phi}_c^g) = -\left(\sum_{n=1}^{N} k_{ny} r_{nx}^2 + \sum_{n=1}^{N} k_{nx} r_{ny}^2\right)\phi_c^r - \sum_{n=1}^{N} k_{ny} r_{nx} U_{cy}^r \tag{5}$$

Equations (4) and (5) can be written as follows:

$$\begin{bmatrix} m & 0 \\ 0 & J_c \end{bmatrix}\begin{Bmatrix} \ddot{U}_{cy}^r \\ \ddot{\phi}_c^r \end{Bmatrix} + \begin{bmatrix} k_{yy} & k_{y\varphi} \\ k_{\varphi y} & k_{\varphi\varphi} \end{bmatrix}\begin{Bmatrix} U_{cy}^r \\ \phi_c^r \end{Bmatrix} = -\begin{bmatrix} m & 0 \\ 0 & J_c \end{bmatrix}\begin{Bmatrix} \ddot{U}_{cy}^g \\ \ddot{\phi}_c^g \end{Bmatrix} \tag{6}$$

where

$$k_{yy} = \sum_{n=1}^{N} k_{ny},\ k_{\varphi\varphi} = \sum_{n=1}^{N} k_{ny} r_{nx}^2 + \sum_{n=1}^{N} k_{nx} r_{ny}^2,\ k_{y\varphi} = k_{\varphi y} = \sum_{n=1}^{N} k_{ny} r_{nx} \tag{7}$$

3 Analytical Solution of Roof Torque and Column Shear

3.1 The Translational Acceleration and the Rotational Acceleration of Rigid Foundations

The acceleration time history of seismic wave can be expressed as follows [13].

$$\ddot{U}_g = \sum_{i=1}^{\infty} A_{gi} \sin \theta_i t \tag{8}$$

where

$$A_{gi} = \frac{2}{T}\int_0^T \ddot{U}_g(t) \sin \theta_i t dt$$
$$\theta_i = \frac{i\pi}{T}, i = 1, 2, 3... \tag{9}$$

In this paper, the translational acceleration and rotational acceleration of rigid foundation with wave passage excitation are obtained, and then the two accelerations are substituted into the equations to solve the response of single-story frame structure with wave passage excitation.

Using the Eq. (8), the translational and rotational acceleration of rigid foundation with traveling wave excitation can be obtained as follows [3, 11]:

$$\ddot{U}_{cy}^g = \frac{1}{2}(\ddot{U}_L^g + \ddot{U}_R^g) = \frac{1}{2}\left[\sum_{i=1}^{\infty} A_{gi} \sin \theta_i t + A_{gi} \sin \theta_i \left(t - \frac{D}{v}\right)\right] \tag{10}$$

$$\ddot{\phi}_c^g = -\frac{\dddot{U}_g}{v} = -\frac{1}{v}\sum_{i=1}^{\infty} A_{gi}\theta_i \cos \theta_i t \tag{11}$$

where, $\ddot{U}_L^g$ is the acceleration on the left side of the foundation, and $\ddot{U}_R^g$ is the acceleration on the right side of the foundation with time delay.

And the translational and rotational acceleration of rigid foundation with uniform excitation can be obtained as follows:

$$\ddot{U}_{cy}^{g} = \ddot{U}_{g} = \sum_{i=1}^{\infty} A_{gi} \sin \theta_i t \tag{12}$$

$$\ddot{\phi}_{c}^{g} = 0 \tag{13}$$

3.2 Analytical Solutions of Natural Frequency, Vibration Mode and Relative Displacement

According to Eq. (6), the natural frequencies of the structure are:

$$\omega_{1,2}^2 = \frac{\left(\omega_y^2 + \omega_\varphi^2\right)}{2} \mp \sqrt{\left(\frac{\omega_y^2 - \omega_\varphi^2}{2}\right)^2 + \frac{k_{y\varphi}^2}{mJ_c}} \tag{14}$$

where

$$\omega_y = \sqrt{\frac{k_{yy}}{m}} \tag{15}$$

And then the matrix expression of the main vibration mode is

$$\phi = \begin{bmatrix} 1 & \varphi_{12} \\ \varphi_{21} & 1 \end{bmatrix} = \begin{bmatrix} 1 & \frac{k_{y\varphi}}{m\omega_2^2 - k_{yy}} \\ \frac{m\omega_1^2 - k_{yy}}{k_{y\varphi}} & 1 \end{bmatrix} \tag{16}$$

Substituting Eqs. (10) and (11) into Eq. (6), and using the mode superposition method, the translational displacement and rotational displacement of the roof centroid relative to the foundation with wave passage excitation can be obtained as follows:

$$\begin{aligned}
U_{cy}^{r} = & - A_1 \left\{ \sum_{i=1}^{\infty} \frac{A_{gi}\left[\sin \theta_i t + \sin \theta_i \left(t - \frac{D}{v}\right)\right]}{2\left(\omega_1^2 - \theta_i^2\right)} - \sum_{i=1}^{\infty} \frac{A_{gi}\theta_i\left[\sin \omega_1 t + \sin \omega_1 \left(t - \frac{D}{v}\right)\right]}{2\omega_1\left(\omega_1^2 - \theta_i^2\right)} \right\} \\
& - B_1\phi_{12} \left\{ \sum_{i=1}^{\infty} \frac{A_{gi}\left[\sin \theta_i t + \sin \theta_i \left(t - \frac{D}{v}\right)\right]}{2\left(\omega_2^2 - \theta_i^2\right)} - \sum_{i=1}^{\infty} \frac{A_{gi}\theta_i\left[\sin \omega_2 t + \sin \omega_2 \left(t - \frac{D}{v}\right)\right]}{2\omega_2\left(\omega_2^2 - \theta_i^2\right)} \right\} \\
& - A_2 \left[-\sum_{i=1}^{\infty} \frac{A_{gi}\theta_i}{v\left(\omega_1^2 - \theta_i^2\right)} \cos \theta_i t + \sum_{i=1}^{\infty} \frac{A_{gi}\theta_i}{v\left(\omega_1^2 - \theta_i^2\right)} \cos \omega_1 t \right] \\
& - B_2\phi_{12} \left[-\sum_{i=1}^{\infty} \frac{A_{gi}\theta_i}{v\left(\omega_2^2 - \theta_i^2\right)} \cos \theta_i t + \sum_{i=1}^{\infty} \frac{A_{gi}\theta_i}{v\left(\omega_2^2 - \theta_i^2\right)} \cos \omega_2 t \right] \\
\phi_{c}^{r} = & - A_1\phi_{21} \left\{ \sum_{i=1}^{\infty} \frac{A_{gi}\left[\sin \theta_i t + \sin \theta_i \left(t - \frac{D}{v}\right)\right]}{2\left(\omega_1^2 - \theta_i^2\right)} - \sum_{i=1}^{\infty} \frac{A_{gi}\theta_i\left[\sin \omega_1 t + \sin \omega_1 \left(t - \frac{D}{v}\right)\right]}{2\omega_1\left(\omega_1^2 - \theta_i^2\right)} \right\}
\end{aligned} \tag{17}$$

$$
\begin{aligned}
&-B_1\left\{\sum_{i=1}^{\infty}\frac{A_{gi}\left[\sin\theta_i t+\sin\theta_i\left(t-\frac{D}{v}\right)\right]}{2\left(\omega_2^2-\theta_i^2\right)}-\sum_{i=1}^{\infty}\frac{A_{gi}\theta_i\left[\sin\omega_2 t+\sin\omega_2\left(t-\frac{D}{v}\right)\right]}{2\omega_2\left(\omega_2^2-\theta_i^2\right)}\right\}\\
&-A_2\phi_{21}\left[-\sum_{i=1}^{\infty}\frac{A_{gi}\theta_i}{v\left(\omega_1^2-\theta_i^2\right)}\cos\theta_i t+\sum_{i=1}^{\infty}\frac{A_{gi}\theta_i}{v\left(\omega_1^2-\theta_i^2\right)}\cos\omega_1 t\right]\\
&-B_2\left[-\sum_{i=1}^{\infty}\frac{A_{gi}\theta_i}{v\left(\omega_2^2-\theta_i^2\right)}\cos\theta_i t+\sum_{i=1}^{\infty}\frac{A_{gi}\theta_i}{v\left(\omega_2^2-\theta_i^2\right)}\cos\omega_2 t\right]
\end{aligned}
\tag{18}
$$

where

$$
A_1=\frac{m}{\left[m+J_c\left(\frac{m\omega_1^2-k_{yy}}{k_{y\varphi}}\right)^2\right]},\quad A_2=\frac{\frac{m\omega_1^2-k_{yy}}{k_{y\varphi}}J_c}{m+J_c\left(\frac{m\omega_1^2-k_{yy}}{k_{y\varphi}}\right)^2}
\tag{19}
$$

$$
B_1=\frac{\frac{k_{y\varphi}}{m\omega_2^2-k_{yy}}m}{m\left(\frac{k_{y\varphi}}{m\omega_2^2-k_{yy}}\right)^2+J_c},\quad B_2=\frac{J_c}{m\left(\frac{k_{y\varphi}}{m\omega_2^2-k_{yy}}\right)^2+J_c}
\tag{20}
$$

Substituting Eqs. (12) and (13) into Eq. (6), the translational displacement and rotational displacement of the roof centroid relative to the foundation with uniform excitation can be obtained as follows:

$$
\begin{aligned}
U_{cy}^r=&-A_1\left[\sum_{i=1}^{\infty}\frac{A_{gi}}{\left(\omega_1^2-\theta_i^2\right)}\sin\theta_i t-\sum_{i=1}^{\infty}\frac{A_{gi}\theta_i}{\omega_1\left(\omega_1^2-\theta_i^2\right)}\sin\omega_1 t\right]\\
&-B_1\phi_{12}\left[\sum_{i=1}^{\infty}\frac{A_{gi}}{\left(\omega_2^2-\theta_i^2\right)}\sin\theta_i t-\sum_{i=1}^{\infty}\frac{A_{gi}\theta_i}{\omega_2\left(\omega_2^2-\theta_i^2\right)}\sin\omega_2 t\right]
\end{aligned}
\tag{21}
$$

$$
\begin{aligned}
\phi_c^r=&-A_1\phi_{21}\left[\sum_{i=1}^{\infty}\frac{A_{gi}}{\left(\omega_1^2-\theta_i^2\right)}\sin\theta_i t-\sum_{i=1}^{\infty}\frac{A_{gi}\theta_i}{2\omega_1\left(\omega_1^2-\theta_i^2\right)}\sin\omega_1 t\right]\\
&-B_1\left[\sum_{i=1}^{\infty}\frac{A_{gi}}{\left(\omega_2^2-\theta_i^2\right)}\sin\theta_i t-\sum_{i=1}^{\infty}\frac{A_{gi}\theta_i}{\omega_2\left(\omega_2^2-\theta_i^2\right)}\sin\omega_2 t\right]
\end{aligned}
\tag{22}
$$

It can be seen from Eq. (22) that even under uniform excitation, the structure may produce torsional vibration, and the vibration amplitude is related to the distance between the center of mass and the center of rigidity.

3.3 Relative Translational and Rotational Displacement of the Roof Centroid of a Symmetric Structure

For symmetric structure:

$$
A_1=B_2=1,\quad A_2=B_1=0,\quad \phi_{12}=\phi_{21}=0
\tag{23}
$$

$$\omega_1 = \omega_y, \quad \omega_2 = \omega_\varphi \tag{24}$$

Substituting Eq. (23) into Eqs. (17) and (18), the relative translational and angular displacement solutions of the roof centroid of the symmetric structure with wave passage excitation are obtained as follows:

$$U_{cy}^r = -\sum_{i=1}^{\infty} \frac{A_{gi}\left[\sin\theta_i t + \sin\theta_i\left(t - \frac{D}{v}\right)\right]}{2\left(\omega_1^2 - \theta_i^2\right)} + \sum_{i=1}^{\infty} \frac{A_{gi}\theta_i\left[\sin\omega_1 t + \sin\omega_1\left(t - \frac{D}{v}\right)\right]}{2\omega_1\left(\omega_1^2 - \theta_i^2\right)} \tag{25}$$

$$\phi_c^r = \sum_{i=1}^{\infty} \frac{A_{gi}\theta_i}{v\left(\omega_2^2 - \theta_i^2\right)} \cos\theta_i t - \sum_{i=1}^{\infty} \frac{A_{gi}\theta_i}{v\left(\omega_2^2 - \theta_i^2\right)} \cos\omega_2 t \tag{26}$$

Substituting Eq. (23) into Eqs. (21) and (22), the relative translational and angular displacement solutions of the roof centroid of the symmetric structure with uniform excitation are obtained as follows:

$$U_{cy}^r = -\sum_{i=1}^{\infty} \frac{A_{gi}}{\left(\omega_1^2 - \theta_i^2\right)} \sin\theta_i t + \sum_{i=1}^{\infty} \frac{A_{gi}\theta_i}{\omega_1\left(\omega_1^2 - \theta_i^2\right)} \sin\omega_1 t \tag{27}$$

$$\phi_c^r = 0 \tag{28}$$

It can be seen from Eq. (28) that under uniform excitation, the symmetric structure does not produce torsional vibration.

When every frequency θ_i of the local seismic wave tends to infinity, Eqs. (25), (26) and (27) tend to 0. When every frequency θ_i tends to 0, the above three equations also tend to zero. It can be seen that when every frequency θ_i tends to 0 or infinity, whether it is traveling wave excitation or uniform excitation, the roof relative to the ground is in a static state.

The translational and torsional responses of symmetric structures are independent, each with only one resonance region, while the asymmetric structures are coupled with each other, and each has two resonance regions.

3.4 Roof Torque

According to Eq. (5), the roof torque is:

$$T = -\left(\sum_{n=1}^{N} k_{ny} r_{nx}^2 + \sum_{n=1}^{N} k_{nx} r_{ny}^2\right)\left[\sum_{i=1}^{\infty} \frac{A_{gi}\theta_i}{v\left(\omega_2^2 - \theta_i^2\right)} \cos\theta_i t - \sum_{i=1}^{\infty} \frac{A_{gi}\theta_i}{v\left(\omega_2^2 - \theta_i^2\right)} \cos\omega_2 t\right] \tag{29}$$

Similar to the results of relative displacement of the roof, the roof torque is 0 when every frequency θ_i tends to 0 or infinity with wave passage excitation.

With uniform excitation, the torque of the roof is 0.

3.5 Column Shear Force

The shear force of column n in the y and x directions can be calculated respectively according to the following equation:

$$F_{ny} = k_{ny}U_{cy}^{r} + k_{ny}r_{nx}\phi_{c}^{r} \tag{30}$$

$$F_{nx} = k_{nx}r_{ny}\phi_{c}^{r} \tag{31}$$

Substituting Eqs. (25) and (26) into Eqs. (30) and (31), we can get:

$$\begin{aligned} F_{ny} =& k_{ny}\left\{-\sum_{i=1}^{\infty}\frac{A_{gi}\left[\sin\theta_i t + \sin\theta_i\left(t-\frac{D}{v}\right)\right]}{2\left(\omega_1^2-\theta_i^2\right)} + \sum_{i=1}^{\infty}\frac{A_{gi}\theta_i\left[\sin\omega_1 t + \sin\omega_1\left(t-\frac{D}{v}\right)\right]}{2\omega_1\left(\omega_1^2-\theta_i^2\right)}\right\} \\ &+ k_{ny}r_{nx}\left[\sum_{i=1}^{\infty}\frac{A_{gi}\theta_i}{v\left(\omega_2^2-\theta_i^2\right)}\cos\theta_i t - \sum_{i=1}^{\infty}\frac{A_{gi}\theta_i}{v\left(\omega_2^2-\theta_i^2\right)}\cos\omega_2 t\right] \end{aligned} \tag{32}$$

$$F_{nx} = k_{nx}r_{ny}\left[\sum_{i=1}^{\infty}\frac{A_{gi}\theta_i}{v\left(\omega_2^2-\theta_i^2\right)}\cos\theta_i t - \sum_{i=1}^{\infty}\frac{A_{gi}\theta_i}{v\left(\omega_2^2-\theta_i^2\right)}\cos\omega_2 t\right] \tag{33}$$

Similar to the results of relative displacement of the roof, the column shear force is 0 when every frequency θ_i tends to 0 or infinity with wave passage excitation.

With uniform excitation, the shear forces of each column are:

$$F_{ny} = k_{ny}\left[-\sum_{i=1}^{\infty}\frac{A_{gi}}{\left(\omega_1^2-\theta_i^2\right)}\sin\theta_i t + \sum_{i=1}^{\infty}\frac{A_{gi}\theta_i}{\omega_1\left(\omega_1^2-\theta_i^2\right)}\sin\omega_1 t\right] \tag{34}$$

$$F_{nx} = 0 \tag{35}$$

It can be seen from Eq. (35) that with uniform excitation, each column does not produce shear in the non excitation direction.

4 Calculation Example

Take the column grid of the structure in Fig. 1 as the two forms shown in the following figure. The roof mass of structure 1 (Fig. 2) is taken as 3.26 × 105 kg, rotational inertia is 1.09 × 108 kg · m^2, and the column height is 4.5 m. The natural frequence on translation degrees is 0.183 Hz and on rotation degrees is 0.181 Hz.; The roof mass of structure 2 (Fig. 3.) is taken as 1.31 × 105 kg, rotational inertia is 2.22 × 107 kg·m^2, the column height is 2.9 m, and the section size of each column is 400 mm × 400 mm. The natural frequence on translation degrees is 16.539 Hz and on rotation degrees is 17.845 Hz. El Centro wave and Taft wave are selected as seismic waves.

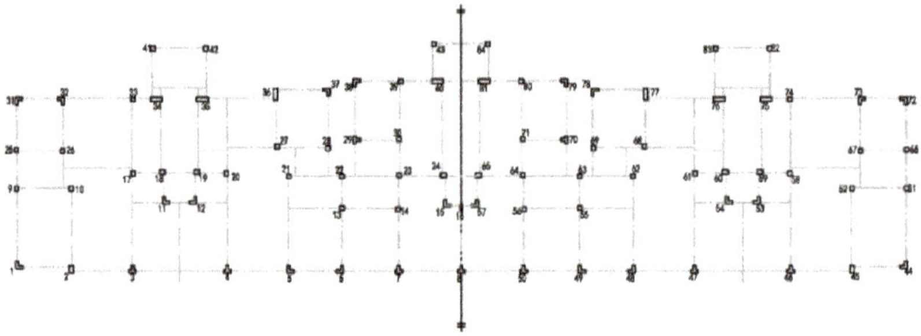

Fig. 2. Column grid of structure 1

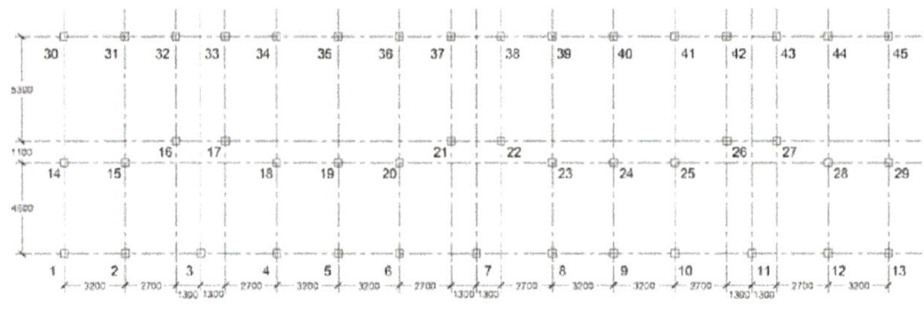

Fig. 3. Column grid of structure 2

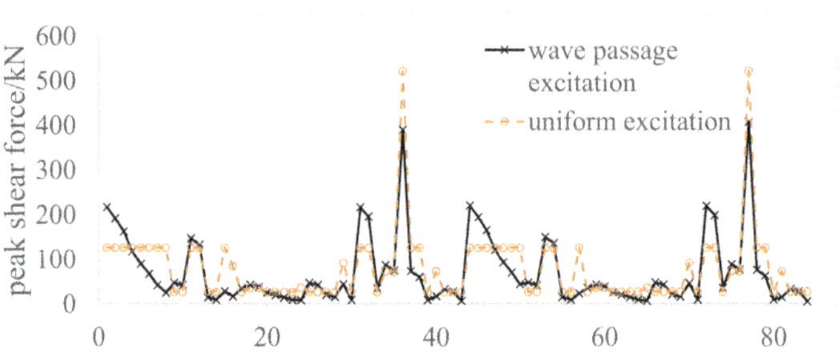

Fig. 4. Structure 1 with El-centro

4.1 Column Shear Force in Y Direction

For the two structures with the excitation of two seismic waves, the comparison of peak column shear in Y direction between wave passage excitation and uniform excitation is shown in Fig. 4, Fig. 5, Fig. 6, Fig. 7.

It can be seen from the above figure that, compared with the uniform excitation, the peak shear forces of the columns near the centroid decrease in the propagation direction with the wave passage excitation, and the reduction amplitude increases with the decrease of distance, which decreases by 79.5%, 81.5%, 23.6% and 31.2% respectively; the peak shear force of columns far from the centroid increases, and the increase amplitude

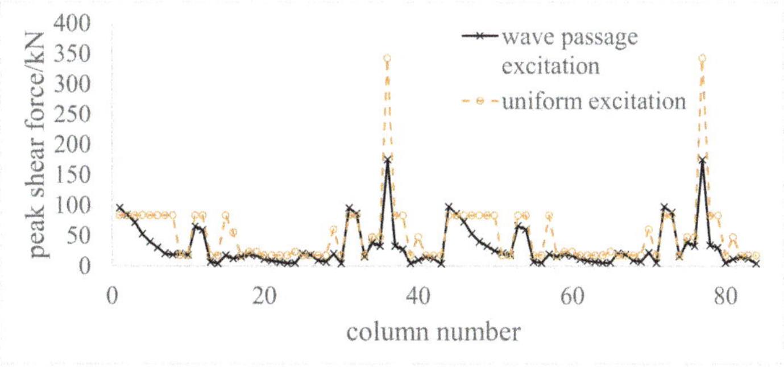

Fig. 5. Structure 1 with Taft

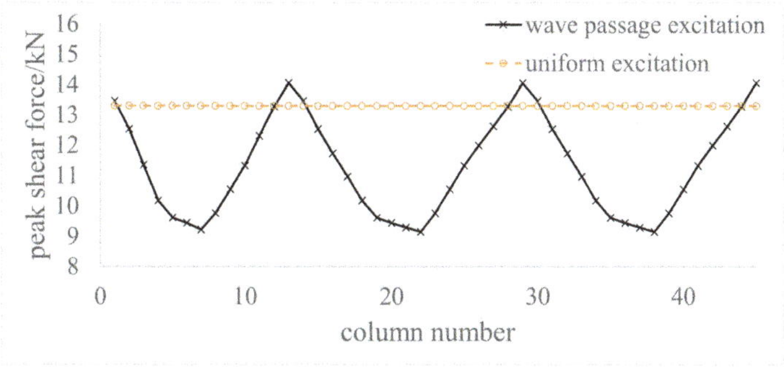

Fig. 6. Structure 2 with El-centro

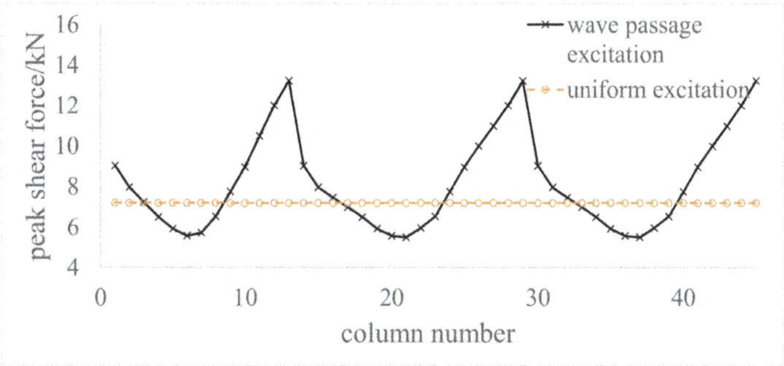

Fig. 7. Structure 2 with Taft

increases with the increase of distance, which decreases by 17.2%, 73.1%, 83.6% and 5.6%.

The influence of wave passage excitation is different for different structures with the same seismic wave and the same structure with different seismic waves.

With wave passage excitation, the peak shear force of the column on the right side of the symmetry axis is higher than that of the corresponding column on the left side up to 60% and 55% higher respectively.

4.2 Column Shear Force in X Direction

For the two structures with the excitation of two seismic waves, the comparison of peak column shear in X direction between wave passage excitation and uniform excitation is shown in Fig. 8, Fig. 9, Fig. 10, Fig. 11.

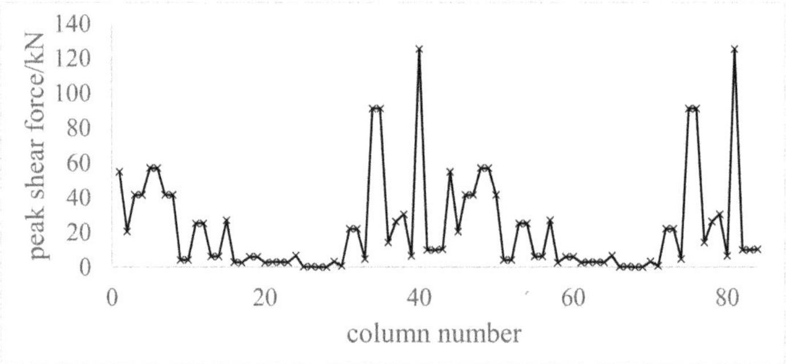

Fig. 8. Structure 1 with El-centro

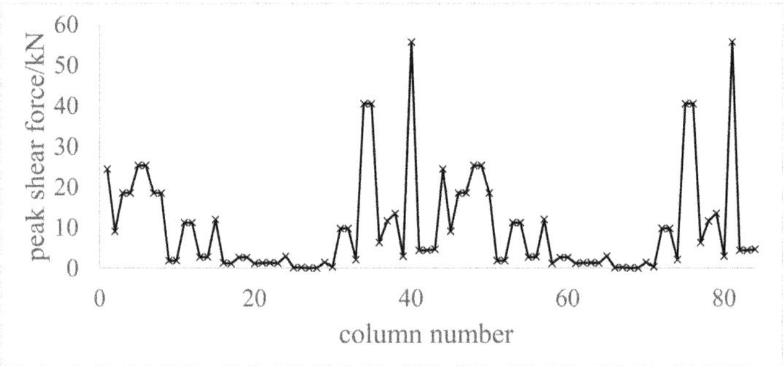

Fig. 9. Structure 1 with Taft

It can be seen from the above figure that with the wave passage excitation, the peak shear forces of the columns near the centroid decreases in the excitation direction, and the reduction amplitude increases with the decrease of distance; The peak shear forces of the columns far from the centroid increases, and the increase amplitude increases with the increase of distance.

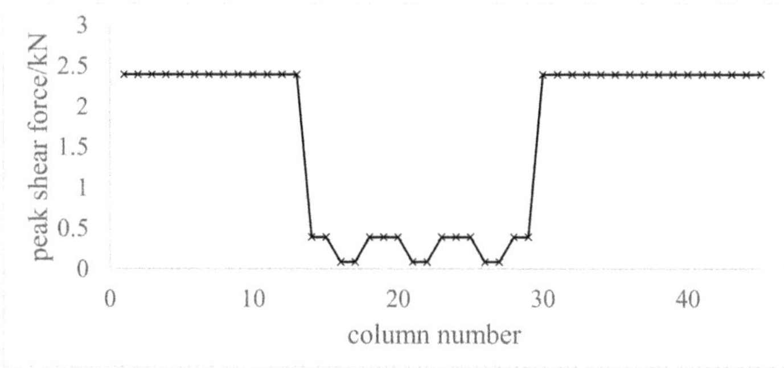

Fig. 10. Structure 2 with El-centro

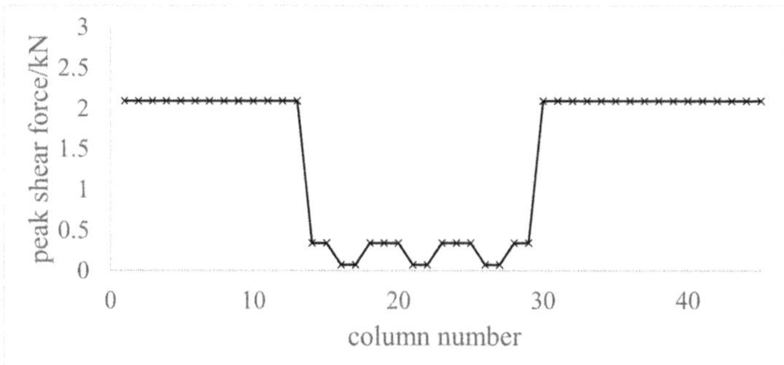

Fig. 11. Structure 2 with Taft

5 Conclusion

In this paper, the dynamic equilibrium equation is established for the single-story frame structure subjected to wave passage excitation.

The translational acceleration and rotational acceleration of rigid foundation with wave passage excitation are substituted into the equilibrium equation, and the analytical solutions of roof torque and column shear force are given.

The roof torque and column shear force of two single-story structures with the excitation of two seismic waves are analyzed, and the following conclusions are obtained:

1. Torsional vibration of the structure is induced by wave passage excitation.
2. The column vibrates along the non excitation direction, and the amplitude of the vibration is related to the distance from the column to the roof centroid in the excitation direction.
3. Compared with the uniform excitation, the peak shear forces of the columns near the centroid decrease and the peak shear force of columns far from the centroid increases.
4. The peak shear forces of the distal columns in the direction of wave passage propagation are bigger than those of the proximal columns (facing the wave input).

5. Compared with the uniform excitation, the shear force in the excitation direction is smaller under the traveling wave excitation, but the resultant force of the shear force may be larger.

References

1. Rosenblueth, E., Meli, R.: The 1985 earthquake: causes and effects in Mexico City. Concr. Int. **8**, 23–24 (1986)
2. Newmark, N.M.: Torsion in symmetrical buildings. In: Proceedings of the 4th World Conference on Earthquake Engineering, Santiago, Chile, pp. A3, 19–32 (1969)
3. De La Llera, J.C.: Accidental torsion in buildings due to stiffness uncertainty. Earthq. Eng. Struct. Dyn. **23**, 117–136 (1994)
4. Cao, Y.: Effects of wave passage on torsional response of symmetric buildings subjected to near-fault pulse-like ground motions. Soil Dyn. Earthq. Eng. **88**(2016), 109–123 (2016)
5. Sarokolayi, L.K.: Modeling of ground motion rotational components for near-fault and far-fault earthquake according to soil type. Arab. J. Geosci. **8**(6), 3785–3797 (2014)
6. Tso, W.K.: Induced torsional oscillations in symmetrical structures. Earthq. Eng. Struct. Dyn. **3**, 337–346 (1975)
7. Hao, H.: Multiple excitation effects on response of symmetric buildings. Eng. Struct. **18**(9), 732–740 (1996)
8. Hao, H.: Seismic response of asymmetric structures to multiple ground motions. J. Struct. Eng. **1995**(121), 1557–1564 (1995)
9. Tso, W.K.: Seismic torsional provisions for dynamic eccentricity. Earthq. Eng. Struct. Dyn. **8**, 275–289 (1980)
10. De La Llera, J.C.: Accidental torsion in buildings due to base rotational excitation. Earthq. Eng. Struct. Dyn. **23**, 1003–1021 (1994)
11. Shakib, H.: Evaluation of accidental eccentricity in buildings due to rotational component of earthquake. J. Earthq. Eng. **6**(4), 431–445 (2002)
12. Hao, H.: Torsional response of building structures to spatial random ground excitations. Eng. Struct. **19**(2), 105–112 (1997)
13. Li, W.: Compositions of time histories of acceleration and velocity and dis-placement of ground motion. Eng. Mech. **37**(Suppl), 164–167 (2020). (in Chinese)

Simulation Study of Fire Spread on Exterior Wall of Twin-Tower High-Rise Building Linked with Connective Structures

Zengchao Song[1](✉), Yansheng Song[1], Jian Su[2], and Yan Zeng[1]

[1] Shenyang Jianzhu University, Shenyang 110168, China
2274205126@qq.com, ceyssong@sjzu.edu.cn
[2] Dalian Ocean University, Dalian 116023, China
sj@dlou.edu.cn

Abstract. When a fire occurs in a Twin-tower high-rise building linked with connective structures, it will spread laterally through the connective structures, causing greater damage. In this paper, PyroSim is used to simulate the whole process of the "8·27" fire in Dalian Kaixuan International Building. The results show that within 50s, the compartment fire overflow occurs, the flame spreads upward at a speed of 0.79 m/s, and spreads to the top of the tower at 86 s. It spreads horizontally at a speed of 0.54 m/s, and 126 s spreads to another tower; The local maximum temperature of the outer surface of the tower fire area is 1200 °C. The maximum temperature of the outer surface of the corridor is 1143.13 °C. The concentration of CO in all rooms of the corridor in the burned area exceeded the critical concentration leading to death. When the flame passes through the corridor to another tower, the maximum temperature is 524.40 °C lower than before.

Keywords: Twin-tower Linked with Connective Structures · High-rise Building · Fire Spread · PyroSim

1 Introduction

Twin-tower high-rise building linked with connective structures has a beautiful architectural appearance, open vision and can save a lot of land. More and more high-rise buildings have begun to adopt this architectural form, such as CCTV Building, Singapore Sands Hotel, Suzhou Oriental Gate and so on. However, the ensuing problem is that once a fire occurs in the connected structure of a high-rise building, it will involve two towers at the same time and it is difficult to extinguish the fire. Therefore, the fire and smoke diffusion law of connected twin-tower high-rise buildings is an important research content of fire safety. The conclusions drawn in this paper can provide reference for the fire management of the connected twin-tower high-rise buildings.

Xu Yanqiu [1] used polyurethane foam to replace the thermal insulation material of the actual high-rise building exterior wall, and simulated a 28-story teacher's apartment by direct ignition. The fire spread trend, air temperature, thermal radiation and smoke concentration were analyzed. The law of flame spread, the influence of thermal radiation

I. Tuns et al. (Eds.): CIBv 2023, LNCE 510, pp. 263–273, 2024.
https://doi.org/10.1007/978-3-031-60765-3_22

on the human body and the surface of the surrounding buildings, and the law of smoke spread were obtained. Dai Changqing [2] studied the high-rise fire smoke spread law, smoke temperature field and other characteristics, obtained the smoke spread rate, carbon monoxide and carbon dioxide concentration variation, ambient temperature distribution.

Quintiere [3] studied the fire characteristics of various building decoration materials, and obtained the influence of material thickness and ignition source characteristics on upward fire spread. Zhang Taolin and Zhou Xiaodong [4] studied the characteristics of flame spread rate, weight loss rate and temperature field distribution of polystyrene foam (EPS), one of the high-rise external wall insulation materials, obtained the flame spread law under single ignition and double ignition conditions, and analyzed the factors of mass loss rate change. Eric [5] obtained the difference between the two by comparing and analyzing the real fire and the simulated fire. Liu [6] explored the law of fire smoke spread in high-rise buildings under the opening and closing conditions of fire corridors and automatic fire extinguishing systems, and obtained the influence of fire corridors and automatic fire extinguishing systems on smoke spread. Wang Hairong [7] explored the law of fire smoke spread and the characteristics of personnel evacuation behavior in high-rise buildings, and obtained the key time period and evacuation method of personnel evacuation in fire. Shi Jianyong [8] explored the fire characteristics of three typical exterior insulation materials and the fire characteristics of high-rise exterior insulation materials, and obtained the fire spread law of the adhesive layer with or without cavity and the fire prevention strategy of high-rise building exterior insulation materials. Hao [9] explored the law of smoke spread in apartment fires and the influence of smoke on personnel evacuation, and obtained the distribution law of smoke under the influence of chimney effect. Wang Weigang [10] explored the environmental changes of CO concentration, smoke distribution and visibility distribution in high-rise residential buildings during fire, and obtained the influence of the three on personnel evacuation and life and health and the law of fire development over time. Chen Shuyang [11] explored the law of smoke spread when a fire broke out in a student dormitory building, and obtained the distribution law of CO concentration, visibility and temperature.

Based on the above basis, the fire diffusion model of the connected twin-tower high-rise building was established in PyroSim, focusing on the tower to study the overall fire spread law on the external wall insulation material, the fire temperature field of the connecting wall and the CO concentration distribution law.

2 Model Establishment

2.1 Establishment of PyroSim Numerical Model

The height of Dalian Kaixuan Building is about 98 m, and the two towers are connected by two steel structure connectors on the top floor. The building area is 63,000 m^2, 2 floors underground and 29 floors above ground. The tower consists of 28 floors, and the 25th to 28th floors of the tower are connected by steel structure connectors. The fire occurred in a household on the 19th floor of Tower B. The cause of the fire was determined to be caused by the poor contact between the power cord plug and the socket of the electric balance vehicle charger, which ignited the surrounding wood wardrobe and other combustibles. After breaking through the window, the fire ignited the aluminum plastic plate and foamed polyurethane insulation layer of the outer curtain wall of Tower B, causing the fire to expand (Fig. 1).

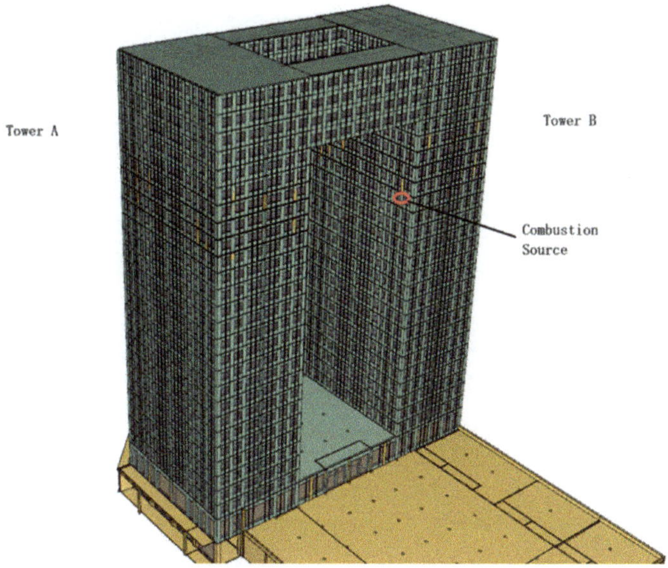

Fig. 1. 3D model diagram

2.1.1 Fire Scene Settings

According to the calculation formula of the characteristic diameter of the fire source [12]

$$D^* = \left(\frac{Q}{\rho c T \sqrt{g}}\right)^{2/5} \tag{1}$$

In the formula: D^* is the characteristic diameter of fire source, m; Q is the heat release rate of the fire source, kW; ρ is ambient air density, kg/m^3; c is the specific heat of air at constant pressure, kJ/(kg·K); K is ambient air humidity, K; g is the acceleration of gravity, m/s^2.

The heat release rate of the fire source is set to 4 MW, and the characteristic diameter of the fire source is calculated to be $D^* = 2.0$. Referring to the paper, the reasonable value range of the grid size is 0.0625 D^* ~ 0.25 D^*. We have chosen the default grid size values of 0.5 m × 0.5 m × 0.5 m, the total number of grids is 922720, and the simulation time is set to 500 s.

The cladding is foamed polyurethane insulation layer and aluminum plastic plate. The thickness of the decorative layer of the aluminum-plastic plate is 4 mm, and the thickness of the foamed polyurethane insulation layer is 40 mm. The initial temperature of the environment is 20 °C, and the southwest wind speed is 4m/s.

2.1.2 Layout of Measuring Points

The vertical interval of the measuring points is 3 m (Fig. 2 and Tables 1, 2).

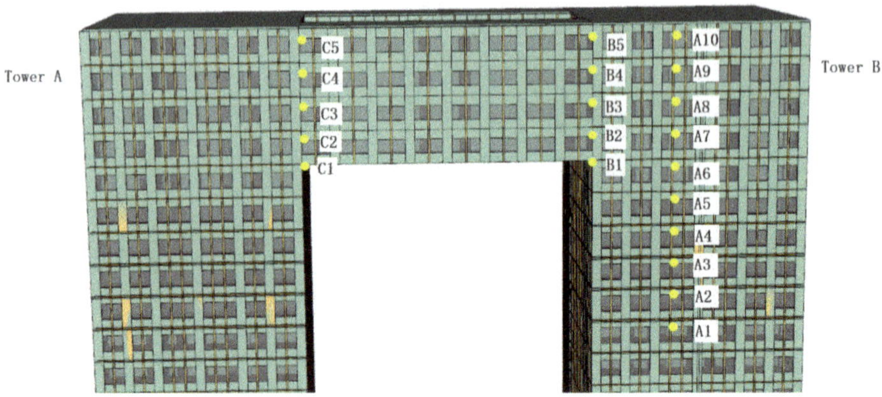

Fig. 2. Thermocouple measuring points layout

Table 1. Measuring points position

Measuring point position	Outer window of 19–28 floors of B tower	The right end of the corridor	The left end of the corridor
Measuring point number	A1–A10	B1–B5	C1–C5

Table 2. Fire scene layout

Fire scene	Fire source location	Fire type	Fire heat release rate (MW)	Tower window open state	Mechanical smoke exhaust open state	Indoor door open state
1	Inside the room on the left side of the 19th floor of Tower B	Fast fire	6	NO	NO	YES

(Note: NO means closed state, YES means open)

3 PyroSim Simulation Results Compared with Reality

According to the failure mode of a connected twin-tower high-rise building in Dalian after fire, the fire area is roughly in the shape of '7'. Due to the burning and falling of external thermal insulation materials, the flame spreads to the lower part of the 19th floor. As shown in Fig. 4, the simulated fire spread pattern is roughly consistent with the reality, lays a foundation for studying the whole process of fire spreading on the external thermal insulation layer of the outer wall of the connected twin-tower high-rise building and the concentration of CO (Fig. 3).

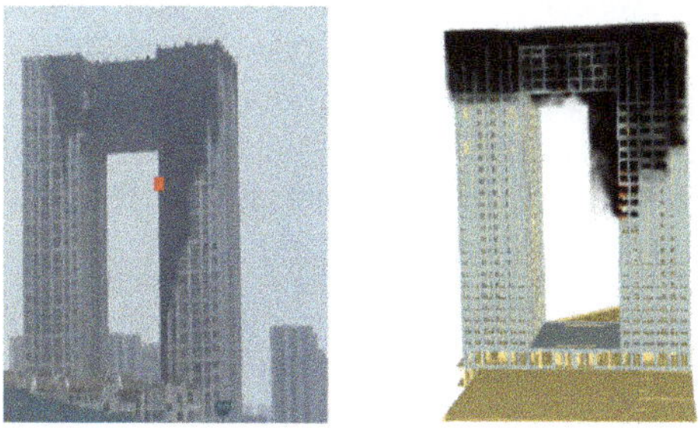

Fig. 3. Comparison of real and simulated fire spread patterns

4 Fire Diffusion Analysis Based on PyroSim

4.1 Fire Spread Law

It can be seen from Fig. 4 and Fig. 5 that the compartment fire overflow occurred within 50 s, igniting the external thermal insulation material. The flame spread to the A1 measuring point of 19 layers at 52 s, and spread to the A10 measuring point at 86 s. The vertical propagation path of the flame is 27 m. Therefore, the vertical upward flame spread speed is about 0.79 m/s. From the rising rate and maximum temperature of each curve, it can be seen that under the influence of natural wind, the maximum temperature of 19 layers is 549.30 °C. From 20 layers to 28 layers, the fire spread speed gradually accelerated, and the maximum temperature remained at about 1200 °C.

From the curves in Fig. 4 and Fig. 5, it can be seen that after about 100s, the outer surface of Tower B has cooled, and then at 225 s, the temperature rises to more than 100 °C, and there may be a secondary reignition phenomenon.

It can be seen from Fig. 6 and Fig. 7 that the time for the fire to spread to the B5 measuring point is 76 s, and the maximum temperature is 1146.13 °C. The time for the fire to spread horizontally to the C5 measuring point is 126 s, and the maximum temperature is 618.73 °C. The distance between B5 and C5 is 27 m, and the speed of fire horizontal spread is 0.54 m/s. The maximum temperature of the fire is reduced by 524.40 °C. It can be seen from all the measuring points on the corridor that the horizontal spread of fire is a process of temperature attenuation.

From the above data, we can see that the vertical spread of fire is faster than the horizontal spread. When the flame spreads vertically upward, the temperature will gradually increase at the beginning, and then stabilize near a maximum of 1200 °C. Moreover, at the measuring point, the time interval between the maximum temperature of the fire and the cooling time will not exceed 100 s, which means that the fire will quickly burn out the external insulation layer and continue to spread.

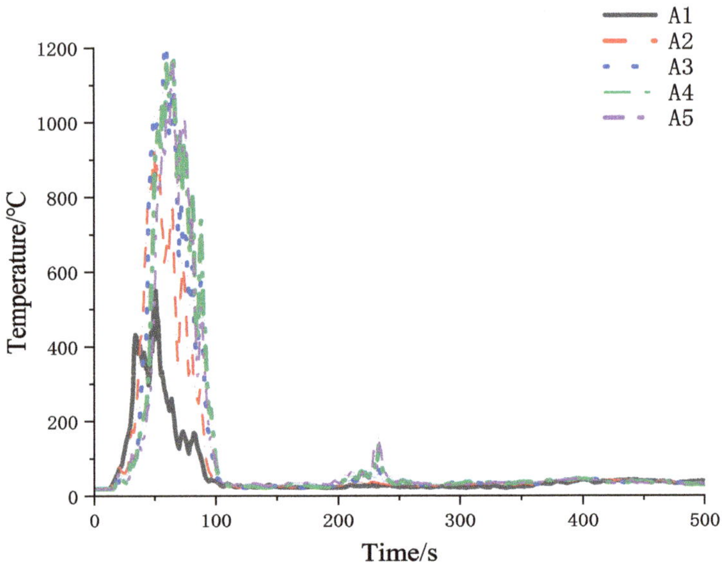

Fig. 4. A1–A5 temperature change curves

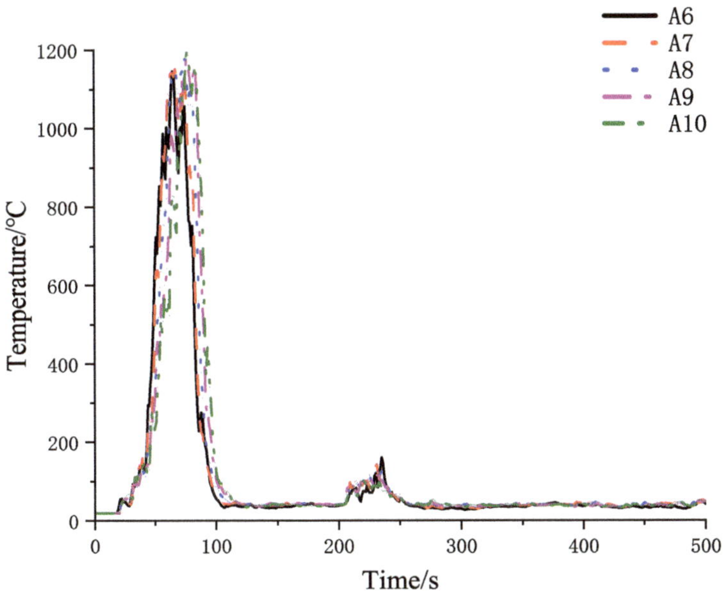

Fig. 5. A6–A10 temperature change curves

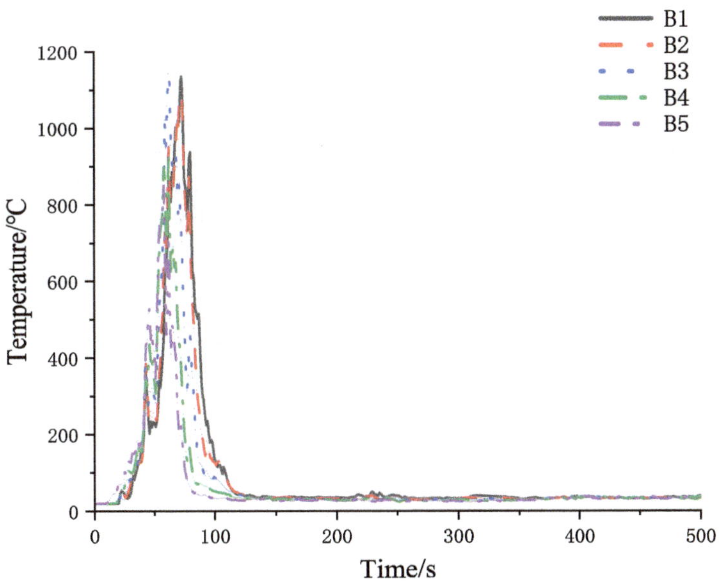

Fig. 6. B1–B5 temperature change curves

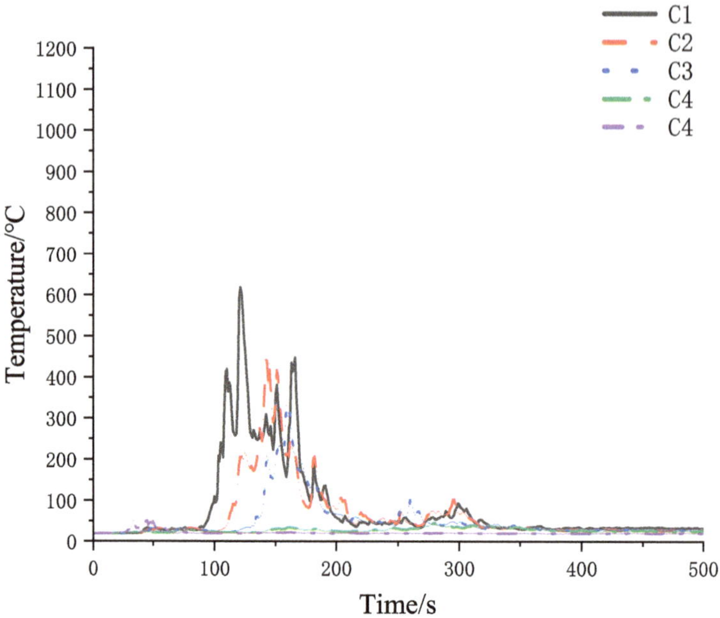

Fig. 7. C1–C5 temperature change curves

Figure 8 shows the mass loss rate in the simulation process.

It reflects the pyrolysis rate and pyrolysis behavior of polymer materials under certain heat flow conditions. It can be seen that the quality loss of external thermal insulation materials experienced heating stage, exponential rising stage, gentle decline stage and secondary rising stage. At 100 s, both the tower and the connecting external insulation material were pyrolyzed, and the mass loss rate reached the first peak. At 300 s, the curve rises again. At this time, the flame has spread to the A tower, and the combustible materials in the fire area are pyrolyzed in a large area. The fire of double-connected double-tower high-rise building should be controlled before the second rise of mass loss rate, otherwise the fire scale is too large and distributed at the height that is difficult to reach by ordinary fire engines, and it is more difficult to extinguish the fire in high-rise rooms.

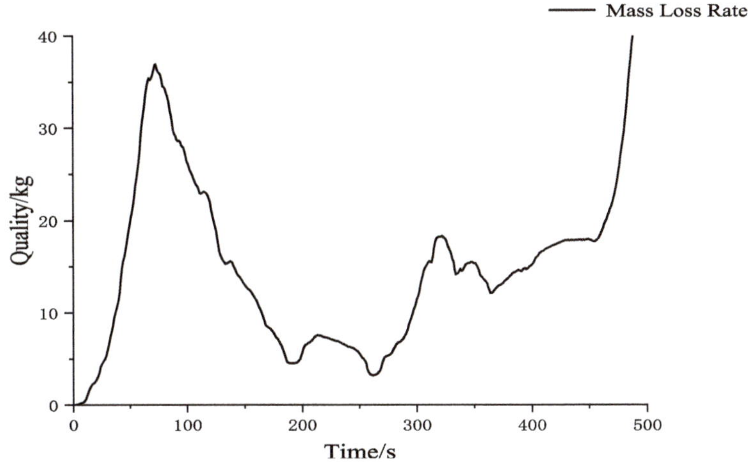

Fig. 8. Mass Loss Rate

5 Distribution of Flue Gas and CO Concentration

5.1 CO Concentration Distribution

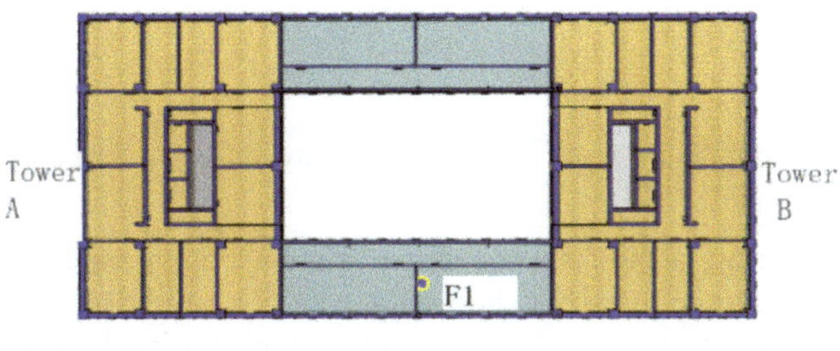

Fig. 9. Distribution of CO concentration measuring point in corridor room

CO is one of the main combustion products that cause death in fires. According to the conditions of dangerous critical state in performance-based fire protection design, the alarm concentration of CO is 300 ppm (1ppm = 1×10^{-6} mol/mol) in China. When the concentration reaches 1000 ppm, the person will lose the ability to escape, and the critical concentration of death caused by human inhalation of CO is 2500 ppm [13]. Here, the CO concentration in the 25–28 story corridor room is compared. The measuring points of each layer are arranged in the same position, the height is 1.6 m from the ground, the measuring point of the 25 layer is F1, the 26 layer is F2, and so on (Fig. 9).

It can be seen from Fig. 10 that the CO concentration in the corridor room far more than 2.5×10^{-3} mol/mol (2500 ppm = 2.5×10^{-3} mol/mol) with the spread of the fire, the maximum value reached 3.1×10^{-2} mol/mol, which exceeded the critical concentration of causing death and lasted for about 100 s. The CO concentration in the room on the first floor of the corridor is lower than that in other floors of the corridor and is unstable.

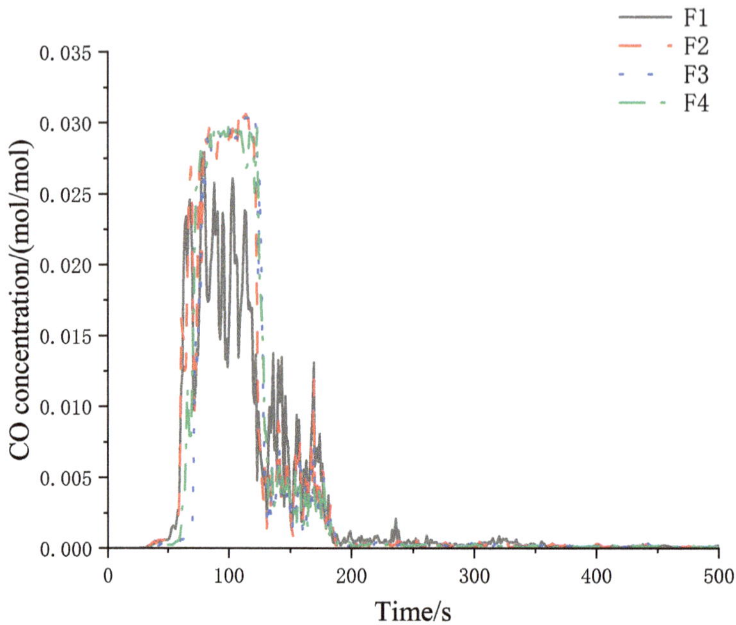

Fig. 10. CO concentration curves in corridor rooms

6 Conclusion

According to the simulation results and the actual fire situation, the conclusions are as follows:

(1) In general, the fire area of the connected twin-tower high-rise building is in the shape of '7', and the fire spread horizontally in the corridor is a cooling process. The vertical spread and upward spread of fire is a process in which the temperature rises first and then stabilizes to the maximum value. Spreading the fire to another tower will cause another large area of burning. The re-ignition phenomenon that may occur at 225 s will pose a great threat to the life safety of firefighters.

(2) The second peak of the mass loss rate of the external thermal insulation material of the connected twin-tower high-rise building is due to the fact that the fire will ignite the combustibles in the room on a large scale after 300 s and spread along the connector to Tower A.

(3) The CO concentration in the corridor room will rapidly reach the critical value of causing death with the spread of the fire and last for about 100 s. The CO concentration in the first floor of the corridor is lower and unstable than that in other floors.

References

1. Xu, Y., Wang, Z., Li, Z.: Numerical simulation and analysis of outdoor fire in high-rise buildings. Secur. Technol. **112**(06), 8–11 (2011)

2. Dai, C., Yuan, H.: Numerical simulation analysis of high-rise residential building fire based on FDS. J. Anhui Jianzhu Univ. **25**(04), 14–18 (2017)
3. Quintiere, J.G., Haynes, G., Rhodes, B.T.: Applications of a model to predict flame spread over interior finish materials in a compartment. J. Fire Protect. Eng. **7**(1), 1–13 (1995)
4. Zhang, T., Zhou, X., Lei, G., Wang, W., Gong, J., Yang, L.: Numerical simulation of fire spread characteristics of typical exterior wall thermal insulation materials for high-rise buildings. J. Disast. Prevent. Mitigat. Eng. **32**(02), 230–234 (2012)
5. Eric, G., Virginie, D.: Reconstruction of Grenfell tower fire. Part 3—numerical simulation of the Grenfell tower disaster: contribution to the understanding of the fire propagation and behaviour during the vertical fire spread. Fire Mater. **44**(1), 35–57 (2020)
6. Liu, X., Zhao, W., Li, X., Chen, J., Xu, X.: Simulation study on fire spread law of high-rise residential buildings with corridors. J. China Jiliang Univ. **32**(02), 260–266 (2021)
7. Wang, H., Chen, Q., Yu, Y., Yan, J., Liang, D.: Fire simulation and escape behavior of high-rise buildings. J. Sun Yat-sen Univ. (Nat. Sci. Edn.) **53**(03), 150–154 (2014)
8. Shi, J., Liu, P., Xu, Q.: Fire characteristics and fire prevention strategies of high-rise residential exterior insulation. Fire Sci. Technol. **32**(11), 1220–1223 (2013)
9. Hao, Y., Zhang, Q., He, F.: Smoke simulation and evacuation of apartment fire based on Pyrosim and Pathfinder. In: Environmental Engineering Branch of Chinese Society of Environmental Sciences. 2022 Annual Conference of Science and Technology of Chinese Society of Environmental Sciences-Environmental Engineering Technology Innovation and Application Branch Field Papers (1), p. 5. ‘Industrial Architecture’ Magazine Co., Ltd., ‘Environmental Engineering’ Editorial Department (2022)
10. Wang, W., Wen, H., Jia, Y.: Fire simulation of high-rise residential buildings based on FDS. J. Xi’an Univ. Sci. Technol. **40**(02), 314–320 (2020)
11. Chen, S., Yu, T.: Numerical simulation of a student dormitory building fire based on Pyrosim. Value Eng. **41**(36), 146–148 (2022)
12. McGrattan, K., McDermott, R., Weinschenk, C., Forney, G.: Fire Dynamics Simulator, Technical Reference Guide, 6tth Edition. Special Publication (NIST SP), National Institute of Standards and Technology (2013)
13. Duan, H., Zhao, Y., Zhao, J., et al.: Numerical simulation of fire spread in external wall thermal insulation system of high-rise residential building. J. Disast. Prevent. Mitiga. Eng. **35**(01), 23–27 (2015)

Seismic Performance Analysis of a Connected Twin-Tower Structure Connection Mode

Jian Gao[1(✉)], Yansheng Song[1], and Jian Su[2]

[1] Shenyang Jianzhu University, Shenyang 110168, China
1148859009@qq.com, ceyssong@sjzu.edu.cn
[2] Dalian Ocean University, Dalian 116023, China
sj@dlou.edu.cn

Abstract. In order to determine the connection scheme of the high-level connection of a connected twin-tower structure, four models of different connection forms are established. Model I is rigidly connected to both ends of the tower. Model II is hinged with tower A and rigidly connected with tower B. Model III is slidingly connected with tower A and rigidly connected with tower B. Model IV is hinged with both ends of the tower. The elastic dynamic time-history analysis of structures of four different connection modes under earthquake action is carried out in ABAQUS. The results show that the connection mode of rigid connection at both ends under earthquake can not only reduce the structural displacement, but also realize the stiffness coordination between towers, and has the smallest the Inter-story Drift Ratio and vertex displacement, and the maximum displacement of key beam-column joints is small, and the damage of each connection mode is similar. Therefore, for the connected twin-tower structure, the rigid connection at both ends of the high-position connection is more favorable to the overall structure.

Keywords: Twin-tower Link with Connective Structures · Connection Mode · Dynamic Time History Analysis · Earthquake resistant behavior

1 Foreword

With the continuous development of China's economy and society, people's requirements for architectural forms and architectural functions are getting higher and higher, which makes the structural forms have made great progress in the development of buildings, and various new structural forms have developed rapidly. Among them, the connected structure has gradually become a common structural form in high-rise buildings. The connected structure is often used to connect the internal space of two or more towers at a certain height. On the one hand, the connected structure has a unique shape and good visual effect, which can make the building more distinctive. On the other hand, the connected part generally has a good lighting effect and a broad vision, so it can also be used as a sightseeing corridor and a leisure place to meet the connection function between the towers. However, at the same time, the interaction between the corridor and the tower has a significant impact on the seismic performance of the whole structure, so the connection scheme of the corridor connector is the key point in the design of the connected structure.

I. Tuns et al. (Eds.): CIBv 2023, LNCE 510, pp. 274–293, 2024.
https://doi.org/10.1007/978-3-031-60765-3_23

2 Connection Mode

The connection between the connected structure and the tower is divided into rigid connection, hinged connection and elastic connection. Rigid connection is a kind of strong link between the connecting body and the rigid connection of the tower. In the rigid connection, the connecting body not only bears its own load, but also transmits the bending moment, shear force and axial force at the connection with the tower. The more important role is to coordinate the effects caused by displacement and vibration between the towers. Because of its unique mechanical properties, articulated connection limits the relative movement between the connector structure and the tower, but releases the local bending moment of the connection part, so it is also a strong connection form. The relative rotation of this connection form reduces the negative bending moment value at both ends of the connection, so as to avoid the failure of the connection due to excessive force at both ends, which leads to a sharp increase in the bending moment value in the middle of the connection, and reduces the integrity of the two to a certain extent. The sliding connection of one or both ends is a weak connection, and the weak connection mode can realize that the connector will not break prematurely when the tower and the connector work together. In the design, it should be considered to avoid the slip of the connector and the collision of the towers under the action of seismic forces [1].

3 Research Status

When the connectors and the main structure are rigidly connected, the calculation, analysis and structure are easy to grasp, and the engineering construction is also easy to realize. Gong Haijun et al. [2] conducted seismic analysis on the long-span connected high-rise structure of Shijiazhuang Gate, and found that the two towers of the building have similar structural forms and the same height, and the tower and the high corridor structures are relatively regular and symmetrical. The analysis results show that the bearing capacity, deformation capacity and ductility of the whole structure are good. Xu Weibin et al. [3] conducted a special analysis of the strong connection scheme of a connected structure in Ningbo, analyzed its adverse impact on the towers on both sides, and took anti-seismic strengthening measures for the connected floor slab and the components on both sides. Li Jing et al. [4] adopted a rigid connection scheme to save the construction and maintenance cost of the complex support structure with sliding connection for the two tower structures with large height difference in Yinlong Tower in Shaanxi Province. Lin Yu et al. [5] took a twin tower conjoined structure in Chengdu as an example to analyze the influence of low rigid connection corridors on the tower's dynamic characteristics under frequent and rare earthquakes, and the results showed that when towers with similar structural forms had different heights, the adoption of corridors with sufficient stiffness could meet the goal of structural safety and reliability. Qu Hong et al. There are references [6, 7] that have calculated and analyzed connected structures with weak connection scheme, and it is found that the stiffness change of the connecter has little influence on the overall natural vibration period and mode, but has obvious influence on the dynamic response of the structure.

Although the strong connection scheme between towers with large differences in height, stiffness or structural form can simplify the connection construction, the increase in corridor stiffness also increases the construction cost. In addition, the strong connection scheme makes the overall mechanical characteristics under earthquake action complicated, the torsional vibration types rich, and the connection parts easy to produce weak layers, which may cause relatively serious earthquake damage [8]. Therefore, for such cases, the corridor weak connection scheme can play its advantages. Xiao Fei et al. [9] set friction pendulum supports at both ends of the corridor of three connected structures with large volume differences to achieve weak influence on the main body of the tower, and the corridor and tower can be separately calculated for seismic design. Lu Qing et al. [10] conducted a simulated seismic shaking table test on a four-tower connected structure with friction pendulum isolation support, and the results showed that the friction pendulum support and the damper had good energy dissipation and shock absorption effect, and the corridor was in elastic working state under rare earthquakes. Feng Ziqiang et al. [11] designed the corridor structure of split-level connection of a certain five ring towers in Shanghai with two-way sliding connection at both ends, and strengthened the anti-fall measures at the same time. This connection scheme can reduce the difficulty of strong connection design. For the sliding connection, it has become a technical difficulty to accurately calculate the sliding distance of the sliding support and the safe distance between the corridor and the main body of the tower to avoid the collision between them under the action of earthquake [12]. In addition, the structural design and bearing capacity calculation of weakly connected supports also need attention.To sum up, the selection of corridor connection scheme is related to the structure form, stiffness distribution, force characteristics and setting conditions of multi-tower connected structures. For the connected structures whose connection mode cannot be determined by qualitative analysis, multi-scheme calculation and analysis are required to determine the optimal connection scheme. There are few cases and results for calculation and comparison of different corridor connection schemes [13–15], and there is a lack of analysis on the influence of irregular corridor connection schemes on the seismic performance of the overall structure [16].

In this paper, a symmetrical twin tower connected structure is taken as an engineering case, the calculation model of four kinds of high level connection schemes is established, the elastic dynamic time-history analysis of structures is carried out under earthquake action, and the design suggestion of high level steel connected scheme is given.

4 Project Profile

The project is located in Dalian, with a total floor area of 997.92 m^2. The project is composed of two towers, Tower A and Tower B, both of which are 29.7 m high. The overall building connects the two towers with each other by placing two horizontal high steel conjunctions at 19.8 m. The size of the connected main beam is I500 × 400 × 28 × 26, the transverse secondary beam is I350 × 200 × 24 × 22, and the interlayer contact member is the column I500 × 400 × 28 × 26, and the steel is Q345 and HRB400, the concrete is C30. The conjunctions are connected to the steel and concrete frame of the tower. As the connectors are two symmetrically left and right, as shown in Fig. 1, it is difficult to determine the strong or weak connection mode through qualitative analysis for the high-rigidly connected structures between towers A and B. Therefore, this paper will establish several analysis and comparison models to determine the optimal connection mode.

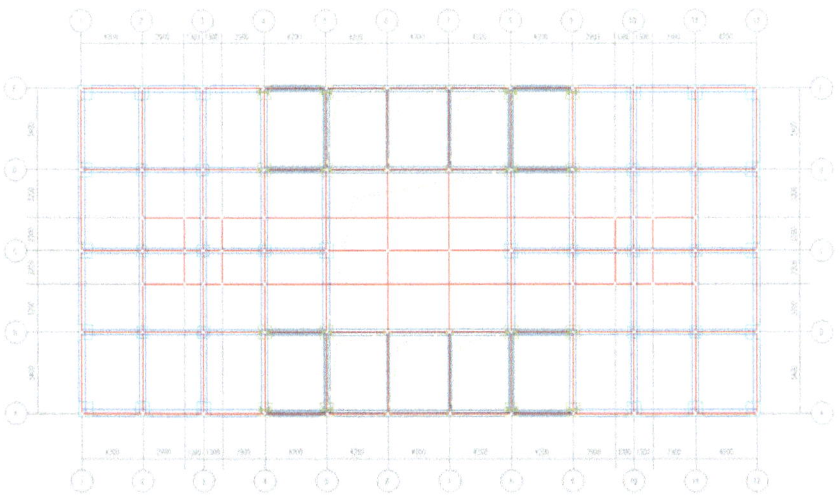

Fig. 1. Architectural plan

5 Calculation Model

PKPM is a project management software developed by the Institute of Architectural Engineering Software, China Academy of Building Research. Is a series, in addition to building, structure, equipment (water supply and drainage, heating, ventilation and air conditioning, electrical) design in one of the integrated CAD system, the current PKPM and building budget series (steel calculation, engineering quantity calculation, engineering valuation), construction series software (series, safety calculation series, construction technology series), construction enterprise information. PKPM has an absolute advantage in the domestic design industry. It has tens of thousands of users and a market share of more than 90%. It has become the most widely used CAD system in China. It keeps up with the needs of the industry and the update of norms, and constantly develops software products that have a huge impact on the industry, so that domestic software with independent intellectual property rights has occupied the dominant position in the application and technology of China's structural design industry for more than ten years. Timely to meet the needs of the rapid development of China's construction industry, significantly improve the design efficiency and quality. PMSAP is a core integrated design of PKPM software architecture design module (Fig. 2).

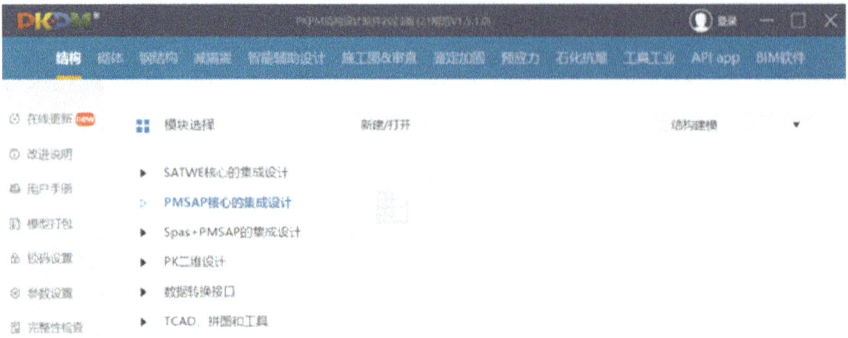

Fig. 2. The interface of PKPM software

PMSAP analysis and design software and ABAQUS CAE software were used to establish four different connection modes of high steel connections between blocks A and B, respectively, as shown in Fig. 3 and 4. Four kinds of models were established by PMSAP, then the models were converted into CAE models by PA-TRANS software, and the elastic dynamic time-history analysis under earthquake was analyzed by ABAQUS software. Model II is hinged with tower A and rigidly connected with tower B. Model III is slidingly connected with tower A and rigidly connected with tower B. Model IV is hinged with both ends of the tower. Geometric nonlinearity is also considered. Among them, the constraint mode of sliding connection support point is only restricted z translational freedom, the constraint mode of hinged connection support point is restricted x, y, z translational freedom, and the bottom node of the structure adopts fixed constraint.

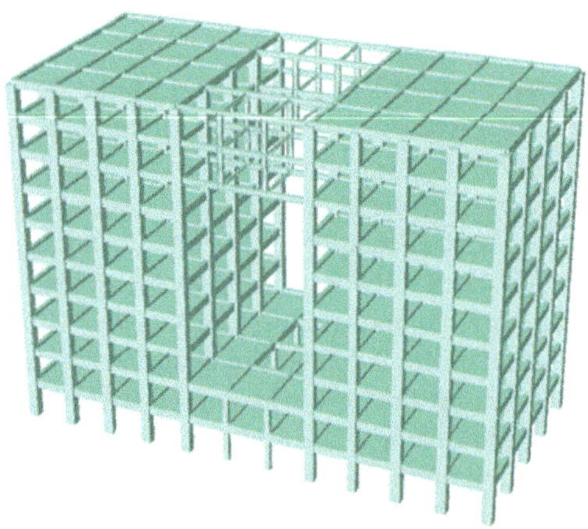

Fig. 3. CAE Model

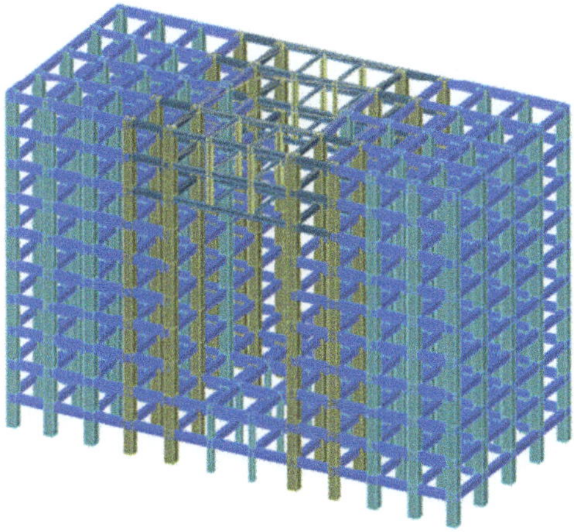

Fig. 4. PMSAP Model

5.1 Model Verification

(See Table 1).

Table 1. Model cycle comparison

Vibration mode number	Period/s		Main formation direction	Period error /%
	PMSAP Model	ABAQUS Model		
1	0.7561	0.7440	X translation	1.62
2	0.6702	0.6886	Y translation	2.67
3	0.6536	0.6621	torsion	1.28

Note: The period error is [PMSAP model period-ABAQUS model period]/ABAQUS model period.

6 Dynamic Time History Analysis

Dynamic time history analysis is a method of structural dynamics analysis, which is used to predict and evaluate the response of a structure under a specific dynamic action. Based on the vibration mode and dynamic frequency of the structure, the frequency response function and power spectral density function of the structure are calculated to predict the acceleration, velocity and displacement response of the structure under dynamic action. The dynamic time history analysis method has the advantages of fast calculation speed, accurate and reliable calculation results, and is suitable for complex structures. It is widely used in the fields of structural seismic response analysis, structural seismic design, earthquake disaster assessment and so on.

At present, the main seismic analysis methods are divided into the following: bottom shear method, mode decomposition method and time history analysis. Time history analysis method, also known as stepwise integration method, is the most intuitive and seismic analysis method with the least assumptions in the calculation process. Starting from the initial state of the structure to the end of the appropriate seismic acceleration time-history curve applied to the structure, the whole process is calculated by stepwise integration method to calculate the response of the structure at any time [17].

6.1 Seismic Wave Selection

According to GB50011-2010 [18] Dalian has set the seismic fortification intensity as 7 degrees (0.15 g).The design earthquake group is the first group, the site category is class I, and the characteristic period Tg is 0.25 s. The selection of ground motion acceleration time history curve needs to meet the three elements of ground motion, namely spectral characteristics, effective peak and duration. In order to meet the spectral characteristics, three seismic waves are preliminarily screened by the characteristic period Tg, as shown in Table 2. In the seven-degree area, the design basic seismic acceleration is 0.15g, and the effective peak value is 55 cm/s^2 according to Table 5.1.2-2 of 'Code for seismic design of buildings' [18]. The effective duration is generally not less than 5 times of the basic period of the structure. The basic period of the model is about 0.8s, and the selected seismic wave meets the requirements (Fig. 5).

Table 2. Ground motion information

Serial number	Type	Name	Time step	Duration	Peak time
1	Natural	Niigata Japan	0.02	30	5.68
2	Natural	Chi-Chi Taiwan-06	0.01	30	7.17
3	Artificial wave	RH1TG025	0.02	20	3.82

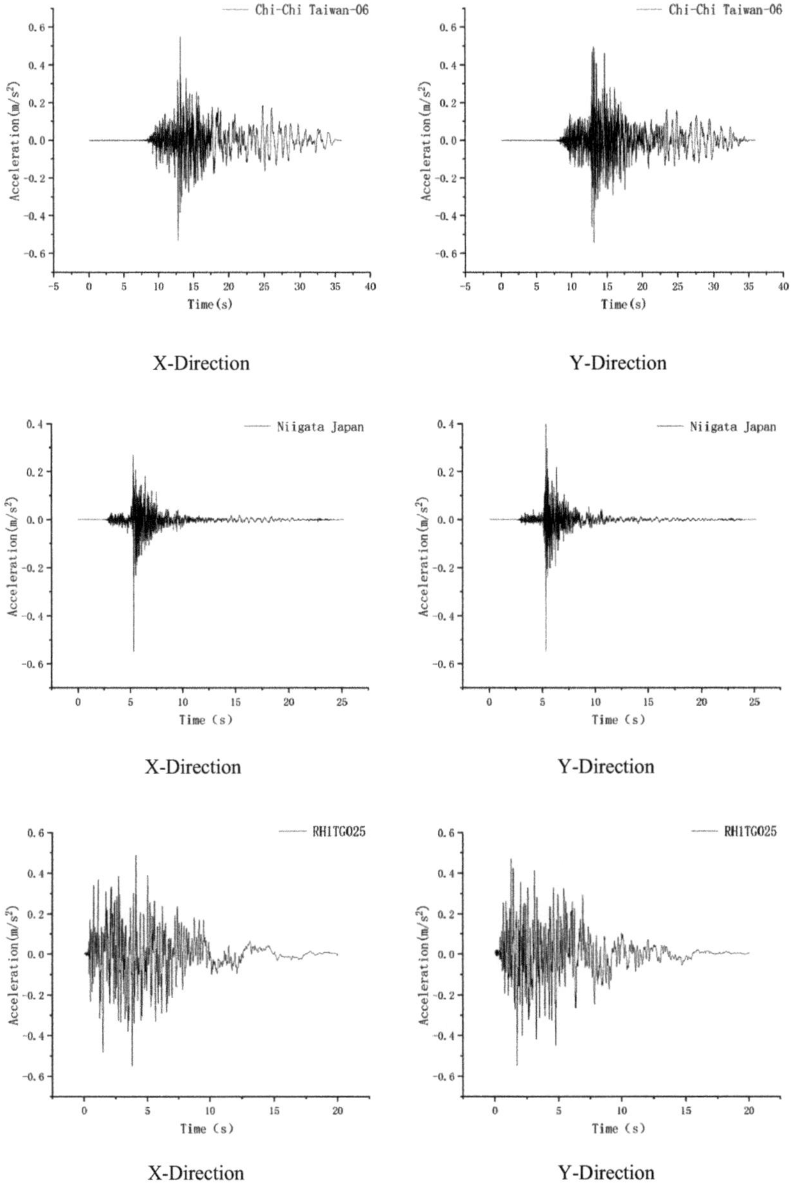

Fig. 5. Ground motion acceleration curve

Usually, the maximum acceleration of ground motion measured is different from the peak acceleration required by the seismic fortification intensity design of the building. Therefore, the maximum acceleration of ground motion should be adjusted according to the GB50011-2010 [18] (see Table 3), the maximum value is adjusted to 55 cm/s^2 by linear scaling (Fig. 6).

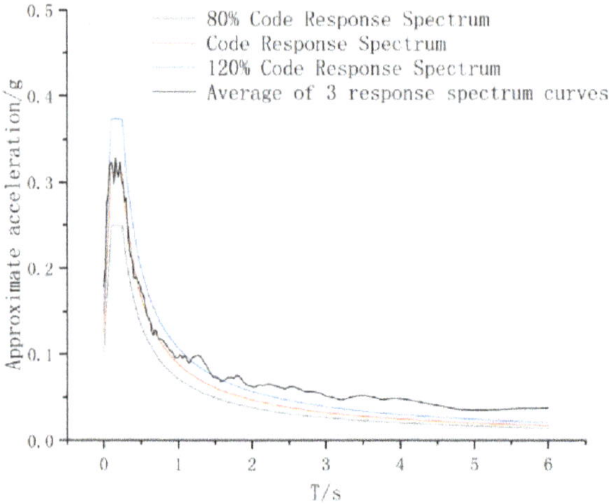

Fig. 6. Comparison of seismic wave response spectrum curve and design response spectrum curve

The difference between the response spectrum of the three seismic waves and the standard response spectrum is between 10% and 30%, and the average difference between the average of the three response spectra and the standard response spectrum is within 20%.

Table 3. Maximum value of seismic acceleration adjustment

Earthquake intensity	Adjusted maximum value of frequent earthquake (cm/s^2)	Adjusted maximum value of rare earthquake (cm/s^2)
6	18	125
7	35 (55)	220 (310)
8	70 (110)	400 (510)
9	140	620

6.2 Analysis of Interlayer Displacement Angle

The Inter-story Drift Ratio is one of the very important indicators in the seismic performance deformation checking calculation of buildings. It is the ratio of the maximum inter-story displacement to the height of the floor under the action of the standard value of multiple earthquakes calculated according to the elastic method, which is mainly used to ensure that the high-rise structure has sufficient stiffness. Under the action of multiple earthquakes, the Inter-story Drift Ratio limits of different structural systems are determined. The following table is the allowable values of the Inter-story Drift Ratio of different structural systems specified in the GB50010-2010 [18] (Table 4).

Table 4. Allowable value of the Inter-story Drift Ratio

Structural type	$[\theta_e]$
Reinforced concrete frame	1/550
Slab column-seismic wall, reinforced concrete frame-seismic wall, frame core tube	1/800
Reinforced concrete anti-seismic wall, tube in tube	1/1000
Reinforced concrete frame supporting layer	1/1000
Multi-storey, high-rise steel structure	1/250

(1) Analysis of the Inter-story Drift Ratio of Chi-Chi Taiwan-06 wave earthquake (Fig. 7)

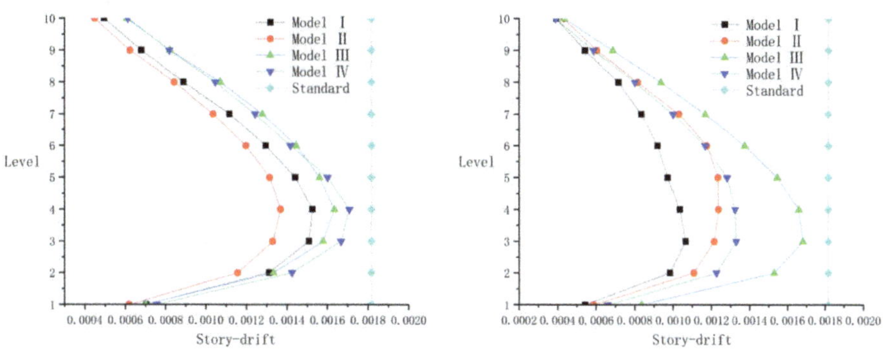

(a) The Inter-story Drift Ratio under the action of Chi-Chi Taiwan-06 wave earthquake in X direction

(a) The Inter-story Drift Ratio under the action of Chi-Chi Taiwan-06 wave earthquake in Y direction

Fig. 7. The Inter-story Drift Ratio of four models under Chi-Chi Taiwan-06 ground motion

The figure above shows the connected structures under four different connection modes. After ground motion is applied to Taiwan in the X and Y directions, it can be seen from the figure that the Inter-story Drift Ratio of the structure in both directions increases first and then decreases, and the IDR in the X direction is larger than that in the Y direction. Among them, the fourth IDR in the X direction is the largest, and the third IDR in the Y direction is the largest, and the maximum IDR of the structure is 656^{-1} and 939^{-1}, respectively. The maximum IDR of the four models is less than the displacement Angle limit of 550^{-1}, indicating that the structure has not been damaged, meeting the criterion of "small earthquake is not bad", and the structure is in the elastic range. The maximum elastic-plastic displacement Angle of the Models I and II with strong connection is smaller than that of the models with weak connection. Synthesize the the Inter-story Drift Ratio in two directions, when the two ends of the high connection between tower A and tower B are strongly connected, especially when the two ends are just connected, the structural integrity can be effectively improved and the elastic-plastic displacement response of the structure under earthquake action can be reduced (Table 5).

Table 5. The Inter-story Drift Ratio analysis of four models under Chi-Chi Taiwan-06 wave ground motion

Floor	The maximum value in X direction	Model	The maximum value in Y direction	Model	Whether out of gauge
10	2135^{-1}	II	2550^{-1}	I	No
9	1596^{-1}	II	1843^{-1}	I	No
8	1195^{-1}	II	1395^{-1}	I	No
7	960^{-1}	II	1198^{-1}	I	No
6	829^{-1}	II	1088^{-1}	I	No
5	755^{-1}	II	1029^{-1}	I	No
4	728^{-1}	II	966^{-1}	I	No
3	752^{-1}	II	939^{-1}	I	No
2	862^{-1}	II	1016^{-1}	I	No
1	1626^{-1}	II	1840^{-1}	I	No

(2) Analysis of the Inter-story Drift Ratio of Niigata Japan wave earthquake (Fig. 8)

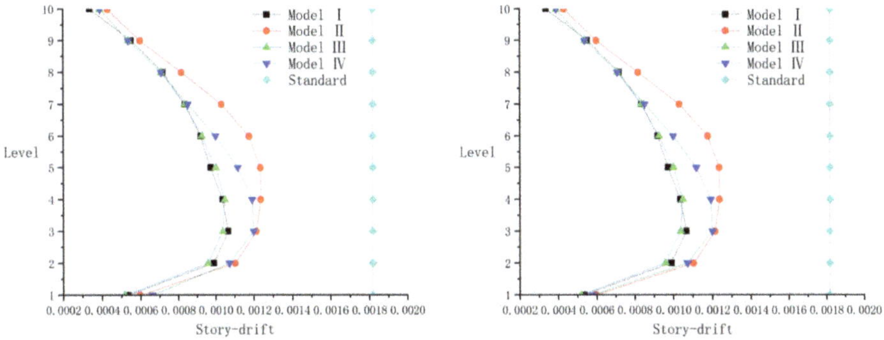

(a) The Inter-story Drift Ratio under the action of Niigata Japan wave earthquake in X direction

(b) The Inter-story Drift Ratio under the action of Niigata Japan wave earthquake in Y direction

Fig. 8. The Inter-story Drift Ratio of four models under Niigata Japan ground motion

The above figure shows the connected structures under four different connection modes. After the application of Niigata Japan ground motion in X and Y directions, it can be seen from the figure that the Inter-story Drift Ratio of the structure in both directions presents a trend of first increasing and then decreasing, and the IDR in X direction is the same as that in Y direction. The fifth inter-layer displacement Angle is the largest, and the maximum IDR of the structure is 810^{-1}. The maximum IDR of the four models is all less than the displacement Angle limit of 550^{-1}, indicating that the structure is not damaged and the lateral stiffness meets the structural requirements. In the third layer, the maximum displacement Angle of Model I is greater than that of Model III by 0.2%; in the fourth layer, the maximum displacement Angle of Model III is greater than that of Model I, and the difference between them is 1.9%. Starting from the fourth floor, the IDR between the two floors of model I is smaller than that of Model III, which proves that the rigid connection at both ends of the connection can strengthen the integrity of the structure, make the tower more closely connected, and the superstructure becomes a whole and the stiffness increases (Table 6).

Table 6. The Inter-story Drift Ratio analysis of four models under Niigata Japan wave ground motion

Floor	The maximum value in X direction	Model	The maximum value in Y direction	Model	Whether out of gauge
10	2972^{-1}	I	2972^{-1}	I	No
9	1820^{-1}	I	1820^{-1}	I	No
8	1329^{-1}	I	1329^{-1}	I	No
7	1198^{-1}	I	1198^{-1}	I	No
6	1088^{-1}	I	1088^{-1}	I	No
5	1029^{-1}	I	1029^{-1}	I	No
4	966^{-1}	I	966^{-1}	I	No
3	964^{-1}	III	939^{-1}	III	No
2	1043^{-1}	III	1016^{-1}	III	No
1	1930^{-1}	III	1840^{-1}	III	No

(3) Analysis of the Inter-story Drift Ratio of RH1TG025 wave earthquake (Fig. 9)

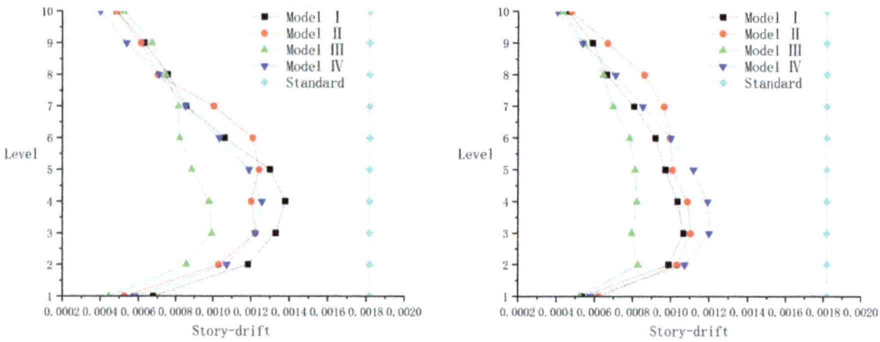

(a)The Inter-story Drift Ratio under the action of RH1TG025 wave earthquake in X direction

(b)The Inter-story Drift Ratio under the action of RH1TG025 wave earthquake in Y direction

Fig. 9. The Inter-story Drift Ratio of four models under RH1TG025 ground motion

The above figure shows the connected structures under four different connection modes. After the ground motion of RH1TG025 is applied in the X and Y directions, it can be seen from the figure that shear force sudden change occurs in both Model II and Model III in the X and Y directions, and the IDR of Model I and Model IV present a trend of first increasing and then decreasing. Among them, the fourth layer displacement Angle in the X direction is the largest, the third layer displacement Angle in the Y direction is the largest, and the maximum IDR of the structure is 725^{-1} and 833^{-1}. The maximum IDR of the four models is less than the displacement Angle limit of 550^{-1}, indicating that the structure is not damaged and the lateral stiffness also meets the structural requirements. The strong connection at both ends can still strengthen the overall stiffness of tower A and Tower B and absorb large earthquake force under the earthquake action dominated by Y direction (Table 7).

Table 7. The Inter-story Drift Ratio analysis of four models under RH1TG025 wave ground motion

Floor	The maximum value in X direction	Model	The maximum value in Y direction	Model	Whether out of gauge
10	2476^{-1}	IV	2976^{-1}	IV	No
9	1826^{-1}	IV	1861^{-1}	IV	No
8	1424^{-1}	II	1551^{-1}	III	No
7	1227^{-1}	III	1433^{-1}	III	No
6	1218^{-1}	III	1275^{-1}	III	No
5	1128^{-1}	III	1230^{-1}	III	No
4	1021^{-1}	III	1216^{-1}	III	No
3	1006^{-1}	III	1257^{-1}	III	No
2	1165^{-1}	III	1208^{-1}	III	No
1	2244^{-1}	III	1916^{-1}	III	No

7 Top Displacement Results

(1) Top displacement analysis of the model under Taiwan wave earthquake

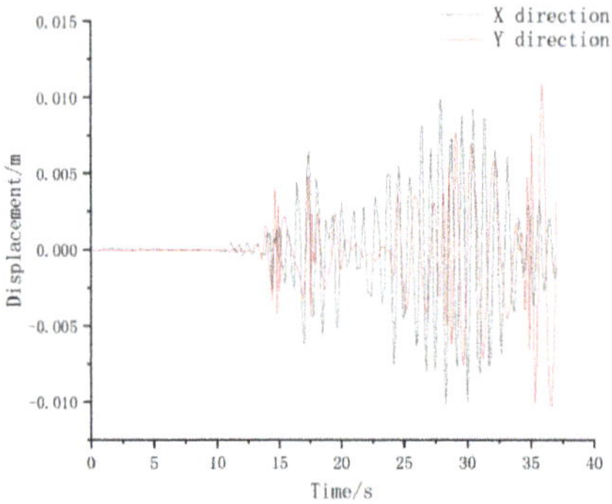

Model I

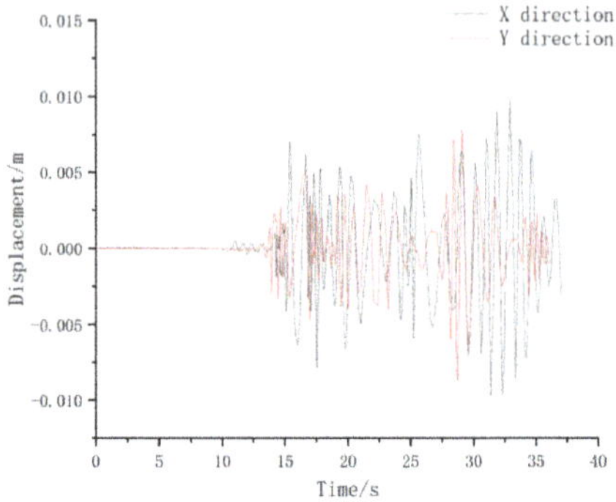

Model II

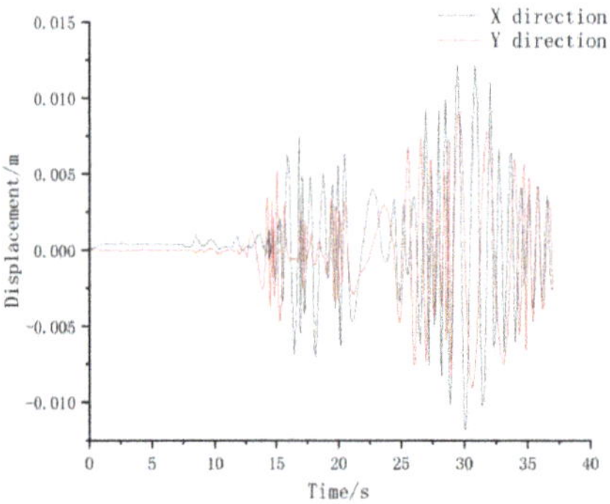

Model III

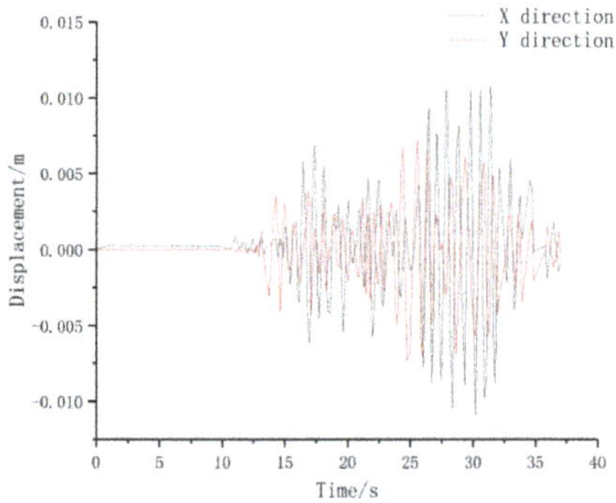

Model IV

(2) Top displacement analysis of the model under Niigata Japan wave earthquake

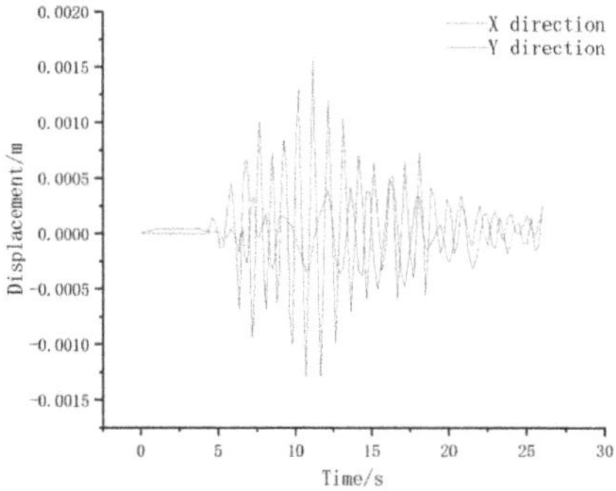

Model I

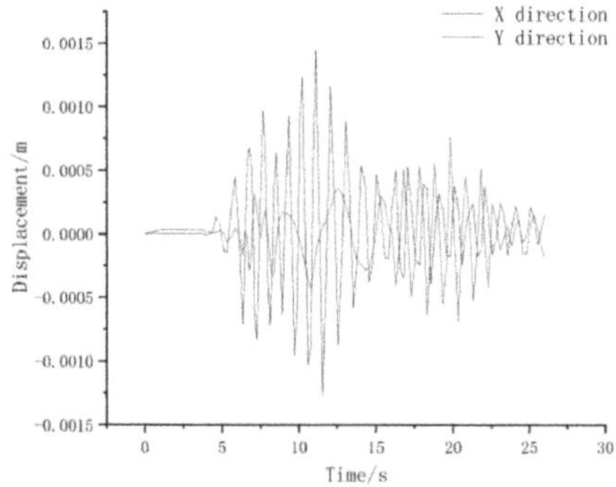

Model II

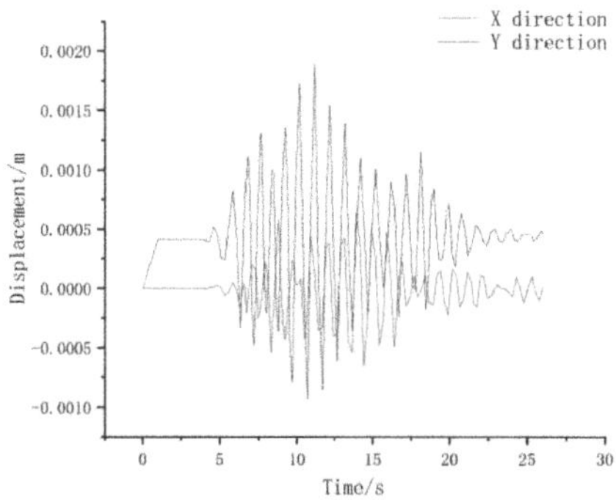

Model III

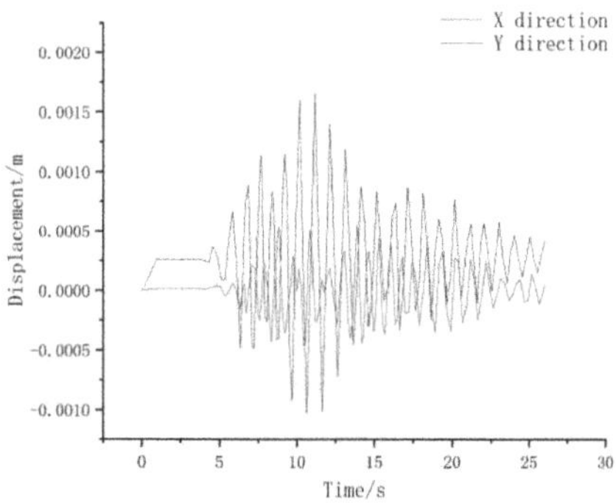

Model IV

(3) Top displacement analysis of the model under RH1TG025 wave earthquake (Table 8)

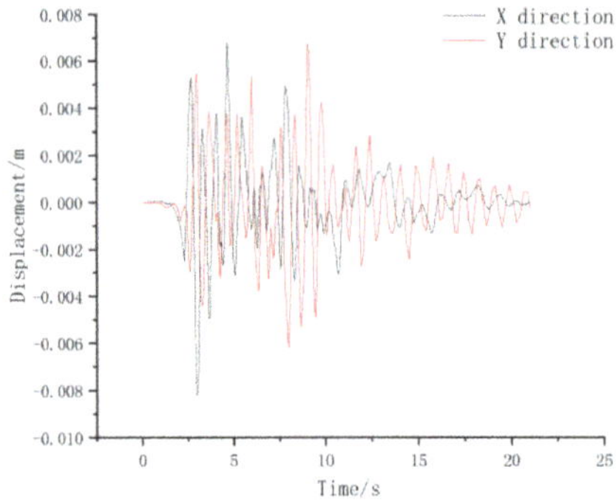

Model I

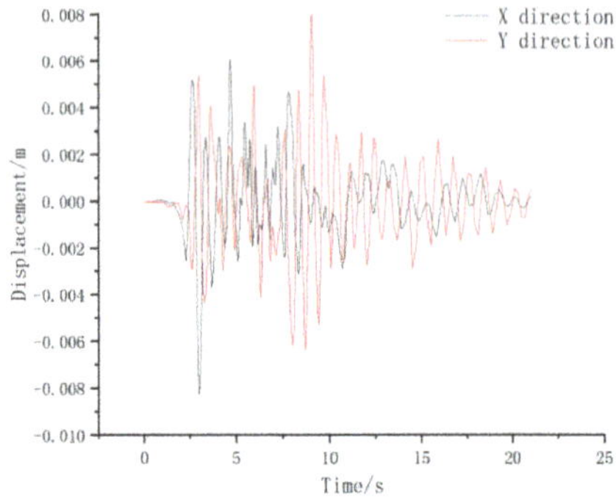

Model II

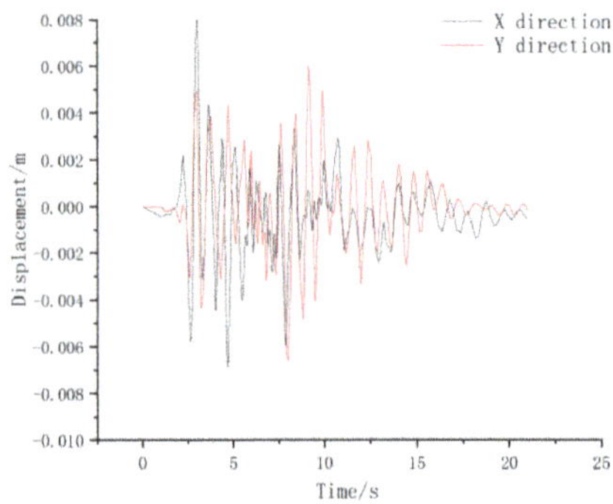

Model III

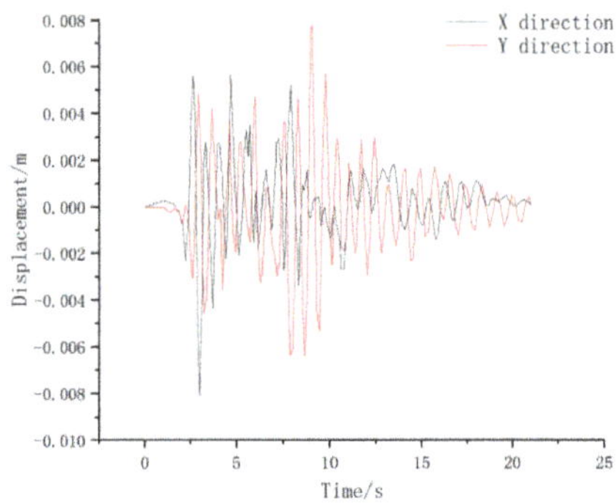

Model IV

Table 8. Maximum displacement at the top/mm

Seismic wave	Model I		Model II		Model III		Model IV	
	X	Y	X	Y	X	Y	X	Y
C-C Taiwan	9.2	9.12	9.8	8.7	12.1	9.18	10.9	7.3
Niigata Japan	1.5	0.5	1.4	0.7	1.8	0.6	1.6	0.5
RH1TG025	8.1	6.7	8.2	8.1	8.1	6.6	8.1	7.8

The total displacement at the top of the model with rigid connection is slightly smaller than that with flexible connection. Under the action of C-C Taiwan wave earthquake, the maximum displacement in X direction increases by 23.9% compared with that of Model I just connected at both ends, and the displacement in Y direction has little difference. There is little difference in displacement in X direction and Y direction under Niigata Japan and RH1TG025 earthquakes. 1) The maximum displacement in the X-direction occurs at node 3722, the maximum displacement in the Y-direction occurs at node 717, and the maximum displacement in the Z-direction occurs at node 3849. The above nodes are located in the middle of the 9-story connected structure, the middle of the 10-story connected structure main beam, and the middle of the 9-story connected structure secondary beam, respectively. 2) The maximum displacement in X direction at the top of the four model structures all appears at node 77, the maximum displacement in Y direction at node 65, and the maximum displacement in z direction at node 5475, which are the middle right side of the top floor of tower A, the right side of the top floor of tower B, and the middle span of the 10-story connected structure. In the direction of z, the maximum displacement of corridors A, B, C and D is the maximum displacement of the whole structure. 3) The maximum displacement in X and Y direction of the conjunctions of Model I, II, III and IV is less than the maximum displacement at the top of the structure. 4) In general, the displacements of the conjunctions and top structures of Model I are similar to those of Models II, III and IV, but they are all reduced. It is shown that the rigid connection between the high link and the two towers improves the stiffness of the connected structure and can effectively reduce the displacement of the structure (Table 9).

Table 9. The maximum internal force value of key components under different seismic waves

Internal force	Name	Model I		Model II		Model III		Model IV	
		X	Y	X	Y	X	Y	X	Y
Axial force (KN)	C-C Taiwan	42.1	43.1	52.0	52.4	26.9	26.3	45.1	46.4
	Niigata Japan	42.0	42.0	51.9	52.0	26.2	26.2	45.1	45.3
	RH1TG 025	42.0	42.1	51.9	52.1	26.8	26.2	45.1	45.5
Shearing force (KN)	C-C Taiwan	70.4	81.2	70.6	69.2	70.6	75.3	70.4	69.7
	Niigata Japan	70.3	70.3	70.3	70.4	70.4	70.3	70.4	70.3
	RH1TG 025	70.3	70.0	70.3	71.1	70.4	70.4	70.3	70.7
Bending moment (KN·M)	C-C Taiwan	98.7	104.1	99.1	96.4	115.4	110.9	115.3	112.4
	Niigata Japan	98.3	98.2	98.3	98.5	110.3	110.3	111.9	112.3
	RH1TG 025	98.3	97.6	98.3	99.6	110.1	110.3	112.6	112.3

For the connected key beam, the axial force, shear force and bending moment values of the four models are basically the same under different seismic waves, so it is only necessary to compare the maximum internal force values of the key members under different models under one seismic wave. Because Model III is a sliding connection at one end, part of the axial force will be released under the action of ground motion, so the axial force of Model III under the action of three seismic waves is the smallest, which is reduced by 49.76% compared with the rigidly connected Model II. The bending moment envelope values of Model III and IV are generally higher than those of Model I and II, and the X and Y directions are about 14.4% and 7.4% higher, respectively. The shear values of Models I, II, III and IV have little difference. According to the displacement envelope analysis at the top of the high-connected structure and the internal force envelope analysis of key components under different connection schemes, it can be seen that the rigid connection (Model I) can effectively reduce the displacement of the structure and the bending moment of the key beam of the corridor under frequent earthquakes. Therefore, the rigid connection of the high-connected structure can realize the stiffness coordination of the connected structure.

8 Conclusion

(1) For the four connection schemes of the high steel connection between tower A and tower B of a connected twin tower structure, the elastic analysis results show that the displacement of the Model I connection and its top is small under frequent earthquakes, and the rigid connection between the high steel connection and the towers at both ends can improve the stiffness of the overall structure and effectively reduce the structural displacement; In addition, the rigid connection scheme can realize the stiffness coordination of the connected structure and reduce the bending moment of the key beam.

(2) The strong connection scheme can also reduce the displacement of the key beam-column joints around the connected joints to varying degrees, which can effectively improve the structural integrity. For the damage of model members, more frame beams enter the damage stage in the model with strong connection, but generally speaking, there is little difference between the four models, which meet the design principle of strong column and weak beam, and the energy dissipation mechanism is reasonable.

(3) To sum up, for the two-tower connected structure in this project case, the rigid connection scheme in strong connection is more beneficial to the overall structure.

References

1. Liang, X., Zhou, Y., Niu, X., et al.: Seismic performance analysis of a multi-tower connected structure high steel corridor connection scheme. Progr. Build. Steel Struct. **25**(08), 16–25 + 37 (2023). https://doi.org/10.13969/j.cnki.cn31-1893.2023.08.003
2. Gong, H., Liu, B., Huang, L., et al.: Seismic analysis of large-span connected out-of-code high-rise structures at the gate of Shijiazhuang. North China Earthq. Sci. **33**(z1), 89–94 (2015)
3. Xu, W., Zhou, Y., Zhang, Y., et al.: Structural design of twin-tower connected high-rise building in Ningbo Block B1–5. Struct. Eng. **34**(3), 24–31 (2018)
4. Li, J., Hu, P., Xu, L., et al.: Structural design of the connected high-rise building of Yinlong building in Shaanxi. Build. Struct. **52**(2), 30–37 (2022)
5. Lin, Y., He, C., Chen, D.: Analysis of the influence of corridors on towers in a low-level connected structure. Build. Struct. **46**(1), 87–91 (2016)
6. Mazars, J., Kotronis, P., Davenne, L.: A new modelling strategy for the behaviour of shear walls under dynamic loading. Earthq. Eng. Struct. Dynam. **31**(4), 937–954 (2010)

7. Chintanapakdee, C., Chopra, A.K.: Evaluation of modal pushover analysis using generic frames. Earthq. Eng. Struct. Dynam. **32**(3), 417–442 (2010)
8. Zhen, W., Zhang, L., Zhang, Y., et al.: Research on connection control method of multi-tower connected high-rise buildings. Build. Struct. **49**(18), 49–54 (2019)
9. Xiao, F., Chen, Y., Liu, F., et al.: Design method and analysis of weak connection between connected buildings with different volume. Eng. Seismic Reinforcement Reconstruct. **40**(3), 145–152, 144 (2018)
10. Lu, Q., Lu, W., Lyu, X., Zhou, Y.: Shaking table test of four-tower connected structure with passive control. J. Build. Struct. **38**(03), 46–57 (2017)
11. Feng, Z., Gong, D., Sun, H., et al.: Structural design of Shanghai Junkang financial Plaza. Build. Struct. **51**(6), 79–85 (2021)
12. Li, Y., Chen, B., Jiang, B., et al.: Design and analysis of connector scheme for a twin-tower connected super high-rise structure. Spec. Struct. **35**(4), 46–51, 124 (2018)
13. Sun, H.S., Liu, M.H., Zhu, H.P.: Connecting parameters optimization on unsymmetrical twin-tower structure linked by sky-bridge. J. Cent. South Univ. **21**(6), 2460–2468 (2014)
14. Luco, J.E., Barros, F.C.P.: Control of the seismic response of adjacent structure. In: World Conference on Structural Control and Monitoring, no. 2, pp. 3–5 (2013)
15. Clough, R., Penzien, J.: Dynamic of Structures. Computers and Structures Inc., Berkeley (2010)
16. Pan, Y., Zhang, B., Liu, Y., et al.: Influence of connection mode on seismic performance of large-span special-shaped steel corridor connected structure. J. Build. Struct. 1–11. https://doi.org/10.14006/j.jzjgxb.2021.0328
17. Pan, Y.: Research on seismic performance of concrete frame after local fire. Shenyang Jianzhu University (2022). https://doi.org/10.27809/d.cnki.gsjgc.2022.000265
18. Chinese national standard GB50010-2010. Code for seismic design of buildings. China Building Industry Press, Beijing (2010)

Comparison and Analysis of Dust Explosion Characteristics Between Bungalow Granary and Silo Granary

Chang Li(✉), Shiguang Shen, and Ruojie Chen

School of Civil Engineering, Shenyang Jianzhu University, Shenyang 110168, China
ssg1102@qq.com

Abstract. In order to investigate the impact of different types of grain storage silos on grain dust explosions during transportation and storage. FLACS software was utilized to construct building models of flat and cylindrical grain silos. The study systematically compared the flame propagation range, maximum explosion overpressure, and temperature variations of different silo types by altering the dust concentration, ignition position, and silo shape. The simulation results indicate that dust concentration, ignition position, and silo shape all influence the explosion process. Flame release in flat grain silos occurs horizontally, whereas in cylindrical grain silos, it occurs from the top. Within the mass concentration range of 200–700 g/m^3, an increase in dust concentration leads to higher maximum explosion overpressure and elevated temperatures at observation points. As the ignition source moves closer to the center of the dust cloud, the flame shape becomes more ellipsoidal, leading to higher maximum explosion overpressure and a shorter time to reach it.

Keywords: Dust Explosion · Numerical Simulation · Granary Type

With the expansion of the grain trade market, the state has increased the construction of grain storage facilities. Grain dust, the second most explosive dust type after metal dust, tends to accumulate during grain processing and transportation and from dust clouds [1]. Friction sparks from grain transportation equipment during operation may cause dust explosions when they come into contact with dust clouds [2]. The confined space of granaries intensifies the seriousness of the dust explosion risk [3].

In China, research on dust explosions mainly focuses on explonsion characteristics, explosion protective measures, and causes of explosions [4, 5]. There is a lack of research data on the specific impact of granary building shapes on grain dust explosions.

Currently, experts researching dust explosions utilize commercial Computational Fluid Dynamics (CFD) software for numerical simulations of natural gas, hydrogen, and other gas explosions, as well as simulations involving grain dust and metal dust. Shen et al. used CFD software FLUENT to simulate the explosion process of aluminum powder in a 20 L spherical device, studying the impact of turbulence on explosion characteristics by altering ignition delay time and dust concentration [6]. Wu et al., based on FLACS, simulated the explosion process of tobacco dust in a silo, conducting

I. Tuns et al. (Eds.): CIBv 2023, LNCE 510, pp. 294–310, 2024.
https://doi.org/10.1007/978-3-031-60765-3_24

a comparative study of flame propagation range and explosion overpressure evolution by varying initial concentration, ignition position, and silo volume [7].

Fire and explosion simulation plays a crucial role in engineering, safety assessment, and various other fields. FLACS (FLame ACceleration Simulator), as fire and explosion simulation software based on computational fluid dynamics principles, boasts a range of advanced CFD theoretical foundations, multiphase flow models, chemical kinetic models, turbulence models, and an array of characteristics. These features provide FLACS with the foundation for achieving a high degree of accuracy in fire simulations, rendering it ideal for simulating complex fire scenarios. Wang et al. simulated a gas explosion incident in a residential building in Shijingshan, Beijing, on December 7, 2015, using FLACS software. They combined this with actual explosion accident data, and the results closely mirrored the development process of the original accident [8].

The FLACS software is grounded in CFD theory, a numerical simulation method employed to solve fluid dynamics problems. FLACS utilizes a multiphase flow model that considers air, flue gases, and other gas components that may be present simultaneously. This model can more realistically depict the complex gas mixing state at the fire scene, thereby enhancing the accuracy of simulation results [9]. FLACS supports various chemical kinetic models to simulate chemical reactions during combustion. Users can select different combustion models based on the actual scenario to accommodate the combustion characteristics of diverse ignition sources. FLACS employs an advanced turbulence model, aiding in revealing crucial processes such as thermals and smoke propagation. Additionally, the FLACS software incorporates a thermal radiation model, enabling consideration of radiation effects on fire development.

The FLACS numerical simulation analysis process is as follows: (1). Define the simulation area, including buildings, equipment, etc. (2). Generate and divide the mesh. (3). Set boundary conditions. (4). Select the combustion model. (5). Perform dust explosion simulation and result analysis [10]. FLACS enables users to conduct parameter sensitivity analyses to evaluate the impact of key parameters on simulation results by modifying them. This assists in determining model reliability and handling uncertainties effectively. FLACS software demonstrates high computational efficiency and real-time performance during the simulation process. Its flexible time steps, efficient numerical algorithms, and real-time interactivity make FLACS a robust and practical tool for fire and explosion simulation.

A literature survey reveals limited research data on numerical simulations of grain dust explosions using FLACS software. The current approach lacks the ability to distinguish explosion overpressure and temperature at different locations within the building during dust explosion numerical simulations. FLACS software, with its capacity to monitor overpressure and temperature by setting observation points at various locations, is employed in this paper to conduct numerical simulations of corn dust explosions in bungalow granaries and silo granaries.

1 Method

The modeling of the bungalow granary and the silo granary is based on a grain storage depot in China. Both the bungalow and silo granaries have almost equal building volumes. Figure 1a illustrates the bungalow granary model with dimensions of 23*15*10.5 m. The bungalow granary has 4 doors and 6 windows on the wall. Door dimensions are 3*4 m, and window dimensions are 1.2*1 m, positioned at 6.8m above the ground. Figure 1b depicts the silo granary model with a radius of 6m and a height of 27 m.

(a) (b)

Fig. 1. (a) Bungalow granary model and (b) Silo granary model

In this paper, corn dust is chosen as the type for dust explosion numerical simulation. Explosion parameters for corn dust are obtained through a 20L spherical explosion test, yielding a maximum explosion overpressure of 0.66 MPa, a maximum explosion index of 17.21 MPa.m/s. Initial conditions for FLACS simulation include gravitational acceleration of 9.8 m/s^2, characteristic velocity of 10 m/s, relative turbulence velocity of 1, turbulence scale of 0.0025 m, initial temperature of 20 °C, initial atmospheric pressure of 100 kPa.

Two groups of simulation experiments are conducted in this paper: ① dust explosion numerical simulation with varying dust concentration parameters; ② dust explosion numerical simulation with different ignition source locations. The experimental analysis focused on the flame propagation characteristics, maximum explosion overpressure, and dust explosion temperature of bungalow and silo granaries.

2 Numerical Simulation Results of Different Dust Concentrations

2.1 Dust Explosion Simulation Parameter Setting

Various factors influence the intensity of dust explosions, with dust concentration being the most crucial. Parameter settings: dust concentrations of 200, 300, 400, 500, 600, 700, 800 g/m^3, a dust cloud volume of 3*3m, and ignition positioned at the center of the dust cloud. Numerical simulations of dust explosions are conducted for both bungalow and silo granaries. Figure 2 illustrates the location of the observation point.

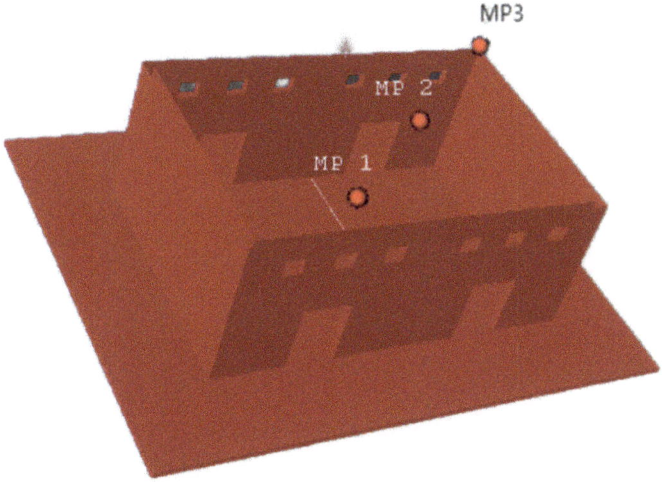

Fig. 2. Location of observation points

2.2 Analysis of Explosion Overpressure Variations for Different Dust Concentrations and Granary Types

In Fig. 3, the curve depicts the maximum explosion overpressure under varying dust concentrations in bungalow and silo granaries. For concentrations of 200 g/m^3–700 g/m^3, the curve shows an upward trend, with the silo granary exhibiting higher overpressure than the bungalow granary. For concentration of 200 g/m^3, the maximum explosion overpressure of bungalow granary and silo granary is 0.07 bar. For concentration of 700 g/m^3, the maximum explosion overpressure of bungalow granary and silo granary is 0.29 bar. In the range of 700 g/m^3–800 g/m^3, the curve shows a downward trend. This phenomenon is attributed to the lack of oxygen during the redox reaction. With the increase of dust concentration, the explosion overpressure increase rate inside the silo granary is higher than that of the bungalow granary.

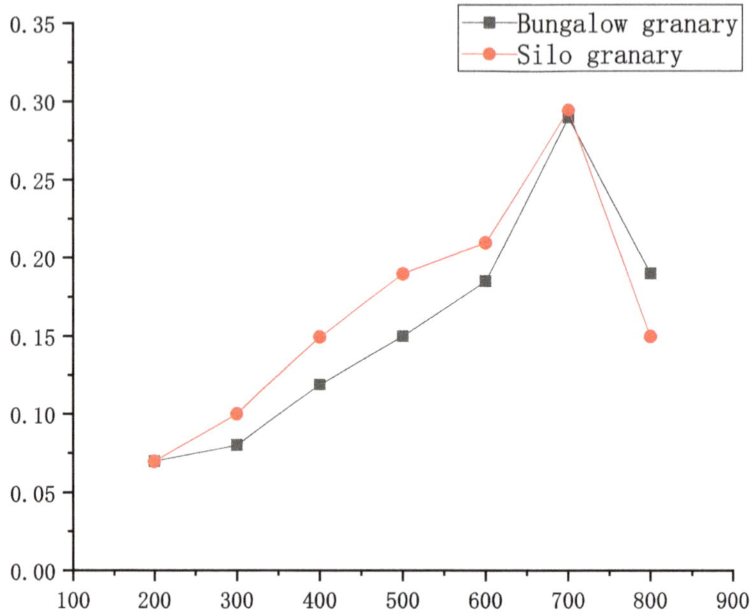

Fig. 3. Comparison of explosion pressure data for different types of granary and dust concentrations

2.3 Analysis of Explosion Temperature Variations for Different Dust Concentrations and Granary Types

The changes in temperature values at the three observation points are shown in the graph (see Fig. 4, Fig. 5 and Fig. 6), with the bungalow granary on the top and the silo granary on the below.

Observation point 1 is located in the center of the dust cloud, and the temperature values change the most. 200 g/m^3, the temperature values of the silo reach about 1500 degrees; 500 g/m^3, the granary dust concentration reaches the limiting temperature of 2000 degrees or more.

The maximum temperature values at observation points 2 and 3, as shown in the figure, are 950° and 625° for the bungalow granary and 1600° and 1000° for the silo granary. When the location of the observation point is far from the center of the dust cloud, the dust explosion produces higher temperature damage to the silo granary than the bungalow granary.

As shown in the figure, the silo granary temperature rises faster than the bungalow granary.

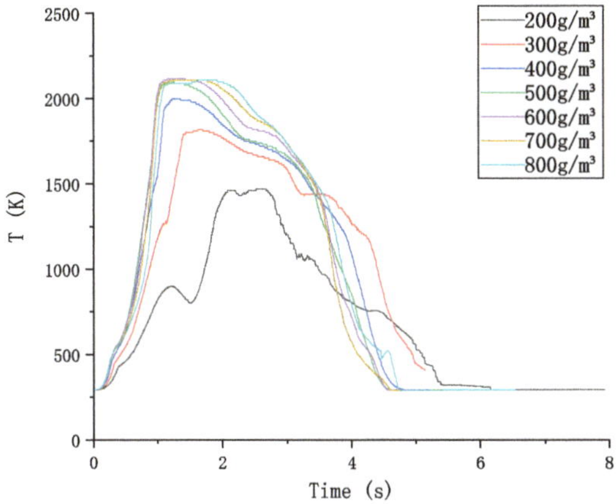

(a) Bungalow granary

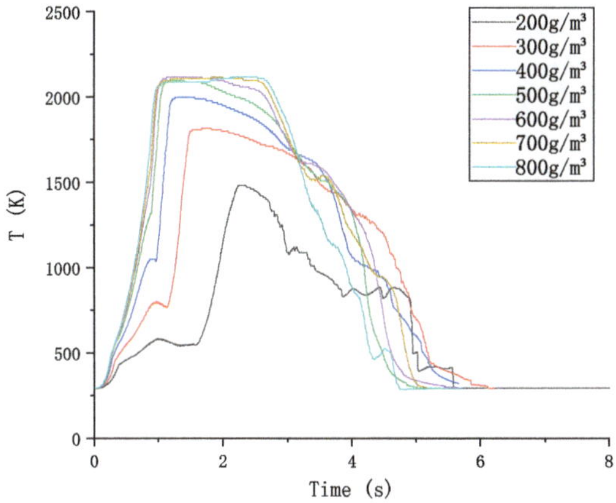

(b) Silo granary

Fig. 4. Observation point 1: temperature curves of bungalow silos and vertical silos at different concentrations

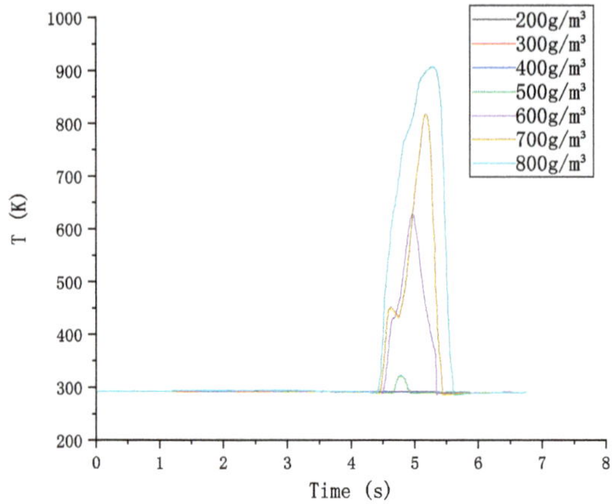

(a) Bungalow granary

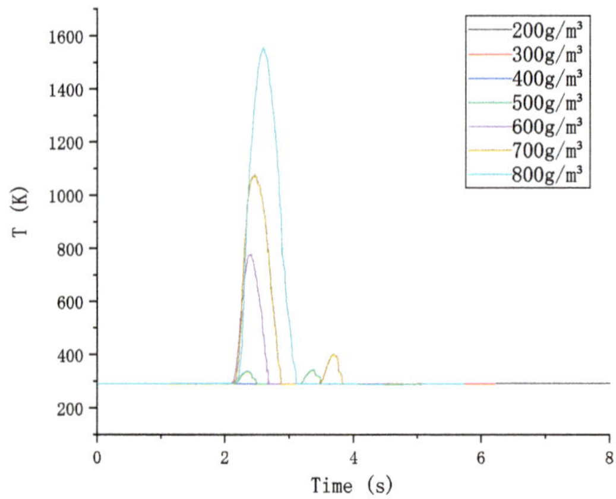

(b) Silo granary

Fig. 5. Observation point 2: temperature curves of bungalow silos and vertical silos at different concentrations

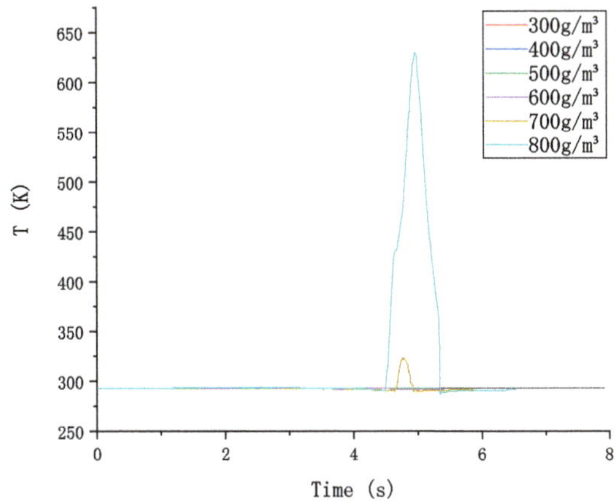

(a) Bungalow granary

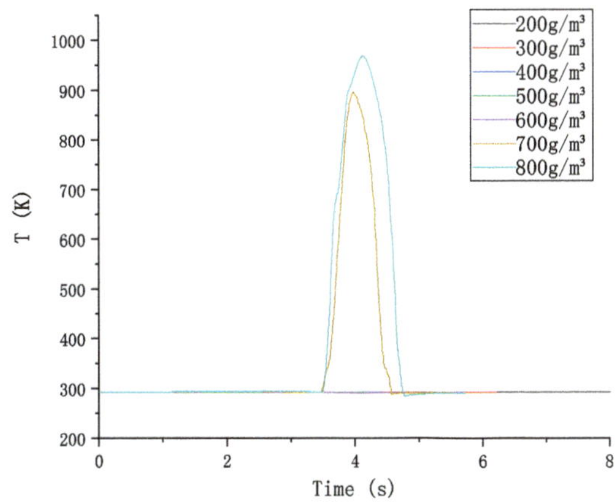

(b) Silo granary

Fig. 6. Observation point 3: temperature curves of bungalow silos and vertical silos at different concentrations

2.4 Dust Explosion Flame Propagation Cloud Map

The flame cloud diagram and overpressure cloud diagram of the explosion that occurred when the dust concentration was 700 g/m^3 are shown in the figure (see Fig. 7).

The flame propagation shape of the bungalow granary and silo granary showed an ellipsoidal shape. When the dust explosion occurred, the explosion overpressure inside the grain silo kept getting bigger. Because of the limitations of the building walls, the later shape of the flame propagation of the bungalow granary and silo granary is different. The diameter of the flame diffusion range at the top of the bungalow granary is 12 m, and the diameter of the flame diffusion range at the top of the silogranary is 10 m.

The flame spread of dust explosions in both the bungalow and silo granaries is ellipsoidal. Different building shapes impose distinct restrictions on flame spread.

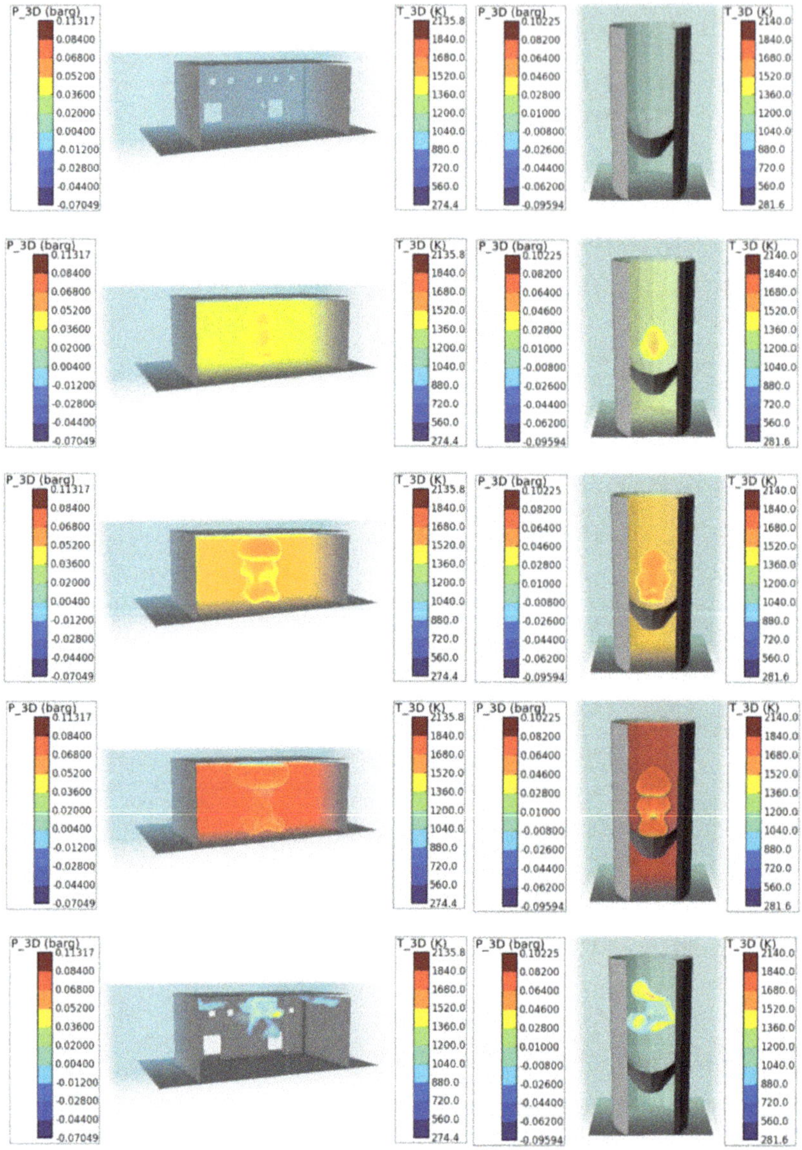

Fig. 7. Comparison of flame propagation cloud maps for bungalow granary and silo granary

3 Simulation Analysis of Dust Explosion with Different Ignition Source Locations and Granary Types

3.1 Dust Explosion Simulation Parameter Setting

The volume of the dust cloud is 3*3 m, dust concentration of 700 g/m^3, Fig. 8 shows ignition source locations 1, 2, and 3, and Fig. 9 shows observation point locations 1, 2, 3, 4, and 5.

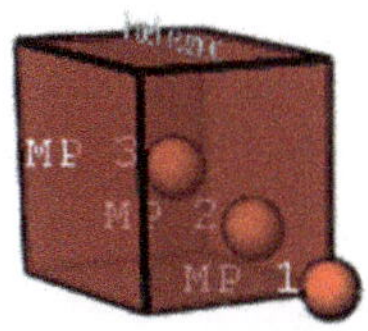

Fig. 8. Ignition source location

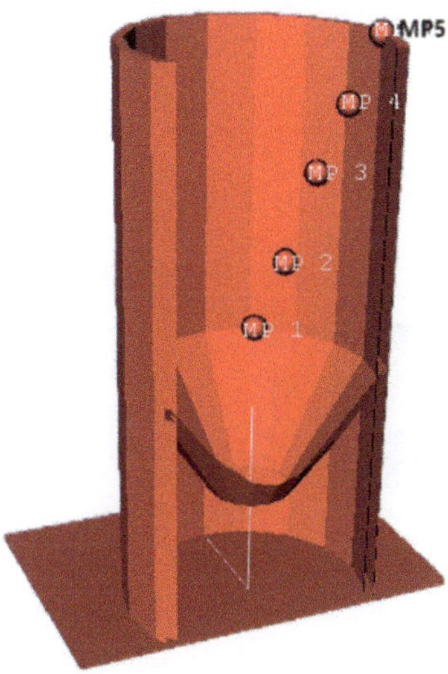

Fig. 9. Location of observation points

3.2 Comparative Analysis of Dust Explosion Flame Propagation Cloud Maps

Figure 10 illustrates flame propagation clouds for different ignition source locations in both the bungalow and silo granaries.

Figure 10a displays the flame propagation cloud when the ignition source is at position 1. The flame starts at this position and spreads towards the dust, with a central gap in the flame extent. Figure 10b shows the flame propagation cloud when the ignition source is at position 2. The flame starts at this position and moves towards an ellipsoidal shape, differing between the bungalow and silo granaries. Figure 10c demonstrates the flame propagation cloud when the ignition source is at position 3. The flame spreads evenly in all directions, forming an ellipsoidal shape in both granaries.

Reasons for differences: Variations in ignition source locations lead to different oxygen distribution around the dust cloud in the bungalow and silo granaries.

The flame begins at the ignition source and spreads towards surrounding dust. The closer the ignition source is to the dust cloud center, the closer the flame propagation shape is to an ellipsoid.

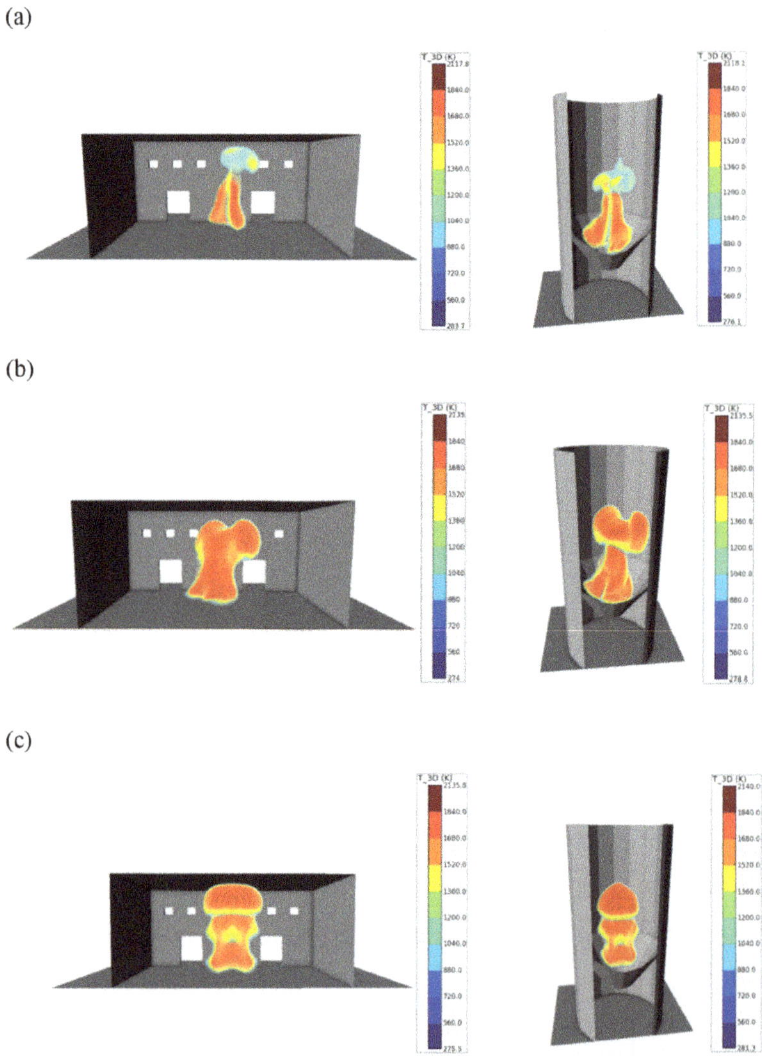

Fig. 10. Comparison of flame propagation cloud Maps for different granary types and ignition source locations (a) Ignition source 1 and (b) Ignition source 2 (c) Ignition source 3

3.3 Analysis of Explosion Overpressure Variations for Different Ignition Source Locations and Grain Bin Types

Figure 11 is the maximum explosion overpressure time - ignition position curve for both granaries. Ignition source 1: Both granaries reach the maximum explosion overpressure time at around 3.5 s. Ignition source 2: Bungalow granary reaches the maximum explosion overpressure time in 3 s, while silo granary reaches it in 2.2 s. Ignition source 3: Bungalow granary reaches the maximum explosion overpressure time in 2.5 s, while silo granary reaches it in 1.8 s. Reason: The closer the ignition source is to the dust cloud center, the more uniform the dust distribution, resulting in a faster dust combustion rate.

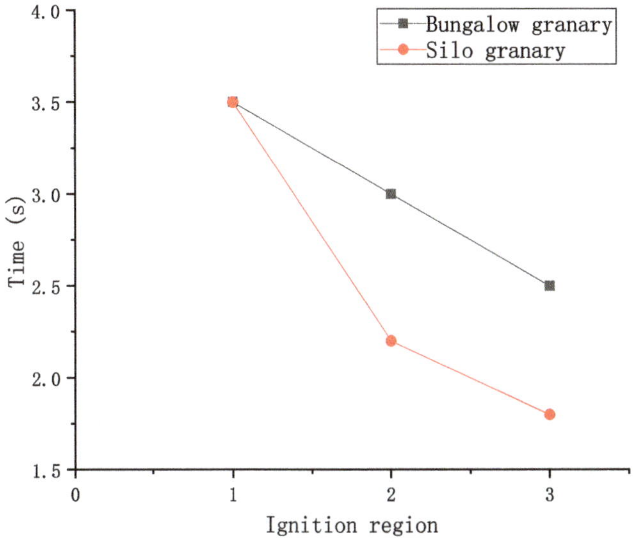

Fig. 11. Comparison of maximum explosive pressure time data for different types of granary and Different Ignition Source Locations

Figure 12 is the maximum explosive overpressure curve for both granaries. Ignition source 1–3: Bungalow granary shows a uniform rising curve from 0.12 bar to 0.27 bar. Silo granary's curve is not uniform, ranging from 0.18bar to 0.3 bar.

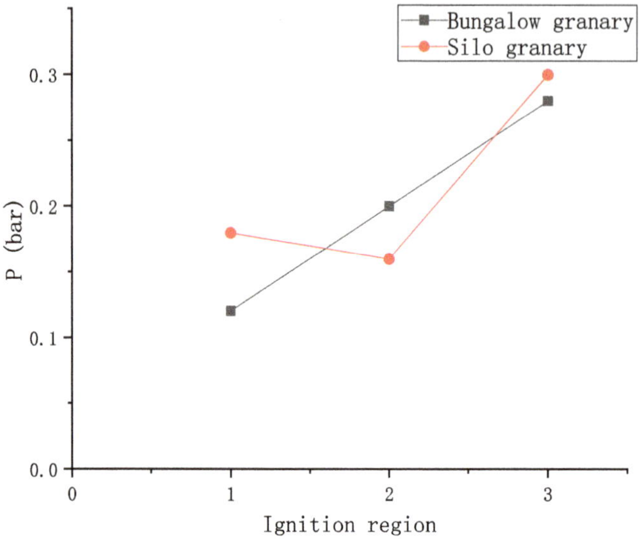

Fig. 12. Comparison of explosion pressure data for different types of granary and different ignition source locations

3.4 Analysis of Explosion Temperature Variations for Different Ignition Source Locations and Grain Bin Types

When a dust explosion occurs, the ignition source is located in different positions, the flame propagation form will also be different from the temperature change inside the building is different.

This section shows temperature-time profiles for different ignition source locations in both granaries, with the bungalow granary on the top and the silo granary on the below. Figure 13 displays temperature variation at observation points when the ignition source is at position 1. Both granaries show fluctuations at observation points 1 and 2, with the bungalow granary reaching a lower maximum temperature near 1600 degrees compared to the silo granary near 1350 degrees. Figure 14 reveals temperature changes at observation points when the ignition source is at position 2. The bungalow granary shows varying degrees of temperature increase, with observation point location affecting the temperature rise. The silo granary exhibits different temperature phenomena. Figure 15 depicts temperature changes at observation points when the ignition source is at position 3. Both granaries show a maximum temperature above 2000 degrees, with observation point location influencing the temperature.

Different ignition source locations lead to varying temperature changes in dust explosions. The closer the ignition source is to the dust cloud center, the higher the explosion temperature in the center of the dust cloud.

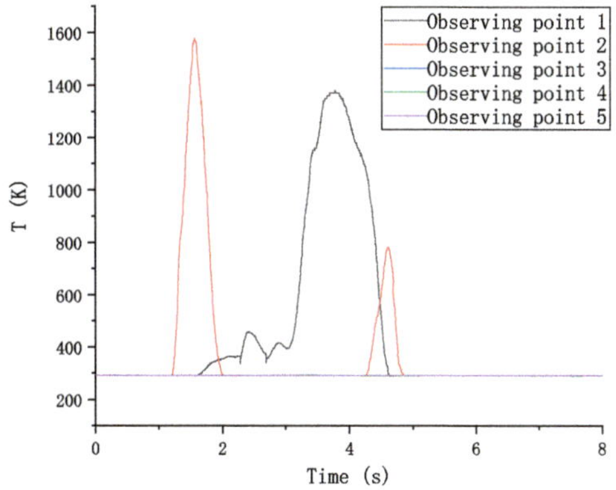

(a) Bungalow granary

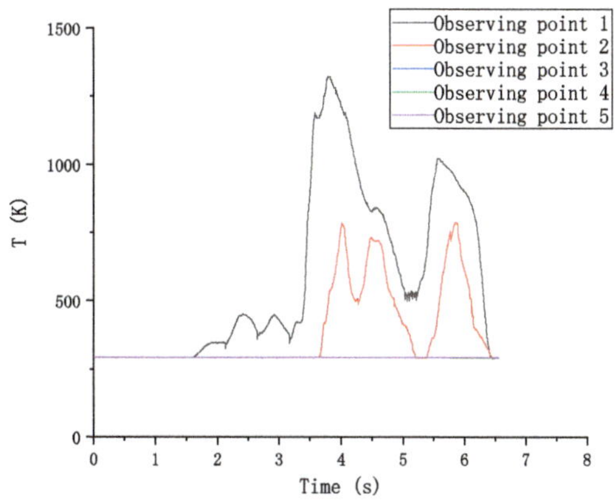

(b) Silo granary

Fig. 13. Ignition source 1: temperature curves at different observation points at the same concentration

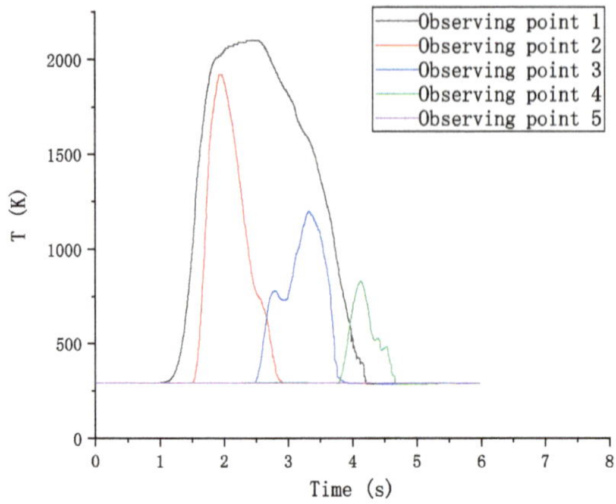

(a) Bungalow granary

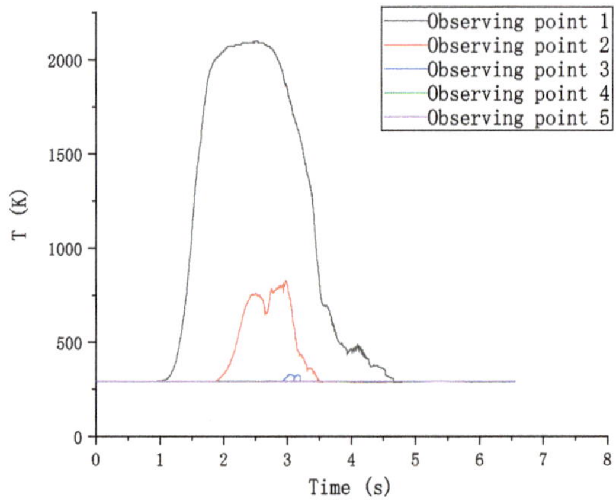

(b) Silo granary

Fig. 14. Ignition source 2: temperature curves at different observation points at the same concentration

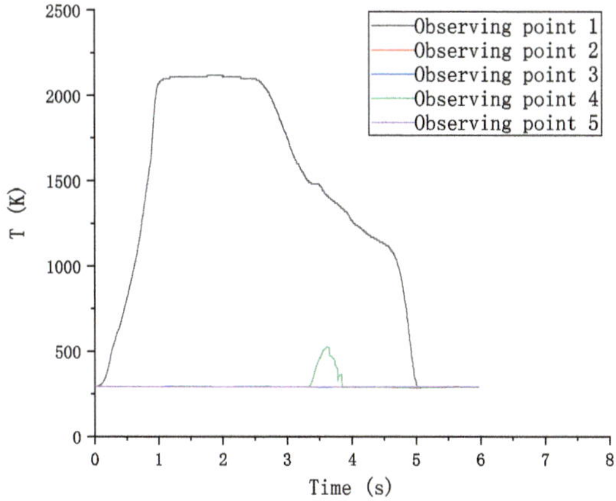

(a) Bungalow granary

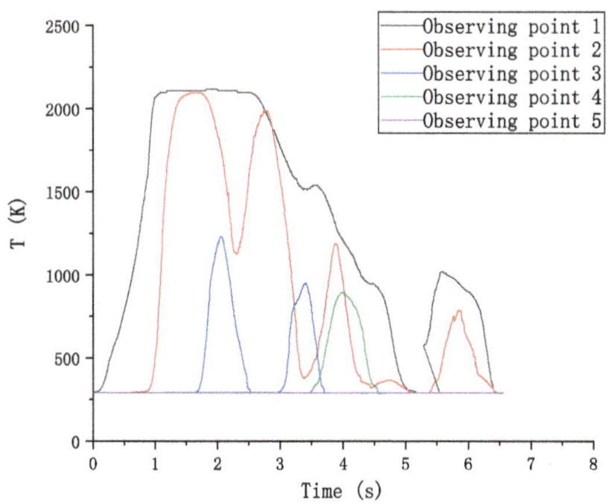

(b) Silo granary

Fig. 15. Ignition source 3: temperature curves at different observation points at the same concentration

4 Conclusion

1. With the increase of dust concentration, the explosion overpressure increase rate inside the silo granary is higher than that inside the bungalow granary, the silo granary temperature rises faster than the bungalow granary. The dust concentration is the same, and the flame diffusion range in the bungalow granary is greater than that in the silo granary.

2. When the ignition source is located at different locations: The closer the ignition source is to the dust cloud center, the closer the flame propagation shape is to an ellipsoid, the shorter the time for a dust explosion to reach the maximum explosion overpressure. (The time for the dust explosion in the silo granary to reach the maximum explosion overpressure is shorter than that of the bungalow granary). The closer the ignition source is to the dust cloud center, the higher the explosion temperature in the center of the dust cloud.

References

1. Qin, X.L., Li, X.Q.: Statistical analysis of grain dust explosion accidents. J. Ind. Saf. Environ. Prot. **46**(05), 78–82 (2020)
2. Xie, Z.H.: Research on dust explosion risk analysis and prevention and control measures in grain storage enterprises. J. Grains Fats. **35**(02), 98–101 (2022)
3. Zou, T.Y., Du, X.J., Zhang, T.: Dust explosion and prevention and control measures. J. Electr. Explos. Prot. **2023**(05), 28–30 (2023)
4. Zhao, J.Z.: Study and numerical analysis of the explosion characteristics of flour in a 20L closed container. Tianjin Univ. Technol. **59** (2021)
5. Cen, K., Song, B., Huang, Y.: CFD simulations to study parameters affecting gas explosion venting in compressor compartments. Math. Probl. Eng. (4), 1–14 (2017)
6. Shen, S.L., Zhang, Q., Ma, Q.J., Li, D., Yan, H.: Effect of turbulence on the explosive characteristics of aluminum powder. J. Armament Ind. **37**(03), 455–461 (2016)
7. Wu, Y.M., Ding, J.X., Wang, X.H., Zhao, Y.X.: Numerical simulation of tobacco dust explosion in silo. J. Ind. Saf. Environ. Prot. **45**(05), 31–35 (2019)
8. Wang, D., Qian, X.M., Yuan, M.Q.: Numerical simulation analysis of explosion process and destructive effect by gas explosion accident in buildings. J. Loss Prev. Process Ind. **49**, 215–227 (2017)
9. Wang, Y.Z., Wu, M., Du, J., Gong, K.: Simulation of the consequences of natural gas leakage explosion accident in canteen based on FLACS. J. Liaoning Shihua Univ. **42**(01), 35–40 (2022)
10. Yan, Z.Y., Liu, Y., Feng, W.: Discuss the use of FLACS in leak explosion simulation. J. Electr. Explos. Prot. **2021**(01), 33–38 (2021)

Study on Smoke Blocking Effect of Vertical Air Curtain

Yansheng Song and Wanwan Zhang(✉)

Shenyang Jianzhu University, Shenyang 110168, China
ceyssong@sjzu.edu.cn, z987996530@163.com

Abstract. To reduce the harm caused by smoke spread in the event of a fire, this article proposes a vertical air curtain. The platform floor of an island style subway station is taken as research background, the impact of four types of air curtains on fire smoke control was investigated using FDS numerical simulation software, including upper air supply (US), downside air supply (DS), upper air supply and downside air exhaust (USDE), and downside air supply and upper air exhaust (DSUE), by analyzing the changes in visibility, CO concentration, and the temperature of smoke. The results show that the smoke control effect of the US air curtain and USDE air curtain are poor, because they cannot prevent the spread of smoke; The DS air curtain and the DSUE air curtain can effectively block the spread of smoke, among which the DSUE air curtain has the best smoke blocking effect. Within 360 s, it can ensure that the visibility at the height of the human eye in the protected area of the air curtain is maintained at 30 m, the CO concentration does not increase, and the maximum smoke temperature is 21 °C. It proves that using the DSUE air curtain can ensure a safe evacuation time of more than 6 min for personnel, providing convenient conditions for personnel evacuation.

Keywords: Subway fire · Air curtain · Air supply vertically · Spread of smoke

1 Introduction

The subway platform is a crowded public gathering place with high pedestrian density. Once a fire accident occurs, it is difficult to evacuate the crowd. The statistical results of numerous subway fire cases indicate that over 85% of deaths are caused by smoke, which becomes a deadly substance for passengers in fires [1]. In recent years, scholars at home and abroad have proposed a new smoke blocking technology based on air curtain, that is, air curtain smoke blocking technology. The air curtain is powered by a fan to shoot out a curtain shaped area with a certain jet flow rate from the nozzle, in order to prevent air, dust, and other substances from invading the protected area [2]. Li Shengtao studied the effects of three factors in the ventilation system: smoke exhaust outlet height, smoke exhaust spacing, and smoke exhaust velocity on the smoke exhaust effect when a fire occurs at a subway transfer station [3]. Wang Meiqi verified and evaluated the operational effect of air curtains in fire accidents. Compared with the traditional smoke prevention mode which uses the hang wall to block the smoke, air curtains can effectively prevent

I. Tuns et al. (Eds.): CIBv 2023, LNCE 510, pp. 311–322, 2024.
https://doi.org/10.1007/978-3-031-60765-3_25

smoke diffusion and facilitate personnel evacuation [4]. L. H. Hu et al. studied the effect of air supply outlet position on smoke exhaust efficiency and discussed the operating time of smoke exhaust fans [5]. Meng Huaning studied the smoke spread of subway fires under different smoke exhaust modes and found that the smoke control effect of lateral smoke exhaust coupled with station hall air supply was the best [6]. Hailing Yue used FDS to simulate the fire at the end and the middle of the platform, and proposed the optimal ventilation method for the subway under different fire conditions [7]. Zhang Wenwu studied the temperature distribution and flow field characteristics of smoke in natural ventilation and mechanical ventilation fires under different fire scenarios [8]. Jian Li et al. found that regardless of whether it is a fully enclosed platform door or a full height platform door system, the removal of rail top smoke exhaust outlets has a lower impact on the temperature, visibility, CO concentration, and other factors of smoke in public areas [9]. He Jiapeng et al. used FDS to simulate the effect of vertical jet and hang wall combination on smoke control in high-rise building fires in corridors [10]. Lou Na et al. used a combination of experimental and numerical simulation methods to study the inhibitory effect of air curtains on smoke spread in high-rise building fires [11]. Guo Jianxiang et al. conducted a fire simulation on a simple building model to study the differences in smoke blocking effects between air curtains and hang wall and at different distances from the fire [12].

This article aims to study the effect of different vertical air curtains on the smoke propagation characteristics when a fire occurs in a subway platform. Schematic diagram of jet flow of different vertical air curtains is shown in Fig. 1.

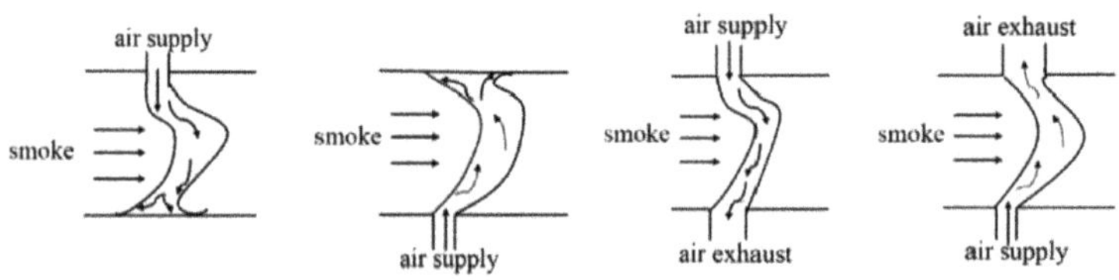

Fig. 1. Airflow of different air curtains

2 Model Establishment and Experimental Conditions

In this paper, FDS fire dynamics numerical simulation software is used to establish a full-scale subway station model. The simulation model is a two-story island subway station, the dimensions of the platform are length × width × height = 114 m × 10.8 m × 4.5 m. The fire source is located 8 m in front of the left staircase, with an area of 0.8 m × 0.8 m × 0.8 m. British scholars carried out statistics on the fire source power of dense public places, and obtained the fire source power size of 2.0–2.5 MW [13]. Therefore, in this paper, the fire source power (HRR) is 2.5 MW. The simulation uses t^2 fast growing fire. The station has three passenger entrances and exits. When a fire occurs on the platform, turn on the smoke exhaust system on the platform level and the air supply system on the concourse level, with a smoke exhaust volume of 72576 m^3/h. The air supply volume is 36432 m^3/h. The air curtain is installed at the left staircase entrance, behind the hang wall. The width of air curtain's air supply outlet is 0.3 m, and the width of the air exhaust outlet is 0.6 m. The blowing and suction speeds of the air curtain are all 2.5 m/s. Without considering the influence of external wind speed, the initial temperature is set to 20 °C and the pressure is set to normal pressure. The grid size near the air curtain is 0.05 m × 0.05 m × 0.05 m, grid size near fire source 0.1 m × 0.1 m × 0.1 m, grid size in other areas 0.2 m × 0.2 m × 0.2 m. The floor plan of the station hall and platform is shown in Fig. 2. The working conditions of this paper are set up as shown in Table 1.

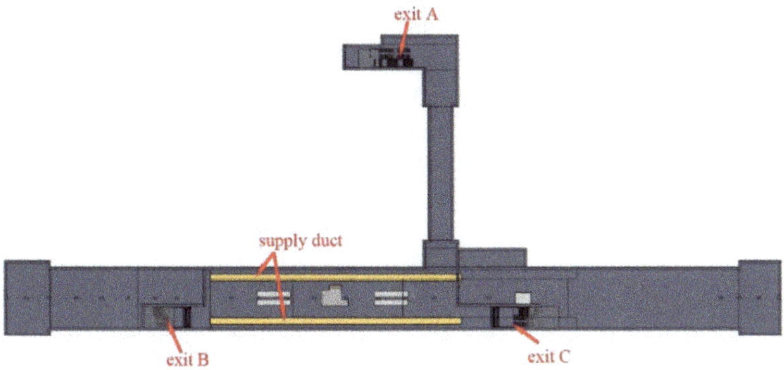

(a) station hall

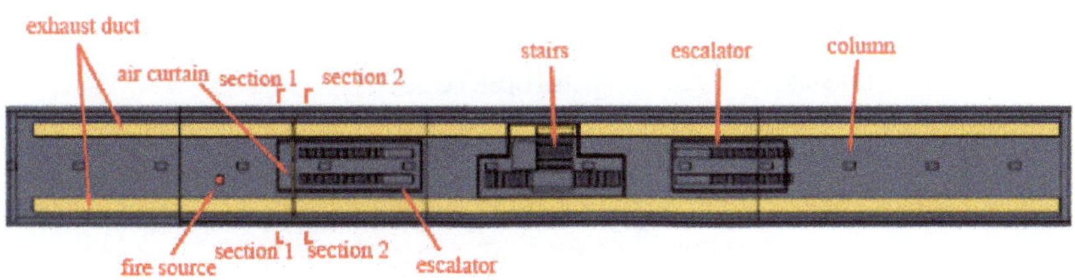

(b)platform

Fig. 2. Plan of the mode

Table 1. Working conditions

Simulated condition	Power of the fire source(HRR)	Form of air curtain
Condition 1	2.5MW	——
Condition 2		US
Condition 3		DS
Condition 4		USDE
Condition 5		DSUE

3 Simulation Results and Analysis

According to the Design Specification for Fire Evacuation and Rescue Engineering in Railway Tunnels [14] and the US NFPA130 [15] regulations, the evacuation time for personnel shall not exceed 6 min, and the numerical simulation duration shall be taken as 6 min. By analyzing the visibility and CO concentration at the staircase exit at 360 s, as well as the temperature changes within 360 s, determine the smoke blocking effect of different air curtains. Considering that the characteristic height of the human eye is 1.8 m, visibility, CO concentration, and temperature measurement points are set at position 1 and position 2, which are 1 m away from the outside and inside of the air curtain at a height of 1.8 m, respectively. The visibility slice is placed at cross-section 1 of the cross-section 1 m away from the inside of the air curtain, and the CO concentration slice is placed at cross-section 2, which is 1.8 m high. The location of the measuring point is shown in Fig. 3. The cross-section is shown in Fig. 4.

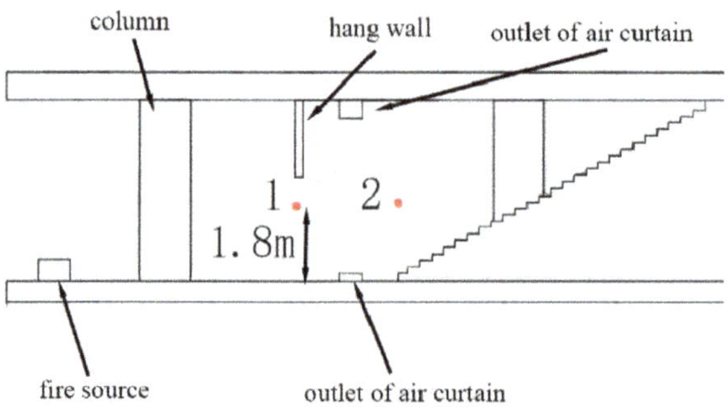

Fig. 3. Location of the measuring point

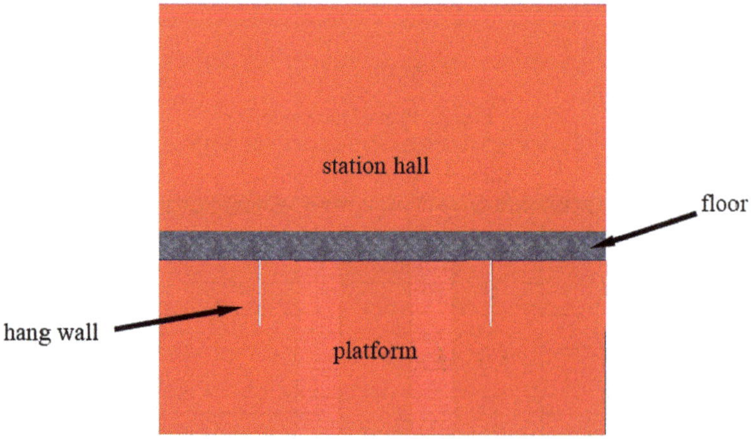

Fig. 4. Section 1

3.1 Impact on Visibility

According to the provisions of the Subway Design Code (GB50517-2013) [16], the acceptable visibility standard in the subway station is set at 10 m to ensure that the subway station personnel can find at least one evacuation sign in the personnel passageway. The visibility distribution on cross-section 2 of each simulated condition is shown in Fig. 5. From the figure, it can be seen that under the same condition of platform fire, different types of air curtains have different smoke control effects. Simulated condition 1 does not have an air curtain installed, on cross-section 1, only a small number of areas in the lower part of the platform layer have reduced visibility to 12 m, while the visibility in the remaining areas has decreased to 0 m. In the station, only the upper part has reduced visibility to 0 m, while the visibility in the middle and lower parts remains at 30 m (normal level). Simulated condition 2 is equipped with an US air curtain, and the visibility on cross-section 2 is basically 0 m. The smoke has spread to the station hall, filling the entire cross-section 2, and the overall visibility is not better than simulated condition 1. In simulated condition 3, a DS air curtain is set up, and the visibility of the lower area on cross-section 2 is low, while the visibility of the upper area is high. Due to the presence of hang wall in the upper part of the platform, smoke accumulation causes the visibility on the outer side of the hang wall to decrease to 0 m. Due to the downward settlement of smoke in the upper part, the visibility in the lower part of the platform decreases to 15 m. At the station hall level, visibility is reduced to 15 m only at the ceiling, while the visibility of the remaining parts remained at 30 m, with overall visibility better than condition 1. In simulated condition 4, an USDE air curtain is set up, and the overall visibility of the platform on cross-section 2 is reduced to 0 m. At the station hall level, the visibility decreased to 6 m at the ceiling, while the visibility of the other areas remained at 30 m, indicating that smoke filled the platform and spread to the ceiling of the station hall floor, and the overall visibility was not better than condition 1. In simulated condition 5, a DSUE air curtain is set up. On cross-section 2, at the platform, the visibility decreases to 0 m in the outer side of the hang wall, while the visibility of the other areas remains at 30 m. The smoke does not settle significantly, nor does it spread to the station hall. The visibility of the station hall has not decreased and remains at 30 m, with overall visibility significantly better than simulated conditions 1–4.

Figure 6 shows the variation curves of visibility over time at positions 1. It can be seen that at position 1, the visibility curve of simulation condition 1 suddenly shows a downward trend at 230 s and then immediately rebounds. The curve fluctuates up and down at 25 m due to the smoke exhaust effect of the exhaust outlet. At 250 s, the visibility curve once again shows a downward trend accompanied by upward and downward fluctuations. At 360 s, the visibility curve drops to 3 m, indicating the presence of a large amount of smoke at position 1. At 40 s, the visibility curve of simulated condition 2 suddenly dropped to 2 m, and the visibility remained at 2 m, indicating that due to the downward airflow of the air curtain, a large amount of smoke had already spread to position 1 at 40 s. The visibility curve of simulated condition 3 and simulated condition 2 shows a downward trend at around 40 s. At 50 s, the visibility decreases to about 10 m, and then the visibility curve rebounds to 30 m. This is because the upward airflow of the air curtain exerts an upward force on the smoke. At 80 s, the visibility

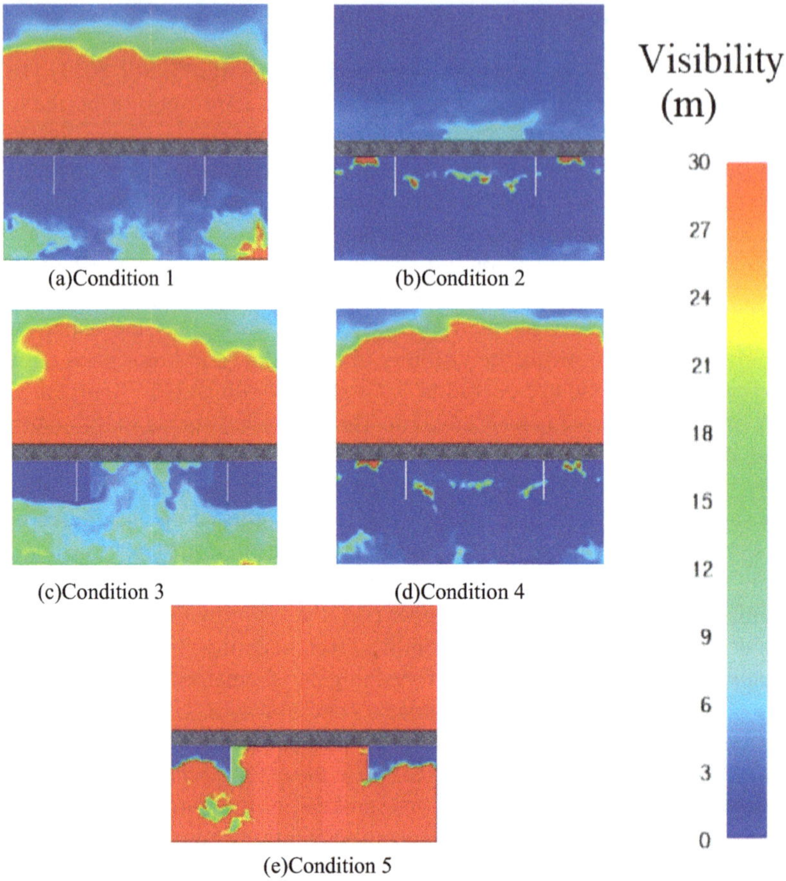

Fig. 5. Visibility distribution on cross-section 1 at 360 s

curve once again decreases, and at 160 s, it drops to about 5 m, indicating that a large amount of smoke is spreading to position 1. After that, the visibility curve slightly rises, but fluctuates violently, This is because the upward airflow of the air curtain has a significant disturbance effect on the smoke. The visibility curve of simulated condition 4 also shows a downward trend around 40 s and suddenly drops to around 3 m. Since then, the visibility has remained at about 3 m. This is because both the downward airflow of the air curtain and the air exhaust outlet on the ground will exert a downward force on the smoke. At 55 s, the visibility curve of simulated condition 5 showed a downward trend, and at 160 s, the visibility decreased to about 10 m. Afterwards, the visibility curve experienced severe fluctuations, similar to the changes in simulated condition 3 curve.

Figure 7 shows the variation curves of visibility over time at positions 2. It can be seen that the visibility curve of working condition 1 suddenly shows a downward trend at about 220 s, and the curve fluctuates significantly up and down at about 20 m. At 250 s, the fluctuation amplitude of the curve decreases, and at 360 s, the curve continues to fluctuate up and down at about 5 m, indicating the presence of a large amount of smoke at position 2. Due to the high concentration of smoke, the disturbance effect of the smoke outlet on the smoke decreases. At 60 s, the visibility curve of condition 2 shows a downward trend, which is 160 s earlier than condition 1. This is because the downward airflow of the air curtain has a suction effect on the smoke, and a large amount of smoke accumulates on the outside of the air curtain and spreads inward across the air curtain. The fluctuation amplitude of the curve is large, and as time increases, the fluctuation decreases. At 170 s, the visibility decreases to 0 m, and then it remains 0 m, which is 3 m lower than the minimum visibility of condition 1. The visibility curve of working condition 3 began to show a downward trend at 210 s, which was 10 s earlier than working condition 1, indicating that the upward airflow has a relatively small suction effect on smoke. At 360 s, the visibility decreases to 7 m, which was 4 m higher than the minimum visibility of condition 1. At 60 s, the visibility curve of condition 4 shows a downward trend, which is 160 s earlier than working condition 1. The curve fluctuated significantly at 15 m, and after 135 s, the visibility remains stable at about 2 m, indicating the presence of a large amount of smoke at position 2. The smoke exhaust outlet at the lower part of the air curtain can not timely discharge the smoke. The visibility curve of condition 5 does not show a downward trend and remained at 30 m, indicating that smoke does not spread to position 2.

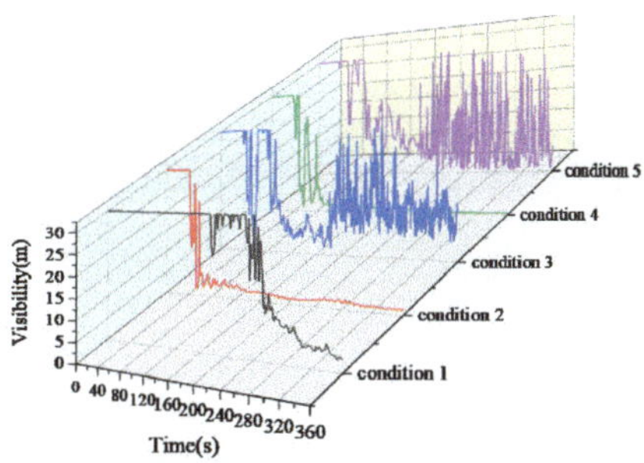

Fig. 6. Visibility -time curves at position 1

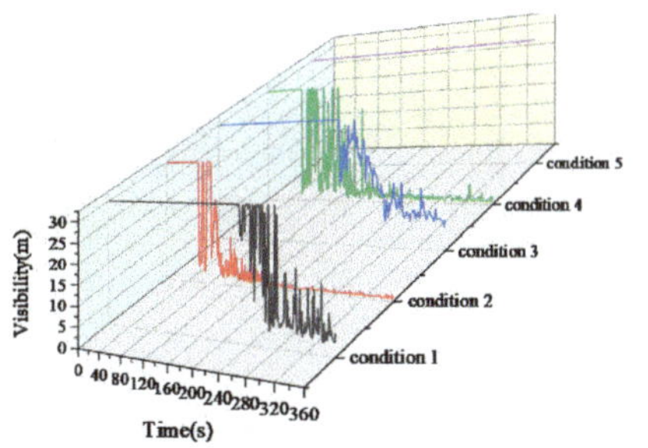

Fig. 7. Visibility -time curves at position 2

3.2 The Impact on CO Concentration

When a subway fire occurs, low oxygen combustion will produce a lot of CO. The CO concentration distribution on Sect. 3 of each simulated condition is shown in Fig. 8. In working condition 1, the CO concentration at the ignition source is 200 ppm, while the CO concentration at the left end and middle staircase is relatively high, reaching 200 ppm and 140 ppm respectively. The CO concentration in the middle is low. In condition 2, the CO concentration at the ignition source is 400 ppm, which is 200 ppm higher than that in condition 1. This is because the downward airflow from the air curtain has a promoting effect on the ignition source's combustion, and smoke accumulates at the left end and middle staircase, resulting in a CO concentration of 240 ppm. In condition 3, the CO concentration at the ignition source is 400 ppm, which is 200 ppm higher than that in condition 1. The CO concentration at the left end and left staircase is higher, reaching

400 ppm and 240 ppm respectively, while the CO concentration in other areas is almost zero. This is because the blocking effect of the air curtain reduces the CO concentration inside the air curtain. In condition 4 the CO concentration at the fire source is 600 ppm, which is 400 ppm higher than condition 1 and 200 ppm higher than conditions2 and condition 3. This is because the downward airflow of the air curtain and the lower air exhaust outlet intensify the settling of the smoke. The CO concentration at the left end

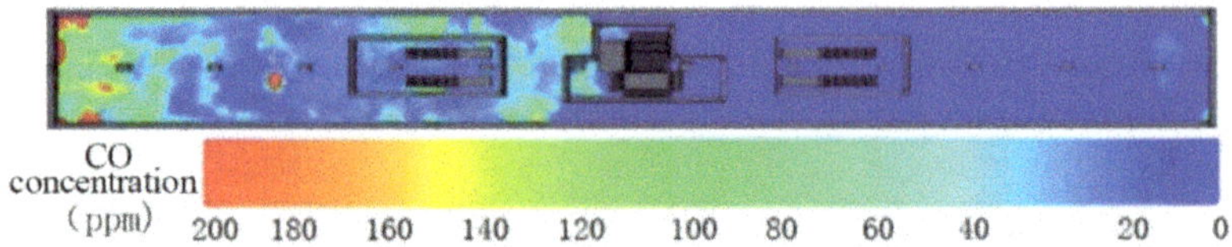

(a) Condition 1

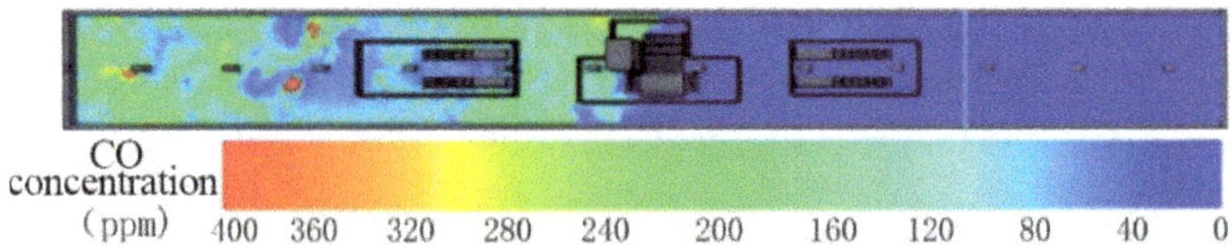

(b) Condition 2

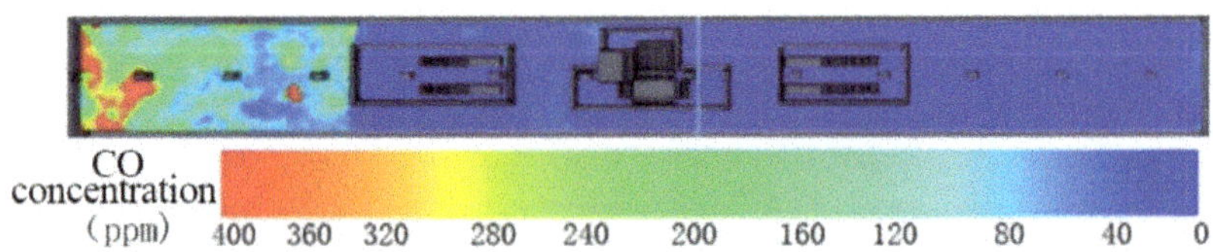

(c) Condition 3

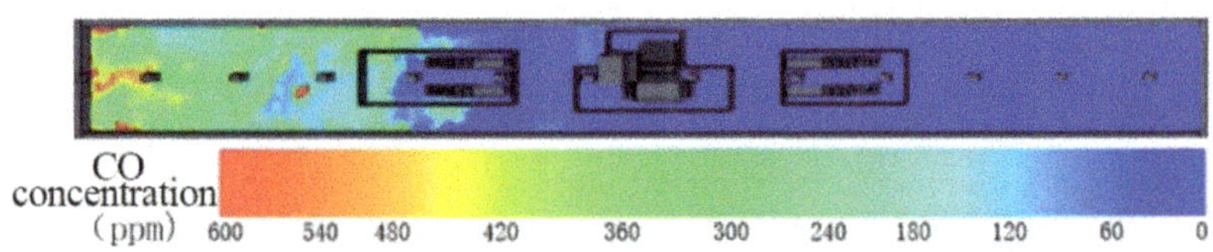

(d) Condition 4

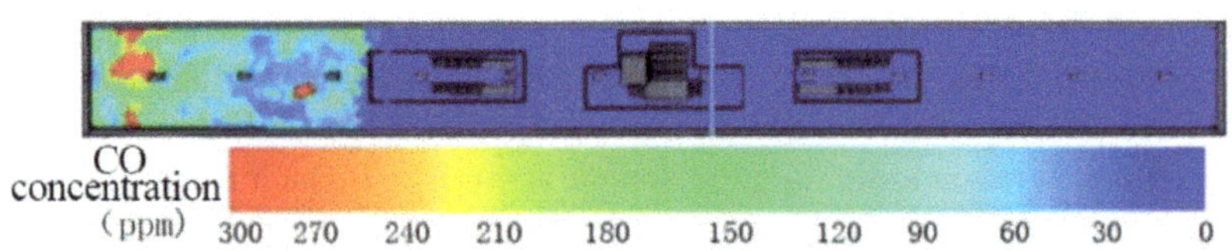

(e)Condition 5

Fig. 8. CO concentration distribution on section 2 at 360 s and left staircase is higher, reaching 360 ppm, while the CO concentration in other areas is almost zero, indicating that the air curtain has played a certain role in blocking smoke. In condition 5 the CO concentration at the fire source is 300 ppm, which is 100 ppm higher than that in condition 1. This is because the upward airflow of the air curtain has a combustion supporting effect on the fire source, but the upper air exhaust outlet of the air curtain can timely discharge some smoke. The CO concentration at the left end and left staircase is relatively high, reaching 300 ppm and 180 ppm, respectively. Due to the smoke blocking effect of the air curtain, the CO concentration in other areas is almost zero.

3.3 Impact on the Temperature of Smoke

Subway fires generate a large amount of high-temperature gas, which radiates to the surrounding area through various heat conduction methods. Figure 9 shows the variation curves of the temperature of smoke over time at positions 1 of each operating condition. In condition 1 the temperature curve at position 1 starts to rise at 80 s, after the temperature slowly rises to 22 °C, the rate of increase accelerates. At 360 s, the highest temperature reaches 37 °C. This is because the smoke first spreads horizontally along the ceiling to the left end, and then begins to spread downwards to the right. Therefore, the temperature at position 1 is initially lower and rises slowly, and then increases at a faster rate; At position 2, the temperature began to slowly rise at 140 s, indicating that smoke has spread to the inside of the air curtain. By 360 s, the highest temperature reached 25 °C; The maximum temperature difference between position 1 and position 2 is 12 °C. In working condition 2 the temperature curve at position 1 began to rise at 35 s, 45 s earlier than working condition 1. At 360 s, the highest temperature reached 42 °C, which was 5 °C higher than the highest temperature in condition 1. This is because the downward airflow of the air curtain plays a role in supporting combustion for the fire source; At 60 s, The curve at position 2 starts to rise at 60 s, 80 s earlier than condition 1. At 360 s, the highest temperature reached 45 °C, which was 20 °C higher than the highest temperature in condition1. The maximum temperature difference between position 1 and position 2 is –3 °C. The highest temperature inside the air curtain is higher than the highest temperature outside the air curtain, indicating that the US air curtain does not have the effect of blocking the spread of smoke. Instead, it will have a combustion promoting effect on the fire source and accelerate the spread of smoke. In working condition 3, the temperature curve at position 1 began to rise at 35 s, 45 s earlier than condition 1. At 260 s, it rises to 33 °C and then began to decline, indicating that a large amount of smoke spreads across the upward blowing airflow towards the inside of the air curtain. The upward airflow of the air curtain played a role in reducing the smoke temperature. At 360 s, the temperature dropped to 23 °C, and the highest temperature was 4 °C lower than the highest temperature in condition 1; The temperature curve at position 2 starts to rise at 100 s, 40 s earlier than condition 1. At 360 s, it slowly rises to 23 °C, 2 °C lower than the highest temperature in condition 1; The maximum temperature difference between position 1 and position 2 is 10 °C, indicating that the downward blowing air curtain can block the spread of smoke towards the inside of the air curtain. In working condition 4 the temperature curve at position 1 also starts to rise at 35 s, and the highest temperature reaches 70 °C at 360 s, which is 33 °C higher than the highest temperature in condition 1. This is because the lower air exhaust outlet of the air curtain causes the upper high-temperature smoke to settle downwards; The temperature curve at position 2 starts to rise at 50 s, 90 s earlier than condition 1. At 360 s, the temperature reaches 35 °C, which is 10 °C higher than the highest temperature in condition 1; The maximum temperature difference between position 1 and position 2 is 35 °C; Overall, the air curtain did not play a role in blocking the spread of smoke; In condition 5 the temperature curve at position 1 starts to rise at 40 s, 40 s earlier than working condition 1. At 160 s, it rises to a maximum temperature of 33 °C and then begins to decrease. The maximum temperature is 4 °C lower than working condition 1. At 350 s, the temperature drops

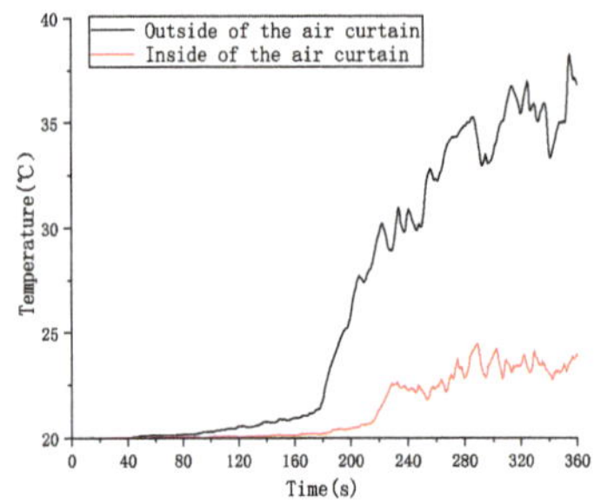

(a)Condition 1

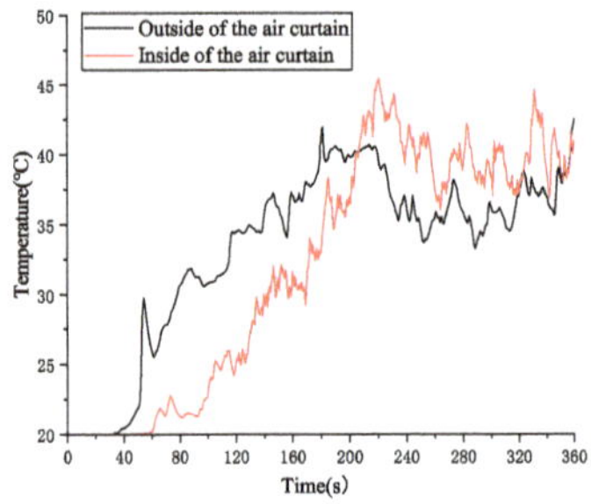

(b)Condition 2

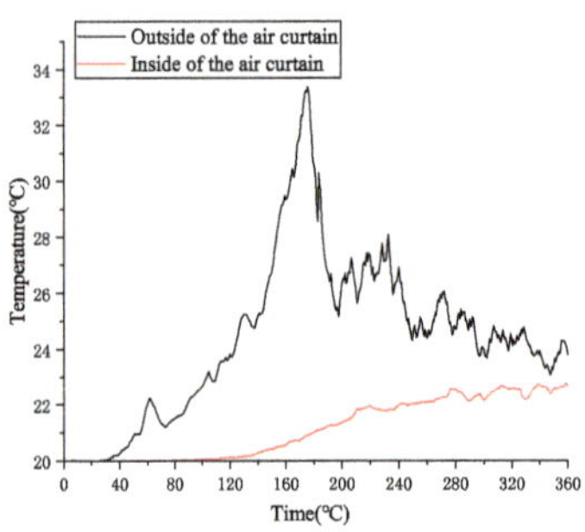

(c)Condition 3

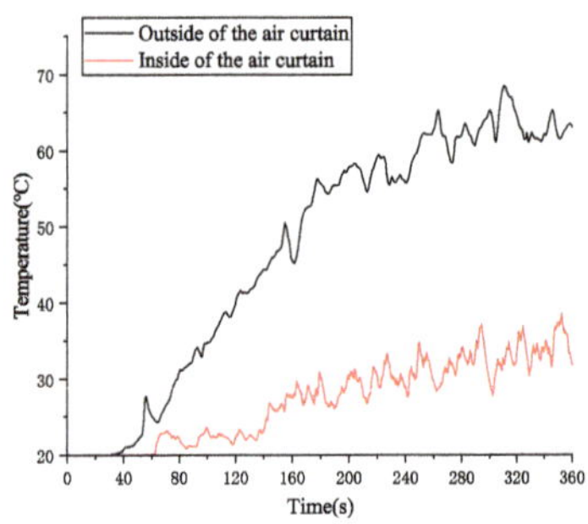

(d)Condition 4

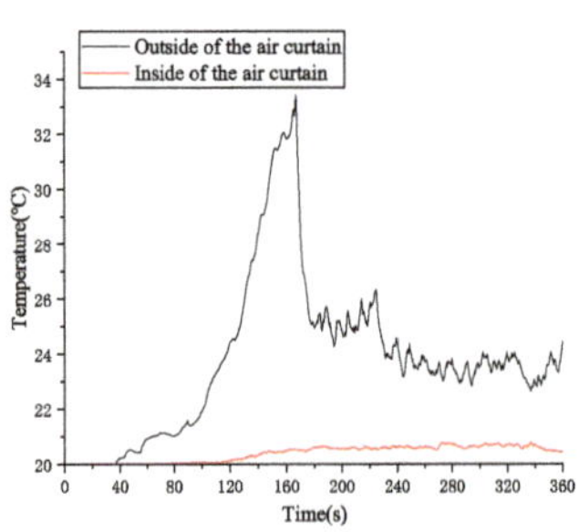

(e)Condition 5

Fig. 9. Temperature – time curves at position 1 and position 2 to 23 °C; The temperature at position 2 started to slowly rise at 110 s, 30 s earlier than condition 1. At 360 s, it rises to 21 °C, 4 °C lower than condition 1. The maximum temperature difference between position 1 and position 2 is 12 °C (Fig. 9).

4 Conclusions

(1) Compared with the absence of an air curtain, the DSUE air curtain has the best smoke control effect, ensuring that the visibility of the protected area by the air curtain is basically maintained at 30 m. Secondly is the DS air curtain, which increases the visibility of the protected area by 4 m. The USDE air curtains have poor effects, and the overall visibility of the area protected by the air curtain decreases.
(2) Compared with the absence of an air curtain, the DSUE air curtain has the best blocking effect on CO. Although the CO concentration in the fire source area has increased, the CO concentration in the area protected by the air curtain has decreased, and the CO concentration in the area protected by the air curtain is almost maintained at a normal level; Secondly, the DS air curtain can effectively reduce the CO concentration in the area protected by the air curtain, but the CO concentration in the fire source area is higher than that of the DSUE air curtain; The USDE air curtain has a certain blocking effect on the CO concentration in the area protected by the air curtain, but the CO concentration in the fire source area increases significantly; The upper supply air curtain has the worst blocking effect on CO concentration.
(3) Compared with the absence of an air curtain, the DSUE air curtain has the best temperature blocking effect. The temperature in the fire source area and the area protected by the air curtain has decreased, and the temperature in the area protected by the air curtain has increased by 1 °C compared to normal temperature. The air curtain has a good temperature blocking effect; secondly is the DS air curtain, the temperature in the fire source area and the area protected by the air curtain has decreased. The temperature in the area protected by the air curtain has increased by 3 °C compared to normal temperature; The temperature in the fire source area and the protected area of the US air curtain and the USDE air curtain have increased, with the US and downside air exhaust air curtain having the highest temperature and the worst barrier effect on smoke.

In future scientific research, the blowing and suction flow ratio of the downside air supply and upper air exhaust air curtain can be simulated and analyzed, as well as the smoke blocking effect of the air curtain when fire occurs at different locations of the subway platform. It is hoped that such research can be carried out in the future to explore whether there are better implementation methods and scenarios for improving the air curtain.

References

1. Hietaniemi, J., Kallonen, R., Mikkda, E.: Burning characteristics of selected substances: production of heat, smoke and chemical species. Fire Mater. **23**(4), 171–185 (1999)
2. Huang, R.: The application of air curtain for smoke confinement in metro station. Chongqing University, Chongqing (2016)

3. Li, S.: Study on factors influencing smoke exhaust effect when platform layer or station hall layer at asubway transfer station in Fire. Xi' an University of Architecture and Technology, Xi' an (2019)
4. Wang, M.: Numerical simulation analysis of smoke control and exhaust system in Shenyang subway. Shenyang Jianzhu University, Shenyang (2020)
5. Hu, L.H., Li, Y.Z.: Full-scale experimental studies on mechanical smoke exhaust efficiency in an underground corridor. Build. Environ. **41**, 1622–1630 (2006)
6. Meng, H.: Research on smoke extraction model of subway stations based on fire simulation and personnel evacuation. Southeast University, 2021
7. Hailing, Y.U.E., Li-ya, X.U., Song, G.A.O.: Simulation of ventilation mode in case of subway platform fire. Res. Explor. Lab. **31**(3), 36–38 (2012)
8. Zhang, W., Liu, G., Ren, F., et al.: Study on fire smoke spread in deep buried station (1): influence of mechanical air supplement at entrance and exit. J. Saf. Sci. Technol. **2021**(17), 13–17
9. Jian, L.I., Congling, S.H.I., Xuan, X.U.: Study on influence of smoke vent at rail top on smoke exhaust effect for fire in underground station of subway. J. Saf. Sci. Technol. **15**(1), 176–180 (2019)
10. Yunchun, Y.A.N.G., Jiapeng, H.E., Ru, Z.H.O.U., et al.: Study on air jet and hang wall combined smoke control. Fire Sci. Technol. **30**(7), 574–576 (2011)
11. Lou, N., Li, A., Gao, R., et al.: An experiment and simulation of smoke confinement utilizing an air curtain. Saf. Sci. **59**, 10–18 (2019)
12. Jianxiang, G.U.O., Zihan, Z.H.O.U., Jinfei, S.U.N., et al.: Influence and effects of different distances of the fire source between the air curtain and hunging wall. J. Saf. Environ. **21**(2), 613–619 (2021)
13. Su, J.K.: Study on smoke flow simulation and smoke exhaust mode of an island subway station in Xi'an. Xi'an University of Science and Technology, Xi'an (2019)
14. National Railway Administration. Design Specification for Fire Evacuation and Rescue Engineering of Railway Tunnels: TB 10020-2017. Beijing: China Railway Press, 2017
15. NFPA130. Standard for fixed guideway transit and passenger rail systems. Quincy: NFPA, 2010
16. Ministry of Housing and Urban-Rural Development, People's Republic of China. GB50157-201, Subway Design Code. Beijing: China Building and Construction Press, 2014

Analysis of Factors Affecting Sleeve Valve Pipe Grouting in Sand Layers

Zhu Gang(✉) and Xu Yan

School of Civil Engineering, Shenyang Jianzhu University, Shenyang 110000, China
1608892565@qq.com

Abstract. The uniform design experiment method was adopted to develop a grouting test plan. By conducting a simulated sleeve valve pipe grouting experiment, the impact of grouting pressure and water-cement ratio on the spreading of grout in sand layers can be analyzed: Under the pressure of grouting, the dominant position of the splitting effect combined with the permeation effect is filling the grout-receiving soil layer in multiple forms and coexisting with the permeation flow portion, which is well bonded to the grout-receiving soil layer. The nonlinear relationship between the diffusion radius of the slurry, grouting pressure, and water-cement ratio during the grouting process is obtained by employing regression analysis for test data design.

Keywords: Sleeve Valve Pipe · Sand Layer · Diffusion Radius · Grouting Simulation Experiment

1 Introduction

In the process of engineering construction and development, there are many problems of engineering foundation: the instability of foundation leads to the collapse of the whole house, the instability of foundation pit, the soil gushing caused by the demolition of shield door, and the collapse of road subgrade. There are more and more problems in foundation treatment, and the difficulty is also increasing, which puts forward higher requirements for foundation treatment. Sleeve valve grouting method was independently developed by Soletanche Foundation Company in France. As a fine-grained grouting process, it was applied in projects such as foundation reinforcement, water sealing curtain and settlement control of buildings, with many successful cases.

At home and abroad, the sleeve valve grouting method is considered to be the most reliable grouting method, the main factors affecting the grouting effect are: sleeve valve tube shell material, grouting material and grouting pressure, many scholars have carried out related research. Through laboratory tests and finite element simulation, Olivier Chupin et al. [1] proved that in terms of grouting volume and grouting pressure during grouting, sand filtration had a significant influence on slurry diffusion in soil layer. Pang Jianyang et al. [2] found in the grouting test of loose water-rich sand pebble layer that the influence of slurry diffusion radius, grouting amount, water content and grouting pressure was particularly significant. Wang et al. [3] designed a series of indoor

I. Tuns et al. (Eds.): CIBv 2023, LNCE 510, pp. 323–332, 2024.
https://doi.org/10.1007/978-3-031-60765-3_26

simulation tests on sandy soil, and conducted comparative analysis on grouting volume and shape of concretions by changing the water-cement ratio of slurry and soil saturation. Li Zhipeng et al. [4] found that the water-cement ratio was the main factor controlling the reinforcement effect of sand layer grouting through the sand layer penetration grouting test. Chupin et al. [5] proved through laboratory tests that the percolation effect of sand has a significant influence on the diffusion of cement grout. YANG et al. [6] found that there was a good power function relationship between the diffusion radius of cement grout in sandy soil and the grouting pressure, permeability coefficient and water-cement ratio, and gave the empirical function formula of the diffusion radius of cement grout.

Currently, there is a lack of in-depth research on the relationship between grouting pressure, slurry ratio and diffusion law in sleeve valve tube grouting. In this paper, by simulating the grouting process of sleeve valve pipe, the splitting and permeation process of grout in sandy soil layer are investigated, and the relationship between grout diffusion radius, grouting pressure and water-cement ratio is established.

2 Grouting Model Experiment

2.1 Experimental Equipment and Apparatus

(1) Simulation experiment box: According to the grouting data of sleeve valve pipe, the slurry diffusion radius is generally 0.5~1.5 m, so the simulation test chamber is set up according to the ratio of 1:5, the specific size is 600 × 600 × 600 mm, and it is built by three-piece wood sheet.
(2) Grouting equipment: According to the data, the grouting pressure is generally 0.6~2 Mpa. In this test, 22 w self-priming pump is selected as the equipment, the grouting pressure is adjusted according to the ratio of 1:100, and the slurry flow rate is adjusted by ball valve to achieve the purpose of changing the grouting pressure.
(3) Grouting pipe: Select a 47 mm diameter sleeve valve as the grouting pipe for the experiment, and set the section spacing to 100 mm due to the size limitations of the experiment box.
(4) Grout drum, grout pipe, grout injector, grout agitator, electronic scale, etc. (Figs. 1, 2 and 3).

2.2 Experimental Materials

The parameters of the sand in its natural state for the experiment are shown in Table 1:

2.3 Experimental Plan

(1) Cement Sodium silicate grout: The scheme for the cement water glass slurry in the grouting experiment is determined as follows: C:S = 1:0.5, and the water glass has a density of 30Bé.
(2) Casing material: The optimal ratio determined for conducting sleeve valve pipe grouting simulation test is as follows: Water: Cement: Bentonite = 2:1:1.

Fig. 1. Grouting device

Fig. 2. Grouting equipment

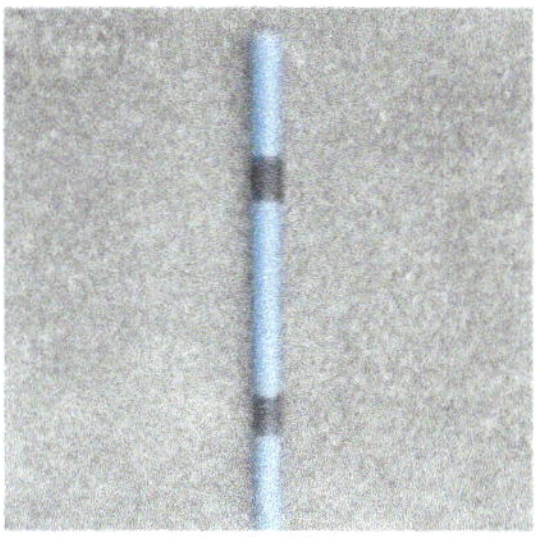

Fig. 3. Grouting pipe

Table 1. Test sand performance index

sample	$P(g/cm^3)$	e	$W(\%)$	$K(10^{-2}/s)$
sand	1.86	0.392	4.4	0.659

(3) Grouting test plan: This experiment considers 2 factors, with s = 2, which are grouting pressure (p) and water-cement ratio (w). The commonly used water-cement ratio for grouting ranges from 0.6 to 2 MPa. Five levels were finally selected as the test points for this experiment, and the specific data are shown in Table 2.

Table 2. Test factor level table

Levels	Factors	
	Grouting pressure(Mpa)	water-cement ratio (w)
1	0.06	0.7
2	0.08	0.8
3	0.1	1
4	0.15	1.25
5	0.2	1.5

2.4 Uniform Experimental Design

The grouting pressure (p) and water-cement ratio (w) are selected as the influencing factors for this grouting simulation experiment. The uniformity schematic diagram is shown in Fig. 4. The $U_{10}(5^2)$ table was used to arrange the test points for this experiment (Table 3).

Table 3. $U_{10}\left(5^2\right)$ uniform design table

Test number	Factors	
	X_1	X_2
1	1	2
2	3	3
3	5	2
4	4	1
5	2	5
6	5	4
7	1	4
8	3	3
9	4	5
10	2	1

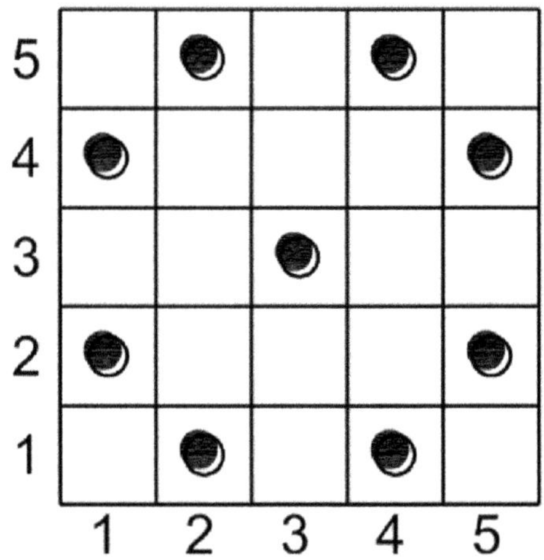

Fig. 4. Schematic diagram of uniformity

2.5 Experimental Procedure

(1) Measure the water, cement, and sodium silicate in the specified proportions by weight. Dilute the sodium silicate to a concentration of 30°Bé.
(2) Adjust the ball valve to the desired grouting pressure and start the grouting process. During grouting, continuously stir the grout mixture. Stop grouting after the grouting time reaches 30 s.
(3) After the completion of grouting, let the simulated test box stand still for 3 to 7 days, then disassemble it. Observe and measure the filling, flow, and diffusion of the grout mixture in the sand layer.
(4) Prepare the experimental box to be used for the next experiment.
(5) Repeat the experiment according to the $U_{10}(5^2)$ test plan, and record the data.

3 Grout Diffusion Shape Analysis

Three days after the completion of the grouting experiment, disassemble the test box, observe the shape of the formed stones, and analyze the diffusion characteristics of the grout mixture.

(1) When the water-cement ratio is greater than 1.25, the cohesiveness of the slurry is poor, and it is prone to breakage at weak points. When the water-cement ratio is equal to 1.25, the slurry has better cohesiveness and is less likely to break.

(a) (b) (c) (d)

Fig. 5. Shape of stone bodies

(2) The upper and lower parts of the stone body diffuse in the form of compaction and infiltration, forming cylindrical shapes with a relatively small diffusion range. The middle part of the rock mass is predominantly fragmented. On the surface, diffusion occurs in the form of infiltration, forming fan-shaped patterns with a larger range.
(3) Within the grouting range, the grout completely fills the pores of the sand, mainly exerting a compaction effect on the soil layer, which can consolidate loose soil into one whole (Fig. 5).

4 Multiple Regression Analysis

4.1 Establishment of Mathematical Models

The relationship between grout flow diffusion radius and grouting parameters is not linear, but a power function relationship [7–9]. Therefore, the linear regression model can be defined as follows:

$$Y = A \cdot X_1^B \cdot X_2^C \tag{1}$$

Taking the logarithm at both ends, we obtain:

$$\log Y = \log A + B \log X_1 + C \log X_2 \tag{2}$$

By substituting $y = \log Y, a = \log A, x_1 = \log X_1, x_2 = \log X_2$ the aforementioned regression model can be transformed into a linear regression model:

$$y = a + bx_1 + cx_2 \tag{3}$$

4.2 Multiple Regression Fitting Effect Analysis

After using the multiple linear regression mathematical model to fit the experimental data, the goodness of fit of the fitting result can be analyzed using the following data:

Residual:

$$res_i = y_i - (A + B_1X_{1i} + B_1X_{2i} + \cdots + B_kX_{ki}) \sim \left(0, \sigma^2\right) \tag{4}$$

Estimated standard deviation:

$$SD = \sqrt{\frac{RSS}{df}} \tag{5}$$

Coefficient of determination:

$$R - Square = 1 - \frac{RSS}{SYY} \tag{6}$$

Multiple regression analysis of variance is shown in Table 4.

Table 4. Decentralized reentry analysis

Content	Degree of freedom	SS	MS	F值
Regression	k	$SSreg = SYY - RSS$	$Msreg = Ssreg/I$	$Msreg/MSE$
Residual	$N - (k + 1)$	RSS	$MSE = RSS/N - k - I$	
Total	N-1	SYY		

Table 5. Grouting simulation test diffusion radius result table

Number	Grouting pressure(Mpa)	Water-cement ratio	Radius (m)
1	0.06	0.9	0.109
2	0.10	1	0.132
3	0.20	0.9	0.139
4	0.15	0.8	0.152
5	0.08	1.5	0.121
6	0.20	1.25	0.159
7	0.06	1.25	0.105
8	0.10	1	0.132
9	0.15	1.5	0.154
10	0.08	0.8	0.118

4.3 Analysis of Grout Diffusion Radius and Influencing Factors

Table 5 below displays the test results for the slurry diffusion radius in the grouting simulation using a sleeve valve tube.

The experimental data was organized according to the basic linear regression model set in the mathematical model, and the data was fitted with a computer. The independent and dependent variables are shown in Table 6.

Table 6. Test data function fitting data table

	1	2	3	4	5	6	7	8	9
X_1	−1.222	−1.000	−0.699	−0.824	−1.097	−0.699	−1.222	−0.824	−1.097
X_2	−0.046	0.000	−0.046	−0.097	0.176	0.097	0.097	0.176	−0.097
Y	−0.963	−0.879	−0.857	−0.818	−0.917	−0.799	−0.979	−0.812	−0.928

Performing linear regression analysis on the experimental data aforementioned can yield the relationship between the injection diffusion radius R and the water-cement ratio W as well as injection pressure P:

$$R = 0.2523P^{0.2598}W^{0.0547} \tag{7}$$

The coefficient of determination is used to judge the goodness of fit of a function, and its value varies between 0 and 1. If the coefficient of determination is close to 1, it means that the fit is good and the goodness of fit is high. The goodness of fit summary of the fitting function in this experiment is shown in Table 7.

Table 7. Goodness of fit table

	Intercept		*P*		*W*		Statistics
	Value	Standard error	Value	Standard error	Value	Standard error	R-square
R	−0.5981	0.0437	0.2958	0.0442	0.0547	0.084	0.8280

According to Table 7, the coefficient of determination R-square is 0.8280, which is close to 1. This indicates that the fitting effect of this experiment is relatively good. The most significant factor affecting the flow and diffusion of grout slurry during grouting is the grouting pressure, and the impact of water-cement ratio on the diffusion radius is significant when it is less than 1.

The flow and diffusion radius R of grouting material under different grouting pressures can be calculated using Eq. (7), as shown in Fig. 6.

From Fig. 6, it can be seen that when the water-cement ratio is between 1 and 0, the grout has a higher diffusion radius growth rate. When the water-cement ratio is greater than 1, the growth rate of diffusion radius gradually flattens. The effect of grouting pressure on the growth rate of grout diffusion radius is relatively stable, which indicates that the diffusion radius has been affected by grouting pressure.

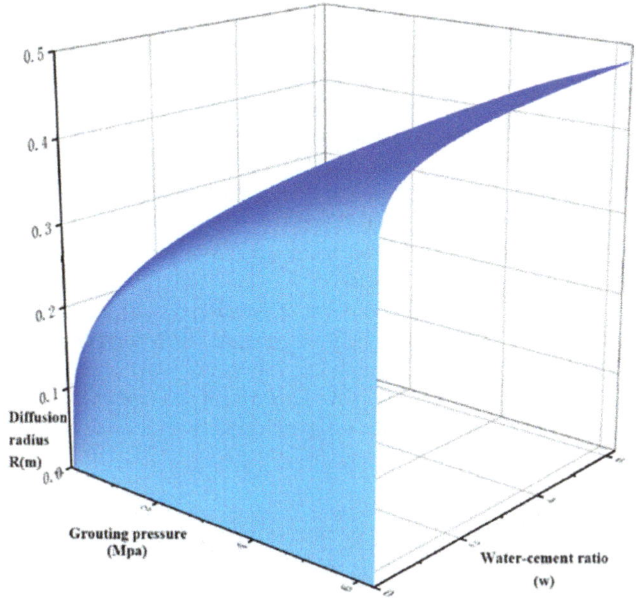

Fig. 6. Fitted function graph

5 Conclusion

(1) Under the action of pressure, the dominant effect on the grout slurry is the fracturing effect, which combines with the penetrating effect. In this process, multiple forms coexist to fill the receiving soil, and the permeable flow part is well bonded with the receiving soil layer.

(2) Through the regression analysis of test data design, the nonlinear relationship between grouting diffusion radius, grouting pressure and water-cement ratio in the sand layer is determined:

$$R = 0.2523P^{0.2958}w^{0.0547} \tag{8}$$

(3) Grouting pressure is the most obvious factor affecting the flow diffusion of grout, and the stability affects the diffusion radius of grout. When the water-cement ratio is less than 1, the growth rate of diffusion radius is greatly affected. In the range of 1~1.25, the diffusion performance of the slurry is better.

References

1. Chupin, O., Saiyouri, N., Hicher, P.Y.: The effects of filtration on the injected of cement-based grouts in sand columns. Transp. Porous Med. **72**, 227–240 (2008)
2. Peng, J., Yao, W.: Grouting simulation test in sandy gravels with loose and water-rich and its engineering application. Chin. J. Underground Space Eng. **15**(4), 1156–1163 (2019)

3. Wang, Q., Wang, S., Sloan, W.S., et al.: Experimental investigation of pressure grouting in sand. Soils Found. **56**(2), 161–173 (2016)
4. Li, Z., Zhang, L., Zhang, Q., et al.: Simulation test for permeation grouting reinforcement effect of sand layer. Meitan Xuebao/J. China Coal Soc. **43**(12), 3488–3497 (2018)
5. Chupin, O., Saiyouri, N., Hicher, P.Y.: The effects of filtration on the injection of cement-based grouts in sand columns. Transp. Porous Med. (2), 72 (2008)
6. Yang, J., Cheng, Y., Chen, W.: Experimental study on diffusion law of post-grouting slurry in sandy soil. Adv. Civ. Eng. **2019**(PT.3), 1–11 (2019)
7. Zhou, F., Xu, Y., Zhu, R., et al.: Diffusion characteristics of grouting slurry in sandy stratum. J. Archit. Civ. Eng. **37**(5), 182–192 (2020)
8. Yu, Y., Zhang, J., Fan, L., et al.: Development of grouting test device for fractured rock and analysis of slurry diffusion law under high-temperature water-rich condition. J. China Coal Soc. **47**(7), 2582–2592 (2022)
9. Hongyuan, F., Kang, L., Xueming, D., et al.: Experiment and numerical simulation of water plugging law of permeable polymer in the water-rich sand layer. Scientia Sinica (Technologica) **53**(3), 457–472 (2023)

Research on Safety Evaluation of Hoisting Engineering in Construction Based on Bayesian Network

Shilong Jia(✉), Yunfu Wu, Hengrui Liu, and Baoquan Peng

School of Civil Engineering, Shenyang Jianzhu University, Shenyang, China
ceslјia@sjzu.edu.cn

Abstract. Hoisting engineering is an important part of building construction, and its safety problems can't be ignored. To effectively assess and reduce the safety risk of hoisting operations in construction engineering, this paper introduces the T-S model based on the traditional accident tree model and combines it with the Bayesian network to establish a kind of safety evaluation model for the project of hoisting operation in construction engineering, and utilizes the model to evaluate the safety risk of the actual hoisting case. The results show that the model can quantitatively calculate the risk probability of the primary risk factors in the hoisting project. It can not only calculate the probability of the final event in a forward reasoning way to determine the critical control order of each risk factor but also carry out reverse reasoning to investigate the causes of safety accidents in the hoisting project. The safety evaluation model is close to the actual construction, which can provide theoretical support for the exemplary safety management of construction hoisting projects.

Keywords: Hoisting Engineering · Safety Evaluation · Bayesian Network

1 Introduction

In recent years, with the continuous expansion of the construction industry, lifting and hoisting, as an important part of construction engineering, is an irreplaceable technical means to effectively solve the vertical transportation and installation of building materials, engineering equipment, components and accessories in construction production. It is widely used in the production and construction of high-rise structures, super-high-rise structures, large-scale factories, and special structures.

While lifting is widely used, the safety problems that follow cannot be ignored. Once an accident occurs in the process of lifting and hoisting, it will not only easily cause a large number of casualties, but also cause significant economic losses and social impact. Therefore, it is necessary to conduct in-depth research on the safety of hoisting engineering.

Safety evaluation is an activity that uses the principles and methods of safety system engineering to identify and evaluate the risk factors existing in production activities, predict the possibility and degree of harm of accidents or occupational hazards,

I. Tuns et al. (Eds.): CIBv 2023, LNCE 510, pp. 333–344, 2024.
https://doi.org/10.1007/978-3-031-60765-3_27

put forward reasonable and feasible suggestions, and make evaluation conclusions. At present, safety evaluation has been widely used in the field of construction safety research.Sertyesilisik et al. investigated the hoisting operations of three construction sites in the UK, and revealed six key points to improve the safety of hoisting operations, namely, planning, training, equipment, communication feedback, personnel arrangement and database improvement [1].

Liu, etc. used the G1 method and the entropy weight method to determine the subjective and objective weights of the safety risk evaluation index of the assembled building hoisting construction, and used the cloud model to evaluate the safety of the assembled building hoisting construction [2]. Zhang, etc. used the analytic hierarchy process to establish the evaluation model of hoisting operation, and analyzed the influencing factors of hoisting operation safety from three aspects: human, machine and environment [3]. Chen, etc. established a safety evaluation model of hoisting machinery based on neural network, and analyzed the whole process of design, use and management of hoisting machinery [4]. Shin, etc. investigated 38 tower crane accidents in South Korea from 2001 to 2011. Through case analysis and group interviews, the safety influencing factors of tower crane installation and demolition stage were summarized [5]. Shapir, etc. used the analytic hierarchy process to establish a system of safety influencing factors in the use stage of tower crane, and quantified the index weight of each level and dimension in the system of safety influencing factors through expert scoring [6].

Fault tree analysis method is an effective tool for reliability analysis and fault diagnosis of complex systems. The traditional fault tree analysis method has been well applied in practice. Its advantages are simple, intuitive, and clear, but there are also certain disadvantages. When the fault mechanism is ambiguous, it cannot be analyzed using accurate fault data, and cannot be accurately described when historical data is scarce. These shortcomings restrict the application of fault tree analysis method in engineering field.

In summary, on the basis of previous studies, this paper uses the analysis model combining T-S fault tree and Bayesian network to evaluate the safety risk of construction hoisting engineering. First of all, through the analysis of the risk factors in the hoisting project, the T-S fault tree with the hoisting safety accident as the top event is established, and the T-S rule is used to describe the causal relationship between the various risk factors in the fault tree and the top event. Secondly, according to the mapping rules, the T-S fuzzy fault tree is transformed into a Bayesian network model of construction hoisting safety accidents. Finally, the model is applied to the risk analysis of practical engineering cases, and the Bayesian network is used to calculate the occurrence probability of each risk state of the top event and the posterior probability of the bottom event. Compared with the existing methods, the use of T-S rules instead of the 'and' gate, 'or' gate in the traditional fault tree to describe the complex relationship of each event in the fault tree can better adapt to the complex construction environment. The analysis results can provide theoretical support for the safety management of construction hoisting engineering.

2 T-S Fault Tree and Bayesian Network Reasoning

2.1 T-S Fault Tree

Fault tree analysis method is an effective tool for reliability analysis and fault diagnosis of complex systems [7]. The traditional fault tree is composed of event symbols and logic gates connected to it. It lists the fault state of the research object as the top event, and then analyzes it layer by layer from top to bottom until the bottom event that leads to the top event occurs, and then uses the logic gate to connect the events into a tree logic diagram, that is, the fault tree.

In the traditional fault tree analysis method, limited logic gates cannot accurately describe the complex relationship between events in the fault tree. Based on the traditional fault tree analysis method, a series of T-S gates with IF-THEN rules are used instead of traditional logic gates to describe the causal relationship between events in the fault tree, and the T-S fault tree is constructed to describe the evolution mechanism of accident risk more accurately. The T-S fault tree is shown in Fig. 1.

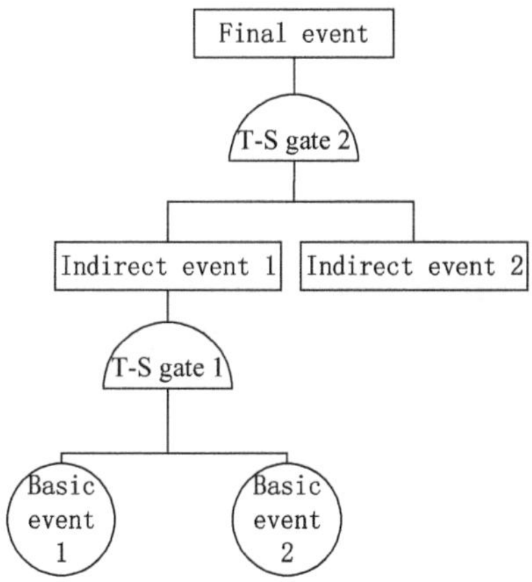

Fig. 1. T-S fault tree

2.2 Bayesian Network

The Bayesian network model consists of three elements: node, one-way arrow and conditional probability table. Each node in the network represents different events, that is, various random variables. The arrows of the one-way connection of the nodes are used to represent the causal relationship between different events or random variables in the Bayesian network. The conditional probability table is used to quantify the causal relationship between nodes in the network. The Bayesian network structure [8] is shown in Fig. 2.

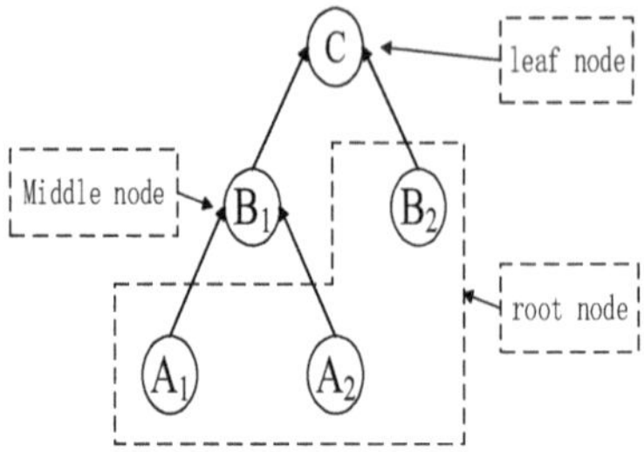

Fig. 2. Bayesian network structure

In the process of risk analysis, Bayesian network provides an uncertain probabilistic reasoning method, which has been successfully applied to practical analysis. It can collect and represent the knowledge of uncertain fields, and can also effectively carry out probabilistic calculus and statistical analysis. It can realize the two-way reasoning of the system, and has been effectively applied in the safety assessment of building construction.

2.3 T-S Fault Tree is Transformed into Bayesian Network

The basic events, intermediate events and top events in the T-S fault tree correspond to the root nodes, intermediate nodes and leaf nodes in the Bayesian network, respectively. The T-S rules in the T-S fault tree are transformed into a conditional probability table of nodes to represent the logical relationship between nodes in the Bayesian network [9]. Thus, the transformation between T-S fault tree and Bayesian network is completed. The conversion process is shown in Fig. 3.

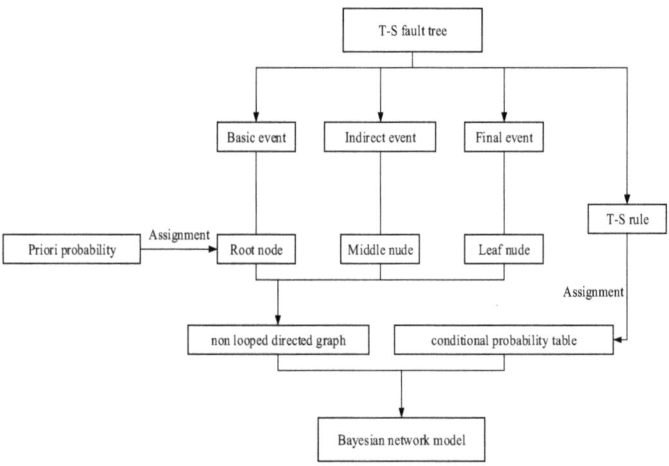

Fig. 3. T-S fault tree to construct Bayesian network process

This method is used to evaluate the safety of hoisting engineering, which is more universal and can meet the requirements of conditional probability and independence.

2.4 Node Probability Calculation

In the Bayesian network model, the probability of occurrence of each risk state of the node is used to describe the security degree of the node. After the occurrence probability of each risk state of the root node is known, the failure probability of the leaf node is calculated positively through the Bayesian network.

When the root node failure state is $x_1\prime, x_2\prime, \ldots, x_n\prime$, the probability of the leaf node failure state T_q is

$$P(T = T_q) = \sum_{x_i\prime; A_i\prime} P(x_i\prime; A_i\prime; T = T_q)$$

$$= \sum\nolimits_{u(A)} P(T = T_q | u(T)) \times \sum\nolimits_{u(b_1)} P(A_1|u(A_1)) \times \sum\nolimits_{u(b_1)} P(A_2|u(A_2)) \times \ldots \times \sum\nolimits_{u(b_1)} P(A_n|u(A_n)) \times \ldots \times P(x_1\prime) \times P(x_2\prime) \times \ldots \times P(x_n\prime) \quad (1)$$

where, $u(T)$ denotes the set of parent nodes of leaf node T, $u(A_n)$ denotes the set of parent nodes of middle node An.

Given the probability of occurrence of leaf node risk state T_q, the posterior probability of root node risk state $x_i\prime$ of Bayesian network can be obtained by using the backward reasoning algorithm of Bayesian network.

$$P(x = x_i\prime | T = T_q) = \frac{P(x = x_i\prime, T = T_q)}{P(T = T_q)} \quad (2)$$

3 System of Assessment

3.1 Identification of Safety Risk Factors for Hoisting Engineering

The hoisting project is mostly a group operation, which requires the cooperation of many people, such as cable signal workers, hoisting machinery drivers, installation personnel, etc., and involves the combined movement of hoisting machinery, hoisting tools and hoisting objects in a certain space. It is a complex interactive system of human, machine and environment, and it is difficult to control safety. Based on the complex safety characteristics of hoisting engineering, combined with the statistics of hoisting safety accident data and related literature analysis and C-HFACS model [10], the T-S fault tree model of construction hoisting engineering safety accident is established as shown in Fig. 4. The event descriptions in the model are shown in Table 1.

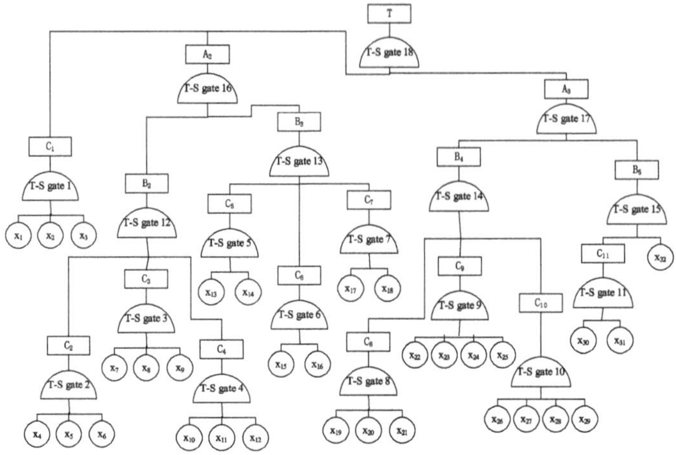

Fig. 4. T-S fault tree model of construction hoisting engineering

Table 1. T-S fault tree model event description

Number	Event	Number	Event
T	Hoisting safety accident	X_8	Lack of safety protection measures
A_2	Manage security risks	X_9	The quality of safety protection facilities is unqualified
A_3	Operation safety risk	X_{10}	Illegal contracting of construction enterprises
B_2	The influence of enterprise and project department organization	X_{11}	Site organization and management confusion
B_3	On-site safety supervision	X_{12}	The construction process is not standardized
B_4	Prerequisites for unsafe behavior	X_{13}	Safety technology disclosure is not in place
B_5	Unsafe act	X_{14}	Inadequate on-site safety inspection
C_1	Poor supervision of relevant parties	X_{15}	The special construction scheme is not perfect
C_2	Lack of cultural atmosphere	X_{16}	Unreasonable staffing
C_3	resource management	X_{17}	Safety responsibilities are not fulfilled in place
C_4	process control	X_{18}	Violate commanding

(*continued*)

Table 1. *(continued)*

Number	Event	Number	Event
C_5	Poor management of construction site	X_{19}	The psychological and physiological state of workers is poor
C_6	Unreasonable design work	X_{20}	Operators have no relevant qualifications
C_7	Violations by regulators	X_{21}	Operators ' safety awareness is weak
C_8	The status of operator	X_{22}	The selection of hoisting equipment does not meet the engineering requirements
C_9	Mechanical equipment	X_{23}	Hoisting equipment safety device failure
C_{10}	Poor environmental conditions	X_{24}	Hoisting equipment maintenance is not in place
C_{11}	Fault	X_{25}	Installation of hoisting equipment is not standardized
C_{12}	Breach of regulation	X_{26}	Bad weather at construction site
X_1	The supervision of relevant government departments is not in place	X_{27}	Insufficient setting of on-site safety warning signs
X_2	The inspection of the third-party testing agency is not in place	X_{28}	Poor night construction lighting conditions
X_3	Relevant safety technical specifications are not perfect	X_{29}	The installation area of hoisting equipment does not meet the operation requirements
X_4	Enterprises do not attach importance to safety	X_{30}	On-site management personnel decision-making mistakes
X_5	The safety management system is inconsistent with the actual situation on site	X_{31}	Operation personnel operation error
X_6	Safety education is not in place	X_{32}	Operation personnel illegal operation
X_7	Insufficient funds for production safety		

According to the characteristics of basic events and the analysis of historical documents, it is determined that the basic events X_5, X_{10}, X_{16}, X_{18}, X_{20}, X_{22}, X_{23}, X_{25}, X_{28} and X_{29} are two risk states, and the remaining basic events are three risk states. The fuzzy numbers 0, 0.5 and 1 are used to represent normal, general and serious risks. According to the transformation rules, the fault tree is transformed into a Bayesian network, as shown in the Fig. 5.

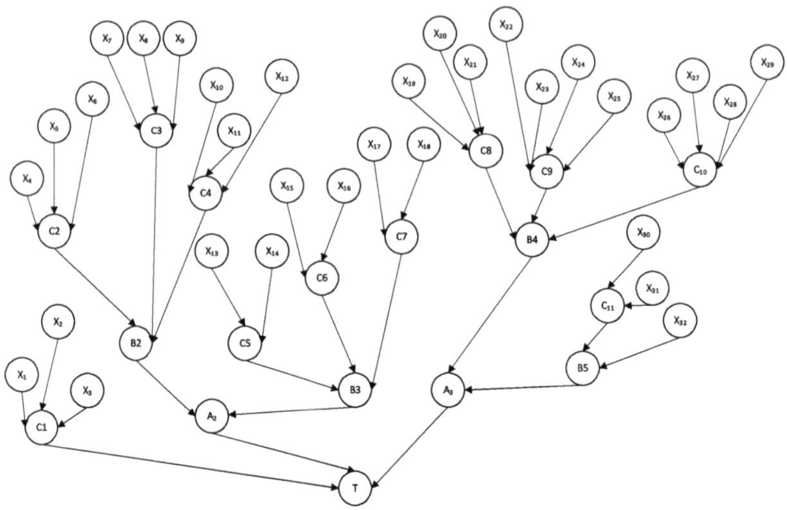

Fig. 5. Safety accident network model of hoisting engineering

4 Instance Analysis

Taking a group of residential projects as the evaluation object, the total land area of the project is 35852 square meters, and the total construction area of the project is about 120704.19 square meters. The building mainly includes 9 25-27-story frame-shear wall structure apartment buildings, 4 one-story frame structure supporting public service buildings, 5 one-story frame structure opening and closing houses, power distribution houses, etc., and a second-story basement. In order to ensure the safety of construction, a special construction scheme for group tower layout has been prepared.A total of four tower cranes are arranged, and the specific distribution area is shown in Table 2.

Table 2. Tower crane distribution area

Tower crane	Model	Covering radius	Application area
C1	QTZ80 (TC6012)	60	7#8#9#
C2	QTZ80 (TC6012)	60	5#6#7#8#
C3	QTZ80 (TC6012)	60	1#2#3#4#5#
C4	QTZ80 (TC6012)	60	1#2#3#

Six engineers with practical engineering experience and university teachers with relevant project research background are invited as experts to evaluate the safety risk of the hoisting project. The analytic hierarchy process is used to assign different values according to the relevant information such as the professional title of the surveyed experts, the time engaged in the construction industry, the level of education, etc., combined with relevant experience. The basic information and questionnaire of experts are shown in Table 3 and Table 4.

Table 3. Basic expert information sheet

Expert	Career	Professional title	Work seniority	Education background
1	Project manager	Engineer	>10 years	Regular college course
2	Chief engineer of project	Engineer	5–10 years	Regular college course
3	Safety director	Engineer	5–10 years	Regular college course
4	Technicist	Assistant engineer	<10 years	Master
5	University teachers	Professor	>10 years	Doctor
6	University teachers	Associate professor	5–10 years	Doctor

Table 4. Questionnaire of probability of occurrence of basic events of safety risk in hoisting engineering

Risk probability description			
Risk code	Risk statement		Probability
1	Very low		0.1
2	Low		0.3
3	Moderation		0.5
4	High		0.7
5	Very High		0.9
Elementary event	Risk state		
	0	0.5	1
X1			
X2			
X3			
...			

(*continued*)

Table 4. (*continued*)

Risk probability description			
X31			
X32			

The data of the accident probability of each basic event under severe risk and general risk obtained through expert investigation are shown in Table 5.

Table 5. Risk probability of basic events

Basic event	$p(d_i = 0.5)$	$p(d_i = 1)$	Basic event	$p(d_i = 0.5)$	$p(d_i = 1)$
X_1	0.062	0.029	X_{17}	0.119	0.034
X_2	0.114	0.034	X_{18}	–	0.110
X_3	0.090	0.045	X_{19}	0.156	0.030
X_4	0.106	0.088	X_{20}	–	0.054
X_5	–	0.062	X_{21}	0.201	0.067
X_6	0.116	0.097	X_{22}	–	0.034
X_7	0.120	0.051	X_{23}	–	0.045
X_8	0.112	0.050	X_{24}	0.105	0.041
X_9	0.056	0.034	X_{25}	–	0.080
X_{10}	–	0.070	X_{26}	0.067	0.029
X_{11}	0.105	0.034	X_{27}	0.055	0.071
X_{12}	0.142	0.078	X_{28}	–	0.129
X_{13}	0.112	0.059	X_{29}	–	0.065
X_{14}	0.101	0.034	X_{30}	0.095	0.045
X_{15}	0.082	0.045	X_{31}	0.198	0.098
X_{16}	–	0.065	X_{32}	0.097	0.041

According to the situation of this project, the Bayesian network model is established, and the probability of leaf node T in safe, general and serious risk state is obtained by using Eq. (1). The calculation results are shown in Table 6.

Table 6. Leaf node T risk state probability

Risk state	Safe	Ordinary risk	Serious risk
Probability	0.9584	0.0284	0.0132

According to the results, it can be concluded that in this project, the probability of serious hoisting safety accident and general hoisting safety accident is small and not much different; the probability of occurrence of the overall safety state of the hoisting project is large, which is consistent with the actual construction situation.

Using the Bayesian network reverse reasoning algorithm, the posterior probability of each basic event can be obtained when the hoisting safety accident occurs, which provides support for the analysis of the cause of the accident. Using Eq. (2) to calculate the posterior probability of the root node when the risk state of the leaf node T is 1, as shown in the Fig. 6.

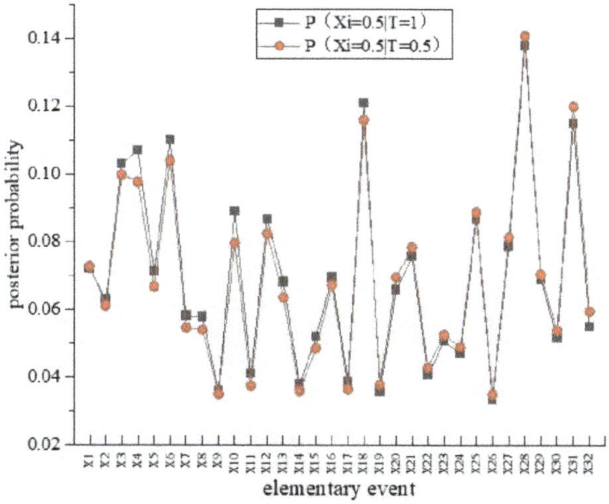

Fig. 6. The posterior probability of the root node T when the risk state of the leaf node is 1

From the diagram, it can be seen that when the hoisting safety accident occurs, the posterior probability of the root node X_3, X_4, X_6, X_{18}, X_{28}, X_{31} is larger. Basic events X_4 (Enterprises do not attach importance to safety), X_6 (Safety education is not in place), X_{18} (Violate commanding) belongs to the management of security risks. In the actual construction process, factors such as unmanned supervision, illegal construction, and lack of professional quality of management personnel directly affect the safety of the hoisting construction process. After the accident, the corresponding risk factors can be investigated and the cause of the accident can be determined.

5 Conclusion

The model can be calculated quantitatively according to the risk probability of the basic risk factors in the hoisting project. It can not only calculate the probability of the final event in a forward reasoning way, so as to determine the key control order of each risk factor, but also carry out reverse reasoning to investigate the causes of safety accidents in the hoisting project. The safety evaluation model is close to the actual construction, which can provide theoretical support for the fine safety management of construction hoisting project.

References

1. Sertyesilisik, B., Tunstall, A., McLouglin, J.: An investigation of lifting operations on UK construction sites. Saf. Sci. **48**(1), 72–79 (2010)
2. Liu, Z., Jiang, J.: Safety risk assessment of prefabricated building hoisting construction based on improved cloud model. J. Qingdao Univ. Technol. **44**(01), 30–38 (2023)
3. Zhang, Z., Wang, J.: Application of AHP evaluation method in safety management of hoisting operation. Constr. Saf. **37**(08), 82–85 (2022)
4. Chen, F.: Safety evaluation method of hoisting machinery based on neural network. Neural Comput. Appl. **33**(2), 565–576 (2021)
5. Shin, I.J.: Factors that affect safety of tower crane installation/dismantling in construction industry. Saf. Sci. **72**, 379–390 (2015)
6. Shapira, A., Simcha, M.: AHP-based weighting of factors affecting safety on construction sites with tower cranes. J. Constr. Eng. Manag. **135**(4), 307–318 (2009)
7. Surui, X.: Safety Systems Engineering. Shanghai Jiao Tong University Press, Shanghai (2015)
8. Yu, M.: Research on safety evaluation system of utility tunnel ancillary facilities based on Bayesian network. Beijing Jiaotong University (2020)
9. Yao, C., Chen, D.: Fuzzy reliability evaluation method based on T-S fault tree and Bayesian network. J. Mech. Eng. **50**(02), 193–201 (2014)
10. Ye, G., Chen, M.: Research on classification of human factors in construction safety accidents. J. Saf. Sci. Technol. **12**(04), 131–137 (2016)

Study on Seismic Performance of Prefabricated Frame Structure with New Beam-Column Joint

Yu Wang, Haille Yue(✉), and Jincheng Huang

No. 9 Hunnan Middle Road, Shenyang City, Liaoning Province, China
2687830556@qq.com

Abstract. In this paper, a new type of prefabricated beam-column joint is proposed, which uses high-strength bolts and embedded steel plates to connect the prefabricated beam-column. The finite element software ABAQUS is used to establish the frame structure model of the new prefabricated beam-column joint, and then two natural ground motion records and one artificial ground motion record are selected to carry out the time history analysis under the action of rare earthquakes. The cast-in-place frame structure model with the same size and loading mode is established for comparative analysis. The core components of the new prefabricated beam-column joints have not yielded to damage under three ground motion records. The damage is mainly concentrated in the beam ends, which meets the design principle of 'strong joints and weak components'. The inter-story shear force, vertex acceleration, and inter-story displacement angle of the prefabricated structure under the action of different ground motion records are slightly lower than those of the cast-in-place structure. The results show that the new prefabricated beam-column joints have good seismic performance, which meets the requirements of seismic code, and its overall seismic performance is slightly better than that of the cast-in-place structure.

Keywords: ground motion record · seismic performance · finite element analysis · prefabricated beam-column joint

With the progress of the times, prefabricated buildings play an increasingly prominent role in the development and construction of cities because of their advantages of low pollution, high efficiency and fast construction. In prefabricated buildings, the connection form of joints is an important factor affecting the seismic performance of structures. In terms of theory and numerical simulation. Li Jun et al. [1] conducted pushover analysis and time-history analysis of cast-in-place frame structure and prefabricated frame structure based on OPENSEES software, and concluded that the seismic performance of prefabricated frame structure is basically the same as that of cast-in-place structure, but its residual strain is smaller than that of cast-in-place structure. Gao Zhen et al. [2] used ABAQUS to establish a fully prefabricated frame structure model with grouting sleeve connection and U-shaped steel bar connection and a cast-in-place frame structure model. Under the action of cyclic reciprocating load, the seismic performance of three different models is compared and analyzed. It is concluded that the ductility initial stiffness and energy dissipation performance of the two connection modes are lower than those

I. Tuns et al. (Eds.): CIBv 2023, LNCE 510, pp. 345–357, 2024.
https://doi.org/10.1007/978-3-031-60765-3_28

of the cast-in-place structure. In terms of experimental research, A. Negoita et al. [3] conducted a quasi-static test on the edge joints of the assembled unbonded prestressed frame. The test results show that the unbonded prestressed prefabricated frame structure has good integrity and ductility. Liu Bingkang et al. [4] conducted a pseudo-dynamic test on a two-story two-span prestressed prefabricated concrete frame to study the failure mechanism, deformation, energy consumption, strength and other properties of the prestressed concrete prefabricated frame. The results show that the plastic hinge first appears at the beam end of the frame, which conforms to the failure mode of 'strong column and weak beam'. The overall lateral stiffness of the frame is improved, and it has good anti-collapse ability.

Most of the above studies focus on the prefabricated structure with wet-connected, while the seismic research on the prefabricated frame structure with dry connection is relatively scarce. On this basis, a new type of prefabricated beam-column joint is proposed. The joint has the advantages of fast construction and less wet work, which can solve the problems of dense reinforcement and more formwork erection of low-rise frame structure joints.

The ABAQUS software is used to establish the prefabricated frame structure model with the new beam-column joint, and then the nonlinear dynamic time history analysis is carried out by inputting the ground motion record, so as to analyze the seismic performance of the new joint prefabricated frame structure.

1 Joint Overview

1.1 Joint Construction

The new prefabricated beam-column joint is embedded with steel plates at the beam and column ends, and the bolt holes and post-pouring holes for screwing bolts are reserved according to the drawings. Finally, the connection is performed by applying preload to the high-strength bolts. The section size of the joint column is 600 mm × 600 mm, the width and height of the beam section is 400 mm × 600 mm, and the steel plate is Q235B steel with a thickness of 25 mm. The bolt adopts 10.9 grade high strength bolt, and 402 kN preload is applied to each bolt. The joint structure is shown in the following figure.

1.2 Structural Description

In order to study the seismic performance of the new assembled beam-column joints applied to the specific structure, a four-story and two-span frame was established by ABAQUS. The longitudinal reinforcement of the column was 8 steel bars with a diameter of 18 mm and 4 steel bars with a diameter of 22 mm. The upper part of the beam used 2 compressive steel bars with a diameter of 25 mm, and the lower part used 4 steel bars with a diameter of tension. The diameter of the stirrup was 10 mm and the spacing was 100/200. All the steel bars in the test piece are HRB400, and the specific size of the frame is shown in the following figure (Figs. 1 and 2).

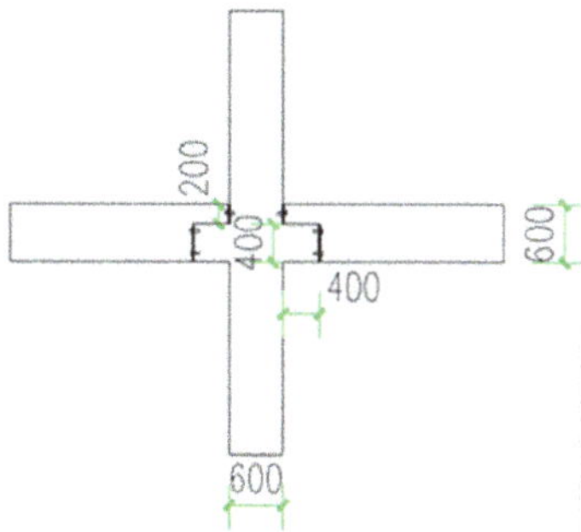

Fig. 1. Fabricated joint facade diagram

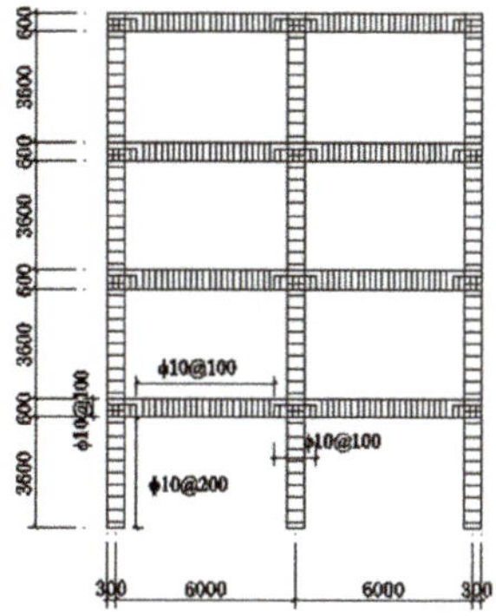

Fig. 2. Frame structure size diagram

2 Selection of Ground Motion Records

In this paper, the selected building is located in Xin Tai, Shandong Province, and the design basic earthquake ground acceleration is 0.1 g [7]. According to the equivalent shear wave velocity of the soil layer and the thickness of the overburden layer of the building site, the site category is class II according to the 'Code for Seismic Design of Buildings'. The peak acceleration of the selected three ground motion records is adjusted to 220 gal, which is input into the structure in the form of acceleration at the bottom of the column.

2.1 Selection of Natural Waves

The selection of ground motion records and the input mode are the basis and premise of dynamic time history analysis, and also the weak link in seismic engineering.

According to Article 5.1.2 of the Code for Seismic Design of Buildings (GB 50011-2010) [5], when the time history analysis method is used for seismic performance analysis, the actual strong earthquake record and the artificially simulated acceleration time history curve should be selected, and the number of actual strong earthquake records should not be less than 2/3 of the total.

In order to ensure the reliability of ground motion records, the ground motion records in this paper are derived from the peer Pacific Earthquake Center database. Considering the factors such as the fortification intensity and site conditions of the analyzed structure, two typical ground motion record curves with different characteristics are selected, as shown in the following chart (Table 1 and Fig. 4).

Table 1. Selected ground motion records

Name of the earthquake	Earthquake number	Earthquake occurrence time	wave velocity	station of record	site classification
Kobe	RSN1107	1995	312 m/s^2	Kakogawa	Category II site
Whittire	RSN599	1987	322 m/s^2	Brea - S Flower Av	Category II site

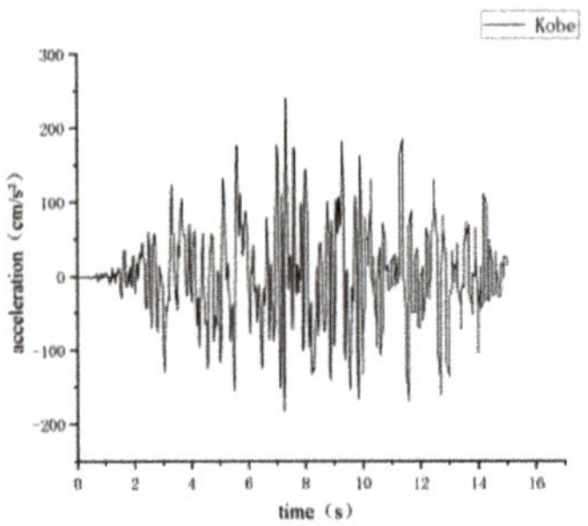

Fig. 3. Kobe ground motion records

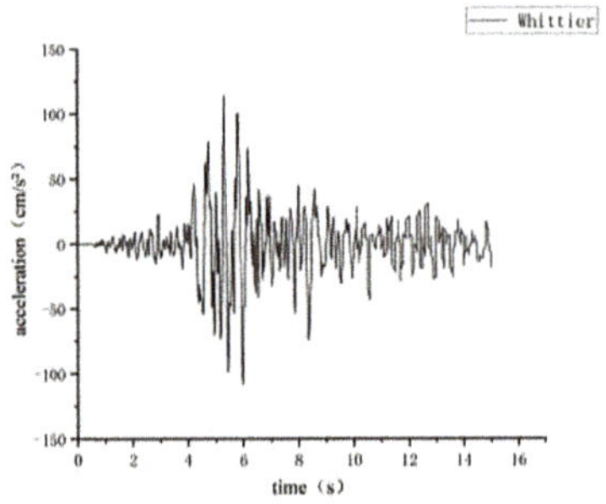

Fig. 4. Whittier ground motion records

2.2 Fitting of Artificial Ground Motion Records

The ground motion record processing software GM tools is used to generate artificial ground motion records by writing input seismic intensity, site category, seismic duration and other conditions.

The difference between the artificial wave earthquake acceleration response spectrum and the standard spectrum at the critical period point is less than 20%, which meets the requirements of the specification, and the fitting degree is good (Figs. 5 and 6).

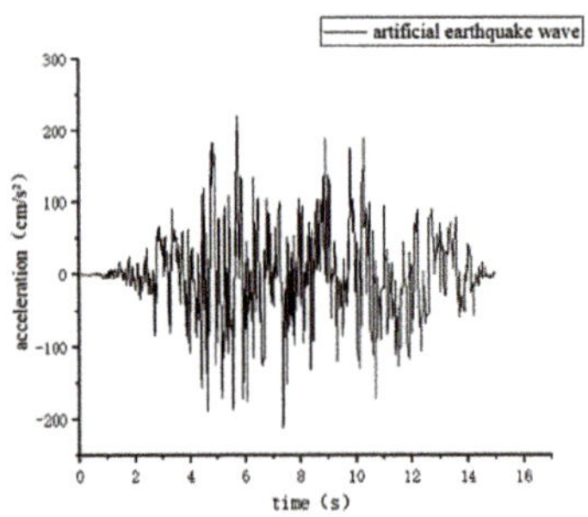

Fig. 5. The fitted artificial ground motion record

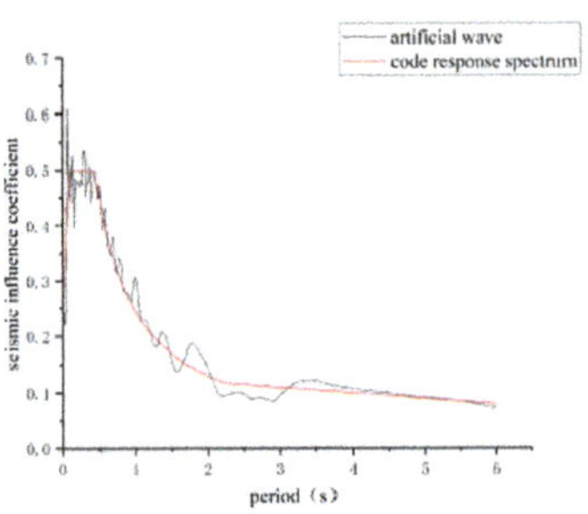

Fig. 6. Seismic ground acceleration response spectrum of artificial ground motion record

3 Finite Element Simulation

3.1 Establishment of Finite Element Model

The prefabricated frame structure model is established by ABAQUS software, and its size is consistent with that shown in Fig. 3. Concrete and bolted steel plates use C3D8R. The 8-node hexahedron linear reduced integral element is divided by T3D2 truss element. The grid size of the beam is 50 mm and the grid size of the column is 100 mm. The mesh of the bolted steel plate is refined, and the mesh size is 10 mm, so that each plate has at least 2 elements along the thickness direction (Fig. 7).

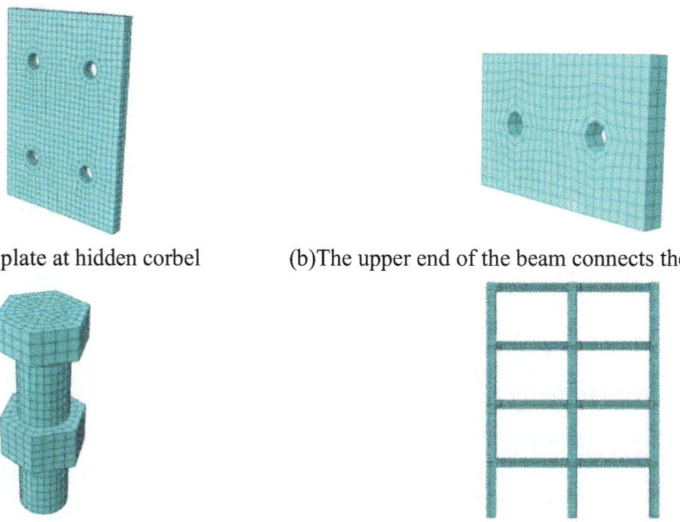

(a)Steel plate at hidden corbel (b)The upper end of the beam connects the steel plate

(c)bolt (d)frame structure

Fig. 7. Component finite element model and meshing

3.2 Material Constitutive Relationship

The formula for calculating the stress-strain curve of concrete materials under uniaxial tension is as follows: Appendix C of 'Code for Design of Concrete Structures' (GB50010-2010) [6] (Fig. 8).

$$d_t = \begin{cases} 1 - \rho_t\left[1.2 - 0.2x^2\right], x \le 1 \\ 1 - \frac{\rho_t}{\alpha_t(x-1)^{1.7}+x}, x > 1 \end{cases} \tag{1}$$

$$\sigma = (1 - d_t)E_c\varepsilon \tag{2}$$

$$x = \frac{\varepsilon}{\varepsilon_{t,r}} \tag{3}$$

$$\rho_t = \frac{f_{t,r}}{E_c\varepsilon_{t,r}} \tag{4}$$

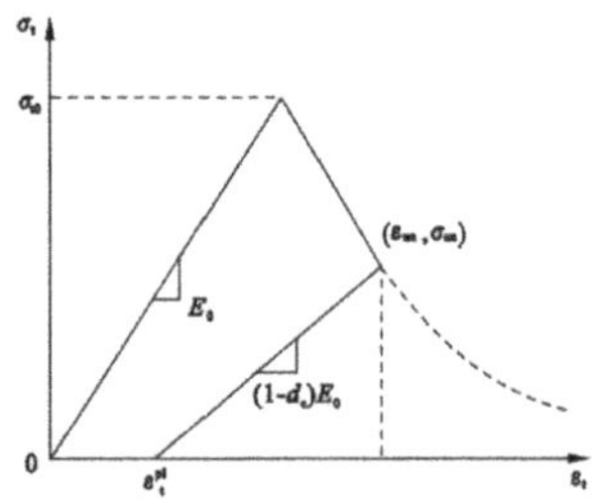

Fig. 8. Tensile stress-strain relationship

In the formula: α_t is the parameter value of the descending section of the concrete uniaxial tensile stress-strain curve; $f_{t,r}$ is the representative value of uniaxial tensile strength of concrete; $\varepsilon_{t,r}$ is the peak tensile strain of concrete corresponding to the representative value of uniaxial tensile strength; d_t is the uniaxial tensile damage evolution parameter of concrete.

The formula for calculating the stress-strain curve of concrete under uniaxial compression is based on the design of concrete structure (GB50010-2010) [6] (Fig. 9).

$$\sigma = (1 - d_c)E_c\varepsilon \tag{5}$$

$$d_c = \begin{cases} 1 - \frac{\rho_c n}{n-1+x^n}, x < 0 \\ 1 - \frac{\rho_c}{\alpha_c(x-1)^2+x}, x \ge 0 \end{cases} \tag{6}$$

$$x = \frac{\varepsilon}{\varepsilon_{c,r}} \tag{7}$$

$$\rho_t = \frac{f_{c,r}}{E_c\varepsilon_{c,r}} \tag{8}$$

$$n = \frac{E_c\varepsilon_{c,r}}{E_c\varepsilon_{c,r} - f_{c,r}} \tag{9}$$

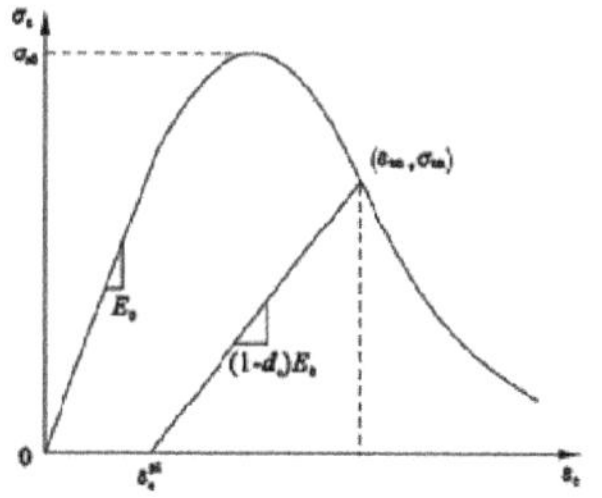

Fig. 9. Compression stress-strain relationship

In the formula: α_c is the parameter value of the descending section of the stress-strain curve of concrete under uniaxial compression; $f_{c,r}$ is the representative value of uniaxial compressive strength of concrete, MPa; $\varepsilon_{c,r}$ is the peak compressive strain of concrete corresponding to the uniaxial compressive strength; d_c is the damage evolution parameter of concrete under uniaxial compression (Table 2).

Table 2. Mechanical parameters of concrete

Concrete model	elastic modulus N/mm^2	poisson ratio	mass density Kg/m^3	compressive strength MPa	tensile strength MPa
C30	30000	0.3	2400	20.1	2.01

The steel adopts the stress-strain constitutive formula of Appendix C of 'Code for Design of Concrete Structures' (GB50010-2010) [6]. The calculation formula is as follows (Fig. 10)

$$\sigma_s = \begin{cases} E_s\varepsilon_s & , \ \varepsilon_s \le \varepsilon_y \\ f_{y,r} + k(\varepsilon_s - \varepsilon_{uy}) & , \ \varepsilon_y < \varepsilon_s \le \varepsilon_u \\ 0 & , \ \varepsilon_s > \varepsilon_u \end{cases} \tag{10}$$

$$k = (f_{st,r} - f_{y,r})/(\varepsilon_u - \varepsilon_{uy}) \tag{11}$$

In the formula: E_s is the elastic modulus of the steel bar; σ_s is the stress of steel bar; ε_s is the steel strain; $f_{y,r}$ is the representative value of yield strength of steel bar; $f_{st,r}$ is the representative value of the ultimate strength of the steel bar; ε_y is the yield strain of steel bar corresponding to $f_{y,r}$; ε_{uy} is the starting point strain of the reinforcement strengthening stage; ε_u is the peak strain of steel bar corresponding to $f_{st,r}$; k is the slope of the reinforced section (Table 3).

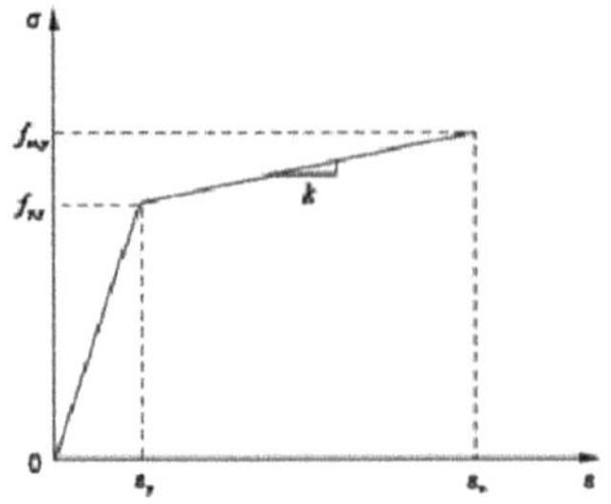

Fig. 10. The constitutive model of steel bar

Table 3. Steel and bolt material parameters

Materials	E_s/MPa	v	σ_y/MPa	σ_u/MPa	$\varepsilon_y/10^{-2}$	$\varepsilon_u/10^{-2}$	$\varepsilon_{st}/10^{-2}$
Q235B	2.06×10^5	0.3	299	436	0.045	19.8	34
Bolt	2.06×10^5	0.3	980	1100	0.47	1.36	—
HRB400	2.00×10^5	0.5	400	540	0.002	0.01	—

In the table, E_s represents the elastic modulus of the material; v denotes the Poisson's ratio of the material; σ_y denotes the yield strength of the material; σ_u represents the ultimate tensile strength of the material; ε_y represents the yield strain of the material; ε_u denotes the ultimate strain of the material; ε_{st} denotes the failure strain of the material.

3.3 Finite Element Model Validation

In order to ensure the accuracy of finite element simulation, this paper selects the pseudo-dynamic test completed by Professor Liu Bing Kang's team [4] of Hefei University of Technology to simulate. In this test, two MTS electro-hydraulic servo actuators fixed to the reaction wall are used to apply lateral horizontal seismic action to the frame layer by layer. EI-centro (1940 NS) was selected as the ground motion record of the pseudo-dynamic test, and the first 10 s was intercepted, and the peak acceleration of the ground motion record was amplitude-modulated into 110 PGA for loading. The corresponding numerical model is established according to the preloaded assembled prestressed concrete frame of the selected test, and the numerical simulation results are compared with the test results. The specific structural model and test device are shown in the Fig. 11.

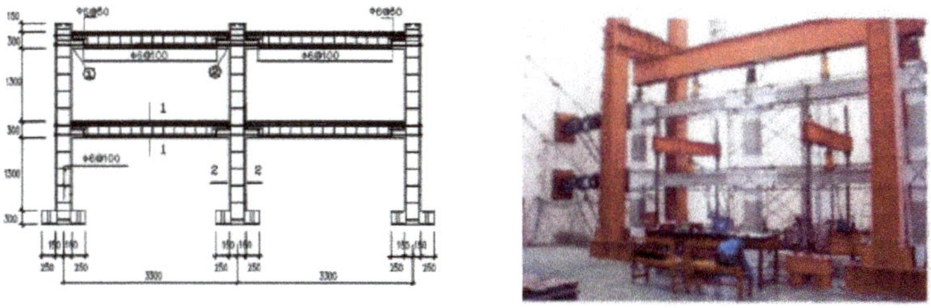

Fig. 11. Prestressed one-bay two-story two-span frame

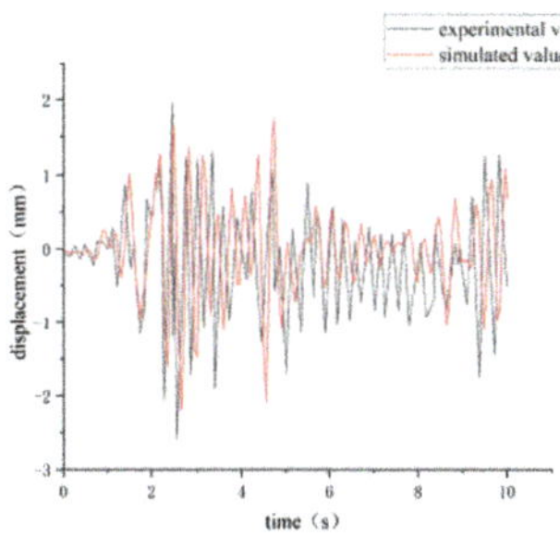

Fig. 12. The displacement time history curve of the first layer under 110 gal condition

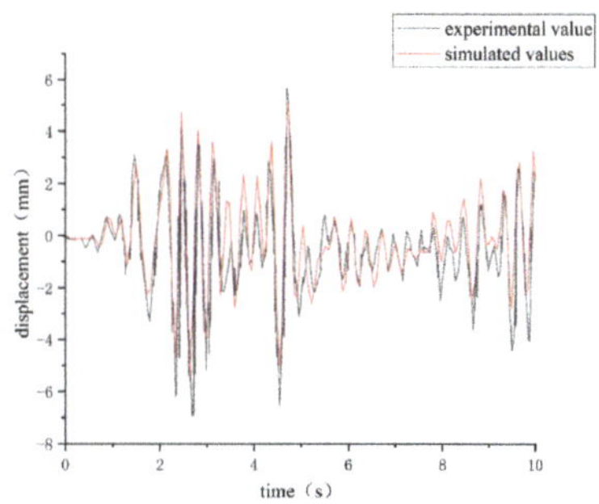

Fig. 13. The displacement time history curve of the second floor under 110 gal condition

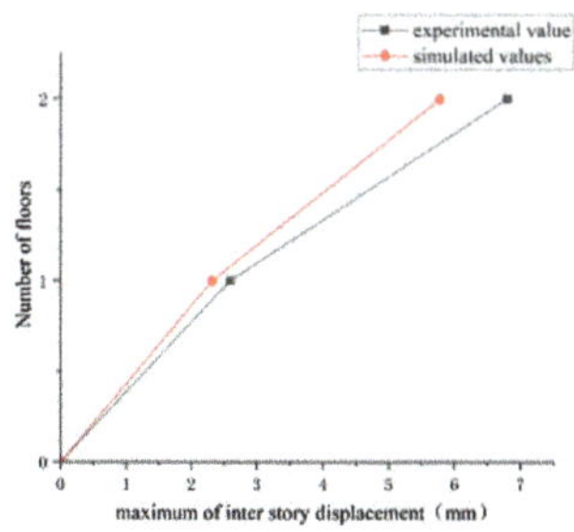

Fig. 14. Maximum displacement of each floor

By simulating the existing test, the displacement time history curve of the pre-stressed assembly frame is obtained. From Fig. 12, 13, 14, it can be seen that the displacement trend of the ABAQUS simulation is close to the test value, and the maximum displacement value of the second layer of the test is 6.8 mm. The simulated value is 5.78 mm (15% error), and the test value of the first layer is 2.6 mm and the test value is 2.32 mm (10.7% error). The error may be due to

(1) the error in the manufacture of test specimens and the damage during transportation.
(2) There may be small cracks between the loading device and the components during installation, resulting in larger test values.

4 Seismic Response of Structures

In order to quantitatively evaluate the seismic performance of the new joint prefabricated frame structure, the cast-in-place frame structure and the new joint prefabricated frame structure are compared and analyzed respectively from the dynamic characteristics and seismic response of the structure. Below is a comparison of the seismic responses of the two structures (Figs. 15, 16 and 17).

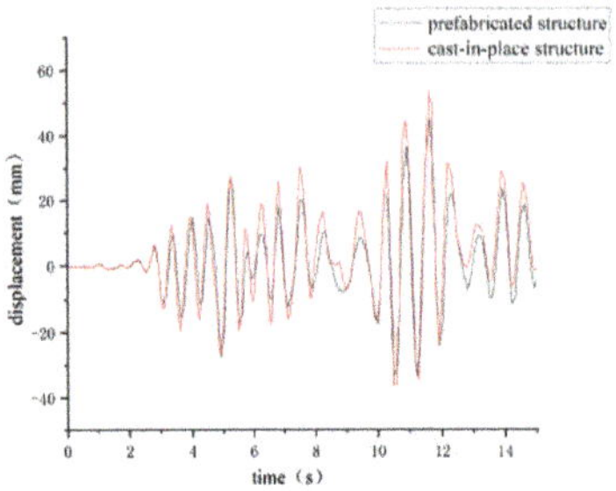

Fig. 15. Time history curve of top layer displacement under Kobe wave

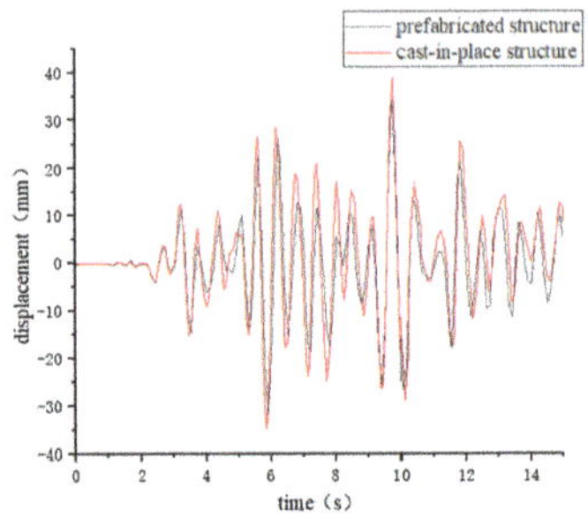

Fig. 16. Time history curve of top layer displacement under whitter wave

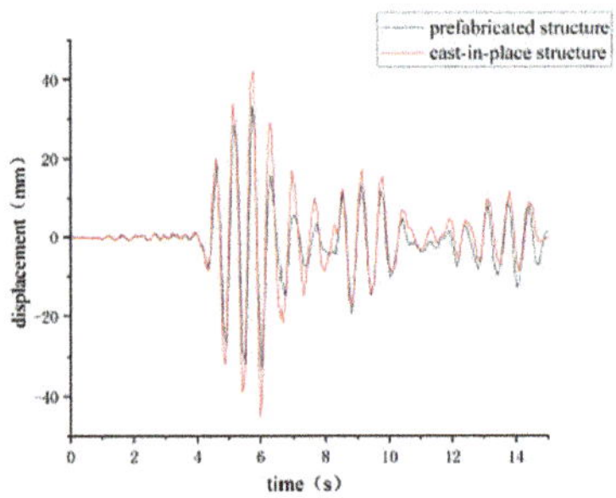

Fig. 17. Time history curve of top layer displacement under artificial wave

Under the action of ground motion records, the displacement time history curve of the cast-in-place structure is enveloped on the prefabricated structure. The peak displacement of the cast-in-place structure under the action of Kobe wave, whitter wave and artificial wave is 39.21 mm, 44.89 mm and 53.83 mm respectively, which is bigger than that of the prefabricated structure of 35.21 mm, 33.25 mm and 44.98 mm. That is to say, under the action of each ground motion record, the interlayer displacement of the new joint prefabricated frame structure is smaller than that of the cast-in-place structure. This shows that the new joint prefabricated frame has better seismic performance.

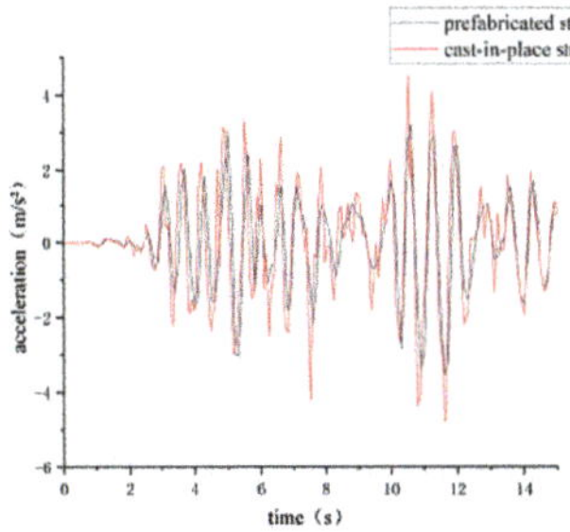

Fig. 18. The vertex acceleration time history curve of frame structure under Kobe wave

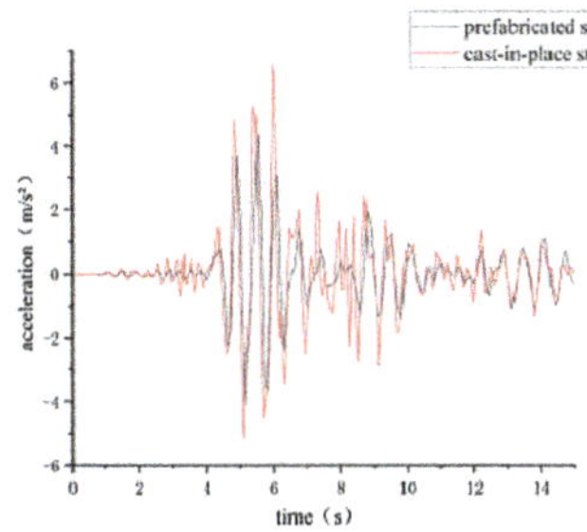

Fig. 19. The vertex acceleration time history curve of frame structure under Whitter wave

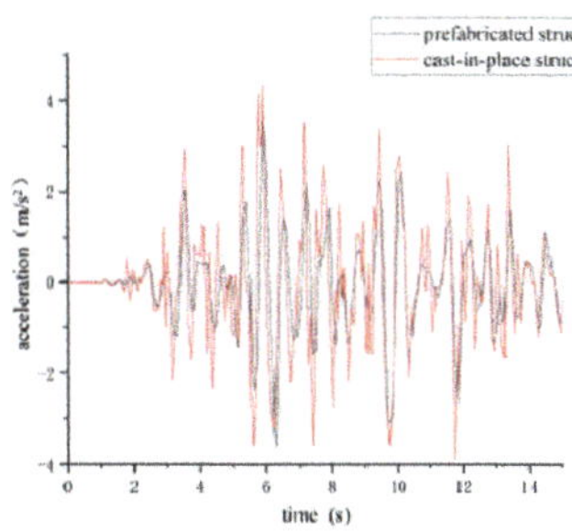

Fig. 20. The vertex acceleration time history curve of frame structure under artificial wave

According to Fig. 18, 19, 20, the peak acceleration of the prefabricated structure under the action of three ground motion records is 3.62 m/s^2 and 4.36 m/s^2 3.52 m/s^2, respectively, which are smaller than the 4.33 m/s^2, 6.54 m/s^2 and 4.76 m/s^2 of the cast-in-place structure. The above shows that the response of the new joint prefabricated frame to the earthquake is similar to that of the cast-in-place structure, and the response to the top acceleration is not as obvious as that of the cast-in-place structure (Figs. 21, 22 and 23).

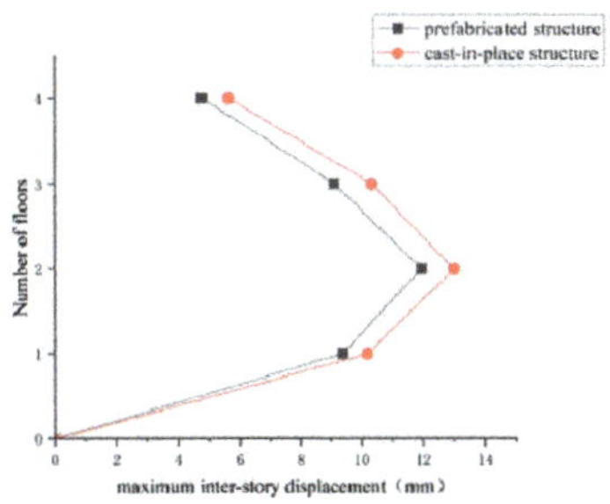

Fig. 21. Maximum interlayer displacement of each layer under Kobe wave action

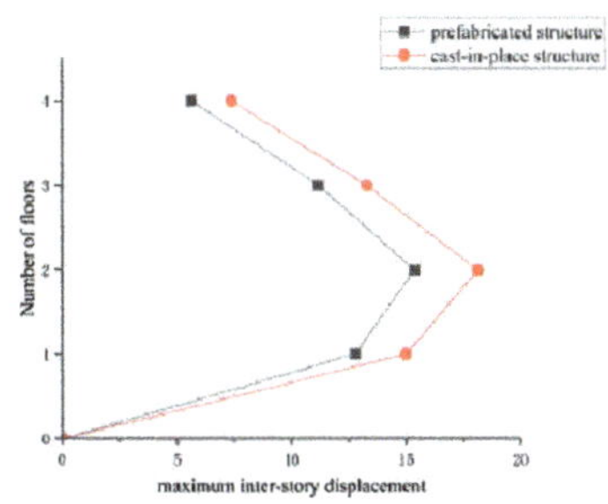

Fig. 22. Maximum interlayer displacement under artificial ground motion record

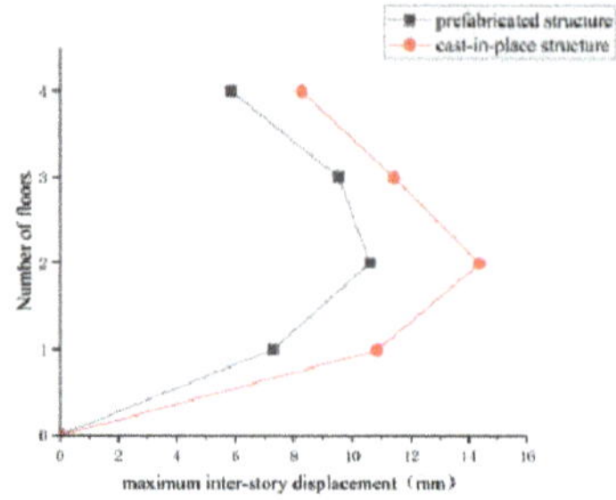

Fig. 23. Maximum interlayer displacement of each layer under Whitter wave action

Because under the respective action of different ground motion records, the maximum inter-story displacement of the prefabricated structure and the cast-in-place structure appears on the second floor. And the interlayer displacement of the cast-in-place structure is greater than that of the prefabricated structure. It shows that the seismic response of the two is close, and the second layer is the weak layer of the structure.

The stress cloud diagram of the component at the joint is taken as an example under the action of Kobe seismic wave.

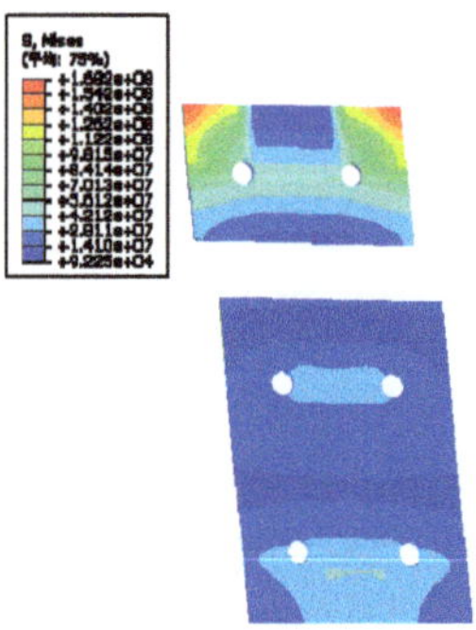

Fig. 24. Steel plate stress cloud diagram

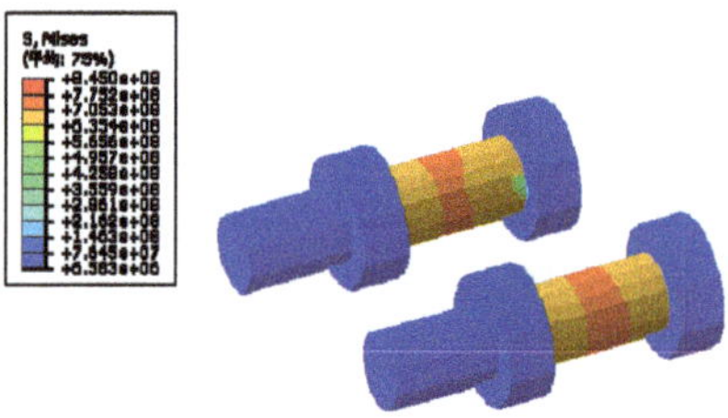

Fig. 25. Steel plate stress cloud diagram

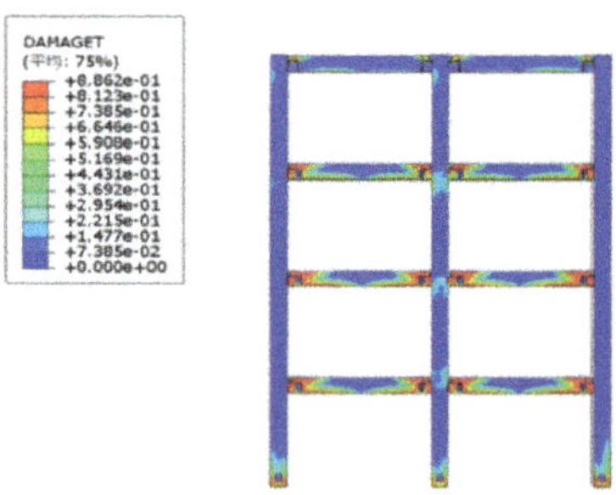

Fig. 26. The tensile damage cloud diagram of the new joint prefabricated frame structure

1. It can be seen from Figs. 24 and 25 that the bolts and steel plates at the joints did not reach the yield strength. It shows that there is no plastic deformation at the joint and there is a large surplus stress, which is in line with the design concept of strong joint and weak component.

2. In Fig. 26, the damage of concrete is concentrated at the bottom of the beam, and the 45° crack extends obliquely to the upper beam end, which conforms to the design concept of strong column and weak beam.

5 Conclusion

1. the new joint prefabricated frame structure shows excellent seismic performance under the action of three ground motion records. The inter-story displacement, inter-story displacement angle and vertex acceleration of the prefabricated frame structure are smaller than those of the cast-in-place frame structure.
2. Under different seismic waves, the seismic response of the new joint prefabricated frame structure is different, and the seismic response of the artificial wave is the strongest. However, under the action of three seismic waves, the maximum inter-story displacement angles of the new joint prefabricated frame structure are 2/703, 2/793 and 1/273, respectively, which are less than the elastic-plastic displacement angle limit of 1/50 specified in the 'resistance code'. This shows that the new joint prefabricated frame structure has good seismic performance under the action of seven degree rare earthquake.
3. Under the action of seven-degree rare earthquake, the new joint prefabricated frame structure shows better seismic performance than the cast-in-place structure, and conforms to the design concept of 'strong column and weak beam' and 'strong joint and weak component'. Because of its many advantages, the new prefabricated joint has great application prospects.

References

1. Li, J.: Master 's degree thesis on seismic analysis method of prestressed prefabricated building structure, Dalian University of Technology (2015)
2. Gao, Z.: Master 's degree thesis on seismic performance analysis of assembled frame structures with different joints, Shandong University of Science and Technology (2019)
3. Negoita, A.: The seismic analysis, theoretical ard experimental of the joints of prestressed concrete frames. CEB Bull. D'inf. **1979**, 53–61 (1979)
4. Bingkang, L., Shenjiang, H., Manrong, S., Honglin, W., Lihua, C.: Experimental study on seismic performance of precast prestressed concrete frame. J. Civ. Eng. **11**, 1–8 (2011)
5. Ministry of Construction of the People 's Republic of China.Code for seismic design of buildings GB50011-2010 (2016). China Construction Industry Press, Beijing
6. Ministry of Construction of the People 's Republic of China. Code for design of concrete structures ' GB50010-2010 (2011). China Building Industry Press, Beijing
7. Seismic ground motion parameter zoning map of China: GB 18306-2015 (2015)

Simulation Study on Wave Attenuation Performance of Rigid Explosion-Proof Wall with Cone-Shaped Blast-Facing Surface

Shilong Jia(✉), Shuo Liu, Chang Li, and Mingzhen Hou

School of Civil Engineering, Shenyang Jianzhu University, Shenyang, China
cesljia@sjzu.edu.cn

Abstract. According to the previous research on the simulation of corrugated plate explosion-proof wall, the corner cone explosion-proof wall is proposed by changing the structure of the explosion-proof wall. ANSYS/LS-DYNA finite element software was used to simulate and analyze the new explosion-proof wall with a cone height of 15, 22.5, 30, 37.5, 45 cm and a bottom edge of 30 cm under the same blasting height and proportional blasting distance. The feasibility of the model is verified by comparing the empirical formula, and then the overpressure peak of each test point is analyzed. Firstly, it is concluded that the triangular cone explosion-proof wall has a stronger weakening effect on the shock wave than the straight wall, and the cone heights of 15 cm and 30 cm are the most obvious. Secondly, the triangular cone wall will change the shape of the shock wave after passing through the explosion-proof wall to a certain extent. Finally, the peak value of superimposed overpressure at 1–1.5 times the height of the wall behind the triangular cone wall will be greater than that of the ordinary straight wall.

Keywords: Explosion Shock Wave · Triangular Cone Explosion-proof Wall · Numerical Simulation · Shock Wave Behind the Wall

In recent years, explosion accidents in factories, mines and warehouses have been frequent [1]. In order to reduce losses, setting up explosion-proof walls has become a common technical measure. Domestic scholars have carried out a lot of research on the special-shaped face of the explosion-proof wall by means of numerical simulation, empirical formula verification, experimental comparison and other methods, such as special-shaped explosion-proof wall, guardrail explosion-proof wall, arc explosion-proof wall, corrugated plate explosion-proof wall, inclined explosion-proof wall and so on [2]. At present, corrugated plate explosion-proof wall has many applications in anti-explosion of offshore platforms. Compared with the corrugated plate explosion-proof wall with one-way slotted on the blasting surface, the corner cone explosion-proof wall with two-way slotted is rarely studied.

The present study employs the ANSYS/LS-DYNA finite element software to simulate ordinary straight blast walls and corner cone blast walls under the same explosive height and scaled distance conditions. The corner cones have a height of 30 cm and a base width of 50 cm. A comparative analysis is conducted to investigate the ability of irregular blast walls to attenuate shock waves compared to straight walls under the action

I. Tuns et al. (Eds.): CIBv 2023, LNCE 510, pp. 358–372, 2024.
https://doi.org/10.1007/978-3-031-60765-3_29

of blast waves. The peak overpressure values at various measurement points behind the walls in both horizontal and vertical directions are analyzed to understand the influence of corner cone blast walls on shock wave protection and to summarize the characteristics of peak overpressure values behind the walls and the propagation patterns of shock waves.

1 Geometric Reflection Principle of Corner Cone to Wave

According to the principle of optical reflection, the inclined plane on the corner cone will greatly increase the reflection area and frequency of the wave [7]. Assuming that the reflection point is *x*, the height of the reflection point is *h*, the height of the corner cone is *H*, the top angle of the corner cone is *2a*, and the incident angle is θ, as shown in Fig. 1.

$$h_i = \frac{hsin(\theta - \alpha)}{cos\beta_i} \tag{1}$$

That is, when $h_i < H$, the *i-1*th reflection will impact the opposite corner cone shock wave and continue to propagate in the corner cone gap. When $h_i > H$, the shock wave is emitted from the gap after the *i-1* reflection, and the shock wave is reflected *i-1* times in the corner cone. As shown in Fig. 1, the two triangles are the corner cone sections, and the line is the reflection path of the explosion shock wave on the corner cone surface.

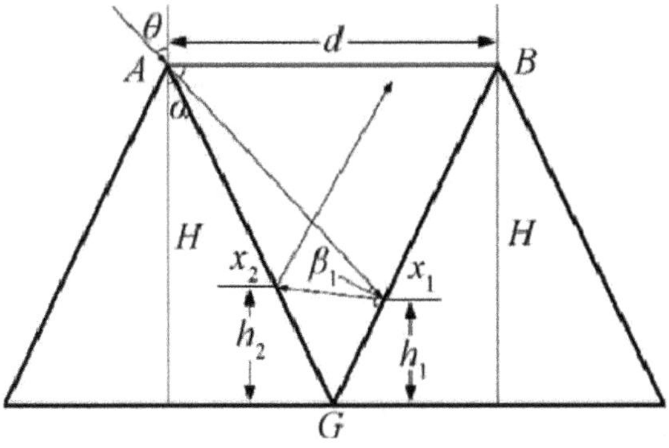

Fig. 1. The shock wave path of corner cone reflection

Experiments have shown that the rational use of passive noise control devices will significantly reduce noise waves with frequencies above 2000 Hz, that is similar optical concepts. In the field of acoustics, it is proved to be equally feasible to reasonably arrange acoustic interference of the same or different sound sources to enhance or weaken the amplitude of sound waves. Therefore, the high-frequency explosion shock wave can weaken the shock wave by interfering with the probability of increasing the opposite phase of the wave through reasonable arrangement of obstacles.

2 Material Parameters and Calculation Model

The finite element analysis software LS-DYNA, which is more mature in recent years, is used to simulate and analyze the improved wall. LS-DYNA program is a simulation program for geometric nonlinearity, material nonlinearity and contact nonlinearity. The LS-DYNA program has two solving functions: explicit and implicit. It can better deal with the nonlinear dynamic analysis in the structure. In the numerical simulation of the anti-explosion effect analysis of the cone-shaped frontal surface of the rigid wall, the model is composed of a rigid wall, air, and explosive. The TNT is used to simulate the explosion of a chemical plant or warehouse [8]. After comparing the previous test and simulation experience, the main parameters are as follows.

2.1 Walls

Wanzhen Wang et al. proved and analyzed the process of the reflection of the detonation wave in the impact of the explosion-proof wall by means of experiments. Although the wall is deformed under the action of the explosion load, the deformation response of the wall is not synchronized with the reflection of the detonation wave after the impact of the wall [9].

At the same time, Chen Shaofan chose the numerical simulation method in the design principle of steel structure, and simulated the deformation of the column under the explosive load and the reflection process of the detonation wave in the process. The results also show that the influence of the detonation wave on the deformation response of the structure is delayed. The process of reflecting overpressure is instantaneously completed and the structure cannot react at the same time, which cannot be coupled with the deformation process of the structure. Therefore, in order to simplify the calculation [10], the wall is assumed to be rigid when calculating and analyzing the load of the explosion-proof wall. The wall is automatically generated by the rigid material MAT RIGID model parameters.

2.2 Air

The air material adopts the ideal air MAT_Null material, adopts the state Eos_Linear_Polynomial, and defines the pressure according to the following formula:

$$p = C_0 + C_{1m} + C_{2m^2} = C_{3m^3} + (C_4 + C_{5m} + C_{6m^2}) \times E_0 \tag{2}$$

In the formula: $m = \rho/\rho_0 - 1$; $C_0 = C_1 = C_2 = C_6 = 0$; $C_4 = C_5 = 0.4$; $\rho_0 = 1.29 \times 10^{-3}$; E_0 is the initial energy of air, taking 2.068×10^5J; ρ is gas density; ρ_0 is the initial density of gas; C_0–C_5 is the calculated coefficient. The values of the air parameters in the material constitutive model are shown in Table 1.

Table 1. Air material parameters

R_0(kg/m^3)	C_0/MPa	C_1	C_2	C_3	C_4	C_5	C_6	E_0/MPa	V_0
1.293e+3	-0.1	0	0	0	0.4	0.4	0	0.25	1.0

2.3 Explosives

The explosive is MAT_High_Explosive_Burn high-energy explosive, and the standard state equation EOS_JWL is used. The expression is:

$$P = A\left(1 - \frac{\omega}{R_1 V}\right)e^{-R_1 V} + B\left(1 - \frac{\omega}{R_2 V}\right)e^{-R_2 V} + \frac{\omega E_0}{V} \tag{3}$$

In the formula: *A*, *B*, *R1*, *R2*, ω are material parameters; E_0 is the initial energy density; v_0 is the initial relative volume. The values of explosive parameters in the material constitutive model are shown in Table 2.

Table 2. Explosive parameters

Density ρ (g/cm^3)	initial energy E_0 (kJ/m^3)	CJpressure P_{cj} (mbar)	detonation velocity D (cm/μs)	A	B	R1	R2	ω
1.63	7.0e + 4	0.27	0.693	4.15	0.1	4.5	1.1	0.3

2.4 Calculation Model

In the model, the Euler algorithm is used for explosives and air, the multi-material ALE algorithm is used for the unit, and the coupling algorithm is used between the concrete wall and the air material. The model does not consider the end effect and is simplified from the central symmetry axis to a 1/2 model. Figure 2 is the plane along the *xoy* coordinate axis. Without considering the end effect, the central symmetry axis is simplified to 1/2 model. The size of the air domain is 8 m × 15 m × 1.5 m × 2. The front, rear, upper and left boundaries are non-reflective boundaries. The material in the air domain flows out naturally, simulating the infinite domain. The lower boundary of the air domain simulates the ground and imposes fixed constraints on the lower boundary. The right boundary of the air domain is a symmetrical boundary, and the *Z*-axis direction symmetry effect is applied to the right boundary. The bottom of the explosion-proof wall is rigidly restrained to simulate the fixed concrete base.

The explosive is a 0.15 m side length cube with symmetrical structure. The center point is 0.5 m high and the vertical distance from the bottom of the corner cone of the explosion-proof wall is 4.5 m. A total of 23 measuring points is set up in the model. The horizontal distance between the measuring points is 1m, and the height from the ground is 1 m. The distance between the measuring points in the vertical direction is 0.5 m below 3 m, 1m from 3 m to 8 m, and 1 m from the explosion-proof wall. The height of explosion-proof wall is 3 m, the height of corner cone is 0.5 m, and the width is 2.4 m (Fig. 3).

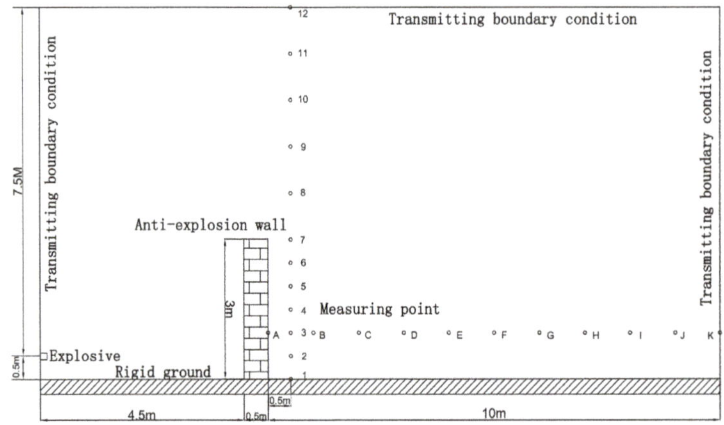

Fig. 2. Explosion-proof wall calculation model

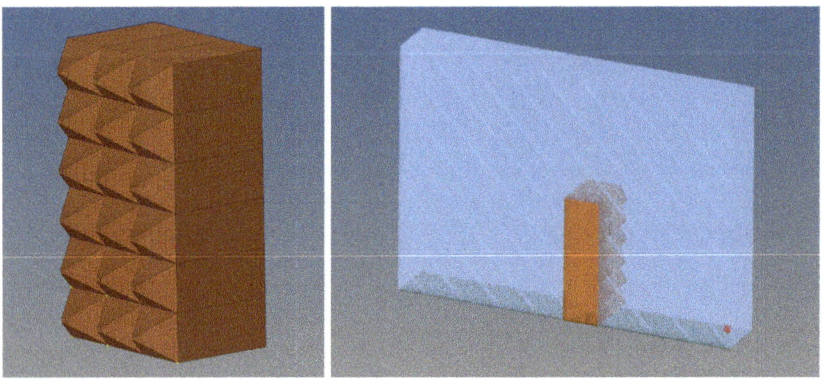

Fig. 3. Corner cone explosion-proof wall model

3 Model Validation Analysis

Because the unit size of the current simulation software still has a certain error effect on the simulation results, it is necessary to verify the accuracy of the simulation results through the empirical formulas summarized in previous experiments. Therefore, the calculation formulas of Henrych, Mills and Sadows are used for comparison [11].

$$Pso = \begin{cases} \frac{1.472}{Z} + \frac{0.554}{Z^2} - \frac{0.357}{Z^3} + \frac{0.00625}{Z^4} 0.1 \le Z < 0.3 \\ \frac{0.619}{Z} - \frac{0.033}{Z^2} - \frac{0.213}{Z^3} 0.3 \le Z < 1 \\ \frac{0.066}{Z} - \frac{0.405}{Z^2} - \frac{0.329}{Z^3} 1 \le Z \le 10 \end{cases} \tag{4a}$$

$$Pso = \frac{0.108}{Z} - \frac{0.114}{Z^2} + \frac{1.772}{Z^3} \tag{4b}$$

$$Pso = \begin{cases} \frac{1.07}{Z} - 0.1Z < 1 \\ \frac{0.108}{Z} - \frac{0.114}{Z^2} + \frac{1.772}{Z^3} 1 \le Z \le 15 \end{cases} \tag{4c}$$

In the formula: P_{so} is the peak overpressure of detonation wave, Mpa; Z is the proportional detonation distance, $Z = RQ^{-1/3}$; R is the distance from the explosion center to the measuring point, m; Q is the equivalent TNT charge, kg.

The verification model is not different from the test model except for no explosion-proof wall and different measuring points. The measuring point is 1m from the ground height. The measuring points are arranged horizontally with a spacing of 0.5 m in the horizontal direction. There are 10 measuring points in the model, the first measuring point starts from 0.5 m, and the last is 5 m.

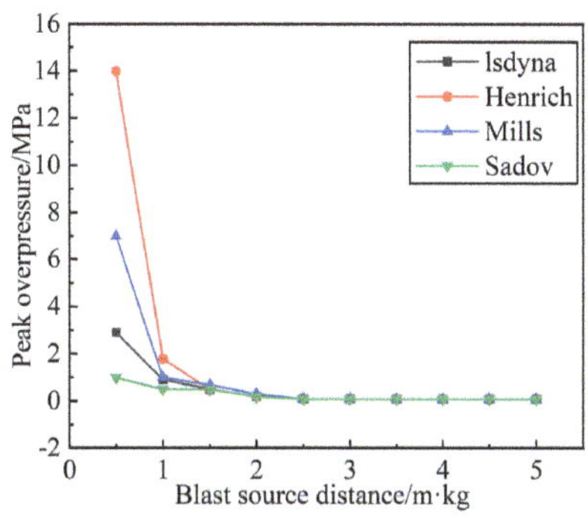

Fig. 4. Comparison of simulation and empirical formula calculation results

According to the calculation experience, the calculation result is the most accurate when the grid is divided into 5–20 mm, but in order to simplify the calculation and improve the calculation speed of the server, the calculation speed and accuracy are more reasonable 75 mm grid is selected. After comparing the simulation results with the above empirical calculation results, it is found that the difference between the simulation results and the empirical calculation results in Fig. 4 is large within the proportional explosion distance of 1.5 m·kg$^{-1/3}$, that is, the explosion source distance is within 2.66 m. This gap is mainly due to the large side length of the calculation model unit, which causes a large error when it is close to the explosion source. When it exceeds 1.5 m·kg$^{-1/3}$, the simulation results of LS-DYNA are in good agreement with the empirical results of Henrich et al. The simulation effect of LS-DYNA is worthy of trust because the minimum explosion source distance of the explosion-proof wall is not less than 3.55 m, that is, the proportional explosion distance is higher than 2 m·kg$^{-1/3}$. Therefore, the effect of numerical simulation is worthy of trust.

4 Simulation Results and Numerical Analysis

The models of straight wall and 15, 22.5, 30, 37.5, 45 cm equiangular cone length were established, and the ratio of cone height to top distance was determined. The final results analyze the trend of overpressure behind the rigid corner cone wall and the causes of this trend, and compare the shock wave weakening rate of each measuring point of the straight wall.

4.1 Morphological Difference of Oblique Reflection

When the angle between the propagation direction of the air shock wave and the vertical line of the reflection surface is φ_0, the oblique reflection of the shock wave will occur, as shown in Fig. 5. *I* is the incident wave, *R* is the reflected wave, and the B, C, D, E, F points on the reflection surface form different incident angles φ_0 with the explosion center.

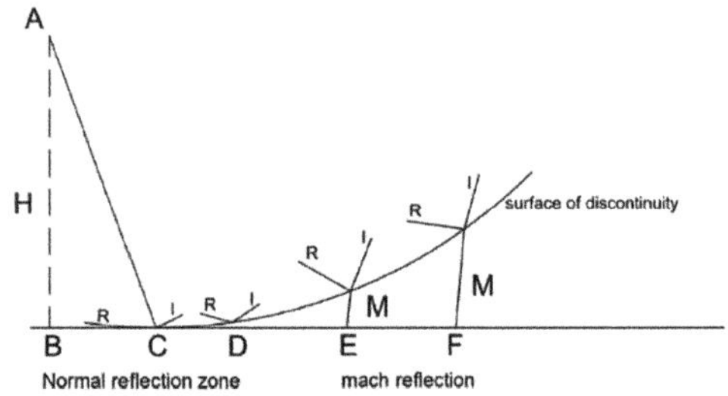

Fig. 5. Diagram of oblique reflection morphology

When point B $\varphi_0 = 0°$, the shock wave produces positive reflection. The φ_0 of the incident wave front at C and D points is less than the critical angle φ_{0c}, so the normal reflection occurs and the shape is close to "V' [12]. The φ_0 of the incident wave front at E, F point is greater than the critical value.

4.2 The Influence of the Cone-Shaped Face on the Shock Wave Shape

Because the influence of the height of each wedge on the shock wave is similar, the pressure cloud analysis of the straight wall and the 45 cm corner cone wall in Fig. 6 is used. It can be seen from Fig. 6: (1) The cone-shaped blasting surface has a certain influence on the overpressure circulation of the shock wave at the top of the wall. After the shock wave hits the ordinary rigid straight wall, a large amount of reflection will be generated on the wall surface. The smooth wall will quickly transmit the shock wave gathered at the bottom of the wall to the top of the wall, thus diffraction occurs, and the overpressure circulation gradually enwraps the back of the wall at the top of the wall. (2) Corner cone wall has an effect on the propagation form of shock wave behind the wall. The diffracted wave is the main component of the shock wave stroke behind the explosion-proof wall. Compared with the corner cone explosion-proof wall, the reflected wave form generated by the diffracted wave behind the straight wall after the impact with the ground is closer to the regular reflection, showing a 'V' type. After the close shock wave behind the corner cone explosion-proof wall is superimposed with the plane wave, the shape of the reflected wave is closer to the Mach reflection, showing a 'Y' type.

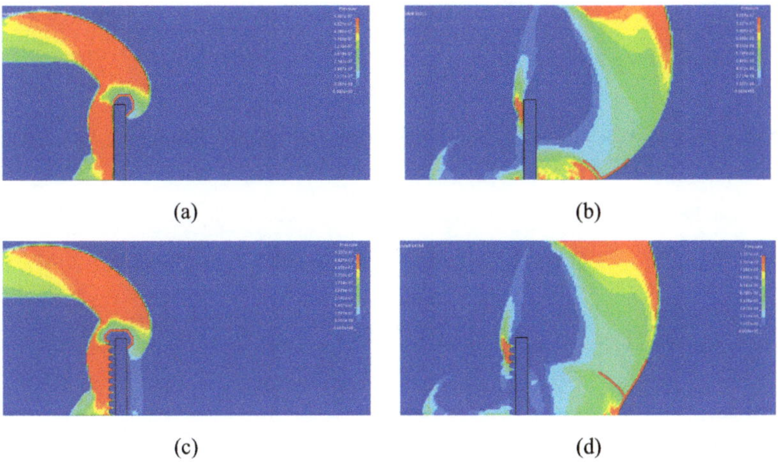

Fig. 6. Detonation wave propagation pressure cloud diagram of straight wall and corner cone wall

4.3 Shock Wave Overpressure Peak Transverse Propagation Characteristics

Figures 7 and 8 show that the overpressure of each measuring point in the horizontal direction of the ordinary straight wall and the 15 cm high angle cone explosion-proof wall changes with time.

In the LS-Prepost post-processing software, the d3plot file generated by the explosion simulation is opened, and the horizontal overpressure peak data option is selected in the history chart analysis option, that is, the horizontal overpressure peak of each measuring point is drawn. The vertical overpressure peak chart in the next section is also made by the same method.

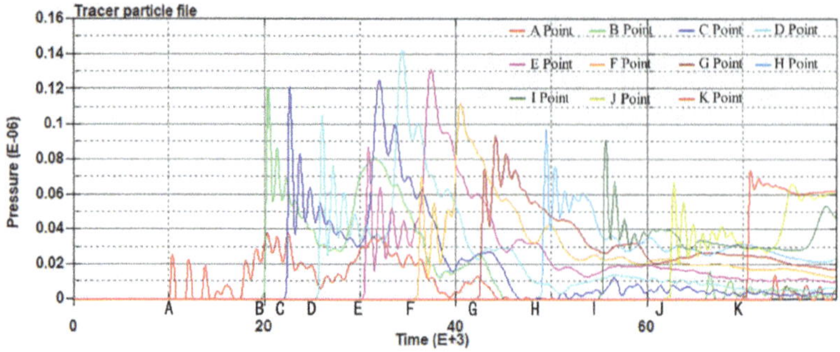

Fig. 7. Pressure of A–K measuring point in horizontal direction behind ordinary straight wall

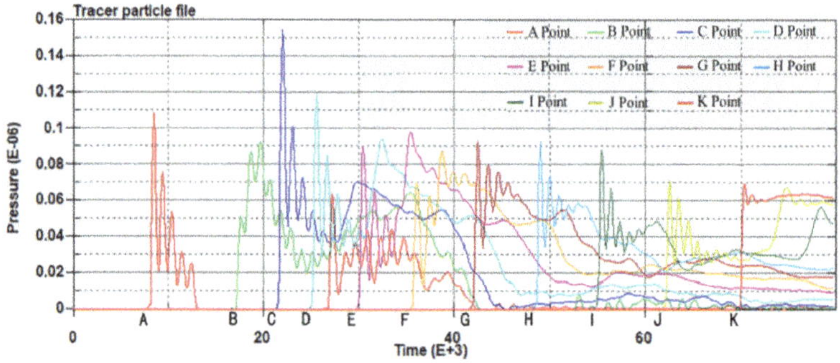

Fig. 8. Pressure of A–K measuring point in horizontal direction behind 15 cm corner cone wall

The maximum overpressure peak behind the corner cone explosion-proof wall is higher than that of the ordinary straight wall. The diffraction wave on the top of the straight wall will completely surround the whole explosion-proof wall, and produce positive reflection or regular reflection when it contacts with the ground. The reflected wave generated at 1.5 to 2.5 times the height of the wall becomes the maximum overpressure behind the whole wall [13], and causes greater damage to the wall within a certain range.

The cone angle increases the number of reflections of the shock wave, resulting in more interference to consume energy. Compared with the ordinary straight wall, the stiffness of the corner cone explosion-proof wall is reduced, so a stronger plane wave will be generated behind the wall.

The top overpressure circulation is affected by the increase in the number of corner cone reflections and the decrease in stiffness when wrapping the back of the entire wall, so it is not as complete as an ordinary straight wall. More shock waves will obliquely impact the ground at an angle of about 45° from the top of the wall, resulting in the generation of Mach reflection.

The maximum overpressure of the corner cone explosion-proof wall appears on the propagation path, which is about 1–1.5 times the height of the ground behind the wall, and the pressure of some paths is higher than the maximum overpressure peak behind the straight wall. Due to the dissipation effect of the corner cone on the shock wave, the peak value of the shock wave overpressure behind the corner cone explosion-proof wall will be slightly lower than that of the ordinary straight wall before and after this point.

Through the results of Fig. 9, it is found that the overpressure of the corner cone explosion-proof wall reaches the maximum at 2 m behind the wall, that is, at the measuring point of 7 m. Within 6 m behind the wall, the cone-shaped explosion-proof wall has a better weakening effect on the shock wave. The peak overpressure of the 15 cm high cone and the 22.5 cm high cone is smaller, indicating that the damage behind the wall is smaller.

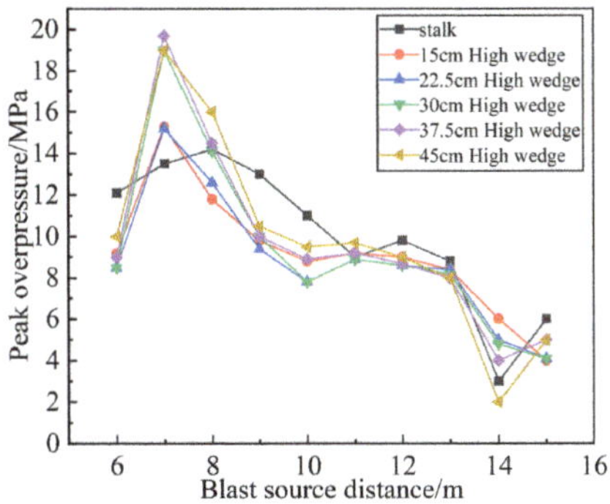

Fig. 9. Peak overpressure in the horizontal direction behind various types of walls

4.4 Longitudinal Propagation Characteristics of Shock Wave Overpressure Peak

Figure 10 and 11 show the variation of overpressure with time at 7 measuring points from the ground to the top of the wall at a horizontal distance of 0.5 m behind the wall.

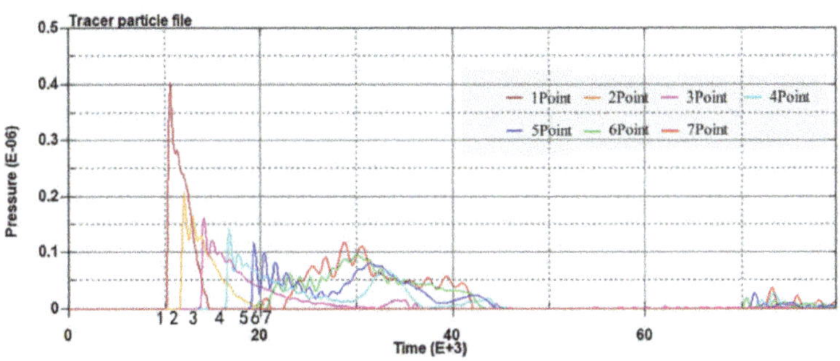

Fig. 10. Pressure of No.1–No.7 measuring points in the vertical direction behind the ordinary straight wall

Due to the overpressure circulation at the top of the wall, the wrapping effect of the ordinary straight wall is relatively complete. Therefore, the time sequence of the

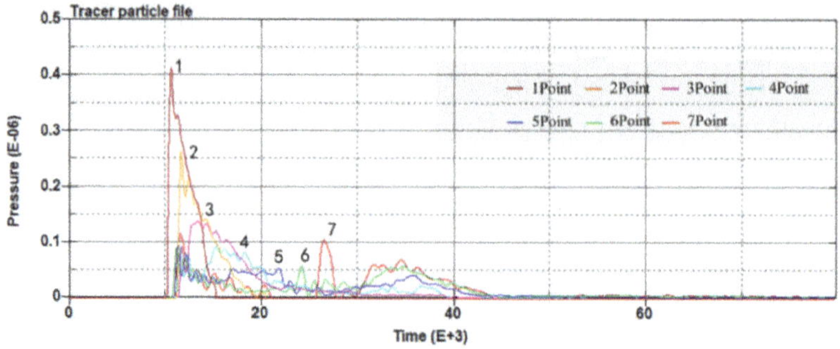

Fig. 11. Pressure of No.1–No.7 measuring points in the vertical direction behind the 30 cm high angle cone wall

overpressure of the measuring point is that the measuring point occurs from top to bottom, and the overpressure peak decreases in turn, which has good regularity. Subsequently, the overpressure circulation, which was disturbed and weakened by the angle cone, passed through the measuring point again from top to bottom, resulting in the second overpressure peak on the image. As shown in Fig. 12, in the range of 0–3 m, except for the 22.5 cm high pyramid wall, the other walls are more effective than the ordinary straight wall to weaken the shock wave. Even if the plane wave is generated behind the wall, it is still lower than the overpressure peak behind the ordinary straight wall. At the same time, due to the interference effect between the corner cones, more shock waves are superimposed and dissipated rather than transmitted to the top of the wall, so the shock wave weakening effect of the corner cone explosion-proof wall is still better than that of the ordinary straight wall in the range of 4–6 m.

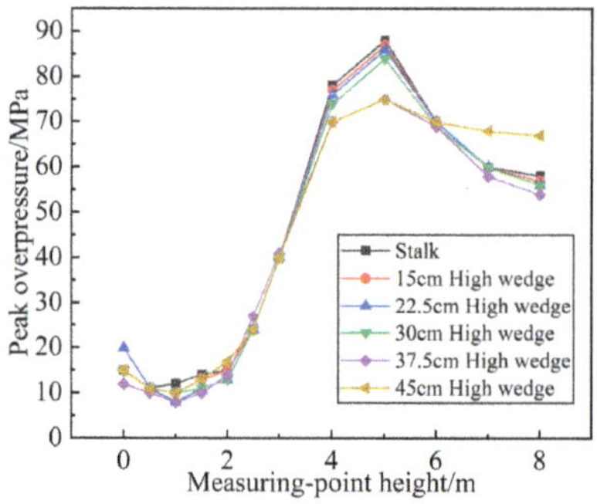

Fig. 12. The peak overpressure in the vertical direction behind each type of wall.

4.5 Shock Wave Weakening of Each Measuring Point

When the detonation wave passes through the corner cone wall, the overpressure behind the explosion-proof wall is reduced by a percentage compared to the straight wall. The effectiveness of reducing the explosion wave is defined as:

$$P_r = (P_b - P_p)/P_b \times 100\% \quad (7)$$

In the formula, P_r is the percentage of overpressure reduction after the guardrail column, that is, the weakening rate; P_b is the overpressure peak behind the straight explosion-proof wall; P_p is the overpressure peak behind the corner cone explosion-proof wall. If $P_b = P_p$, then $P_r = 0\%$, that is, the weakening effect of the corner cone explosion-proof wall on the shock wave is 0 compared with the straight wall.

As shown in Fig. 13, all the cone-shaped explosion-proof walls reach the overpressure peak at 7 m measuring point, that is, 2 m behind the wall, and are higher than the ordinary straight wall. The 15 cm high cone has the smallest increase in the superposition of shock waves here, about 13.67%; the 37.5 cm corner cone wall has the greatest effect on the power of the shock wave, reaching about 45.31%, which will cause greater danger to the target behind the wall.

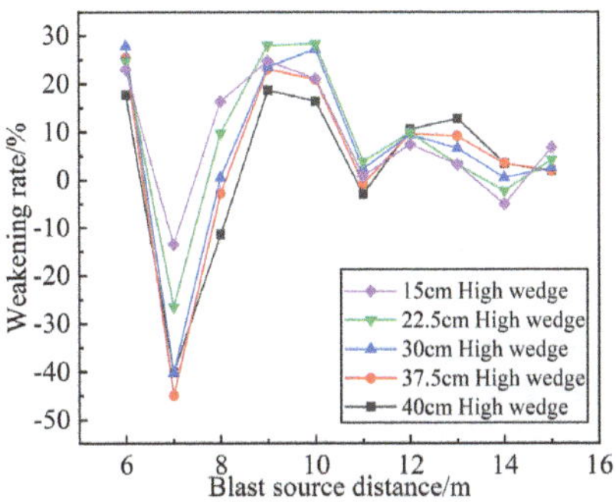

Fig. 13. The weakening rate of each measuring point in the horizontal direction

The weakening rate curve set can be divided into two parts, with 10 m as the dividing line. The average weakening rate of the measuring points within 10 m is higher than that after 10m, and the weakening effect decreases with the height of the cone from low to high. The 15 cm high cone with the best weakening effect can weaken the shock wave by 14.73%. Contrary to that within 10 m, the highest average weakening rate of the measuring point outside 10 m is the 45 cm high pyramid which reduces 5.34% overpressure, and the weakening effect decreases with the height of the pyramid from high to low.

Figure 14 shows the shock wave weakening effect of the longitudinal measuring point at 0.5 m behind the wall, which can also be divided into two parts, with the wall top height of 3 m as the dividing line. The chaotic shock wave weakening curve within 3 m is caused by the superposition of plane wave and incomplete overpressure circulation behind the wall. Although the regularity of the cone height is not strong, the overall effect is still better than that of the ordinary straight wall. The average weakening rate of the 30 cm high cone is the highest in this range, reaching 10.44%. The average weakening rate of the 30cm high cone is the highest in this range, reaching 6.00%.

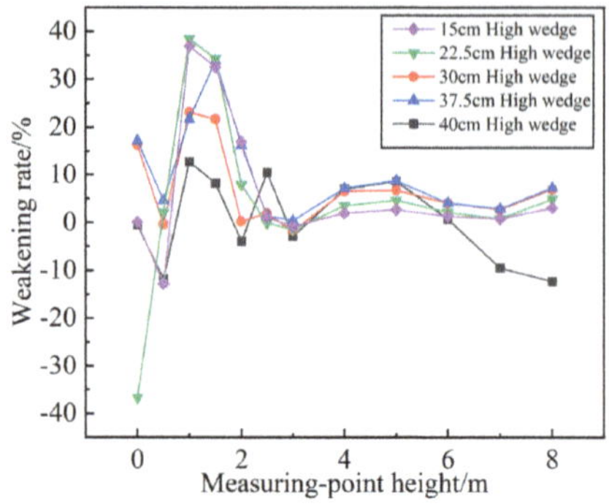

Fig. 14. The weakening rate of each measuring point in the vertical direction

Figure 15 shows the weakening effect of each cone height on the shock wave in different directions and intervals. In general, the best types of blast-facing surfaces for shock wave weakening are 15 cm and 30 cm high pyramids, but the advantages of the two walls are opposite. In the horizontal direction, the weakening rate of the 15 cm pyramid is higher than that of the 30 cm pyramid, and the average weakening rate of the 15 mm pyramid in the horizontal short distance (within 5 m behind the wall) can reach 14.74%, and the average weakening rate in the 10 m distance is reduced but still 8.71%. In the vertical direction, the shock wave weakening effect of the two kinds of walls is opposite. In the vertical direction, the average weakening rate of the 30 cm corner cone within the wall height (0–3 m) reaches 10.44%, while the weakening rate at the height of 0–8 m can still reach 7.97%.

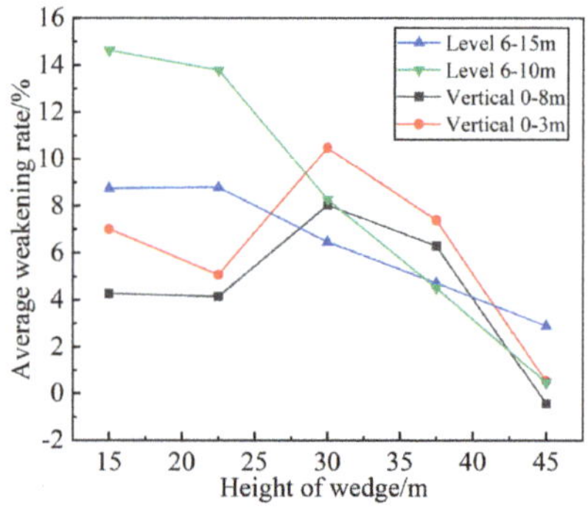

Fig. 15. The average weakening rate of various types of corner cone wall

5 Conclusion

The study of conical explosion-proof wall not only helps to improve the performance of the explosion-proof wall itself, but also provides an important reference for the overall anti-explosion design of buildings and facilities. By reasonably setting and arranging the conical explosion-proof wall, the overall anti-explosion performance of buildings and facilities can be improved, and their survivability and stability in explosive events can be enhanced.

(1) Compared with the ordinary straight wall, the cone-shaped blasting surface strengthens the ability to weaken the detonation wave to a certain extent. In particular, taking the square pyramid with a side length of 30 cm at the bottom of the pyramid as an example, it is found that the explosion-proof surface with a cone height of 15 cm has a better explosion-proof effect in the horizontal direction, while the explosion-proof surface with a cone height of 30 cm performs better in the vertical direction. However, trade-offs need to be made in practical applications.
(2) There is a certain difference in the propagation form of the shock wave behind the wall between the cone-shaped blast-facing surface and the ordinary straight wall, which is mainly affected by the plane wave generated by the stiffness reduction and the blocked top overpressure circulation. The shock wave behind the cone-shaped blast-facing wall is more manifested as Mach reflection, while the shock wave behind the ordinary straight wall is closer to the regular reflection. At the same time, the peak value of reflected overpressure behind the straight wall is higher than that of Mach reflection behind the corner cone wall.
(3) At a distance of about 1–1.5 times the distance from the wall after the cone-shaped front surface, due to the stronger plane wave and the overpressure circulation at the top of the cone, the shock wave will obliquely impact the ground from the top of the wall at a large angle and produce Mach reflection. This phenomenon will lead to the pressure exceeding the maximum overpressure peak behind the ordinary straight wall, which will cause greater damage to the protection target behind the wall, so special attention should be paid to control.

In summary, the research of conical explosion-proof wall is of great significance and value for the field of explosion-proof wall. It can not only improve the efficiency and performance of explosion-proof wall, but also reduce the impact of explosion shock on the surrounding environment, promote the development of explosion-proof technology, and provide important reference for the anti-explosion design of buildings and facilities, so as to provide more reliable protection for people's life and property safety.

References

1. Fang, Q., Yang, S., Chen, L.: Tianjin port '8·12' extraordinary fire and explosion accident building. J. China Civ. Eng. J. **50**(3), 12–18 (2017)
2. Ma, Y., Zhao, L., Nie, J.: Numerical model of anti-air shock wave of special-shaped explosion-proof wall. J. Explosion **27**(1), 26–30 (2010)
3. Zong, R., Bai, P.: Design and numerical simulation of explosion-proof wall with new guardrail structure. J. Explosion **33**(3), 140–145 (2016)
4. Hou, J., Zhuang, L.Y., Feng, S.: Arc explosion-proof wall under the impact of coal mine gas explosion. J. Henan Polytech. Univ. (Nat. Sci. Ed.) **38**(4), 32–38 (2019)
5. Yu, W.J., Zhao, J.C., Shi, J.Y.: Corrugated plate explosion-proof wall under blast load. J. Sichuan Build. Sci. Res. **38**(2), 78–81 (2012)
6. Hong, W., Fan, H.L., Jin, F.N.: Study on the calculation method of the load on the blast face of rigid explosion-proof wall. J. Eng. Mech. **29**(11), 30–31 (2012)

7. Yi, M.D., Wang, B., Wang, D.: The absorbing mechanism of absorbing materials in microwave darkroom design. J. Inf. Commun. **164**(8), 30–31 (2016)
8. Mu, Z.M., Ren, H.Q., Li, Y.C.: Research on diffraction effect of blast shock wave on wall. J. Mech. Pract. **31**(5), 35–40 (2009)
9. Wang, W.Z.: Hysteretic behavior, failure mechanism and resistance of beam-to-column bolt-welded rigid connections in steel frames. Dissertation of Xi'an University of Architecture and Technology (2003)
10. Chen, S.F.: Steel Structure Design Principle. Science and Technology Press, Beijing (1998)
11. Yang, X.H.: Study on explosion-proof mechanism and anti-explosion performance of shaped explosion-proof wall. Beijing Jianzhu University (2022)
12. Luo, X.B., Zhang, Y.L., Ding, Y.K.: Explosive Mechanics Tutorial, pp. 265–266. National Defense Industry Press (2016)
13. Zhang, Q.L., Zhang, Y., Nian, X.Z.: The influence of concrete explosion-proof wall on the propagation of explosion shock wave. J. Vibr. Shock **32**(24), 192–197 (2013)

Seismic Performance Analysis of Precast Frames Considering Joint Defects

Ma Yanqiang and Jin Qiao(✉)

School of Civil Engineering, Shenyang Jianzhu University, Shenyang 110118, NJ, China
1422840226@qq.com

Abstract. To investigate the influence of partial reinforcement failure in FRP-confined rebar splicing joints on the seismic performance of precast portal frames, based on the experimental results of anchorage performance and splicing performance of the joints, this paper established a macroscopic analysis model of precast concrete portal frames considering partial longitudinal reinforcement failure in the joints using OpenSees finite element analysis software. By combining fiber section modeling technique and strain penetration effect simulation technique, under low-cycle reversed loading condition, the influence of different positions and numbers of failed splicing rebars on the overall seismic performance of portal frames was analyzed. The results show that when the joints have defects, the area of the envelope curve of the structural hysteresis loop will significantly decrease, and it will increase with the increase of the number of failed rebars; the joint defects will slightly reduce the initial stiffness of the overall frame structure, but have a greater impact on the stiffness degradation in the middle and late stages of loading; compared with the defect-free precast frames, the cumulative energy dissipation capacity of the precast frames with edge-failed rebar connections is decreased by 6.40%–24.45%, and their ductility coefficient is decreased by 16.49%–24.45%, while the cumulative energy dissipation capacity of the precast frames with middle-failed rebar connections decreased by 5.26%–11.60%, and their ductility coefficient decreased by 14.11%–20.36%. In summary, for the precast frame structures with defective joints, their hysteresis performance, stiffness degradation, cumulative energy dissipation, ductility coefficient and other evaluation indicators are lower than those of the defect-free precast frames, and the position and number of failed longitudinal rebars in the joints have an impact on the overall seismic performance of the structure.

Keywords: Joint defects · Frame structure · OpenSees · Seismic performance · Fiber reinforced polymer (FRP) confinement

1 Introduction

A prefabricated concrete frame structure is a building structure system that uses factory-made components to assemble and connect on site, which has the advantages of saving resources, improving efficiency, and protecting the environment. It has received wide attention and application in China in recent years [1]. However, most of the existing

I. Tuns et al. (Eds.): CIBv 2023, LNCE 510, pp. 373–386, 2024.
https://doi.org/10.1007/978-3-031-60765-3_30

studies follow the "equivalent cast-in-place" principle, assuming that the dimensional accuracy of the prefabricated components and the quality of the construction assembly are ideal, ignoring the possible construction quality problems such as grouting defects, eccentricity, etc. in the actual engineering, which may cause the failure of the local connection joints of the structure, thereby affecting the overall performance of the structure [2, 3]. In fact, these "problem joints" are not uncommon in the early engineering practice of prefabricated structures in China, and their impact on the seismic performance of the structure needs to be further evaluated and analyzed.

The defect problem of prefabricated structure joint connection is one of the hotspots that scholars have paid attention to in recent years. J. Chen et al. [4] conducted experimental studies on the failure mode of semi-grouted sleeve connection rebar, and found that when the ratio of bond failure length to embedment length of rebar in the sleeve (SR) increased from 0 to 100%, the failure mode changed from rebar fracture to rebar pullout; Zheng Qinglin et al. [5] analyzed the influence of different forms of grouting defects on the failure mode of grouted sleeve connection by uniaxial tension test, and the results showed that the failure mode of grouted sleeve connection with grouting defects was mainly rebar pullout failure and rebar yield failure; Li Xiangmin et al. [6] set grouting defects at the bottom of the lower rebar anchorage section in the sleeve, and studied the influence of different defect rates on the uniaxial tensile strength of grouted sleeve, and found that the larger the defect rate, the lower the connection strength. In addition, Zheng Qinglin [7], Jiang Youbao [8], Li Weiwei [9] and other scholars further studied the correlation between grouting sleeve defects and the seismic performance of prefabricated concrete columns and frames, and the results showed that grouting defects reduced the bearing capacity, stiffness and ductility performance of components or structures. The above studies mainly focus on grouted sleeve connections, and the defect problems of other connection methods have not been involved.

In the research methods of prefabricated assembly structure, OpenSees is a common analysis and simulation tool. J. Zhao et al. [10] used the zero-length section element to simulate the end rotation of components at the foundation and component interface, and established a longitudinal reinforcement slip constitutive model considering strain penetration effect, which reflected the mechanical response of components well; Girgin et al. [11] used the force-based beam-column element and the fiber section model to consider the influence of geometric nonlinearity on column response, and used the uniaxial reinforcement model to consider the influence of longitudinal reinforcement buckling, and analyzed the influence of longitudinal reinforcement buckling and strain penetration on structural performance; L. Zhao et al. [12] used the zero-length element and the Pinching4 bilinear uniaxial material to simulate the strength degradation, unloading and loading stiffness degradation characteristics under cyclic load, which had high simulation accuracy, but required experiments to determine the parameters of the Pinching 4 material.

This paper takes a portal frame as the object, adopts the joint connection method of "vertical reinforcement lap connection with FRP confinement ring", and establishes the numerical model of the structure using the OpenSees finite element software. It focuses on the analysis of the influence of joint connection performance failure on the seismic performance of the structure, and provides a reference basis for evaluating the seismic performance of prefabricated structures with joint connection defects.

2 Finite Element Modeling

2.1 The Numerical Model Experiment Design

In order to explore the correlation between the joint connection performance and the overall seismic performance of the prefabricated structure, we established a portal frame structure model with prefabricated joints. The joint form adopts the new connection structure of "reinforcement lap connection with FRP confinement ring" proposed by the author, as shown in Fig. 1. The joint connection has completed the anchorage and lap basic test [13, 14], and has been verified in the corresponding quasi-static test [15]. In the above experiment, the grout was provided by Shenyang Branch of Beijing Saigu Weiye Technology Co., Ltd., which was a dry material prepared by mixing cement, concrete admixture, fine aggregate and other auxiliary materials in a certain proportion. The grout was mainly composed of cement, concrete admixture, fine aggregate and other auxiliary materials. The grout had the characteristics of early strength, high strength and micro-expansion, and could play a good role in connecting the structure.

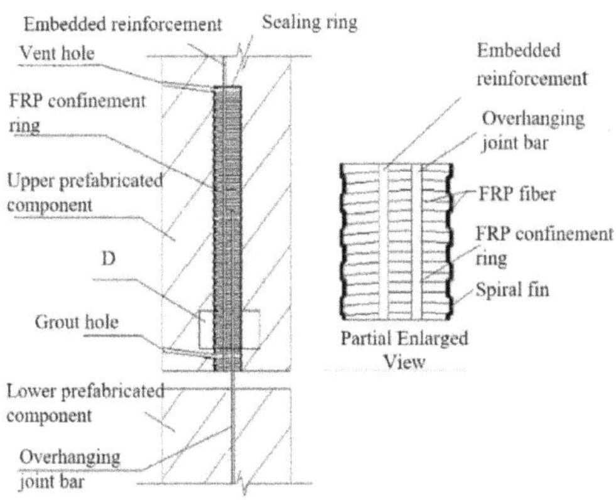

Fig. 1. Schematic diagram of bond connection of reinforcing bars with FRP restraint ring

The overall structure adopts a portal frame form, with a size of 3000 mm × 4200 mm. The two end columns are assembled and connected at the bottom, and the connection joints are set in the column foot area; the beam-column components in the frame are all C40 concrete, the longitudinal reinforcement section is HRB400 grade steel bar with a diameter of 14 mm, and the stirrup is HRB400 grade steel bar with a diameter of 8mm; the protective layer thickness of the beam-column section is 30 mm; the axial compression ratio is 0.6, and the axial compression force is applied to the top of the two edge columns respectively. The specific dimensions and reinforcement are shown in Fig. 2.

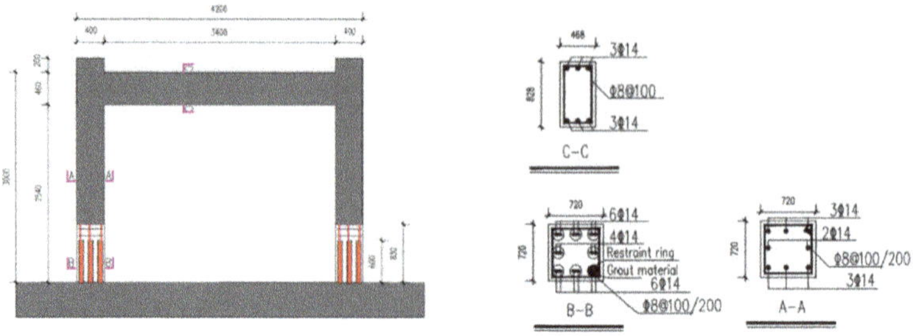

Fig. 2. Size and reinforcement of test piece

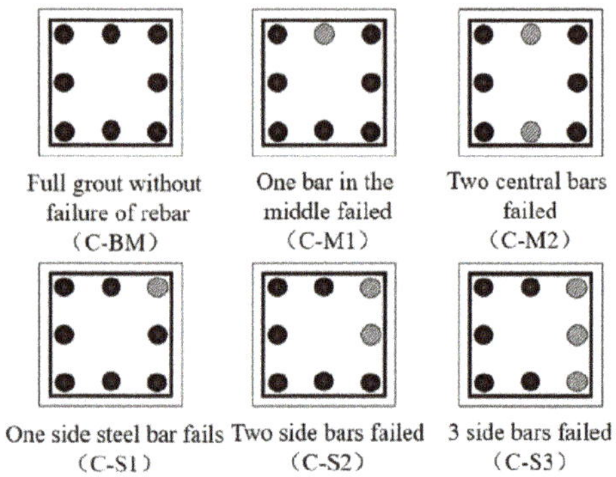

Fig. 3. Defect conditions of the study

In order to simulate the performance failure of the connection joint, five defect conditions and one non-defect condition were designed (see Fig. 3). For the joint defects, the failure of the middle and edge reinforcement of the column section were considered respectively, and the specimen numbers were C-M1, C-M2, C-S1, C-S2, C-S3, respectively, where the symbol "C" represents the column end connection, the symbol "M1" means the failure of the reinforcement connection within one FRP confinement ring in the middle of the section, the symbol "S1" means the failure of the reinforcement connection within one FRP confinement ring at the edge of the section, and so on.

2.2 The Macro Analysis Model of Prefabricated Frame Structure

According to the numerical model experiment design scheme, the structural analysis model was established based on the OpenSees platform. The model assumed that the column foot of the frame structure was fixed to the foundation, and the column and beam used the force-based beam-column elements (i.e. the Element NonlinearBeam-Column), with three Gauss integration points in the element; the beam-column section used the fiber model, with the Concrete01 constitutive model for concrete and the hysteretic constitutive model for longitudinal reinforcements; the model considered the two confinement effects of stirrup on core concrete and FRP confinement ring on grout, using the Kent-Park model for the former and the FRPConfined Concrete constitutive model for the latter. The element and its constitutive model of the model are shown in Fig. 4.

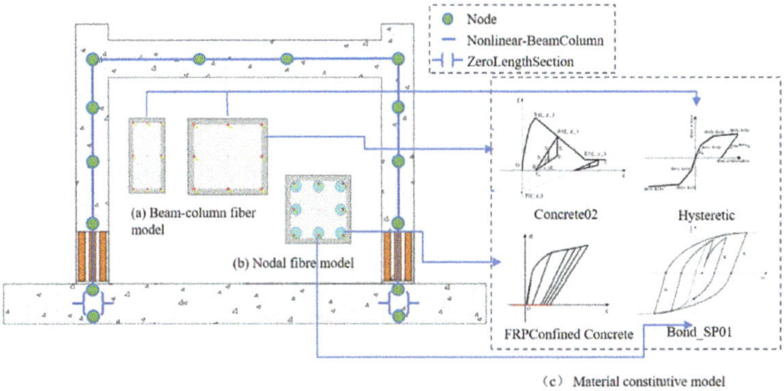

Fig. 4. Schematic diagram of model elements and their constitutive curves

2.3 The Constitutive Model of Fiber Section

In the macro analysis model of prefabricated frame structure, the section of the joint connection area is the most complex area in terms of material types and constraint conditions (see Fig. 5). Taking this as an example, the constitutive model of fiber section is explained.

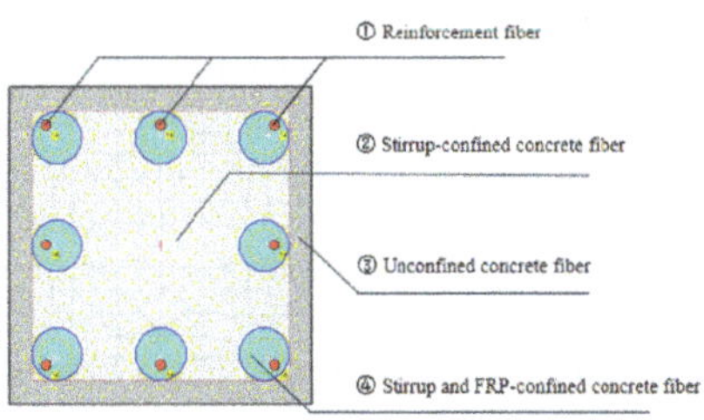

Fig. 5. Precast concrete column fiber section

As can be seen from Fig. 5, the fiber section of this area is divided into four types: ① Reinforcement fiber section (for the longitudinal reinforcements), using Hysteretic constitutive model that considers the effects of pinching, ductility, etc.

② Stirrup-confined concrete fiber section (for the core concrete), using the Concrete04 model (i.e., the Mander confined concrete constitutive model) that considers the hoop effect of stirrup on the core area; ③ Unconfined concrete fiber section (for the protective layer), using the Concrete01 model (i.e., the modified Kent-Scott-Park constitutive model) that neglects the tensile behavior of concrete;

④ Stirrup and FRP-confined concrete fiber section (grout), using FRPConfined-Concrete model (i.e., uniaxial Megalooikonomou-Monti-Santini concrete constitutive model); The constitutive relationships of some key materials are shown in Fig. 6. In addition, for the reinforcement fiber on the zero-length element section in the column foot area, the Bond_SP01 uniaxial material constitutive model is used to consider the strain penetration effect of anchored reinforcement.

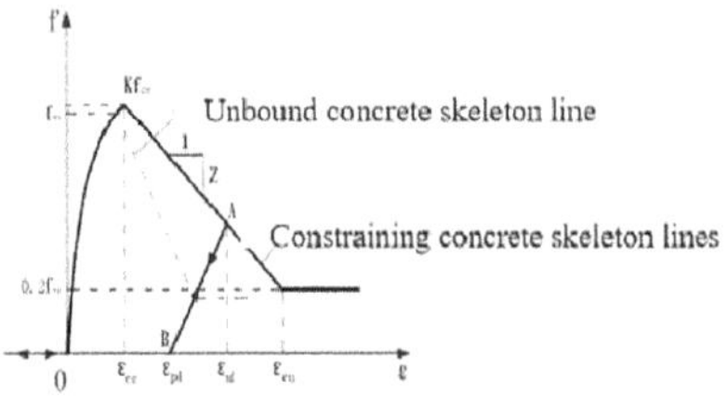

(a) the Concrete01 constitutive model

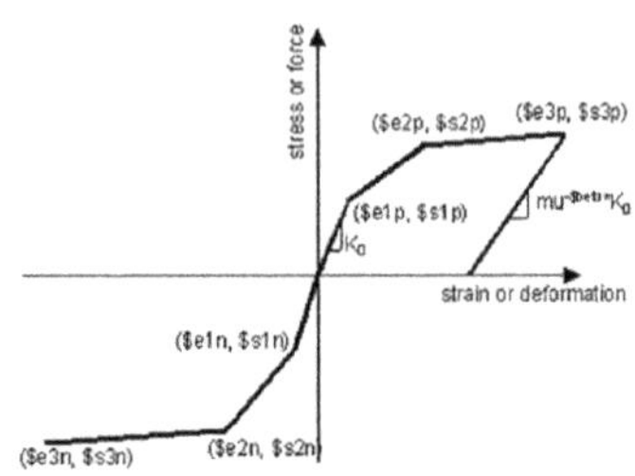

(b) the Hysteretic constitutive model

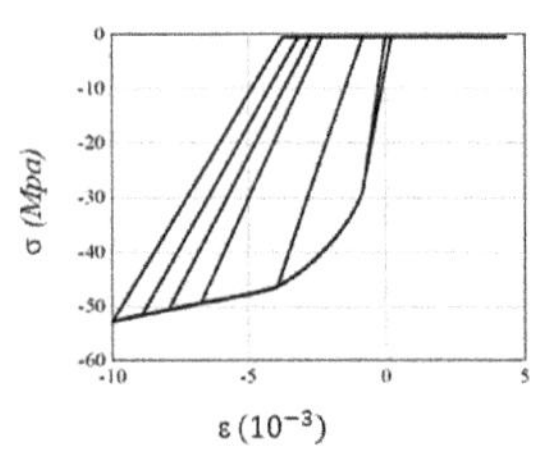

(c) the FRPConfinedConcrete model

Fig. 6. Typical material constitutive models. (Note:$s1p and $e1p represent the stress and strain (or force & deformation) at first point of the envelope in the positive direction respectively; $s2p and $e2p represent the stress and strain (or force & deformation) at second point of the envelope in the positive direction respectively; $s3p and $e3p represent the stress and strain (or force & deformation) at third point of the envelope in the positive direction respectively; $s1n and $e1n represent the stress and strain (or force & deformation) at first point of the envelope in the negative direction respectively; $s2n and $e2n represent the stress and strain (or force & deformation) at second point of the envelope in the negative direction respectively; $s3n and $e3n the stress stress and strain (or force & deformation) at third point of the envelope in the negative direction respectively;)

2.4 Loading Regime

This paper adopts the quasi-static test method, simulating the biaxial loading of the structure subjected to the axial load on the top of the two columns and the horizontal displacement on the top of the left column. According to the "Code for Seismic Design of Buildings" -GB 50011–2010, the limit value of the axial compression ratio of the frame structure is 0.3–0.7, and 0.6 is taken as the axial compression ratio of this study. The axial load is controlled by force, and the horizontal displacement is controlled by displacement. The maximum interlayer displacement angle is determined according to the elastic-plastic interlayer displacement angle limit value of the reinforced concrete frame structure (i.e., 1/50). The loading system is shown in Fig. 7.

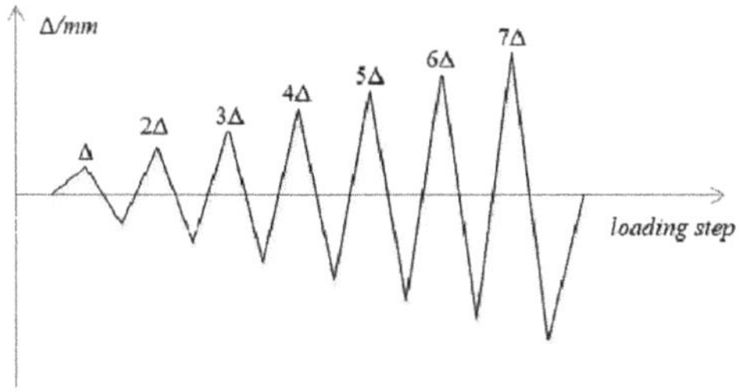

Fig. 7. Loading system

3 Performance Analysis of Prefabricated Frame Without Connection Defects

In order to compare the seismic performance difference between the defect-free prefabricated frame and the cast-in-place frame, this paper carried out a simulation analysis based on the macro analysis model for the two frames. The cast-in-place frame used for comparison was taken from the F25 frame in reference [16]. The size of the frame was 1870 mm × 2750 mm, the compressive strength of concrete was 34.5 Mpa, the longitudinal reinforcement of the beam-column section was HRB400 grade steel bars with a diameter of 16mm, the stirrup was HRB400 grade steel bars with a diameter of 8mm, and the protective layer of the beam-column section was 50 mm.

In the simulation, the prefabricated frame used the reinforcement lap connection with FRP confinement ring, located at the bottom of the column. The material constitutive model, node, element type, fiber section division and loading method of the cast-in-place frame were the same as those of the prefabricated frame. The only difference was that the bottom of the column of the cast-in-place frame did not contain FRP confinement ring, but used the ordinary concrete constitutive model, to reflect its "cast-in-place" construction method.

The hysteresis curves of the two frames are shown in Fig. 8. It can be seen from the figure that the two frames have similar initial stiffness and peak bearing capacity. The bearing capacity of the prefabricated frame with column bottom splicing is slightly higher than that of the cast-in-place frame, which is because the FRP confinement ring and the high-strength grout in the plastic hinge area significantly improve the stiffness and strength of the column section at the connection. At the same time, the pinching effect of the prefabricated frame is more obvious, and the overall difference from the cast-in-place frame is not significant. It can be seen that the prefabricated frame without connection defects has similar seismic performance to the cast-in-place frame.

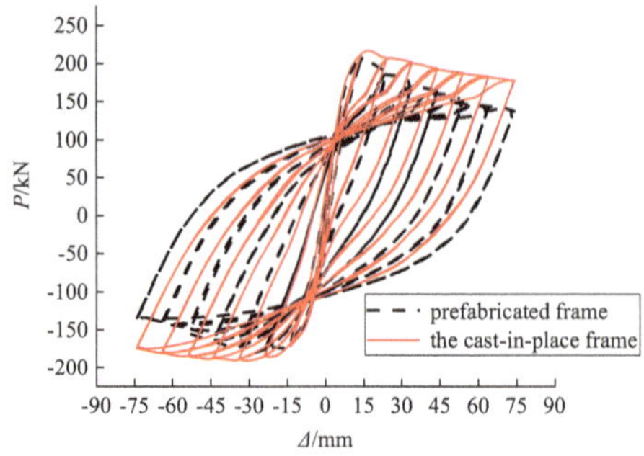

Fig. 8. Comparison of hysteresis curves

4 Performance Analysis of Prefabricated Frame with Connection Defects

Based on the macro analysis model of the prefabricated single-bay frame, this paper analyzed the influence of the five connection joint defects (see Fig. 3) on the seismic performance of the overall frame structure in detail, from the aspects of hysteresis behavior, stiffness degradation, cumulative energy dissipation, bearing capacity, displacement and ductility.

4.1 Hysteresis Curve Analysis

Figure 9 and Fig. 10 show the comparison of the hysteresis curves of the overall frame structure under different positions and numbers of connection joint defects. As shown in Fig. 9, when the connection joint defect is one edge reinforcement failure for C-S1, compared with the defect-free joint for C-BM, the initial stiffness and initial bearing capacity peak of the overall frame structure are both reduced, but the shape of the hysteresis curve is basically consistent, only the area of the hysteresis loop is slightly reduced. With the increase of the number of edge reinforcement failures in the connection joint for C-S2 and C-S3, the shape of the hysteresis curve also changes, the area of the hysteresis loop further shrinks, indicating that the energy dissipation capacity and seismic performance of the overall frame structure are constantly declining. Figure 10 shows the hysteresis curve results of the overall frame structure when the connection joint defect is the middle reinforcement failure for C-M1 and C-M2. Compared with Fig. 8, the middle reinforcement failure has a smaller impact on the hysteresis capacity of the overall frame structure, which is because the middle reinforcement is close to the neutral axis of the section, and has a lower contribution to the bending stiffness and bearing capacity of the frame column. Table 1 shows the envelope area of the hysteresis curves for all frames and the percentage reduction of the envelope area compared with the C-BM frame.

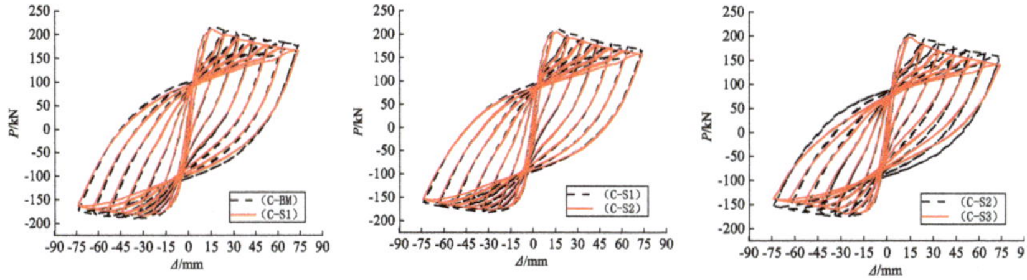

(a) C-S1 & C-BM (b) C-S2 & C-S1 (c) C-S3 & C-S2

Fig. 9. Comparison of hysteresis curves(specimens C-S1, C-S2, C-S3 and C-BM)

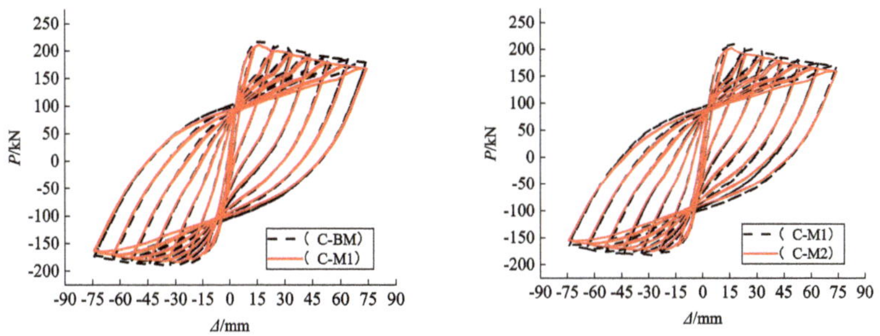

(a) C-M1 & C-BM (b) C-M2 & C-M1

Fig. 10. Comparison of hysteresis curves(specimens C-M1, C-M2 and C-BM)

Table 1. Total envelope area and corresponding reduction percentage of hysteresis curves for all frames

Specimen number	Total envelope area ($kN \cdot mm$)	Reduction percentage
C-BM	153496.3	-
C-S1	143213.7	6.7%
C-S2	135581.5	11.7%
C-S3	123253.8	19.7%
C-M1	145102.7	5.5%
C-M2	135171.8	11.9%

4.2 Stiffness Degradation Analysis

Figure 11 shows the stiffness degradation curves of the overall frame structure under different positions and numbers of connection joint defects. As shown in Fig. 11 (a), when the connection node defect is the edge reinforcement failure for C-S1, C-S2 and C-S3, compared with the defect-free frame for C-BM, the initial stiffness of the overall frame structure decreases slightly with the increase of the number of failed reinforcements. In the early stage of loading, the stiffness degradation curves of the four specimens are almost coincident, indicating that the stiffness degradation rate is similar. In the middle and late stage of loading, the stiffness of the defect-free specimen is significantly higher than that of the defective specimens, and the stiffness degradation rate of the four specimens is still close, resulting in the stiffness degradation curves being basically parallel. As shown in Fig. 11 (b), when the connection joint defect is the middle reinforcement failure for C-M1 and C-M2, compared with the defect-free frame for C-BM, the initial stiffness difference of the overall frame structure is not significant, but the initial stiffness of the specimen C-BM is still slightly higher. In the early stage of loading, the stiffness degradation curves of the three specimens are almost coincident, indicating that the stiffness degradation rate is similar. In the middle and late stage of loading, the stiffness of the defect-free specimen is still higher than that of the defective specimens, but the stiffness degradation rate of the three specimens is different, resulting in the stiffness degradation curves showing a branching trend, among which the specimen C-M2 has the largest stiffness degradation rate, the specimen C-M1 has the medium stiffness degradation rate, and the specimen C-BM has the smallest stiffness degradation rate. Table 2 shows the average stiffness of the softening segment for all frames and the corresponding reduction percentage of the stiffness compared with the C-BM frame when the loading displacement is between 10 mm and 75 mm.

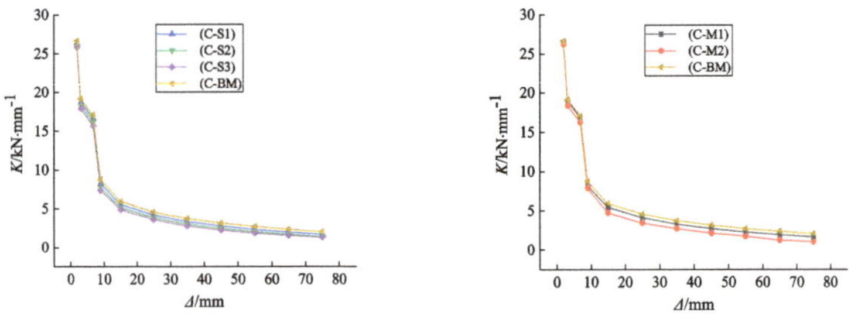

(a) Specimens of C-S1、C-S2 and C-S3 (b) Specimens of C-M1 and C-M2

Fig. 11. Comparison of stiffness degradation curves

Table 2. Average stiffness of the softening segment and corresponding reduction percentage for all frames

Specimen number	Average stiffness at 10 mm–75 mm loading disp. (kN/mm)	Reduction percentage
C-BM	4.22	-
C-S1	3.83	9.24%
C-S2	3.47	17.78%
C-S3	3.26	22.75%
C-M1	3.71	12.09%
C-M2	3.18	24.64%

4.3 Cumulative Energy Dissipation Analysis

As shown in Fig. 12 (a), when the connection joint defect is the edge reinforcement failure for C-S1, C-S2 and C-S3, compared with the defect-free frame for C-BM, the cumulative energy dissipation of the overall frame structure decreases significantly with the increase of the number of failed reinforcements. The maximum cumulative energy dissipation of specimens C-S1, C-S2 and C-S3 are 6.40%, 11.22% and 24.45% lower than that of specimen C-BM, respectively. This is because the failure of the reinforcement joint connection at the frame column foot reduces the number of energy dissipation reinforcements, lowers the actual reinforcement ratio, and affects the energy dissipation capacity of the overall frame.

As shown in Fig. 12 (b), when the connection joint defect is the middle reinforcement failure for C-M1 and C-M2, compared with the defect-free frame for C-BM, the cumulative energy dissipation of the overall frame structure also decreases, but the decrease is smaller. The maximum cumulative energy dissipation of specimens C-M1 and C-M2 are 5.26% and 11.60% lower than that of specimen C-BM, respectively.

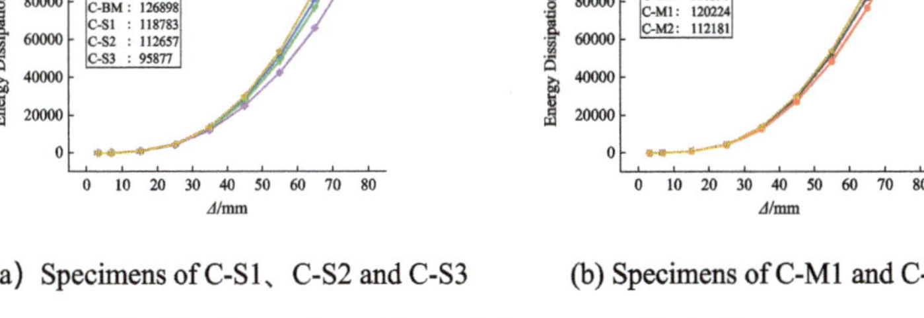

(a) Specimens of C-S1、C-S2 and C-S3 (b) Specimens of C-M1 and C-M2

Fig. 12. Comparison of cumulative energy dissipation curves

4.4 Analysis of Bearing Capacity and Ductility

As shown in Fig. 13 (a), when the connection joint defect is the edge reinforcement failure for C-S1, C-S2 and C-S3, compared with the defect-free frame for C-BM, the yield load and ultimate load of the overall frame structure decrease significantly with the increase of the number of failed reinforcements. The yield loads of specimens C-S1, C-S2 and C-S3 are 2.36%, 5.13% and 8.44% lower than that of specimen C-BM, respectively. The ultimate loads of specimens C-S1, C-S2 and C-S3 are 2.59%, 5.72% and 9.07% lower than that of specimen C-BM, respectively. This is because the failure of the reinforcement joint connection at the frame column foot reduces the number of energy dissipation reinforcements, lowers the actual reinforcement ratio, and hence affects the bearing capacity of the overall prefabricated frame. As shown in Fig. 13 (b), when the connection joint defect is the edge reinforcement failure for C-S1, C-S2 and C-S3, compared with the defect-free frame for C-BM, the ductility coefficient of the overall frame structure also decreases significantly with the increase of the number of failed reinforcements. The ductility coefficients of specimens C-S1, C-S2 and C-S3 are 16.49%, 21.40% and 24.45% lower than that of specimen C-BM, respectively.

Similarly, as shown in Fig. 14, when the connection joint defect is the middle reinforcement failure for C-M1 and C-M2, compared with the defect-free frame for C-BM, the bearing capacity and ductility coefficient of the overall frame structure also decrease, but the decrease is smaller. The yield loads of specimens C-M1 and C-M2 are 2.67% and 6.55% lower than that of specimen C-BM, respectively. The ultimate loads of specimens C-M1 and C-M2 are 4.25% and 6.79% lower than that of specimen C-BM, respectively. The ductility coefficients of specimens C-M1 and C-M2 are 14.11% and 20.36% lower than that of specimen C-BM, respectively.

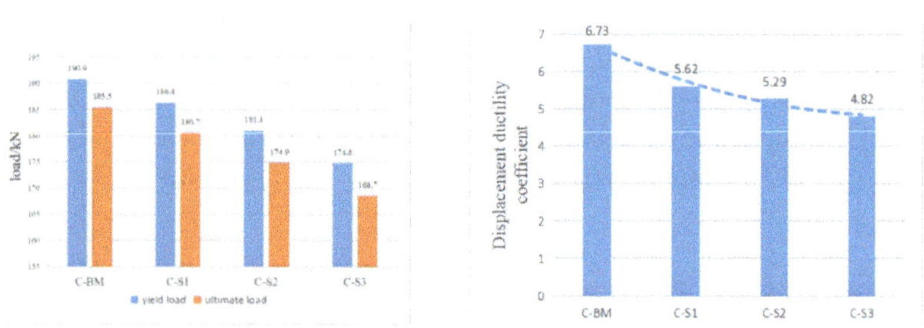

(a) Bearing capacity (b) Ductility coefficient

Fig. 13. The bearing capacity and ductility coefficient of specimens C-S1, C-S2 and C-S3

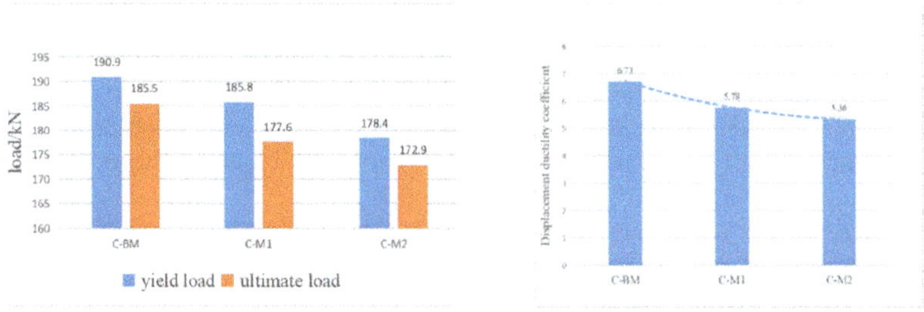

(a) Bearing capacity (b) Ductility coefficient

Fig. 14. The bearing capacity and ductility coefficient of specimens C-M1 and C-M2

5 Conclusions

(1) The prefabricated concrete frame structure with FRP-confined vertical reinforcement lap connection has comparable seismic performance to the cast-in-place frame when the connection joint performance is defect-free; however, its seismic performance will decrease significantly when the connection joint performance has defects. Therefore, in the seismic performance evaluation of prefabricated engineering, the possibility and influence of partial failure of joints should be considered.

(2) The defect of connection joint performance will affect the overall performance of the prefabricated frame structure, and this effect is related to the position and number of failed reinforcements. Generally speaking, the edge reinforcement failure has a greater impact on the overall performance than the middle reinforcement failure, and the more the number of failed reinforcements, the more obviously the overall performance decline.

(3) The defect of connection joint performance has a small impact on the initial stiffness and initial bearing capacity of the prefabricated frame structure, but has a large impact on the hysteresis behavior, cumulative energy dissipation capacity and ductility performance of the prefabricated frame structure, and these indicators are more indicative of the seismic capacity and seismic performance of the prefabricated frame structure.

References

1. Kai, L., et al.: Comparison of the development status and countermeasures of prefabricated buildings at home and abroad. Eng. Constr. **53**(07), 19–24 (2021)
2. Zhao, L., et al.: Numerical studies on the seismic responses of bridge structures with precast segmental columns. Eng. Struct. **151**, 568–583 (2017)
3. Liu, H., et al.: Seismic performance of precast column connected with grouted sleeve connectors. J. Build. Eng. **31**, 101410 (2020)
4. Chen, J., et al.: Experimental investigation of mechanical properties of steel half-grouted sleeve splice with rebar bonding defects. J. Build. Eng. **50**, 104–113 (2022)

5. Zheng, Q., Wang, N., Tao, L., Xu, W.: Experimental study on the effect of grouting defects on the performance of steel bar sleeve grouting joint. Build. Sci. **33**(05), 618 (2017).https://doi.org/10.13614/j.cnki.11-1962/tu.2017.05.010
6. Xiangmin, L., Gao Rundong, X., Qingfeng, W.Z., Fuwen, Z.: Grouting defects on the experimental study of steel sleeve joint grouting connection strength. J. Build. Struct. **13**(07), 52–56 (2018). https://doi.org/10.19701/j.jZJG.2018.07.010
7. Qinglin, Z., Ni, W., Li, T., Xin, T.: Experimental study on effects of grout defects on seismic performance of assembled concrete columns. Chin. Civil Eng. J. **51**(5), 75–83 (2018)
8. Jiang, Y., et al.: Seismic behavior of fabricated concrete columns and frames with sleeve grouting defects. J. Central South Univ. (Sci. Technol.) **51**(05), 1361–1370
9. Weiwei, L., et al.: Experimental study on seismic performance of prefabricated concrete frames under the influence of grouting defects. J. Build. Struct. **43**(07), 123–130 (2022)
10. Zhao, J., Sritharan, S.: Modeling of strain penetration effects in fiber-based analysis of reinforced concrete. ACI Struct. J. **104**, 133–141 (2007)
11. Girgin, S.C., at al.: Nonlinear beam-based modeling of RC columns including the effect of reinforcing-Bar buckling and rupture. Earthq. Spectra **34**, 1289–1309 (2018)
12. Zhao, L., et al. Numerical studies on the seismic responses of bridge structures with precast segmental columns. Eng. Struct. **151**, 568–583 (2017)
13. Xuebin, W.E.N.: Experimental Study on Anchorage Performance of Steel Bars with FRP Confined Ring. Shenyang University of Architecture, Shenyang (2017)
14. Meng, G.A.O.: Study on the Overlapping Connection of Reinforcements in Prefabricated Shear Wall with FRP Confined Rings. Shenyang jianzhu University, Shenyang (2018)
15. Jinglong, L.I.: Seismic Behavior of Assembled Shear Wall with FRP Constrained Vertical Reinforcement Connections. Shenyang University of Architecture, Shenyang (2019)
16. Zhen, M., Weimin, S., Zhanggen, G.: Experimental study on seismic performance of recycled concrete frame. World Earthq. Eng. **27**(01), 22–27 (2011)

Study on Explosion Resistance of a New Type of Prefabricated T-Joint

Yu Wang, Jincheng Huang(✉), and Haile Yue

School of Civil Engineering, Shenyang Jianzhu University, Shenyang 110168, China
15542196265@163.com

Abstract. A new type of prefabricated beam-column connection T-joint is proposed, which connects prefabricated beams and columns through components such as high-strength bolts and embedded steel plates. The deformation and mechanical properties of this new fabricated T-joint under explosion load were studied, and its reliability was analyzed. The finite element analysis software LS-DYNA was used to establish respectively the finite element models of cast-in-place and prefabricated T-joint under the same conditions of the same size and loading mode. The failure characteristics, stress and displacement analysis of T-joints with two different construction techniques under blast load are compared and analyzed. The results show that compared with the cast-in-place joint, the local damage caused by the explosion shock wave to the new prefabricated T-joint is more serious, but there is no large area of concrete collapse. In addition, under the two different construction technologies, the change rules of the steel bar stress and the joint displacement of the T-joint are basically the same. Therefore, it can be concluded that the new prefabricated structure has good explosion resistance.

Keywords: prefabricated · T-joint · finite element analysis · explosive load

Prefabricated structure refers to the structure formed by connecting scattered prefabricated components into a whole through mechanical hoisting and certain connection means when the components are prefabricated in the processing plant or construction site. Its advantages are good quality, high production efficiency, low labor intensity, environmental protection, and is conducive to social sustainable development [1]. The use of prefabricated structures can effectively save resources and energy, improve the efficiency of materials in real building energy conservation and structural performance, reduce the requirements of site construction on site conditions, reduce construction waste and adverse effects on the environment, improve building functions and structural performance, and achieve green development requirements of "four conservations and one environmental protection". Realize the construction process of low energy consumption and low emission, promote the overall development of China 's construction industry, and achieve the predetermined goal of energy conservation and emission reduction:

Y. C. Loo et al. [2] took the lead in carrying out comparative tests of prefabricated concrete beam-column joints with and without corbels. The results show that the ductility and energy consumption of prefabricated concrete beam-column joints without corbels are lower than those of prefabricated concrete beam column joints with Ming corbels.

I. Tuns et al. (Eds.): CIBv 2023, LNCE 510, pp. 387–396, 2024.
https://doi.org/10.1007/978-3-031-60765-3_31

E. Onur et al. [3] tested the assembled beam-column joints using the bright bracket connection method. The results show that the prefabricated beam-column joints using the bright bracket connection method have better ductility and energy dissipation capacity than using the dark bracket. Guo et al. [4] used high-strength bolts and anti-wire sleeves to connect dark corbel beam-column joints and carried out full-scale frame edge joint tests. The results show that the new prefabricated joints have better ductility and energy dissipation capacity. The new prefabricated joints can effectively realize the out-ward movement of plastic hinges and protect the core area of joints. In the above literature, the mechanical properties of beam-column joints with bright corbel and dark corbel were tested and numerically simulated, and the feasi-bility of dark corbel connection was verified.

Wu C and Hao H [5] studied the combined effect of shock wave and ground vibration on the surrounding structure under near-ground explosion by numerical simulation. The variation law of parameters such as overpressure peak, particle velocity and overpressure duration under the action of near-ground explosion is given. Morill et al. [6] carried out three sets of full-scale RC column explosion tests, studied the column resistance and related reinforcement measures under explosive load, and proposed a simplified reinforcement design procedure. Zhao W J [7] carried out the low-cycle reciprocating test of prefabricated concrete frame shear wall structure. The finite element software was used to calculate the low-cycle reciprocating loading of prefabricated and cast-in-place structures, and the design method of strengthening structural measures in weak parts was proposed. Considering the influence of model grid on the results of explosion analysis, Shi Y C [8] proposed a correction method to eliminate the grid size effect, which is beneficial to improve the reliability of explosion numerical simulation. Sun H P [9] analyzed the design and construction of new prefabricated beam-column bolt joints, proposed a new connection technology for prefabricated structural joints, and calculated and analyzed the bearing capacity of bolts. ANSYS finite element analysis software is used to numerically calculate the assembled beam-column bolts under load, and the full simulation expression is carried out. The key links and difficulties of the modeling process are effectively explained through a detailed modeling process, which provides an effective numerical model for the stress analysis of new prefabricated beam-column joints.

In view of this research status, the author proposes a new type of dark corbel connection beam-column joint, which has the advantages of convenient construction and saving formwork. It is a new type of connection method for assembled concrete members, that is, the assembled beam-column is connected by high-strength bolts and embedded steel plates.

In this study, the direct simulation method is used to establish the explosive entity and the air medium, and simulate the whole process of explosive explosion, shock wave propagation and action to the structure. Using ANSYS / LS-DYNA finite element software, the finite element models of cast-in-place and assembled T-joints were established, and the cast-in-place and new dark corbel connection prefabricated T-joints were analyzed under explosion load. We will simulate the propagation process of the explosion and compare the prefabricated T-joint with the cast-in-place. The research shows that the new assembled structure has good anti-explosion performance.

1 Establishment of Model

1.1 Structural Model

The length × width of the prefabricated concrete column is 600 mm × 600 mm. The longitudinal steel bars of the column are four with a diameter of 25 mm. The stirrups are 100/200 with a diameter of 8 mm. The width × height of the assembled concrete beam is 300 mm × 600 mm. Four tension steel bars with a diameter of 28 mm are used in the lower side of the prefabricated beam, two compression steel bars with a diameter of 25 mm are used in the upper side, and the stirrups are 100/200 with a diameter of 8 mm. The longitudinal reinforcement and stirrups of the bracket are the same as those of the column. The steel plate type of the joint connection in the specimen is Q235 B, and the steel bar type is HRB400. The concrete adopts strength grades of C35. The geometric dimensions are shown in Fig. 1.

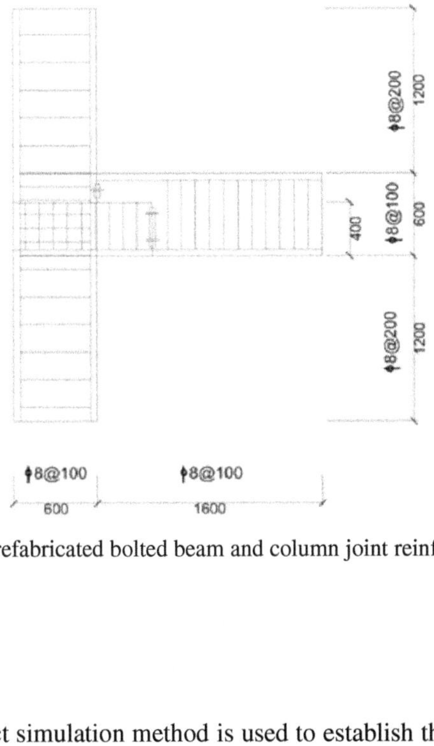

Fig. 1. Prefabricated bolted beam and column joint reinforcement.

1.2 Material Model

In this study, the direct simulation method is used to establish the explosive entity and the air medium, and to simulate the whole process of explosive explosion, shock wave propagation and action to the structure. The numerical simulation uses the * MAT _ HIGH _ EXPLOSINE _ BURN model and the state equation * EOS _ JWL provided by the LS-DYNA material library to simulate the explosive material. In this paper, the air is assumed to be a nonviscous ideal gas, and the * MAT _ NULL material model and the state equation * EOS _ LINEAR _ POLYNOMAL are used to simulate the air.

1.2.1 Concrete Material Model

The material model * MAT _ JOHNSON _ HOLMQUIST _ CONCRETE (HJC) is used to simulate the concrete material. The HJC model fully considers the material properties of concrete materials affected by large strain, high strain rate and high pressure. The damage constitutive relation model is based on the cumulative plastic strain, and considers the pressure effect and strain rate sensitivity. It is suitable for Lagrange and Euler element algorithm [10]. The equivalent yield strength is a function of pressure, strain and damage. The pressure is a function of volumetric strain and contains the compaction effect. The cumulative damage is a function of plastic volumetric strain, equivalent volumetric strain and pressure [11]. The yield expression of the material is

$$\sigma^* = \left[A(1-D) + BP^{*N}\right]\left[1 - cln\left(\varepsilon^*\right)\right] \tag{1}$$

where σ^* is the equivalent yield strength; P^* is the standardized pressure; ε^* is the dimensionless strain rate; A, B, N and C are the constant parameters of concrete, where A is the standard cohesive strength, B is the standard strength increase index, N is the pressure increase index, and C is the strain rate sensitivity coefficient. D is the damage factor in the model, $0 \le D \le 1$. HJC model discusses damage accumulation and damage coefficient D from two aspects of equivalent plastic strain and plastic volume strain.

By accumulating the damage caused by equivalent plastic strain and plastic volumetric strain, it is obtained that:

$$D = \sum \frac{\Delta\varepsilon_P + \Delta\mu_P}{\varepsilon_P^f + \mu_P^f} \tag{2}$$

$\Delta\varepsilon_P$ and $\Delta\mu_P$ are equivalent plastic strain increment and plastic volumetric strain increment; $\varepsilon_P^f + \mu_P^f$ is the plastic strain of fracture at normal pressure:

$$\varepsilon_P^f + \mu_P^f = D1\left(P^* + T^*\right)^{D2} \tag{3}$$

D1 and D2 are damage constants. T^* is the normalized maximum hydrostatic tension, $T^* = T/f_c\prime$; T is the maximum hydrostatic tension that concrete can withstand. The relevant parameters of HJC material model are shown in Table 1.

1.2.2 Steel Material Mode

The steel bar is a kind of isotropic material, and the strain rate is very sensitive. Compared with the static load, the steel bar will deform rapidly under the action of high strain rate explosion load. Therefore, the dynamic description of the steel bar is * MAT _ PLASTIC _ KINEMATIC model. The model is used for beam, shell and solid elements, which is suitable for isotropic kinematic strengthening materials considering strain rate effect. In this paper, the steel bar is a beam element, which meets the requirements of the material model. The Cowper-Symonds model is introduced into the kinematic hardening model to consider the strain rate effect of the material. The yield stress is

$$\sigma^* = \left[1 + \left(\frac{\varepsilon}{C}\right)^{\frac{1}{P}}\right]\left(\sigma_0 + \beta E_P \varepsilon_P{}^{eff}\right) \tag{4}$$

C and P are Cowper-Symonds constants; ε is the strain rate; σ_0 is the initial yield stress; β is hardening parameter; when β = 0, it is a plastic kinematic hardening model; when β = 1, it is an isotropic hardening model; when 0 < β < 1, it is a mixed model of plastic kinematic hardening and isotropic hardening. $\boldsymbol{\varepsilon_P}^{eff}$ is the effective plastic strain; $\boldsymbol{E_P}$ is a plastic hardening modulus (Table 2).

Table 1. Parameters of concrete model

ρ/kg.m^{-3}	A	B	C	PL/MPa	SFMAX	T/MPa	PC/MPa
2400	0.79	1.6	0.007	8.0	7.0	4.0	16.7
EPSO/S^{-1}	EFMIN	N	D1	D2	K1	K2	K3
1E-06	0.01	0.61	0.04	1.0	8.5E10	−1.71E11	2.08E11

Table 2. Parameters of steel material model

Material	ρ/kg.m^{-3}	E/MPa	SIGY/MPa	ETAN/MPa	FS
Reinforcement	7850	2.1×10^5	400	2000	0.12
Steel plate	7850	2.1×10^5	350	6100	0.2
Bolts	7850	2.1×10^5	980	6100	0.2

1.2.3 Unit Type and Grid Division

In the modeling process of this paper, ordinary steel bars are simulated by BEAM161 element, and other materials are modeled by SOLID164 element. The grid model of the assembled T-joint is shown in the following figure. The mesh size of the model element is 50 mm, and the node area is encrypted to 25 mm. In the modeling process, a large number of air domain models will greatly increase the calculation time. In order to solve this problem and improve the computational efficiency. The existing main method is to set the artificial boundary on the limited air domain boundary, which can make the explosion shock wave propagate to the air domain boundary without reflection through the boundary, so as to simulate the environment of the infinite air domain. The non-reflective boundary condition is realized by defining the corresponding keywords for the boundary surface nodes, and the EDNB command is used to define the non-reflective boundary (Fig. 2).

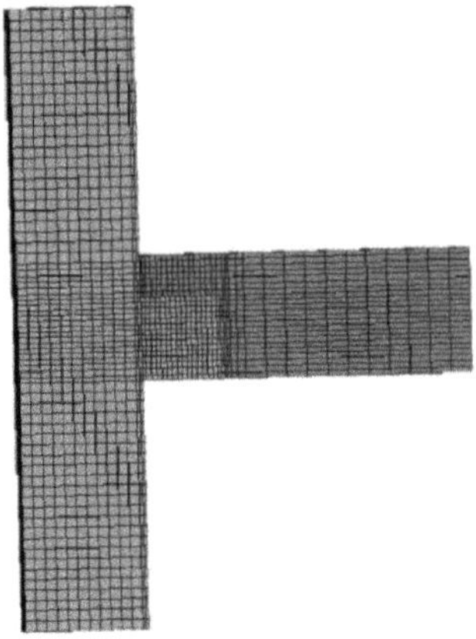

Fig. 2. The finite element meshing of prefabricated nodes.

2 Finite Element Analysis Results

2.1 Failure Characteristics of Joints

2.1.1 Failure Characteristics of Cast-in-Place Edge Joint

TNT explosive with the volume of 200 mm × 200 mm × 200 mm was placed 200 mm in front of the beam T-joint. In the analysis and calculation, the explosive was set to detonate at 0 ms, and the shock wave hit the surface of the beam and column at 0.5 ms. At the impact surface, the concrete far away from the corner was squeezed by the surrounding coagulation, forming a three-way compression state, but the concrete was not crushed. At this time, the concrete at the corner first appeared in the tensile stress state during the extrusion process, and was crushed first (see Fig. 3). Under the protection of dense steel bars in the joint area, the damage range of concrete is not extended. At 5 ms, the shock wave caused the maximum peak damage to the surface of the T-joint (see Fig. 4), and the concrete near the intersection of the beam and the column was crushed by the shock wave.

2.1.2 Failure Characteristics of Fabricated Joints

The TNT explosive was placed 200 mm × 200 mm × 200 mm in front of the T-joint of the beam. The explosive was set to detonate at 0ms, and the shock wave hit the surface of the beam column at 0.5 ms. At the same time as the edge of the beam column was first damaged, due to the weak bolt connection, the local compression protection of the concrete on the impact surface was lost, and the outer concrete far away from the corner did not form a three-dimensional compression state. Therefore, extrusion failure also occurs (see Fig. 5). Compared with the RC cast-in-place T-joints in the same period, the new prefabricated T-joints have greater stress at the side columns, and the deformation of the T-joints is slightly larger than that of the RC cast-in-place T-joints. At 5 ms, the shock wave caused the maximum peak damage to the surface of the T-joints, and the concrete near the beam-column junction was crushed by the shock wave (see Fig. 6). Compared

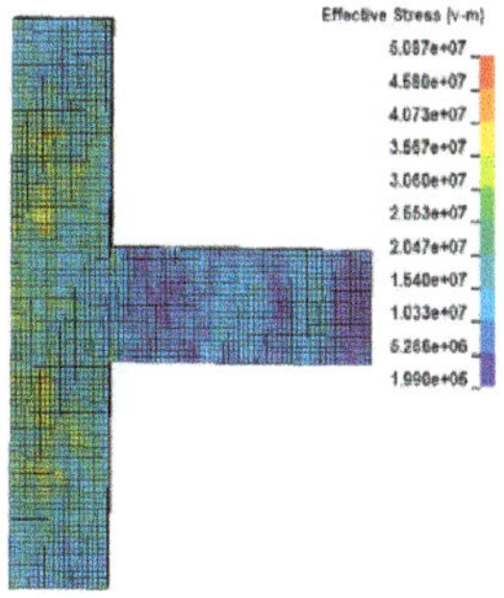

Fig. 3. The T joint stress distribution of concrete at 0.5 ms.

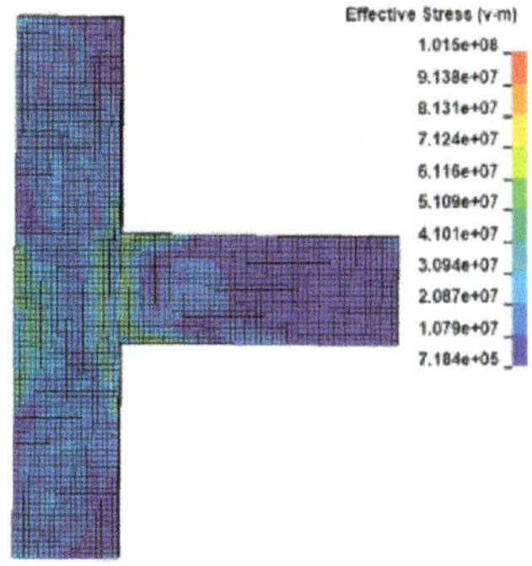

Fig. 4. The T joint stress distribution of concrete at 5 ms.

with the RC cast-in-place T-joints at the same time, the upper part of the beam-column junction is more seriously damaged. Due to the weak connection between the concrete and the steel plate, the concrete and the steel plate are separated by the erosion of the shock wave, and the bolt connection is not working, so the upper connection is more seriously damaged.

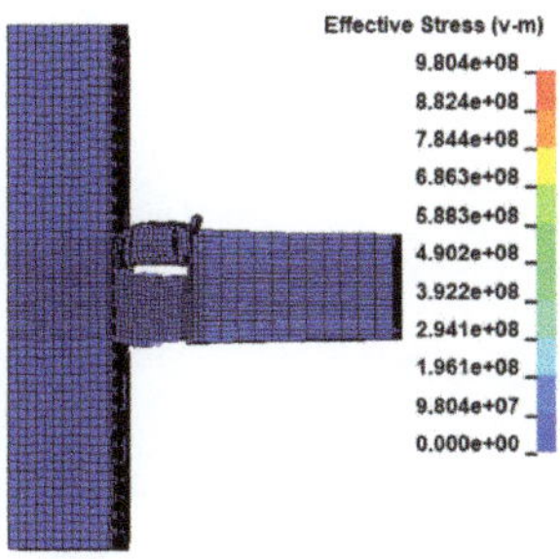

Fig. 5. The T joint stress distribution of concrete at 0.5 ms.

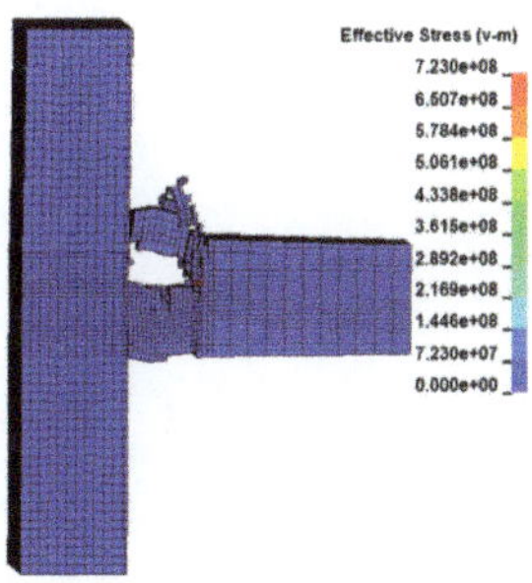

Fig. 6. The T joint stress distribution of concrete at 5 ms.

2.2 Stress Development Characteristics of Beam-Column Main Reinforcement and Stirrups

2.2.1 Cast-in-Place T Node

Within 2 ms after the initiation of the explosive, the peak stress (0.22 MPa) appeared in the main reinforcement elements on the column near the T-joint, and the shear failure did not occur. The elastic strain of most steel bars gradually recovered within 2–10 ms, and the steel bars did not yield under the protection of coagulation (see Fig. 7). At the same time, the main reinforcement elements on the beam near the T-joint and the stirrup elements in the joint range began to recover the elastic deformation before reaching the yield strain, and no shear failure occurred under the influence of the shock wave.

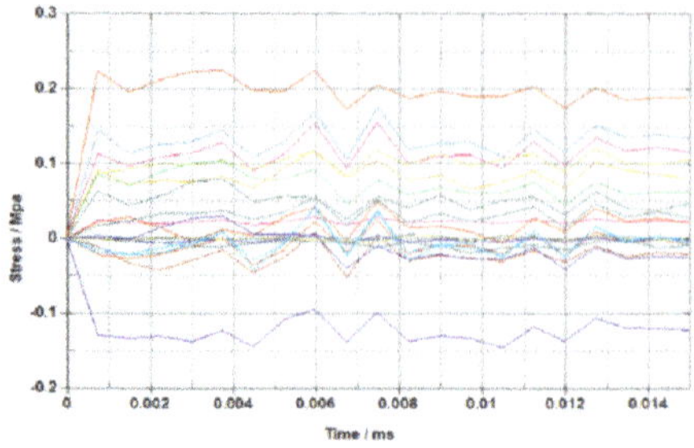

Fig. 7. The time-stress curves of column main steel cage.

2.2.2 Prefabricated T Node

The peak stress (0.21 MPa) of each main reinforcement element on the column near the T-joint appears within 2 ms after the detonation of the explosive, which is close to the peak stress of the cast-in-place T-joint, and the shear failure of the steel bar does not occur. The elastic strain of the steel bar gradually recovers within 2–10 ms, and the steel bar does not yield (see Fig. 8). At the same time, the main reinforcement element on the beam near the T-joint and the stirrup element within the joint range begin to recover the elastic deformation before reaching the yield strain. The peak stress is close to the cast-in-place T-joint, and the shear failure does not occur under the influence of the shock wave.

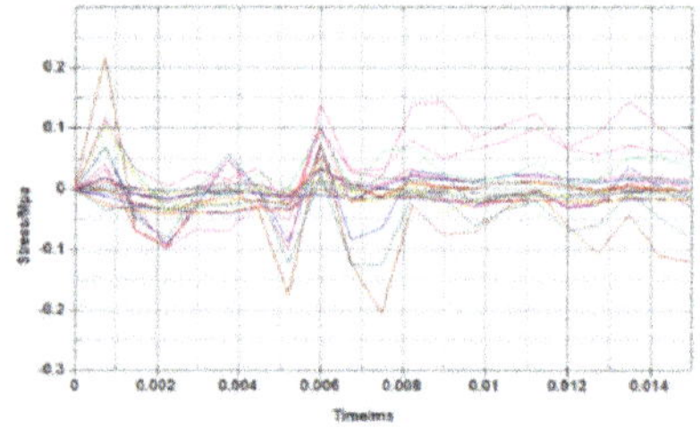

Fig. 8. The time-stress curves of column main steel cage.

2.3 Displacement Variation Characteristics of Beam-Column Main Reinforcement and Stirrups

The same node position of the finite element model is selected in the inter-section range of cast-in-place beam-column and new assembled beam-column, and the displacement-time history curve is obtained the displacement-time history curve of cast-in-place T node is shown in Fig. 9, and the displacement-time history curve of assembled T node is shown in Fig. 10.

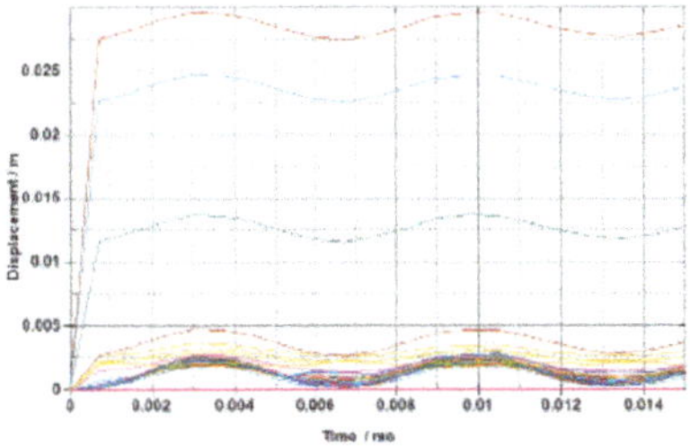

Fig. 9. The time-displacement curves of column main steel cage in cast-in-place concrete joint.

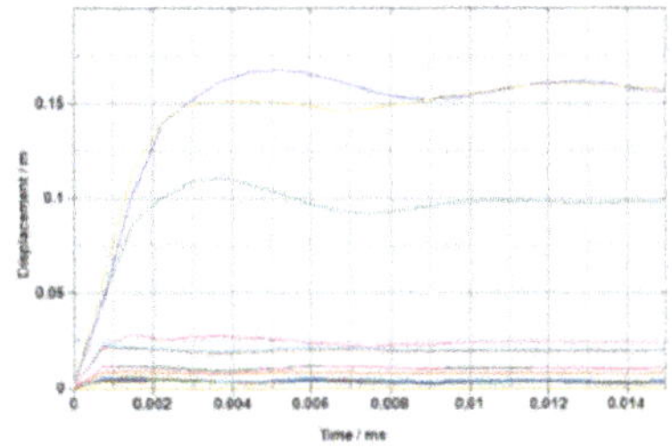

Fig. 10. The time-displacement curves of column main steel cage in fabricated concrete joint.

It can be seen from Fig. 9 and Fig. 10 that under the same explosive load, the displacement of the new prefabricated beam-column T-joint at different times is slightly larger than that of the cast-in-place T-joint, and the bolt connection method improves the ductility of the concrete structure under the impact load to a certain extent. The curve of Fig. 9 is concentrated in 0 - 0.03 m, which indicates that the surface of concrete beam-column joints is damaged by shock wave. The distribution of the curves in Fig. 10 is relatively loose, most of the curves are concentrated in 0 - 0.03 m, and 0.05 - 0.15 m appears locally, indicating that the shock wave erosion causes serious local damage to the new concrete beam-column joints, and other damage is similar to that of the cast-in-place joints.

3 Conclusion

(1) After the explosion, the local concrete shedding area at the bolt connection of the new assembled T-joint is large, and the damage is serious compared with the cast-in-place T-joint. The other damage degree is close to that of the cast-in-place joint, which proves the feasibility of the new prefabricated joint and is of great significance to the anti-explosion performance of the prefabricated joint.
(2) The binding force between the bolt connection and the concrete beam-column is weaker than that of the cast-in-place beam-column, and the post-pouring area needs to be guaranteed in the project. The new assembly connection method improves the ductility of the concrete structure under impact load to a certain extent.
(3) The shock wave erosion has serious damage to the upper part of the bolt connection. The reason is that the connection between the concrete and the steel plate is not strong, and the bolt connection does not work. Concrete is separated from the steel plate. It is suggested to increase the thickness of concrete cover to enhance the connection between steel plate and concrete.

References

1. Architectural Technology Teaching and Research Group of Beijing Institute of Architectural Engineering (1983) Prefabricated building design. China building industry press, Beijing
2. Loo, Y.C., Yao, B.Z.: Static and repeated load tests on precast concrete beam-to-column connections. PCI J. **40**(2), 106—115 (1995)
3. Onur, E., Sevket, O., Turan, O.: Ductile connections in precast concrete moment resisting frames. PCI J. 51 (3), 66–76 (2006)
4. Deng, X.S., Cao, K.F.: Study on FRP bar enhance the ability to resist progressive collapse for reinforced concrete frames. Int. J. Sci. **1**(1), 86–92 (2015)
5. Hu, Y., et al.: Characteristics of confined blast loading in unvented structures. Int. J. Protect. Struct. **2**(1), 21–43 (2011)
6. Morrill, K.B., Malvar, L.J., Crawford, J.E.: Blast resistant design and retrofit of reinforced concrete columns and walls. Struct. Congr. **2004**, 1–8 (2004)
7. Zhao, W.J., Zhong, Q., Jia, L.G.: Low cyclic loading analysis of prefabricated concrete frame shear wall structure. J. Shenyang Jianzhu Univ. **31**(02), 276–285 (2015)
8. Shi, Y., Li, Z.X., Hong, H.: Mesh size effect in numerical simulation of blast wave propagation and interaction with structures. J. Tianjin Univ. **14**(6), 396--402 (2008)
9. Sun, H.P.: Research on the performance and numerical modeling of new prefabricated beam-column connection bolt joints. Build. Technol. Dev. **49**(12), 85–87 (2022)
10. Wang, W.: Influence of shear connectors on seismic behavior of internal steel plate shear walls. J. Xi'an Univ. Archit. **46**(6), 810–813 (2014)
11. Li, G.Q., Zhai, H.Y., Yang, T.C.: The explosive load studying of concrete filled steel tube column. Build. Struct. J. **34**(12), 89–96 (2018)

Research on Damage Identification and Safety Assessment Methods for Nearby Existing Tunnel Structures Due to Vibration from Metro Shield Construction

Jiaxi Liu(✉), Chuanmeng Wei, Kun Zhao, and Rui Zhang

25 Hunnan Middle Road, Shenyang, Liaoning, China
2859399783@qq.com

Abstract. In the field of urban underground engineering, the identification of damage and the safety assessment of nearby existing tunnel structures due to vibrations from Metro Shield Construction are of paramount importance. With the continuous expansion of urban subway systems, the vibrations generated by shield excavation can have varying degrees of impact on existing underground structures, posing threats to their safety. To address this concern, researchers have conducted in-depth studies and proposed an advanced method based on wavelet packet analysis and neural networks. This method involves monitoring vibration response signals from multiple measurement points, extracting feature vectors, and utilizing a genetic algorithm for neural network training, successfully achieving damage identification. Additionally, the application of Dempster-Shafer evidence theory is anticipated to play a crucial role in dealing with uncertainties. This research provides essential technical support for ensuring the safety of urban underground transportation and infrastructure, contributing to a better understanding and assessment of the potential risks associated with vibrations caused by Metro Shield Construction.

Keywords: Vibration from Metro Shield Construction · Damage Identification · D-S Evidence Theory · Structural Safety Assessment

1 Introduction

In recent years, the rapid development of urban underground transportation and infrastructure has become a significant hallmark of modern city development. Meeting the requirements of transportation and accommodating urban expansion underscores the utmost importance of constructing and expanding subway systems. However, the construction and upkeep of subway tunnels typically occur in close proximity to existing underground structures, posing a series of challenges. One particularly significant challenge is the potential impact of vibrations caused by Metro Shield Construction on nearby existing tunnels. These vibrations, arising from excavation and propulsion activities, can result in varying degrees of damage and safety concerns. Accurately identifying and assessing these issues is crucial in the realm of underground engineering research.

I. Tuns et al. (Eds.): CIBv 2023, LNCE 510, pp. 397–411, 2024.
https://doi.org/10.1007/978-3-031-60765-3_32

2 Damage Identification Methods

2.1 Current State of Damage Identification Methods

To safeguard underground engineering structures, it is crucial to identify potential damage to nearby existing tunnel structures. In recent years, researchers have diligently explored advanced technologies to furnish precise health status information. Among these, the approach that integrates wavelet packet analysis and neural networks has garnered considerable attention and is positioned to emerge as an effective technique for damage identification.

Li Hong nan [1] (2003) and others proposed the "energy-damage" principle framework, which integrates wavelet packet analysis and neural networks. It is used for satisfactory diagnosis of damage location and extent using time-domain signals and node energy.

Sun [2] (2007) and others proposed a structural damage identification method that combines wavelet analysis with neural networks. They utilized wavelet packet energy to determine damage indicators and investigated the influence of measurement noise on the results.

Meng Fankong [3] (2013) and others investigated crack damage in flexible cantilever beams. They employed wavelet packet decomposition to extract energy spectral damage indicators and improved the BP neural network for step-by-step identification of damage location and severity in the building sector.

Liu Haiping [4] (2013) and others introduced a damage identification method for marine platform structures. They employed wavelet packet analysis and a BP neural network to decompose acceleration signals and extract energy, facilitating damage detection in the field of construction and architecture.

Wang Qi [5] (2018) investigated cable-stayed structures, employing a combination of wavelet packet analysis and BP neural networks. They explored efficient cable damage indicators and utilized the total energy change rate (RES) as an output to predict damage indicators under various conditions, contributing to the assessment of structural health in the field of construction and architecture.

Yang Linkang [6] (2022) researched on rolling bearings. They employed wavelet packet analysis to extract fault characteristics, calculated energy ratios in various frequency bands, and successfully predicted damage location and severity using a combination of particle swarm optimization and BP neural networks. This approach overcame limitations associated with traditional methods, offering valuable insights into the field of construction and architecture.

Wavelet packet analysis and neural network recognition methods are propelling rapid advancements in the field of structural health monitoring. However, there is still a need to enhance practicality and validate the feasibility of adapting to various engineering environments. Improving these methods will not only enhance structural safety and reliability but also bring forth more opportunities and value to the engineering domain.

2.2 The Framework for the Damage Identification Process

Based on the analysis of domestic and international research, this paper comprehensively considers the damage to existing tunnel structures caused by construction vibrations. It summarizes the damage identification method that combines wavelet packet analysis and neural networks, integrating finite element analysis software. The framework for damage identification in this paper is as follows:

(1) Use finite element analysis software to establish intact and damaged models of existing tunnel structures. Apply actual monitored construction vibration loads to obtain vibration response signals from multiple measurement points.
(2) Perform wavelet packet decomposition on the vibration response signals for both intact and damaged conditions. Calculate and analyze the energy in each frequency band by incorporating relevant parameters to determine the energy values in each frequency band. The energy value is represented by $s_i(i = 1, 2, \cdots, n)$: $s_i = \sum_{j=1}^{n} |x_{ij}|^2$, where $x_{ij}(i = 1, 2, \cdots, n; k = 1, 2, \cdots, n)$ represents the wavelet decomposition coefficients of discrete points in different frequency bands.
(3) Extract the energy values from wavelet packet analysis at multiple measurement points to construct feature indicators, which serve as inputs for a neural network. Train the neural network optimized by a genetic algorithm to achieve damage identification.

This paper presents a research framework for the identification of damage in existing tunnels. It introduces a genetic algorithm to optimize the weights and thresholds of the BP neural network, addressing issues related to training speed and sensitivity. The application of this method holds the potential to effectively overcome challenges faced by neural networks, thereby advancing research and development in this field.

The damage identification framework is illustrated (see Fig. 1).

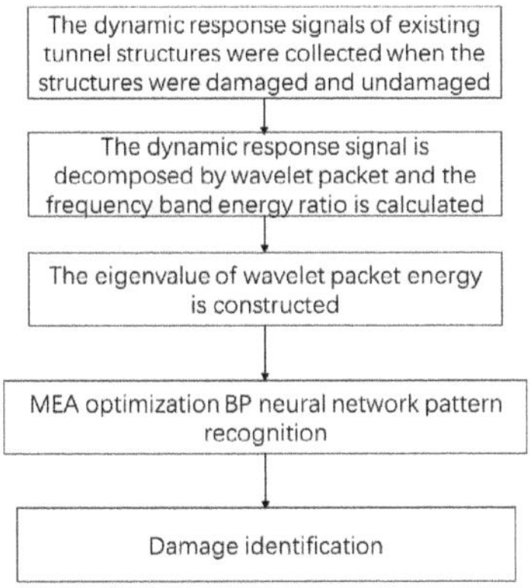

Fig. 1. Damage identification framework flow chart.

In summary, the combination of wavelet packet analysis and neural networks has been successfully applied to structural damage identification and assessment. Liu Jiamei

[7] (2018) and others demonstrated that by employing a 3-level wavelet packet decomposition to extract an 8-dimensional feature vector and utilizing the Mind Evolution Algorithm (MEA) to optimize the weights and thresholds for training a BP neural network, they were able to successfully identify voids, knots, and decay in wooden structures. This indicates the wide-ranging potential applicability of this method in the context of damage identification for nearby existing tunnels under the influence of construction-induced vibrations.

3 Existing Tunnel Structure Safety Assessment Methods

3.1 Current Methods for Structural Safety Assessment

The safety assessment of existing tunnel structures has attracted much attention and is of critical importance to urban transportation and infrastructure. However, changing factors such as time, environment and load can cause safety issues. Domestic and foreign scholars have conducted a large number of researches on existing tunnel structure safety assessment methods, including single index and multi-index assessment, using different methods such as Bayes, expert system, random set theory, neural network, support vector machine, D-S evidence theory, etc. The research method can help to understand the health status of the building structure, find the problems in time, and take the corresponding maintenance measures.

Gan Wenjuan [8] (2021) constructed a deformation prediction model for underground engineering structures using the WA-TCN-LSTM structure based on deep learning algorithms. Furthermore, they developed a multi-attribute fusion-based safety assessment model for underground engineering structures utilizing a fuzzy neural network.

Wu Bohan [9] (2020) conducted research on the overall structural deformation of subway shield tunnels under seismic conditions using Vague sets and set-pair analysis theory. They established a comprehensive seismic level assessment model to more accurately evaluate their safety.

Zhang Huan [10] (2020) and others introduced a Pearson correlation coefficient to address the issue of traditional evidence conflict. They modified the evidence body weights and then applied the D-S combination rule to calculate the fusion results.

Zhang Bin [11] (2023) and others utilized the S11-M-500/35 transformer prototype to acquire information from normal and faulty windings. They quantified and assigned membership degrees, constructing electrical and vibration parameter subevidence. The D-S evidence theory was employed to assess the transformer's condition.

Li Hexuan [12] (2023) and others employed the correlation coefficient weighting algorithm to address evidence conflicts. They introduced the Lance distance function and SAC similarity to develop conflict correction techniques, enhancing the discrimination accuracy of multi-sensor fusion in electrical spindles.

Jia Renfeng [13] (2016) and others introduced a conflict measurement method based on decision probability, improved evidence conflict resolution, and addressed fusion issues in the Dempster-Shafer evidence theory. They established safety assessment indicators for the marine environment and applied enhanced fusion rules to evaluate the safety status of the marine environment.

In summary, the development of safety assessment methods for existing tunnel structures is fraught with challenges. While technological advancements and in-depth research have led to progress, practicality, validation, and adaptability remain areas of concern. Continuous improvement of these methods to enhance structural safety and reliability will contribute significantly to urban infrastructure.

3.2 Traditional Dempster-Shafer (D-S) Evidence Theory and Its Limitations

In intelligent decision systems, when multiple pieces of evidence describe the same event, it becomes necessary to integrate these pieces of evidence to derive a comprehensive judgment result. In order to accomplish this objective, Dempster introduced a fusion rule for calculating basic probability assignments associated with two or more pieces of evidence, consequently yielding a fused result. This rule is recognized as Dempster's fusion rule and has played a significant role in advancing the broad application of the Dempster-Shafer (D-S) evidence theory.

In a given recognition framework θ, when there are two mutually independent items of evidence, E_1 and E_2, each characterized by their basic probability assignment functions m_1 and m_2 respectively, the fusion rule of Dempster for combining these two pieces of evidence is defined as follows:

$$m(A) = \begin{cases} 0 & A = \emptyset \\ \frac{1}{1-k} \sum_{A_i \cap B_j = A, A_i, B_j \subseteq 2^\theta} m_1(A_i) m_2(B_j) & A = \emptyset \end{cases} \quad (1)$$

Among these, $k = \sum_{A_i \bigcap B_j = \emptyset} m_1(A_i) m_2(B_j)$.

$k \in [0,1]$ is referred to as the conflict coefficient, representing the magnitude of the conflict between two items of evidence.

The coefficient "k" in the aforementioned Dempster-Shafer (D-S) evidence theory reflects the degree of conflict among pieces of evidence. While small conflicts may produce satisfactory fusion results, in instances of substantial conflict or complete opposition, traditional D-S theory may prove inadequate in delivering accurate fusion results.

3.3 Improvements to the Dempster-Shafer (D-S) Evidence Theory

Currently, various uncertainty reasoning methods have their advantages and disadvantages. However, the Dempster-Shafer (D-S) evidence theory shows advantages in dealing with multi-level, multi-attribute uncertainty problems. Nevertheless, the classical Dempster combination rule has limitations, especially when the conflict coefficient k equals 1. To address this, this paper proposes an improvement to the D-S evidence theory and presents the safety state assessment for existing tunnel structures. The specific process is as follows [10]:

(1) The formula for calculating the correlation coefficient P_{12} between evidence bodies m_1 and m_2 based on the Pearson correlation coefficient is as follows:

$$P_{12} = \frac{cov(m_1, m_2)}{\sigma_{m1}\sigma_{m2}} = \frac{E((m_1 - \mu_{m1})(m_2 - \mu_{m2}))}{\sigma_{m1}\sigma_{m2}} \quad (2)$$

Among these,

$$\mu_{mi} = E(m_i) \tag{3}$$

$$\sigma_{mi}^2 = E\left(m_i^2\right) - E^2(m_i) \tag{4}$$

In the context of multiple pieces of evidence, calculate the correlation coefficients between each pair of evidence bodies. Then, construct an evidence body correlation matrix P_{ij}, which contains the correlation coefficients between the various evidence bodies.

Calculate the correlation coefficients between each pair of two evidence bodies, forming a correlation matrix.

$$P_{ij} = \begin{bmatrix} p_{11}\ p_{12} & \cdots p_{1n} \\ p_{21}\ p_{22} & \cdots p_{2n} \\ \vdots\ \vdots & \ddots \vdots \\ p_{n1}\ p_{n2} & \ldots p_{nn} \end{bmatrix} \tag{5}$$

(2) Since the range of Pearson correlation coefficient is [−1,1], the negative value indicates negative correlation, and the smaller the negative value is, the higher the degree of negative correlation. In order to control the weight proportion and credibility calculation of non-positively correlated evidence bodies in the recognition framework, and consider overcoming the 0 confidence problem in the combination rules and minimizing the basic probability impact on the overall recognition framework, Here we assign 0.001 to correlation results less than or equal to 0. This approach can be used to define the credibility of the evidence body m_i, denoted as $cred(m_i)$, $cred(m_i) \in [0, 1]$, and evidence body $\sum_{i=1}^{n} cred(m_i) = 1$ as well.

$$cred(m_i) = \frac{\sum_{j=1,i\neq 1}^{n} P_{ij}\left(m_i, m_j\right)}{\sum_{i=1}^{n}\sum_{j=1,i\neq 1}^{n} P_{ij}\left(m_i, m_j\right)} \tag{6}$$

(3) Using the obtained credibility of evidence bodies, the original Basic Probability Assignment (BPA) can be adjusted, and the modified BPA is denoted as $m_i^{\#}(X)$.

$$m_i^{*}(X) = m_i(X) * cred(m_i) \tag{7}$$

(4) To avoid zero-confidence issues, we can replace each zero value in the evidence bodies with 0.001. In the case of multiple maximum probability values, their average can be taken. Similarly, when dealing with multiple zero-confidence items, 0.001 can be extracted from the highest propositions. This results in $m_i^{\#}(X)$, which adheres to these value rules and ensures the invariance of each evidence body $\sum_{X=A,B,\cdots} m_i(X) = 1$.

(5) Following the D-S combination rule of Eq. (1) along with the conflict coefficient formulas of Eqs. (8), (9), and (10), the modified $m_i^{\#}$ is used for computation to obtain the final proposition support values.

Introduce a conflict coefficient k as follows [13]:

Scenario 1: If two items of evidence are in agreement, meaning they have the same maximum decision probability in support of the same proposition, then the conflict coefficient k ∈ (0,0.5].

$$k = y_1 = \frac{n}{4n-4}\sum\nolimits_{A_i}|P_{m1}(A_i) - P_{m2}(A_I)| \tag{8}$$

Scenario 2: If two items of evidence are in disagreement, meaning they support different propositions with different maximum decision probabilities, then the conflict coefficient k ∈ (0.5,1].

$$k = y_2 = \frac{1}{4}\sum\nolimits_{A_i\in\theta}|P_{m1}(A_i) - P_{m2}(A_I)| + \frac{1}{2} \tag{9}$$

Scenario 3: If two pieces of evidence both support the same proposition and also support different propositions, the conflict coefficient can be defined as follows:

$$k = \alpha y_1 + (1-\alpha)y_2 \tag{10}$$

Among these,

$$\alpha = \sum\nolimits_{A_i\in\theta} min(P_{m1}(A_i), P_{m2}(A_i)) \tag{11}$$

α represents the consistency between evidence m1 and evidence m2. If α is relatively large, indicating a high degree of consistency between the two pieces of evidence, then the magnitude of the conflict coefficient k should lean more towards y1; otherwise, it should lean more towards y2.

3.4 Framework for Safety Assessment of Existing Tunnel Structures Based on D-S Evidence Theory

First, the initial weight vector $\omega = \{\ \omega_1\ \omega_2\ \cdots\ \omega_n\ \}$ for damage locations of existing tunnels is calculated using the entropy method. Subsequently, the vector $\omega = \{\ \omega_1\ \omega_2\ \cdots\ \omega_n\ \}$ is normalized to obtain the weight vector for damage at various locations within the tunnel:

$$\overline{\omega_i} = \frac{\omega_i}{\sum_{i=1}^{k}\omega_i} \tag{12}$$

Secondly, the weight vectors for damage locations in the existing tunnel can be represented as matrices A and B in terms of basic probabilities:

$$A = [\overline{\omega_1}, \overline{\omega_2}, \cdots \overline{\omega_k}] \tag{13}$$

$$B = \begin{bmatrix} m_1(e_1) & m_1(e_2) & \cdots & m_1(e_n) \\ m_2(e_1) & m_2(e_2) & \cdots & m_2(e_n) \\ \vdots & \vdots & \ddots & \vdots \\ m_k(e_1) & m_k(e_1) & \cdots & m_k(e_n) \end{bmatrix} \tag{14}$$

The basic probability assignment for each element is determined using a weighted average approach, yielding the following results:

$$\begin{aligned} \bar{m} &= A \cdot B \\ &= [\overline{\omega_1}, \overline{\omega_2}, \cdots \overline{\omega_k}] \times \begin{bmatrix} \mathrm{m}_1(\mathrm{e}_1) & \mathrm{m}_1(\mathrm{e}_2) & \cdots & \mathrm{m}_1(\mathrm{e}_n) \\ \mathrm{m}_2(\mathrm{e}_1) & \mathrm{m}_2(\mathrm{e}_2) & \cdots & \mathrm{m}_2(\mathrm{e}_n) \\ \vdots & \vdots & \ddots & \vdots \\ \mathrm{m}_k(\mathrm{e}_1) & \mathrm{m}_k(\mathrm{e}_1) & \cdots & \mathrm{m}_k(\mathrm{e}_n) \end{bmatrix} \\ &= [m_1, m_2, \cdots, m_n] \end{aligned} \tag{15}$$

Finally, the improved fusion method is employed to fuse $\overline{m}$ for $k - 1$ iterations, resulting in the safety assessment of the existing tunnel denoted as $C = [r_1, r_2, \cdots r_n]$.

To present the safety assessment results of the existing tunnel structure more intuitively, it is necessary to quantify the assessment levels by mapping them to numerical ranges. For example, the corresponding numerical ranges for the assessment levels {"Normal Usage", "Temporary Usage", "Maintenance and Reinforcement", "Life Safety", "Structural Damage"} are {[0, 0.1), [0.1, 0.3), [0.3, 0.45), [0.45, 0.95), [0.95, 1)}, and then their averages are calculated to obtain the quantified levels as follows: {0.05, 0.2, 0.375, 0.7, 0.975}. This mapping assigns numerical values to the assessment levels, making the safety status values more understandable [17]. The assessment levels and their corresponding quantified levels are shown in Table 1.

Table 1. T Assessment levels and quantified intervals are as follows.

Assessment levels	Quantitative Range (C)	Average Value
Normal Usage	$0 \le c < 0.1$	0.05
Temporary Usage	$0.1 \le c < 0.3$	0.2
Maintenance and Reinforcement	$0.3 \le c < 0.45$	0.375
Life Safety	$0.45 \le c < 0.95$	0.7
Structural Damage	$0.95 \le c < 1$	0.975

The mean values of the quantified assessment levels can be represented as the following matrix:

$$D = \begin{bmatrix} 0.05 \; 0.2 \; 0.375 \; 0.7 \; 0.975 \end{bmatrix}^T \tag{16}$$

By multiplying the assessment results with the quantified mean values, the safety state values can be obtained. A higher safety state value indicates a more dangerous assessment level.

The definition of safety state value is as follows:

$$I = C \cdot D = \begin{bmatrix} r_1 \; r_2 \; \cdots \; r_n \end{bmatrix} \cdot \begin{bmatrix} 0.05 \; 0.2 \; 0.375 \; 0.7 \; 0.975 \end{bmatrix}^T \tag{17}$$

Among these, $I \in [0, 1]$.

The flowchart of the safety assessment framework for existing tunnel structures is illustrated (see Fig. 2).

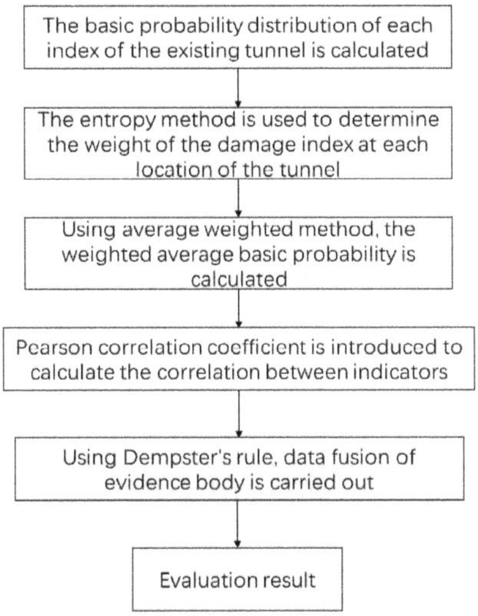

Fig. 2. Framework process of safety assessment of existing tunnel structure.

In summary, this method can integrate information from various sources, including qualitative and quantitative data, and employs the D-S evidence theory to handle information inconsistency and conflicts, thus enhancing decision accuracy. It holds promising prospects for the safety assessment of existing tunnel structures subjected to construction-induced vibrations.

4 Numerical Verification

Based on a subway excavation project, the dynamic response of adjacent existing tunnels under construction disturbance is studied. The wavelet packet energy of each frequency band is obtained by extracting the acceleration dynamic response of the adjacent tunnel, and the eigenvalue of the wavelet packet energy is obtained by data processing. According to the damage presented by the samples, they are divided into four different degree levels:

D1: Case(1), (2) - (1000) basically intact;

D2: Case(3) - (0100) Minor damage;

D3: Case(4) - (0010) Moderate damage;

D4: Case(5) - (0001) Serious damage;

Taking the damaged structure as an example, because the shield tunneling construction has little influence on the adjacent existing tunnel, only the first three levels of damage state exist. 40 groups of energy eigenvalue data were used to select the wavelet packet under each damage state, a total of 120 groups of data. In the BP neural network, 80% of the data is used for training, 10% of the data is used for verification and 10% of the data is used for testing. In order to compare the performance of the three algorithms, the final error, training steps and training time of each algorithm are listed in Table 2, and the three algorithms are compared.

Table 2. Training results of different algorithms

Network type	Training error	Training steps	Training time
Maximum descent method	2.78×10^{-4}	1000	3 s
LM algorithm	6.28×10^{-6}	8	1 s
MEA-BP	1.24×10^{-6}	7	1 s

From the comparison results, it can be seen that in terms of training time, the fastest descent method has not met the accuracy requirements, so the performance is relatively low; LM algorithm in training 8 times to reach a high level of accuracy, MEA method in training 7 times to meet the requirements. By comparing and analyzing the prediction effect, it can be seen that the error of the BP neural network model trained by the fastest descent method is 2.87×10^{-4}, the error of the LM algorithm is 6.28×10^{-6}, and the error of the thinking evolution algorithm is only 1.24×10^{-6}.

After the training of MEA algorithm rules, the structure of the network is 1-8-5, and the convergence state is reached after 41 iterations, and the accuracy is 10^{-6}. Table 3 lists some of the training results. The test samples have been input into the trained network, and the test results are shown in Table 4.

Table 3 and Table 4 show that the network can make correct diagnosis of unknown samples after good training, showing high damage recognition performance.

Table 3. Results of MEA optimization neural network training

sample	Damage degree	Training objective	Training result	Error (%)
1	D1	1000	0.9463 0.0042 0.0435 0.0060	0.02
2	D2	0100	0.0063 0.9538 0.0323 0.0076	0.01
3	D3	0010	0.0085 0.0148 0.9546 0.0221	0.03

Table 4. Test results of MEA optimization neural network

sample	Correct target	Training objective	Damage degree	Error (%)	Diagnostic conclusion
1	1000	0.9125 0.0033 0.0654 0.0188	D1	0.02	correct
2	0100	0.0002 0.9640 0.0312 0.0046	D2	0.01	correct
3	0010	0.0078 0.0134 0.9440 0.0348	D3	0.03	correct

The BP neural network was optimized to identify the damage degree of adjacent existing tunnels by thinking evolutionary algorithm, and the improved D-S evidence theory was used to evaluate adjacent existing tunnels. The damage conditions of 200 locations in the tunnel are selected to represent the damage of the whole tunnel. The compression damage and tensile damage are divided into two frames. According to the difference of damage degree, it was divided into 5 grades, labeled as M1, M2, M3, M4 and M5. It is necessary to conduct quantitative statistics on the sub-damage of each level, as shown in Table 5.

Table 5. Number statistics of sub-evidence bodies

frame	M1	M2	M3	M4	M5
Frame 1	136	54	0	10	0
Frame 2	168	17	3	12	0

For an evidence body whose number of child evidence bodies is 0, the basic probability assignment is denoted as 0.001. The basic probability values and weights of the evidence bodies were calculated, and the specific results were shown in Table 6.

Table 6. Basic probability and weight of fused evidence bodies

evidence	Frame 1	weight	Frame 2	weight
M1	0.030	0.246	0.062	0.189
M2	0.131	0.178	0.186	0.138
M3	0.300	0.139	0.001	0.250
M4	0.538	0.150	0.750	0.173
M5	0.001	0.287	0.001	0.250

After weighted processing of the above data, we can get:

$$\overline{m} = \begin{pmatrix} 0.048\ 0.160\ \ 0.244\ 0.279\ \ 0.001 \\ 0.070\ 0.165\ \ 0.001\ 0.355\ \ 0.001 \end{pmatrix} \tag{18}$$

By using different fusion methods to carry out probability fusion of weighted basic probability assignments, the safety evaluation result $C = \begin{bmatrix} r_1\ r_2\ \cdots\ r_n \end{bmatrix}$ of the existing tunnel is obtained, as shown in Table 7.

Table 7. Safety assessment results of existing tunnels

method	M1	M2	M3	M4	M5
Textual method	0.025	0.159	0.144	0.671	0.001
Reference 10	0.066	0.219	0.333	0.381	0.001
Reference 13	0.089	0.246	0.184	0.479	0.002
Reference 14	0.079	0.234	0.248	0.438	0.001
Reference 15	0.066	0.219	0.333	0.381	0.001
Reference 16	0.169	0.399	0.320	0.109	0.002

The data fusion error rate calculated for the above literature evaluation methods is shown in Fig. 3:

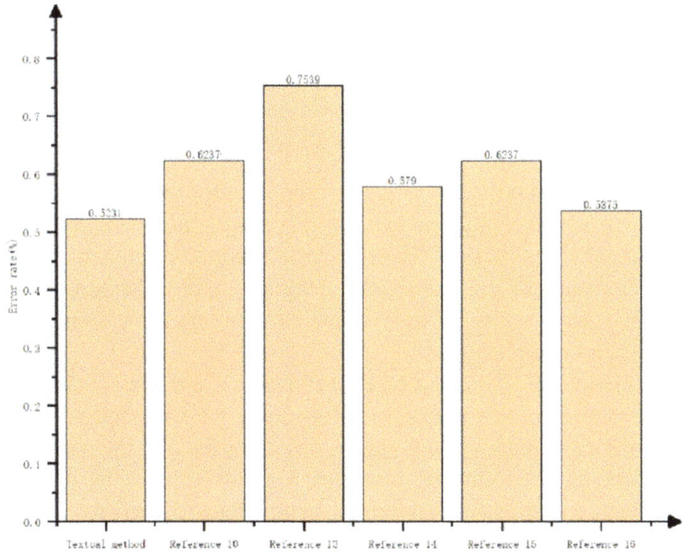

Fig. 3. Basic probability assignment fusion error rate.

From the above results, it can be seen that the method in this paper provides a more accurate and reliable fusion result through accurate calculation of errors in the process of evidence fusion. Compared with traditional methods, the proposed method shows a lower error rate in experiments, which means that it is more robust in the face of uncertainty and conflict. Therefore, the method presented in this paper proves to be the best choice, can provide a more credible basis for decision making, and is expected to achieve significant advantages in various application fields.

Similarly, the basic probability of final fusion is obtained through the above method:

$$C = \begin{pmatrix} 0.025 & 0.159 & 0.144 & 0.671 & 0.001 \end{pmatrix} \tag{19}$$

After calculation, the security status value I is obtained:

$$I = C \cdot D = 0.558 \tag{20}$$

Through the above analysis, for the whole tunnel, the life safety state when the subway shield passes through the existing tunnel is obtained. However, serious cracking and partial detachment of the existing tunnel were found at local locations, and subsequent maintenance and reinforcement were recommended. This conclusion provides an important reference for ensuring the safety of construction and tunnel operation, and it is necessary to take timely maintenance measures to ensure the reliability and safety of rail transit system.

5 Conclusion

In summary, based on the research and analysis presented above, the following conclusions can be drawn regarding the study of Shield Construction Vibration on the identification of damage to adjacent existing tunnel structures:

(1) Previous research has provided valuable references for the identification of damage and safety assessment of adjacent existing tunnel structures caused by Vibration due to Metro Shield Construction. These methods utilize traditional structural health monitoring techniques and empirical assessment approaches, laying the foundation for safety assessments. However, they have certain limitations in dealing with uncertainties and complexities.

(2) The new approach based on wavelet packet analysis and neural networks has introduced innovative ideas for damage identification. This method combines the feature extraction capabilities of wavelet packet analysis with the pattern recognition abilities of neural networks. By monitoring multiple vibration response signals, extracting feature vectors, and optimizing neural network weights using the thought evolution algorithm, efficient damage identification is achieved. Additionally, the proposed framework for damage identification in existing tunnel structures provides feasibility for more precise safety assessments.

(3) The application of D-S evidence theory offers a new approach to dealing with uncertainty issues and integrating multi-source information. By improving the D-S evidence theory, it becomes possible to better manage conflicts between items of evidence, thereby enhancing the accuracy and reliability of safety assessments. Based on this foundation, the proposed safety assessment framework provides a more comprehensive approach and more reliable tools for the safety assessment of existing tunnel structures.

References

1. Li Hongnan, F., Sun, H.S.: Damage diagnosis of frame work structure based on wavelet packet analysis and neural network. Earthq. Eng. Eng. Dyn. **05**, 141–148 (2003)
2. Sun, Z.F., Chang, C.C.S.: Vibration based structural health monitoring: wavelet packet transform based solution. Struct. Infrastruct. Eng. **3**(4), 313–323 (2007)
3. Meng, F.F., Qiu, Z.S.: Damage identification of beam using wavelet packet analysis and neural network. Noise Vibr. Control **33**(01), 197–200+207 (2013)
4. Liu Haiping, F., Diao, Y.S.: Study on structural damage detection based on wavelet packet analysis and BP neural network. Eng. Constr. **45**(04), 13–16 (2013)
5. Wang, Q.F.: Cable Damage Identification Based on Wavelet Packet Analysis and Neural Network. Fuzhou University (2018)
6. Yang, F.: Research on fault diagnosis of rolling bearing based on wavelet packet and machine learning. Beijing Jianzhu University (2022)
7. Liu Jiamei, F., Xu Kaihong, S., Wang, L.T.: Study on the diagnosis of internal defects of wood based on MEA-BP neural network. Forest Prod. Ind. **45**(02), 19–24 (2018)
8. Gan, W.F.: Research on Structure Deformation Prediction and Security Assessment Model of Underground Engineering. Chang'an University (2021)

9. Wu, B.F.: Research on Seismic Safety Evaluation of Subway Shield Tunnel construction. Xi'an University of Technology (2020)
10. Zhang Huan, F., Lu Jianguang, S., Tang, X.T.: An improved DS evidence theory algorithtm for conflict evidence. J. Beijing Univ. Aeronaut. Astronaut. **46**(03), 616–623 (2020)
11. Zhang Bin, F., Song Jinliang, S., Wang, T.T., Wang Chen, N., Li Jingjing, F.: Mechanical condition assessment of transformer winding based on D-S evidence theory. Northeast Electr. Power Technol. **44**(04), 33–37 (2023)
12. Li, H.F.: Research on Rotating Machinery Operation State Recognition Method Based on Improved D-S Evidence Theory. Jilin University (2023)
13. Jia, R.F.: Application of D-S evidence theory in ocean environment safety situation evaluation. Harbin Engineering University (2016)
14. Pan, H.F., Mcmichael, D.S., Lendjel, M.T.: Inference algorithms in Bayesian networks and the probanet system. Digit. Sign. Process. **8**(4), 231–243 (1998)
15. Kam, M.F., et al.: Performance and geometric interpretation for decision fusion with memory. Syst. Man Cybern. Part A Syst. Hum. IEEE Trans. **29**(1), 52–62 (1999)
16. Su, Y.F.: Research and Application of Artificial Neural Network data fusion Method. Zhejiang University (2002)
17. Zhao, F.Y., Jia, S.R., Shi, T.P.: A novel combination method for conflicting evidence based on inconsistent measurements. Inf. Sci. **2016**, 367–368:125–142 (2016)

Study on Hysteretic Performance of Orthogonal-Installed Double Steel Corrugated Plate Wall

Yurong Dai(✉) and Yansheng Song

Shenyang Jianzhu University, Shenyang 110168, China
84744230@qq.com, ceyssong@sjzu.edu.cn

Abstract. In this paper, a set of new orthogonal-installed double steel corrugated plate wall structure is designed based on the stress characteristics of steel corrugated plate and stiffened steel plate. The finite element software ABAQUS was used for numerical simulation to study the effects of 30°, 45° and 90° on the hysteretic performance. The results show that the initial lateral stiffness of the specimens is higher than that of the other two groups when the bending Angle is 30°. When the bending Angle is 45°, the resistance of the specimen to out-of-plane buckling is strong, and its energy dissipation capacity is higher than that of the other two groups. When the bending Angle is 90°, the specimen can still be carried with a symmetrical and complete "X"-shaped tension band after yielding. After the specimen reaches the peak load, the load value drops slowly, which is conducive to the stability of the bearing capacity of the specimen in the later period.

Keywords: Double Steel Corrugated Plate Wall · Hysteretic Performance · Bending Angles · Finite Element Analysis

1 Introduction

Steel corrugated plate wall (SCPW) can improve the lateral stiffness of the structure and act as a shock absorber, and the thickness of the wall plate is small, and the demand for the surrounding frame is low. When the structure is irregular, the control torsional displacement ratio of SCPW can be added to the core tube area of the building, and domestic and foreign scholars have carried out a lot of theoretical and experimental studies on it [1–5]. Hosseinzadeh, L. et al. [6] studied the influence of trapezoidal plate Angle on the mechanical properties of SCPW through experiments. After cyclic loading test of three groups of one-layer single-span 1/2 scale specimens, comparing the test results of the three groups, it is found that with the increase of corrugation Angle, the stress concentration at the corner of the steel plate increases, the energy dissipation capacity decreases, the strength of the structure decreases and the deformation resistance weakens. Qiuhong Zhao et al. [7, 8] systematically studied the lateral resistance of SCPWs with different folding forms, placement methods and geometric sizes through

I. Tuns et al. (Eds.): CIBv 2023, LNCE 510, pp. 412–425, 2024.
https://doi.org/10.1007/978-3-031-60765-3_33

tests and numerical simulations, and found that SCPWs had higher initial lateral stiffness and ultimate bearing capacity, but with the increase of aspect ratio, the lateral resistance curve showed a significant downward trend.

Due to the limitation of corrugated steel plate thickness, double steel corrugated plate wall (DSCPW) has also become the research object of scholars. Ghodratian-Kashan SM, Maleki S et al. [9–11] considered the influence of the direction of corrugated plates (horizontal or vertical), the connection between corrugated plates, and the connection between filled plates and columns on the cyclic performance of DSCPWs. Through the experimental research and numerical simulation, the initial stiffness and ultimate strength of the DSCPW are proposed. Yanlin Guo, Jingzhong Tong et al. [12, 13] revealed the influence of geometric parameters on the shear properties of DSCPWs through tests and numerical simulations. The results show that the ultimate shear resistance of DSCPW increases with the increase of length-diameter ratio, bolt column number and ripple amplitude, which also provides a valuable reference for practical engineering design.

According to the current research, corrugated wall panels are anisotropic and can exert the accordion effect to avoid the vertical load and mainly bear the shear force when placed horizontally, and can effectively share the vertical load when placed vertically. However, it is rarely considered that double steel corrugated plates should be installed orthogonal and used in steel plate shear wall structures. Based on the advantages of the two placement methods of corrugated steel plate, this paper designs three new types of orthogonal-installed double steel corrugated plate wall (OD-SCPW) specimens. Based on the previous studies [14], the influence of the bending Angle on the hysteretic performance of the OD-SCPW is further explored.

2 Orthogonal-Installed Double Steel Corrugated Plate Wall

The specimen of orthogonal-installed double steel corrugated plate wall studied in this paper is a structure composed of two corrugated steel plates crossed and connected with cross-tension bolts at the trough joint, and then welded on a two-layer single-span steel frame through the fishplate. The special structure of the orthogonal surface can realize the effect of improving the mechanical properties of the steel plate shear wall with vertical and horizontal stiffeners, effectively improve the side resistance and seismic performance under appropriate conditions, and reduce the work of stiffening ribs and weld defects caused by welding. The three specimens were named ODCW-1, ODCW-2, and ODCW-3 in turn, and only the twist angles changed. ODCW represents the orthogonal-installed double steel corrugated plate wall. 1, 2 and 3 respectively represent the three bending Angle of 90°, 45° and 30°, and also correspond to the three corrugated sections of rectangle, trapezoid and triangle, respectively. The main view and size of the three test pieces are shown in Fig. 1.

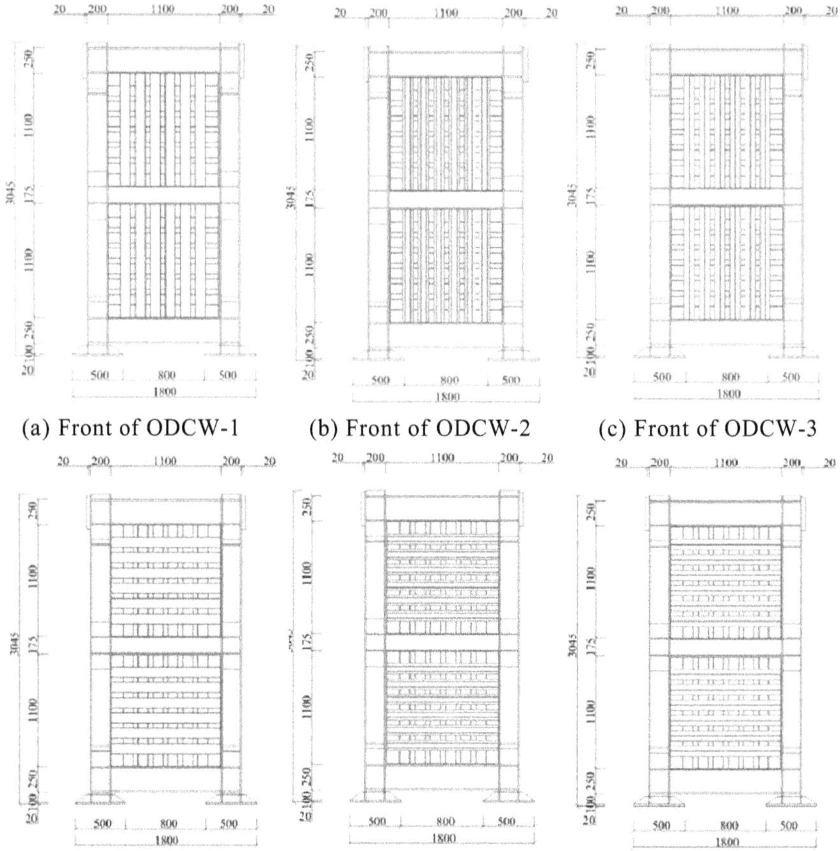

(a) Front of ODCW-1 (b) Front of ODCW-2 (c) Front of ODCW-3

(d) Back of ODCW-1 (e) Back of ODCW-2 (f) Back of ODCW-3

Fig. 1. Orthogonal-installed double steel corrugated plate wall (Unit: mm).

2.1 Numerical Simulation and Verification of Existing Experiments

The experiment of single-span 2-story steel frame-steel plate shear wall structure [15] conducted in Structural Seismic Laboratory of Tianjin University was selected for simulation verification. The specimen size and equipment are shown in Fig. 2. The frame column is selected as HW 200 mm × 200 mm × 8 mm × 12 mm, the section of the side beam HN 250 mm × 200 mm × 12 mm × 14 mm, the section of the middle beam HN 175 mm × 175 mm × 8 mm × 10 mm, and the thickness of the embedded steel plate is 2 mm.

The failure mode of the embedded steel plate was mainly studied. Q355B was selected as the outer frame material, Q235B was selected as the inner steel plate, and the steel two-fold model was adopted for the constitutive relationship [16]. In order to simulate the out-of-plane buckling of steel in the test, the upper and lower edge nodes of the inlaid steel plate were bound to the contact surface of the outer frame (Tie), and the left and right edge nodes of the inlaid steel plate were bound to the contact surface

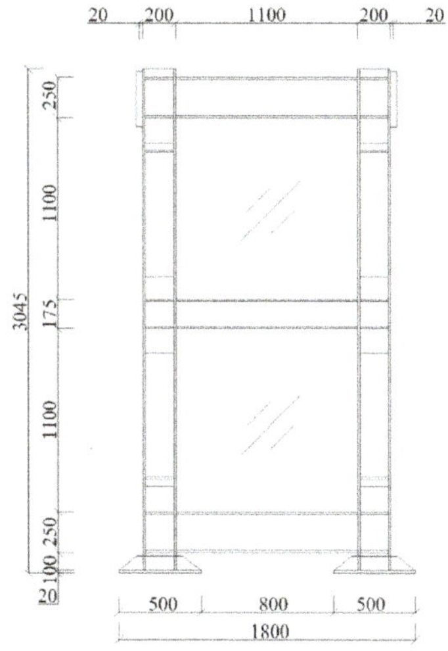

Fig. 2. Front view of steel plate shear wall (Unit: mm).

of the outer frame (Tie). Shell elements (S4R) were selected to simulate the outer frame and the inlaid plate, and the unit size was 30 mm × 30 mm. The total number of units was 8, 993 and 9, 325. Implicit dynamic analysis was used for simulation. In the initial analysis step, boundary conditions are set. In order to simulate the real test condition, the frame column feet are fixed, and the constraint planes of the top beam and middle beam are moved outwardly, that is, Uz = 0, URx = 0, URy = 0. The top of the column is coupled to a point on the cross section centroid line of the top beam of the frame and is used as the loading point. The finite element model is established as shown in Fig. 3.

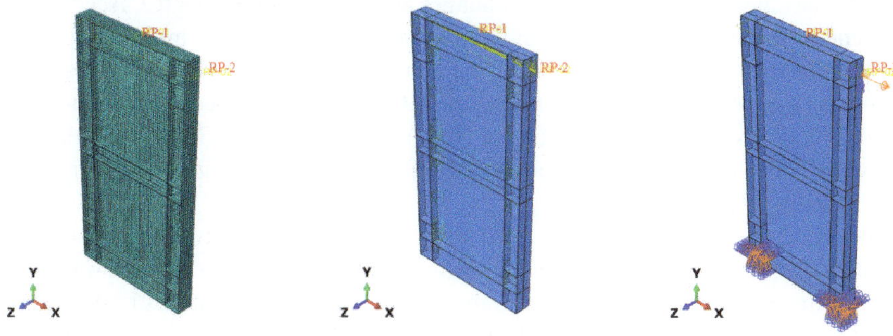

(a) Meshing (b) Interaction (c) Boundary conditions

Fig. 3. The establishment of finite element model.

Horizontal and reciprocating loading is carried out, and the loading system is shown in Fig. 4. First of all, load control is carried out according to 100 kN, 200 kN, 300 kN, 400 kN, 500 kN, 550 kN, 600 kN, 650 kN each cycle. Then the displacement control was carried out, in which the standard loading was divided into 8 stages, corresponding to the lateral displacement Angle of 0.25%, 0.5%, 0.75%, 1%, 1.25%, 1.5%, 1.75%, 2%. After yielding the specimen, each stage was repeated twice. The additional loading is divided into 6 stages, corresponding to the side shift Angle of 2.5%, 3%, 3.5%, 4%, 4.5%, 5%, and each stage is cycled twice.

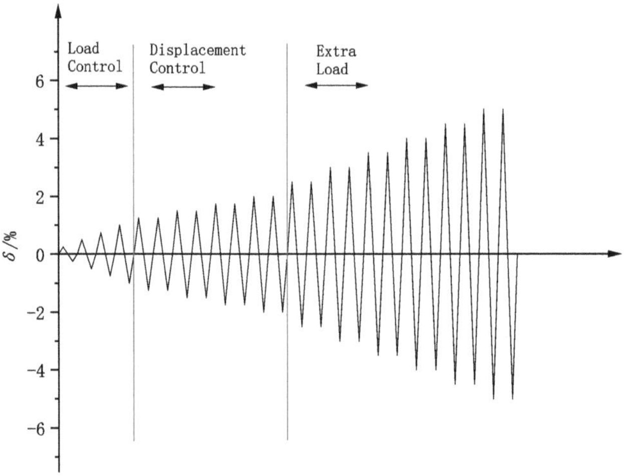

Fig. 4. Loading system.

Figure 5 shows the comparison between the hysteretic curve obtained from the simulation test and the skeleton curve extracted from the hysteretic curve. As can be seen from Fig. 5, the hysteresis curves of the finite element simulation test have a high coincidence, and the change trend of the two is consistent. In each stage of loading, the load-displacement curves have similar changes and high matching degree. According to the trend of the skeleton curve, it can be seen that the load-displacement change trend of the simulation test is consistent, and the values are relatively close. The maximum error point corresponds to the displacement of 65 mm, and the peak load value of the test is 796.11 kN, and the finite element simulation value is 842.39 kN, and the error of both is 5.8%. Meet the requirements of Code for Seismic Design of Buildings (GB 50011-2016) [17].

When the specimen entered the yield state, the comparison between the test failure morphology of the specimen and that of the finite element simulation analysis was shown in Fig. 6. It can be seen that the failure morphology of the specimen was basically consistent with that of the test specimen. Diagonal tension bands appeared on the embedded steel plate, and diagonal outward bulging appeared along the diagonal respectively, and the residual deformation showed an "X" shape. When the damage and material defects at the joint are not considered, the change trend and failure pattern effect of the simulation test meet the expectations, and the data error is within the acceptable range. Therefore, the finite element software ABAQUS can be used for numerical simulation of the steel plate shear wall, and the simulation results can be used as the control group for reference in the following.

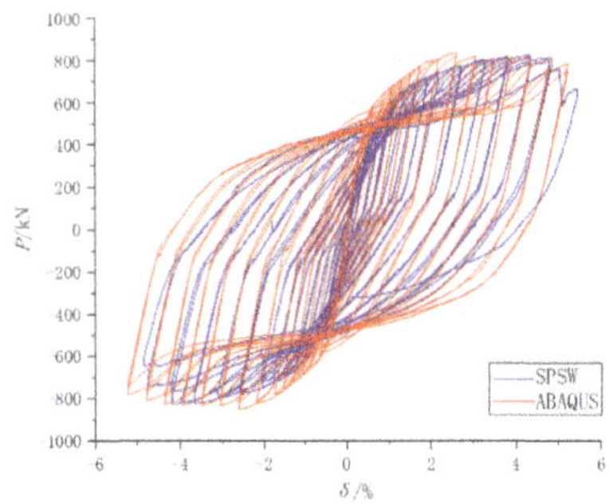

(a) Hysteretic curves

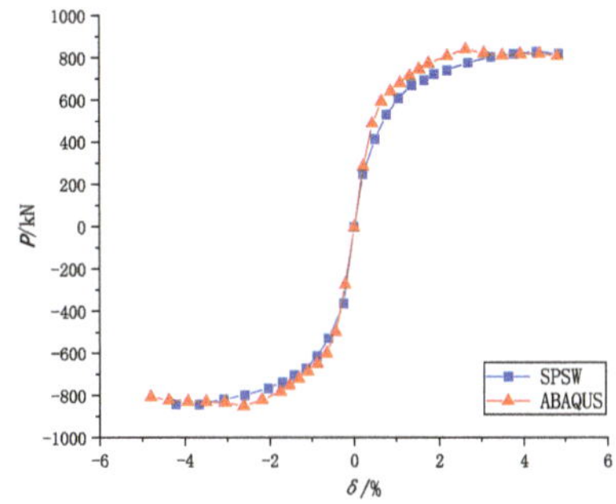

(b) Skeleton curves

Fig. 5. Comparison of hysteretic performance between test and FEM.

(a) Failure pattern

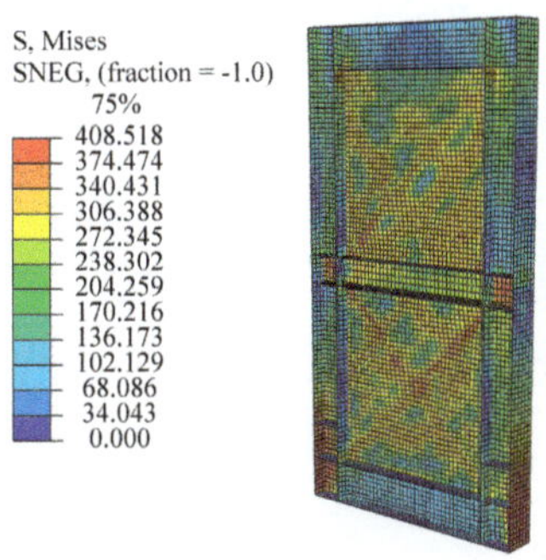

(b) Stress nephogram

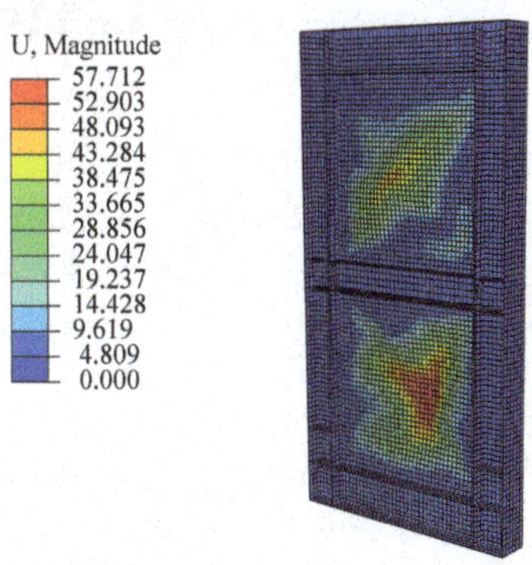

(c) Out-of-plane displacement nephogram

Fig. 6. Comparison of failure mode between test and FEM.

2.2 Establishment of Numerical Model of Specimen

Based on the model verified by numerical simulation, the embedded steel plate in the group was replaced with orthogonal-installed double steel corrugated plate during component assembly. The pre-treatment work such as steel plate material constitutive relationship, plate-frame interaction, boundary conditions of specimens, and loading regime of specimens were adjusted accordingly with reference to the control group. Then the shear wall specimens were subjected to reciprocating loading to further analyze the hysteretic performance of the orthogonal-installed double steel corrugated plate wall. Three sets of finite element models, including ODCW-1, ODCW-2 and ODCW-3, were established, as shown in Fig. 7.

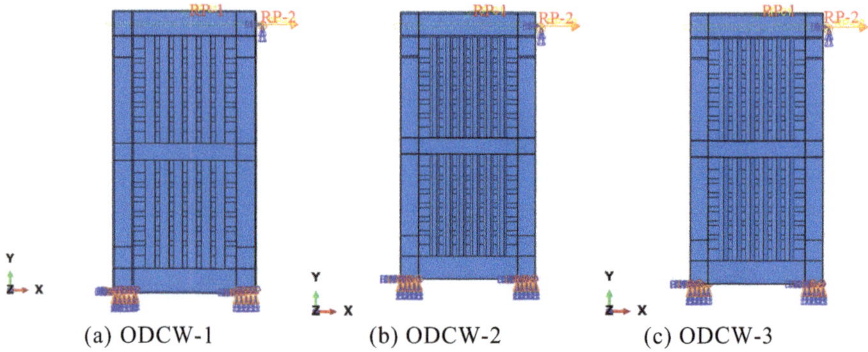

Fig. 7. Model building.

3 The Results of Numerical Simulation and Its Analysis

3.1 Distribution of Stress

Figure 8 shows the stress distribution of the specimen in failure state. It can be seen that the maximum stress of each specimen is at the end of the column foot, and the maximum stress of ODCW-1, ODCW-2 and ODCW-3 at yield is 527.90 kN, 530.50 kN and 540.20 kN respectively. The stress at the end of the column foot in the failure state of the specimen initially presents a negative correlation with the size of the selected bending Angle. The specimen ODCW-1 embedded steel plate also showed diagonal tension bands as in the control group, and the residual deformation showed a symmetrical "X" shape. The orthogonal distribution of out-of-plane ripples hindered the uniform development of the diagonal tension band, and also reduced the size of the "X" shaped distribution stress of the embedded steel plate. The specimen ODCW-2 embedded steel plate does not appear diagonal tension band, and the residual deformation does not show an "X" shape. The steel corrugated plate with 45° bending Angle has improved external integrity after orthogonal lamination, and the embedded steel plate presents an "O"-shaped external drum from the center to the surrounding area, and the stress on the "O" shaped ring is less than the surrounding area. The specimen ODCW-3 panel steel plate showed diagonal tension band and "X" shape residual deformation, but the "X" shape was not symmetrical like the control group and ODCW-1. The "X" intersection point of the upper steel plate was shifted downward, and the "X" intersection point of the lower steel plate was shifted diagonally to the upper right. In addition to the greater stress at the end of the column, ODCW-3 also has greater stress at the joint between the middle beam and the column.

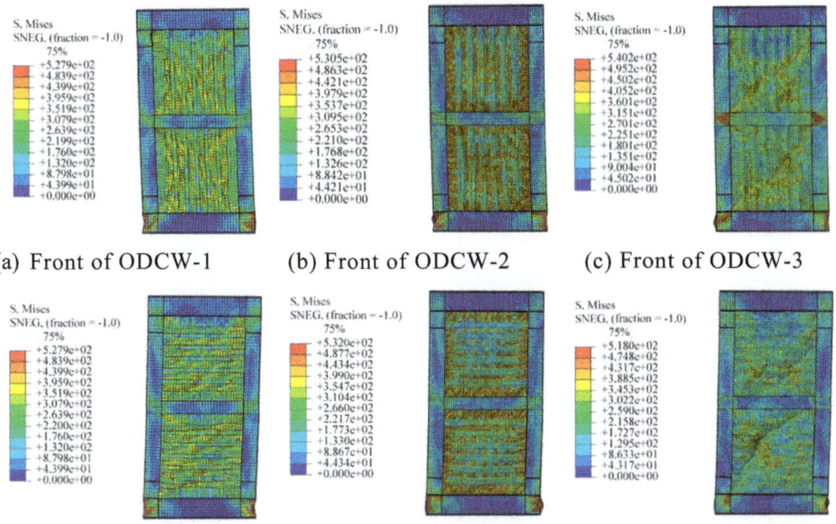

(a) Front of ODCW-1 (b) Front of ODCW-2 (c) Front of ODCW-3

(d) Back of ODCW-1 (e) Back of ODCW-2 (f) Back of ODCW-3

Fig. 8. Stress contours of specimens at failure state (Unit: MPa).

3.2 Out-of-Plane Deformation

Figure 9 shows the out-of-plane deformation diagram of the failure state of the specimen. According to the comparison of out-of-plane deformation values carved in the failure state of the specimen in the figure, it can be seen that ODCW-2 has the smallest out-of-plane deformation, followed by ODCW-1, and ODCW-3 has the largest out-of-plane deformation value. The out-of-plane deformation of the specimen ODCW-1 inlaid steel plate was mainly concentrated on the diagonal line, and the diagonal wavy surface of ODCW-1 appeared along the diagonal. However, the special structure of 90° corrugation outside the orthogonal plane was similar to that of the control group, which inhibited the external buckling of the inlaid steel plate, and the bending range outside the wavy surface was smaller than that of the control group. The center of the ODCW-2 embedded steel plate presents a circle of "O" shape outside the drum, and the drum of the lower steel plate and its range are larger than the upper steel plate. The out-of-plane deformation of ODCW-3 embedded steel plate is also mainly concentrated on the diagonal, with asymmetric distribution of "X" shape outside the plane bulge, and sharp out-of-plane deformation at the "X" intersection. The triangular corrugated section is not conducive to the resistance of the embedded steel plate to out-of-plane buckling.

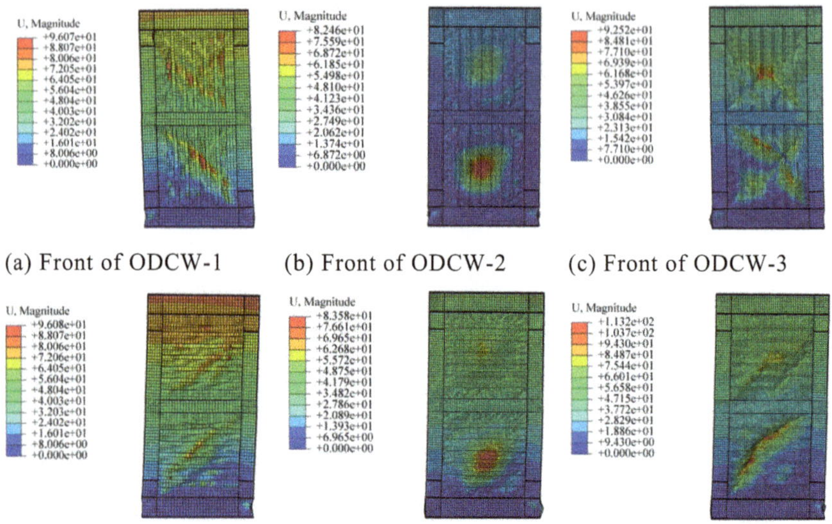

(a) Front of ODCW-1 (b) Front of ODCW-2 (c) Front of ODCW-3

(d) Back of ODCW-1 (e) Back of ODCW-2 (f) Back of ODCW-3

Fig. 9. Out-of-plane deformation of specimens at failure state (Unit: mm).

3.3 Comparison of Hysteresis Curve and Energy Consumption

Under cyclic loading, the hysteretic curve of lateral load-lateral deflection Angle of the specimen is shown in Fig. 10. It can be seen that all specimens have good hysteretic performance, among which ODCW-1 and ODCW-2 have relatively full hysteretic curves, ODCW-2 has the fullest and highest ultimate load, and ODCW-3 has the most obvious hysteretic curve pinching phenomenon. It can be preliminarily seen that the energy dissipation capacity of the specimens is ODCW-2, ODCW-1 and ODCW-3 in order from large to small. ODCW-2's 45° corrugated steel plate orthogonal superposition improves the out-of-plane integrity and improves the seismic performance of the specimen. After the inelastic buckling of ODCW-1 and ODCW-3, the load drops gently. This is because the external deformation of the embedded steel plate causes a certain degree of unloading, but the load is still carried in the form of "X" tension band after yielding.

The energy dissipation capacity curves of the specimens are shown in Fig. 11 and Fig. 12. The hysteretic area of the specimen under each loading stage represents the energy dissipation of the specimen under each loading stage, and is represented by the equivalent viscous damping coefficient (h_c). The hysteresis loops of the specimens before yield are not much different. Compared with each loading stage after yield displacement and combined with accumulated energy consumption, it can be seen that the energy dissipation capacity of the specimens is ODCW-2, ODCW-1 and ODCW-3 in order from large to small.

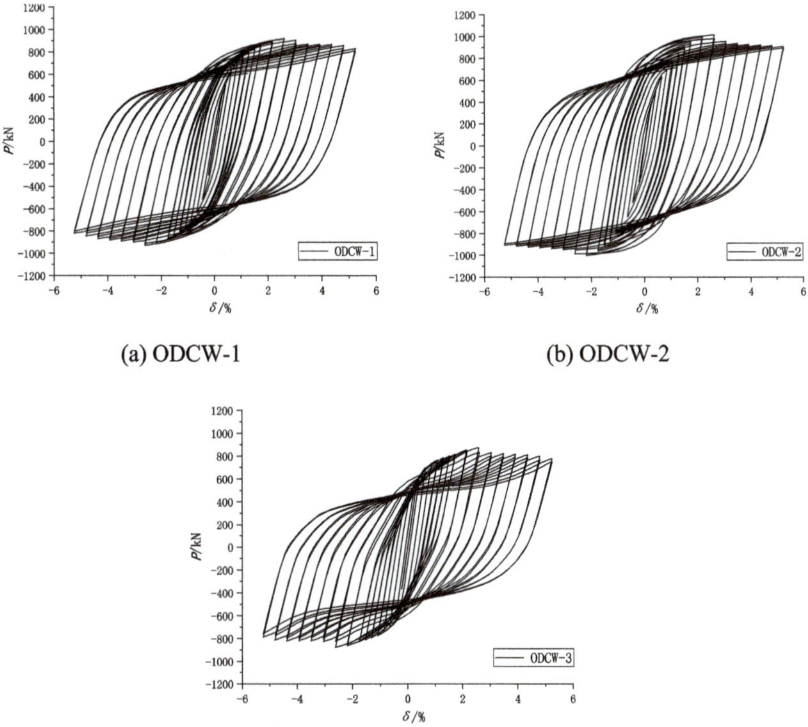

(a) ODCW-1 (b) ODCW-2

(c) ODCW-3

Fig. 10. Hysteretic curves of specimens.

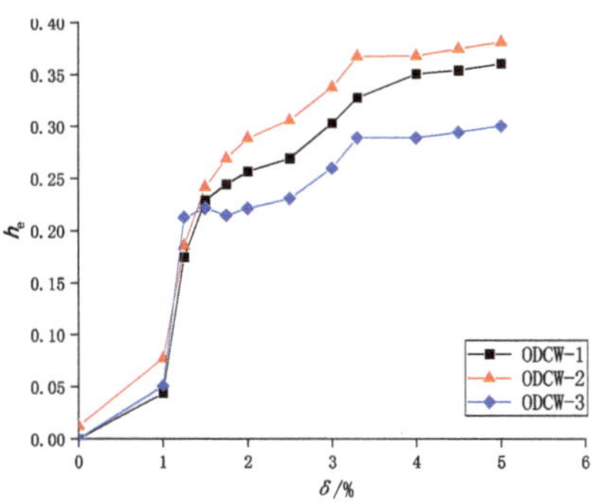

Fig. 11. Equivalent viscous damping coefficient.

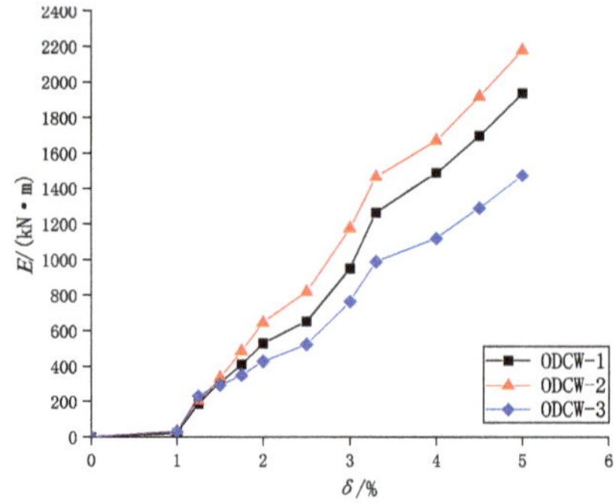

Fig. 12. Cumulative energy dissipation.

3.4 Contrast of Skeleton Curves

The skeleton curve of the specimen was extracted from the hysteresis curve, and the comparison was shown in Fig. 13. At the initial loading stage, the load-displacement curves of the specimens have little difference, among which ODCW-3 has the largest initial lateral stiffness, but the smallest peak load. The triangular corrugated section is conducive to improving the lateral stiffness of the embedded steel plate in the early stage, but the bearing capacity decreases due to the serious out-of-plane deformation in the later stage. The lateral displacement corresponding to the peak load of all specimens is basically the same, and the peak load is ODCW-2, ODCW-1, and ODCW-3 in order from large to small. When the specimen reached the peak value, the load value of the specimen ODCW-1 decreased slowly, while the bearing capacity of ODCW-2 and ODCW-3 decreased relatively quickly, because ODCW-1 could still be carried with a symmetrical and complete "X" -shaped tensile band after yielding.

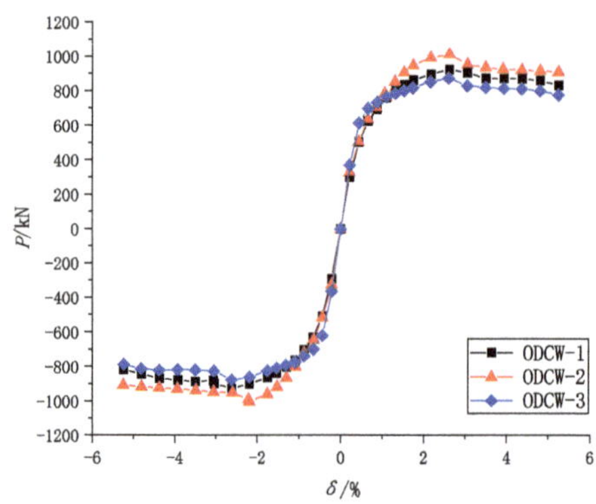

Fig. 13. Comparison of skeleton curves of specimens.

3.5 Comparison of Strength and Stiffness Degradation

The degradation curve of secant stiffness (k) of each specimen with lateral bearing capacity (P) and lateral shift Angle (δ) is shown in Fig. 14. As can be seen from the figure, The initial stiffness of each specimen was ODCW-3, ODCW-2 and ODCW-1 from large to small, the stiffness degradation trend of specimens was consistent, and the stiffness peak value of specimens was similar in the late loading period. As the loading continues, the bearing capacity degradation of ODCW-3 is significantly faster than that of the other two specimens, indicating that the strength degradation of ODCW-3 is faster, and the serious out-of-plane deformation of the triangular corrugated section in the later stage is not conducive to the bearing capacity stability of the member. The bearing capacity degradation of ODCW-1 and ODCW-2 is little different and relatively slow, indicating that the strength degradation of both is slow. The bearing capacity stability of

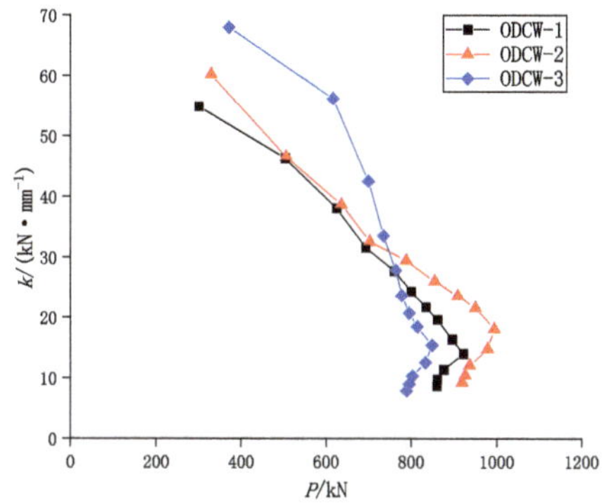

(a) k-P curves

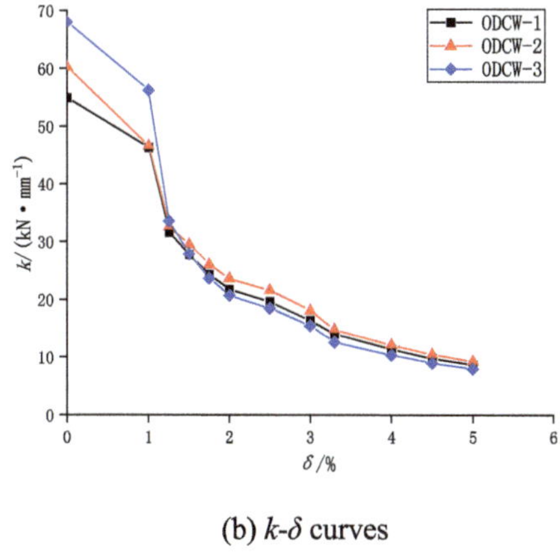

(b) k-δ curves

Fig. 14. Stiffness degradation of specimens.

orthogonal-installed double steel corrugated plate wall with rectangular and trapezoidal corrugated section is better.

4 Conclusion

In this paper, the finite element software ABAQUS is used to simulate three kinds of orthogonal-installed double steel corrugated plate walls explore and compare their mechanical properties under cyclic load, and the following main conclusions can be drawn:

(1) The initial lateral stiffness of the orthogonal-installed double steel corrugated plate wall is related to the selection of the corrugated steel plate's bending Angle, in which the initial lateral stiffness is selected in order of 30°, 45° and 90°, and there is a negative correlation between the two. However, although the triangular corrugated section is beneficial to improve the lateral stiffness of the embedded steel plate in the early stage, the bearing capacity decreases seriously in the late stage due to out-of-plane deformation.
(2) The energy dissipation capacity of orthogonal-installed double steel corrugated plate wall is related to the selection of bending Angle of corrugated steel plate, of which orthogonal-installed double steel corrugated plate with a bending Angle of 45° has good integrity and strong resistance to out-of-plane buckling, and its energy dissipation capacity is higher than the other two groups, followed by the bending Angle of 90°, and the bending Angle of 30° is relatively weak. However, for the orthogonal-installed double steel corrugated plate wall with 45° bending Angle, the distribution of the upper and lower layers of the embedded steel plate is uneven, and the lower layer is larger than the upper layer.
(3) The structure of the orthogonal-installed double steel corrugated plate with a bending Angle of 90° is similar to that of a flat steel plate with vertical and horizontal stiffening ribs. While resisting the out-of-plane buckling, it can still be carried by a symmetrical and complete "X"-shaped tension band after yielding. After the specimen reaches the peak load, the load value declines slowly, which is conducive to the stability of the bearing capacity of the specimen in the later period.

References

1. Zand, S.M.D., Kalatjari, V., Attari, N.K.A.: Cyclic performance of new steel plate shear wall reinforced by trapezoidal corrugated plate. Period. Polytech-Civ. **68**(1), 337–348 (2024)
2. Yadollahi, Y., Pakar, I., Bayat, M.: Evaluation and comparison of behavior of corrugated steel plate shear walls. Lat. Am. J. Solids Struct. **12**, 763–786 (2015)
3. Dou, C., Yang, X., Xie, C.: Reciprocating loading test and lateral resistance analysis of flat-wave steel plate shear wall. Eng. Mech. **39**, 1–13 (2022)
4. Zheng, H., Xie, Y.X., Wu, H.W.: Parameter influence analysis of composite shear walls with cross ribbed flat steel plates. Pro Steel Build. Struct. **24**(09), 76–85 (2022)
5. Zheng, H., Wang, W., Yang, R.P.: Study on hysteretic properties of a new type of cross-stiffened corrugated steel plate shear wall. Pro Steel Build. Struct. **23**(06), 36–43 (2021)

6. Hosseinzadeh, L., Emami, F., Mofid, M.: Experimental investigation on the behavior of corrugated steel shear wall subjected to the different angle of trapezoidal plate. Struct. Des. Tall Spec. **26**(17), e1390 (2017)
7. Zhao, Q.H., Qiu, J., Hao, B.C.: Lateral resistance of vertical corrugated steel plate shear walls with two sides connected. J. Tianjin Univ. (Nat. Sci. Eng. Technol.) **52**(S2), 46–53 (2019)
8. Zhao, Q.H., Gao, J.X., Qiu, J.: Axial buckling analysis of corrugated steel plate shear wall based on actual vertical load distribution. J. Tianjin Univ. (Nat. Sci. Eng. Technol.) **55**(04), 391–401 (2022)
9. Ghodratian-Kashan, S.M., Maleki, S.: Cyclic performance of corrugated steel plate shear walls with beam-only-connected infill plates. Adv. Civ. Eng. **2021**, 1–18 (2021)
10. Ghodratian-Kashan, S.M., Maleki, S.: Numerical investigation of double corrugated steel plate shear walls. J. Civ. Eng. Constr. **10**(1), 44–58 (2021)
11. Ghodratian-Kashan, S.M., Maleki, S.: Experimental investigation of double corrugated steel plate shear walls. J. Constr. Steel Res. **190**, 107138 (2022)
12. Tong, J.Z., Zhang, J.X., Pan, W.H.: Ultimate shear resistance and ultimate performance of double corrugated plate shear wall. J. Constr. Steel Res. **165**, 105895 (2020)
13. Tong, J.Z., Guo, Y.L., Zuo, J.Q.: Experimental and numerical study on shear performance of double corrugated plate shear walls. Thin Wall Struct. **147**, 106485 (2020)
14. Zhu, B.L., Tong, J.Z., Zhang, J.X.: Shear properties and design of orthogonal double corrugated plate shear walls. Thin Wall Struct. **187**, 110759 (2023)
15. Zhao, Q.H., Qiu, J., Li, N.: Experimental study on seismic performance of trapezoidal corrugated steel plate shear wall. J. Build. Struct. **39**(S2), 112–120 (2018)
16. Wang, M., Qian, F.X., Yang, W.G.: Study on the constitutive relationship of LYP 160 steel with low yield point. J. Build. Struct. **38**(2), 55–62 (2017)
17. Ministry of Housing and Urban-Rural Development of the People's Republic of China 2016. Code for Seismic Design of Buildings: GB 50011-2016. China Building and Construction Press, Beijing

Failure Envelopes of Tripod Pile Foundation Under Combined Load in Non-homogeneous Clay

Zihao Zhao[1], Jingchun Han[1], and Hao Zhang[1,2](✉)

[1] School of Civil Engineering, Shenyang Jianzhu University, Shenyang 110819, China
zh18843973567@163.com

[2] Shen Kan Engineering and Technology Corporation MCC, Shenyang 110169, China

Abstract. Tripod pile foundations are an alternative foundation for offshore wind turbines. Three-dimensional finite element limit analyses are carried out to determine the bearing capacities of tripod pile foundations in non-homogeneous clay under combined vertical (*V*), horizontal (*H*) and bending moment (*M*) load. The undrained shear strength of the clay is considered to increase linearly with depth. Three-dimensional failure envelopes in *V-H-M* loading space are established through a simplified loading strategy considering the reality of the loading condition. The obtained results are finally presented in terms of normalized and dimensionless capacity factors. Moreover, the effects of length-to-diameter ratios (*L/D*), spacing-to-diameter ratios (*s/D*) for the tripod piles, and the gradient of undrained strength in terms of soil depth (*k*) on failure envelopes are further discussed. Several failure mechanisms are identified to be associated with different load combinations.

Keywords: Piles · Tripod Pile Group · Bearing Capacity · Clays · Finite Element Limit Analyses · Failure Envelope

1 Introduction

The evaluation of the bearing capacity of the pile foundation subjected to external loads is a critical issue that must be considered in pile design. However, in the traditional design practice, the pile capacity is always considered under independent vertical or lateral loadings. It is not true for the pile foundation when it is used in offshore industry to support upper structures such as oil platforms or wind turbines. Quite significant horizontal loads will be applied on these structures, which in turn transferred to the underlying piles as a combination of vertical (*V*), horizontal forces (*H*) and moment (*M*). On this context, many researches on pile foundations under combined loads have been performed. Liu [1] and Reese [2] studied the lateral bearing capacity of single pile by model test. Murff [3], Randolph [4] and Wang [5] studied the bearing capacity of single pile under unidirectional load through theoretical analysis. Zhao [6] studied the influence of small deformation assumption on the lateral ultimate resistance of single pile

I. Tuns et al. (Eds.): CIBv 2023, LNCE 510, pp. 426–441, 2024.
https://doi.org/10.1007/978-3-031-60765-3_34

by finite element analysis. Zhou [7] studied the seepage mechanism and impermeability of a single screw pile and compared it with the existing centrifuge results. For pile group foundation, Zhao [8] used the limit analysis method to systematically study the failure mechanism of the twin pile foundation under the combined loading mode. Matlock [9] conducted a lateral loading field test of pile groups in clay. Georgiadis [10, 11] studied the ultimate soil resistance of a group of side-by-side piles under lateral load in clay by numerical method and analytical method, and gave the calculation formula of the ultimate soil resistance of double pile groups under general load direction.

In recent years, the tripod foundation is increasingly used for supporting offshore structures in deep sea. Liu [12] studied the coupling bearing capacity of three-pile foundation in *VH* plane. Wang [13, 14] captured the failure envelope of tripod bucket foundation under combined vertical-horizontal loads using three-dimensional finite element analysis method, and further discussed the influence of the structure-soil adhesion factor, aspect ratio, bucket spacing and strength feature of the soil. Similarly, for tripod bucket foundations, Kim [15] considered the loading combination of horizontal load and moment and analyzed its bearing capacity under various geometries and layouts of the buckets in non-uniform clay. Zhao [16] studied the influence of pile group geometry and loading direction on the development of ultimate lateral soil resistance on a three-pile foundation using numerical and analytical upper bound methods.

Nevertheless, it can be found that existing researches of tripod foundations mostly focused on buckets, while the investigation to the tripod pile is rather limited. Otherwise, most studies assumed the soft clay uniform and homogeneous, which ignore the variation of the shear strength along the depth and the heterogeneity feature of the clay. In fact, these features are commonly seen in reality and worth to be considered. At present, only Gourvenec [17] analyzed the influence of clay heterogeneity on the failure envelope of single pile foundation under unidirectional load.

To bridge this gap, this study aims to evaluate the impact of combined *VHM* load interaction on the bearing capacity of a tripod pile foundation in non-homogeneous clay. Three-dimensional failure envelopes in the *VHM* loading space are established following a simplified loading strategy considering the reality of the loading condition. The effects of pile length, pile spacing and the most important heterogeneity of the clay are further analyzed. Finally, an artificial neutral network (ANN) model is constructed to predict the combined capacity of tripod piles in non-homogeneous clay.

2 Problem Statement

Figure 1 shows a tripod pile foundation, which is arranged as an equilateral triangle with equivalent spacing s, embedded in the non-homogeneous clay. The piles, each having embedded length L, diameter D are connected by a rigid cap on their head. The undrained shear strength of the clay is denoted by s_{u} and the soil bulk density is γ. The non-homogeneous profile of s_{u} is expressed as:

$$s_u = s_{um} + kz \tag{1}$$

where $s_{\mathrm{u}0}$ is the undrained shear strength of the soil on the ground surface, k is the gradient of undrained strength in terms of soil depth and z is the soil depth. All loads (the vertical load V, the horizontal load H and the moment M) are applied to the pile cap and act on the same plane. The load and sign conventions used in this paper are consistent with those in Butterfield [18], as shown in Table 1. Note that the value of undrained shear strength at the quarter of the depth is selected for dimensionless use. The sign superscript represents the direction of the load: "+" represents the positive direction along the coordinate axis, "–" represents the negative direction along the coordinate axis.

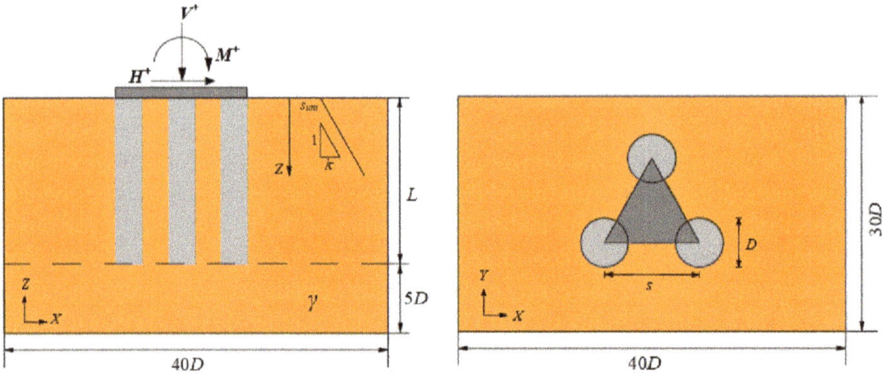

Fig. 1. Tripod pile foundations under combined *V*-*H*-*M* loading.

Table 1. Nomenclature for loads.

	Vertical load	Horizontal load	Moment load
Load	V	H	M
Uniaxial bearing capacity	V_0	H_0	M_0
Normalized load	V/V_0	H/H_0	M/M_0
Dimensionless load	V/s_uD^2	H/s_uD^2	M/s_uD^3

3 Finite Element Limit Analysis Model

Finite element limit analysis (FELA) is an efficient computational tool that combines the capacities of the finite element discretization and the principles of limit analysis. In this study, the capacities of tripod pile foundations are studied using the software program OptumG3. A total number of 36 cases with various combinations of length-diameter ratio, spacing-diameter ratio and soil non-homogeneity are considered to analyze the failure envelopes. The FELA model established is detailed below.

3.1 Model Dimension and Material Properties

As shown in Fig. 2, the tripod pile foundation is simulated using three rigid cylinders with diameter D of 1 m and a rigid plate on their top with thickness of $0.01D$. The geometry is varied within the ranges of practical interest as length-to-diameter ratio $L/D = 3, 6, 9$ and spacing-to-diameter ratio $s/D = 2, 4, 6$. The dimension of the FELA model is determined through a sensitivity analysis, and finally described by a cube with the length of $40D$, the width of $30D$, and the height of $L + 5D$. The boundary condition of the model follows the standard constraint, where the top surface is free, and the side and bottom surfaces are constrained in the normal direction. The rigid behaviors of the piles and their cap are simulated using a "rigid model" available in OptumG3 program. While the soil mass is regarded as a perfectly elastoplastic material and obeys the Tresca yield criterion. The material parameters of the soil are detailed as follows. The unit weight $\gamma = 20$ kN/m^3, the Young's modulus $E_u = 30$ MPa, the Poisson's ratio $\mu = 0.495$, the undrained shear strength on the ground $s_{\mathrm{um}} = 15$ kPa and the gradient of strength in terms of soil depth $k = 0$–2 kPa/m. The interfaces between the pile and the soil are described by shear joint elements with tension cut-off, which is capable of simulating the possible crack between the pile and the soil mass.

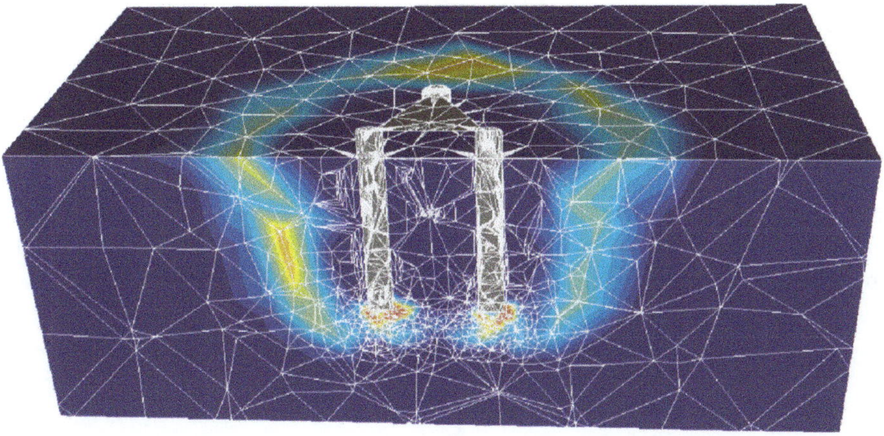

Fig. 2. Tripod pile foundation model with shear dissipations after adaptive meshing.

3.2 Adaptive Meshing Technique

The adaptive meshing technique is another prominent feature in program OptumG3. The scheme of this meshing technique is based on the shear dissipation, where much finer meshes are allocated to the area with large values of shear dissipation, whilst much larger meshes for those with zero shear dissipation. Therefore, with this advanced adaptive scheme, reliable results can be obtained by applying an appropriate number of elements and adaptive steps. Table 2 shows a pilot analysis (a typical tripod pile foundation with $L/D = 9$ and $s/D = 6$) of the influence of the mesh number on the results obtained. It can be observed that the accuracy of the results is not significantly improved when the number of initial meshes reaches 3000 and the number of optimized meshes reaches 15000. So, the mesh density of this is used in all analyses afterwards.

Table 2. Effect of mesh number on ultimate bearing capacity ($L/D = 9$ and $s/D = 6$).

Mesh number	Initial mesh	2000	3000	4000	5000
	Optimized mesh	10000	12000	15000	20000
Vertical load/v_0		6439.1	6389.4	6381.2	6366.7
Horizontal load/h_0		4811.5	4763.5	4760.3	4753.3
Moment load /m_0		34335.7	34109.0	34110.9	34089.4

3.3 Loading Strategy

In order to construct the three-dimensional failure envelope of a tripod pile foundation under the combined *VHM* load, it is necessary to determine a certain loading strategy within the *VHM* loading space. Thus, a typical probe test process is adopted as depicted in Fig. 3. The procedure includes: (1) the vertical load V of a fixed value is set at the center of pile cap, (2) the horizontal load H and the moment M are then applied as multiplier loads under a constant *M/H* ratio ($M/H = \tan\alpha$) to the limit state, (3) the ends of the loading trajectories corresponding to the limit states at various *M/H* ratios are connected to obtain the failure locus on *HM* plane, (4) repeat steps 1 to 3 under different levels of V to attain the final three dimensional failure envelope in *VHM* space. It should be noted that the loading strategy used in this study takes more account of the load feature in reality, since the vertical load is generally from the self-weight of the upper structure and does not change much while the collapse of the foundation is mostly caused by the grow up of the horizontal load.

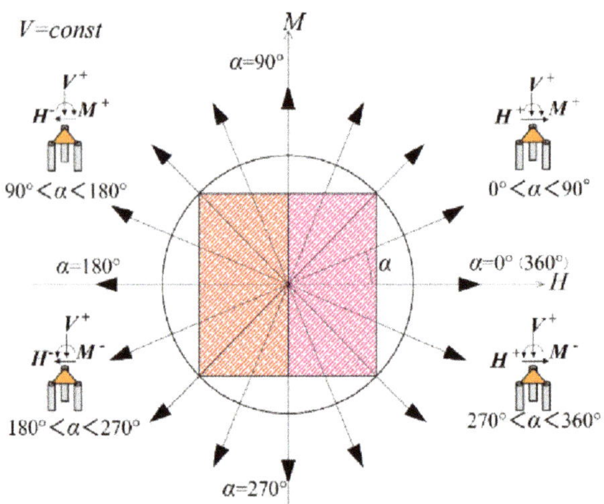

Fig. 3. Load paths on *M-H* plane under a fixed level of *V*.

3.4 Validation

To verify the reliability of the present three-dimensional finite element model, the comparisons with the results reported by the previous work [19–23] for single piles and Liu [12] for tripod pile foundations are performed, as it can be seen in Table 3 and Table 4. The results are shown in the term of dimensionless ratio or normalized ratio based on the original form documented. Herein, V_{ult} and H_{ult} represent the ultimate bearing capacity of single pile foundation under pure vertical and horizontal loading condition, respectively. It is obviously observed from Tables that the numerical solutions in this study generally show good agreement with the results reported in the previous studies for certain pile length and spacing cases. The difference between the results in Table 3 and Table 4 is mainly due to the different research methods. Among them, Liu [12] used the three-dimensional finite element analysis method, Vesic [19] used the field test method, Meyerhof [20], Georgiadis [21] and Hansen [23] used the theoretical analysis method, Jian Yu [22] used a total-displacement-loading approach based on the extended mobilizable strength design (EMSD) method. In this paper, the finite element analysis method is used, so the results are closer to Liu [12]. And it can be clearly seen from the table that the error between the results is acceptable.

Table 3. Comparison of vertical and horizontal bearing capacity of a pile foundation with the previous literature.

L/D	$V_{\mathrm{ult}}/s_u D^2$			$H_{\mathrm{ult}}/s_u D^2$			
	This study	Vesic [19]	Meyerhof [20]	This study	Georgiadis [21]	Jian Yu [22]	Hansen [23]
6	29.37	28.98	25.92	18.72	18.00	23.94	13.11
8	38.25	35.89	32.20	27.31	26.00	34.16	19.07
10	44.64	42.80	38.48	36.38	35.00	45.50	25.67

Table 4. Comparison of vertical and horizontal bearing capacity of tripod pile foundation with the previous literature.

L/D		22		
s/D		4.8	6	8
V/V_{ult}	Liu [12]	2.95	3.07	3.03
	This study	3.13	3.12	3.1
H/H_{ult}	Liu [12]	3.91	4.5	5.34
	This study	3.73	3.96	4.38

4 Three-Dimensional Failure Envelope of the Tripod Pile Foundation

4.1 Failure Locus on the HM Plane

Following the loading strategy stated above, the influence of the vertical load *V* is not considered in this section. Figure 4 shows the *HM* failure locus of the tripod pile foundation with $L/D = 6$ and $s/D = 4$ when $V = 0$. The shape of the locus on *HM* plane is approximately observed as a tilted ellipse with the central symmetry. It shows linear straight lines in the first and third quadrants, and curves with turning points (P2 and P2' in Fig. 4) in the second and fourth quadrants. Variations of the curves in different quadrants reflect the changes of the coupling effect between the horizontal load *H* and moment *M*.

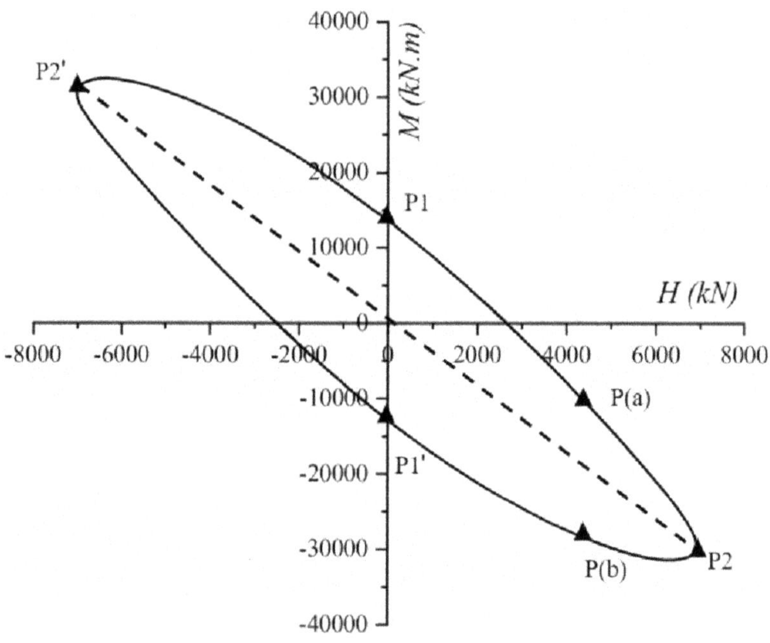

Fig. 4. Failure locus on the *HM* plane for $L/D = 6$ and $s/D = 4$.

When $HM > 0$ (viz, the first and third quadrants), the impacts of the horizontal load and the moment drive the tripod pile foundation leaning to the same direction. In the meantime, the increase of moment will accelerate the overturning failure of pile-soil system, and then reduce the ultimate horizontal force that the system can bear. When $HM < 0$ (viz, the second and fourth quadrants), the horizontal load and the moment potentially impel the tripod pile foundation with the movement trends in the opposite directions. In this case, the ultimate horizontal force that the pile-soil system can withstand increases first and then decreases with the clockwise rotation of the loading path, and the maximum horizontal bearing capacity of the system is obtained at the turning points on the curve

(P2 and P2'). Before the turning points, the horizontal load is the main component that determines the overturning failure of the tripod pile foundation and the existence of the moment will resist the potential movement trend of the pile-soil system, thus the increase of the moment will improve the horizontal bearing capacity of the system. While after the turning point, the moment will gradually become the dominant component that may cause the failure, so the maximum horizontal force that the system can withstand will be reduced. From the perspective of the failure mechanism, before the turning point, the movement direction of the tripod pile foundation is the same as the loading direction of horizontal load components, and more shear dissipations are generated in the soil in front of the foundation (Fig. 5a). After the turning point, the movement direction of the tripod pile foundation is reversed due to the increase of the moment and the shear dissipations are relatively even distributed around the tripod pile foundation in the limit state (Fig. 5b).

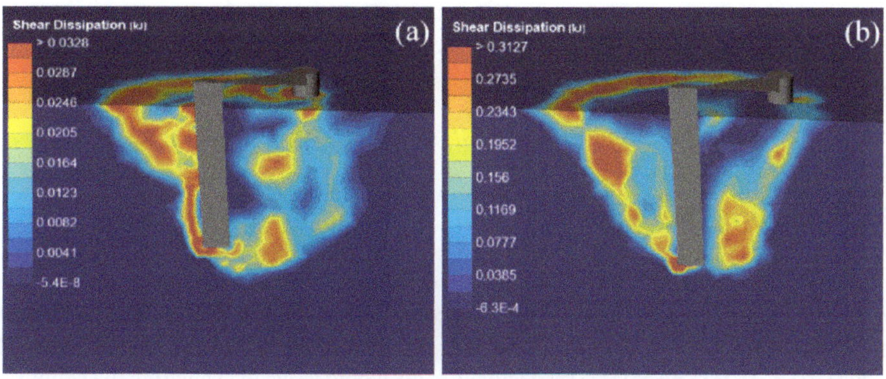

Fig. 5. Shear dissipation patterns for $L/D = 6$ and $s/D = 4$: (a) before P2; (b) after P2.

4.2 Failure Loci for Different Levels of the Vertical Loads

The effect of vertical loads on the capacity of tripod pile foundation is subsequently considered, so as to attain the 3D failure envelope in the *VHM* loading space. Figure 6 shows the *HM* failure loci under different levels of vertical loads ($V = 0$, $0.25V_0^+$, $0.5V_0^+$, $0.75V_0^+$, $0.85V_0^+$, $0.9V_0^+$, $0.95V_0^+$). The selected exemplar case is for $L/D = 6$, $s/D = 4$. In general, the size of the failure locus reduces with the increase of the vertical load. All planar loci corresponding to various V levels are eventually integrated into the 3D coordinate system of *VHM* to obtain a complete 3D failure envelope surface, which is observed as a half of a super-ellipsoid (Fig. 6).

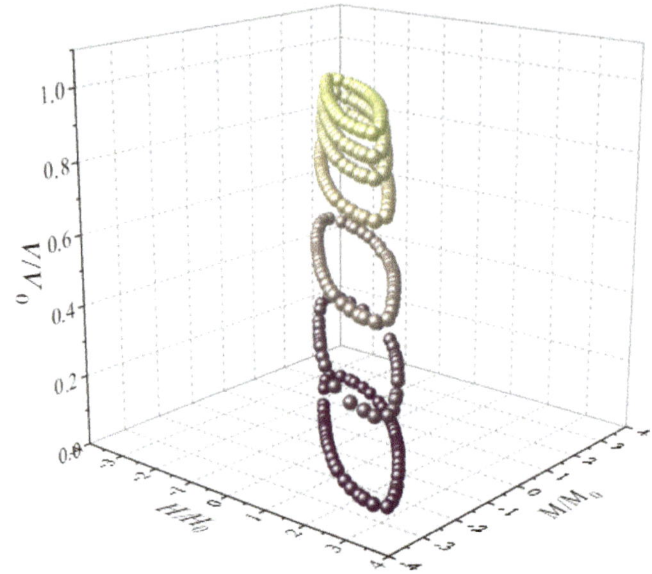

Fig. 6. 3D failure loci in the *V-H-M* space (*L/D* = 6 and *s/D* = 4).

5 Parametric Studies

In this section, the effects of pile length-diameter ratio *L/D*, pile spacing-diameter ratio *s/D* and non-homogeneous coefficient *k* are comprehensively considered. The results are presented in the form of dimensionless (V/s_uD^2, H/s_uD^2 and M/s_uD^3) and normalized loads (V/V_0, H/H_0 and M/M_0) respectively. This would assist in evaluating the coupling effects both quantitatively and qualitatively. It should be noted that the values of uniaxial capacities (V_0, H_0 and M_0) of tripod pile foundations were determined and detailed in our previous work (Zhang [24]), and hence it is not reiterated here anymore.

5.1 Effect of the Geometry Parameters of the Tripod Pile Foundation

Figure 7 shows the dimensionless and the normalized failure loci of the tripod pile foundation in non-homogeneous clay (k = 1.5 kPa/m) under combined *VHM* loads. Therein, various geometries of the pile foundations (the length-diameter ratio *L/D* = 3, 6, 9 and the spacing to diameter ratio *s/D* = 2, 4, 6) are taken into account. Note that the geometry parameters do not impact the evolution law of failure locus on *HM* plane with *V* levels (referred to our previous work of Zhang [24]), the analyses only elaborate on the plane of *H* versus *M* for a typical case of $V = 0.25V_0^+$. It can be observed from the dimensionless diagrams (Fig. 7a, c and e) that all failure loci are similar inclined elliptical shape. The size of the dimensionless failure locus generally increases with the increase of L/D and s/D. Also observe that for the large L/D, the maximum horizontal load the pile can sustain occurs under the path with a higher absolute value of M/H. In the normalized presentations (Fig. 7b, d and f), the size of the locus increases with the increase of s/D, but does not change much when L/D varies. The M/H ratio corresponding to the maximum load components roughly stays at a constant value when the geometry parameters change.

With the help of shear dissipation contour plots, the failure mechanisms of the tripod pile foundation under different pile lengths and pile spacings are identified, as shown in Fig. 8. Comparing the figures related to the variation of the pile length (Fig. 8a, c and e), the increase of the embedded length gradually expands the volume of the conical wedge in front of the piles, and the extension of the wedge is restrained within certain depth. Below this depth (Fig. 8e), a spoon-shaped plastic zone is generated around the bottom of the foundation. Whilst considering the effect of the pile spacing, the tripod piles act as a whole system for closed spacing cases (Fig. 8b), and there is no shear dissipation generated in the soil surrounded by the piles. For larger spacing cases (Fig. 8d and f), the tripod pile foundation tends to act as three individual piles, and there are similar collapse patterns that can observed adjacent to the three piles.

5.2 Effect of Non-homogeneity of the Clay

Figure 9 shows the normalized failure locus of tripod pile foundation considering the variation of non-homogeneous coefficient k of the clay from 0 to 2. The geometry cases of depth-diameter ratio $L/D = 3, 6, 9$ and spacing-to-diameter ratio $s/D = 4$ are considered in the analyses. It can be seen from the figure that when the vertical load level $V/V_0{}^+ = 0.25$ (Fig. 9a, c and e), the shapes of the normalized failure locus are basically the same under different values of non-homogeneous coefficient k, but the increase of the k value generally expands the size of the locus. The law is totally reversed when considering higher vertical load level (for $V/V_0{}^+ = 0.95$, as shown in Fig. 9b, d and f). In these cases, the size of the failure locus obviously shrinks with the increase of the non-homogeneous coefficient k.

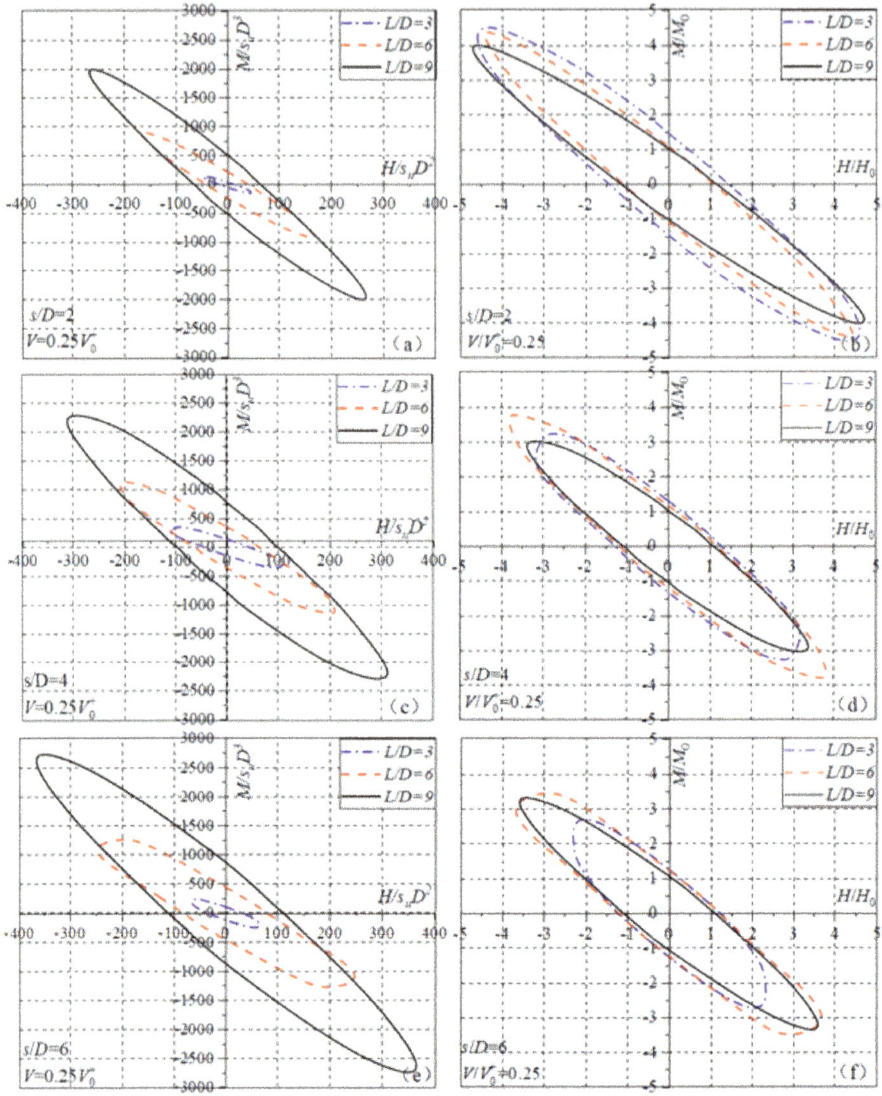

Fig. 7. Dimensionless and normalized failure envelopes of tripod pile foundation in non-homogeneous clay for $V/V_0^+ = 0.25$ and various geometries.

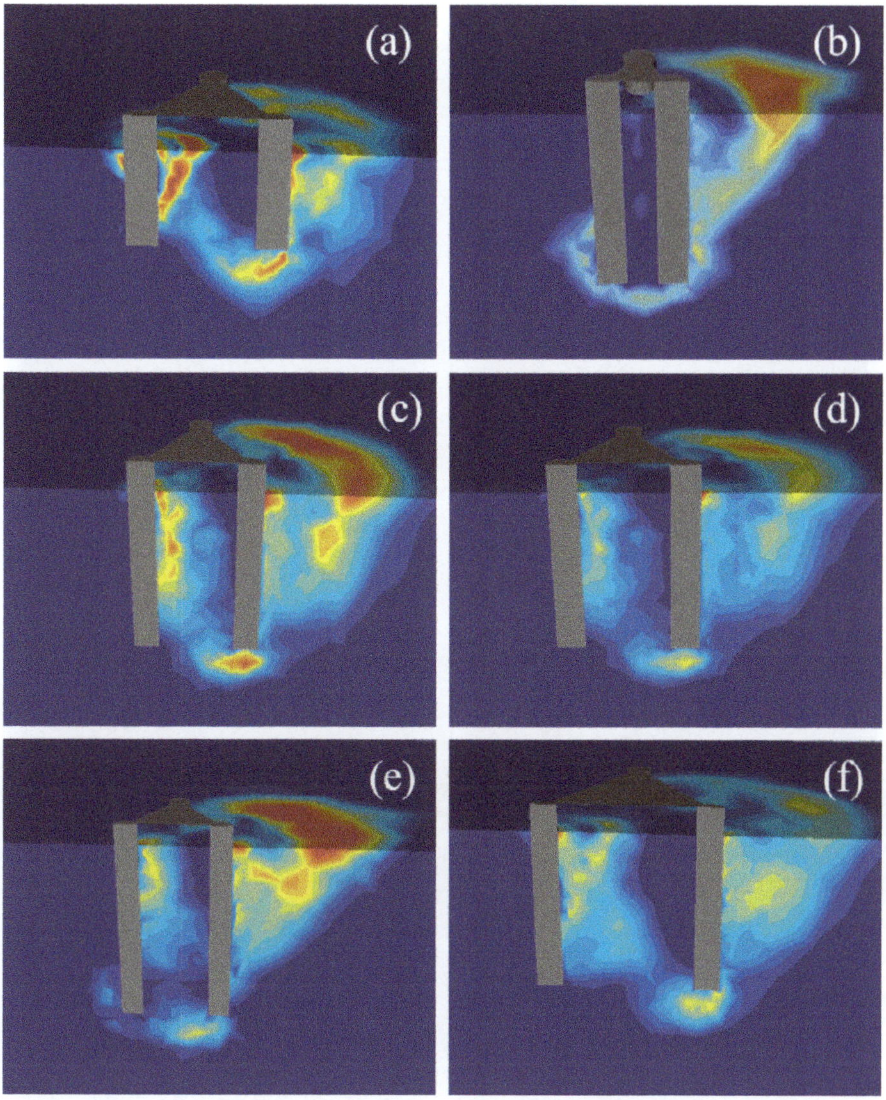

Fig. 8. Failure patterns under $V = 0.25V_0{}^+$ for $s/D = 4$ and different pile lengths: (a) $L/D = 3$; (c) $L/D = 6$; (e) $L/D = 9$ and for $L/D = 6$ and different pile spacings: (b)$s/D = 2$; (d) $s/D = 4$; (f) $s/D = 6$.

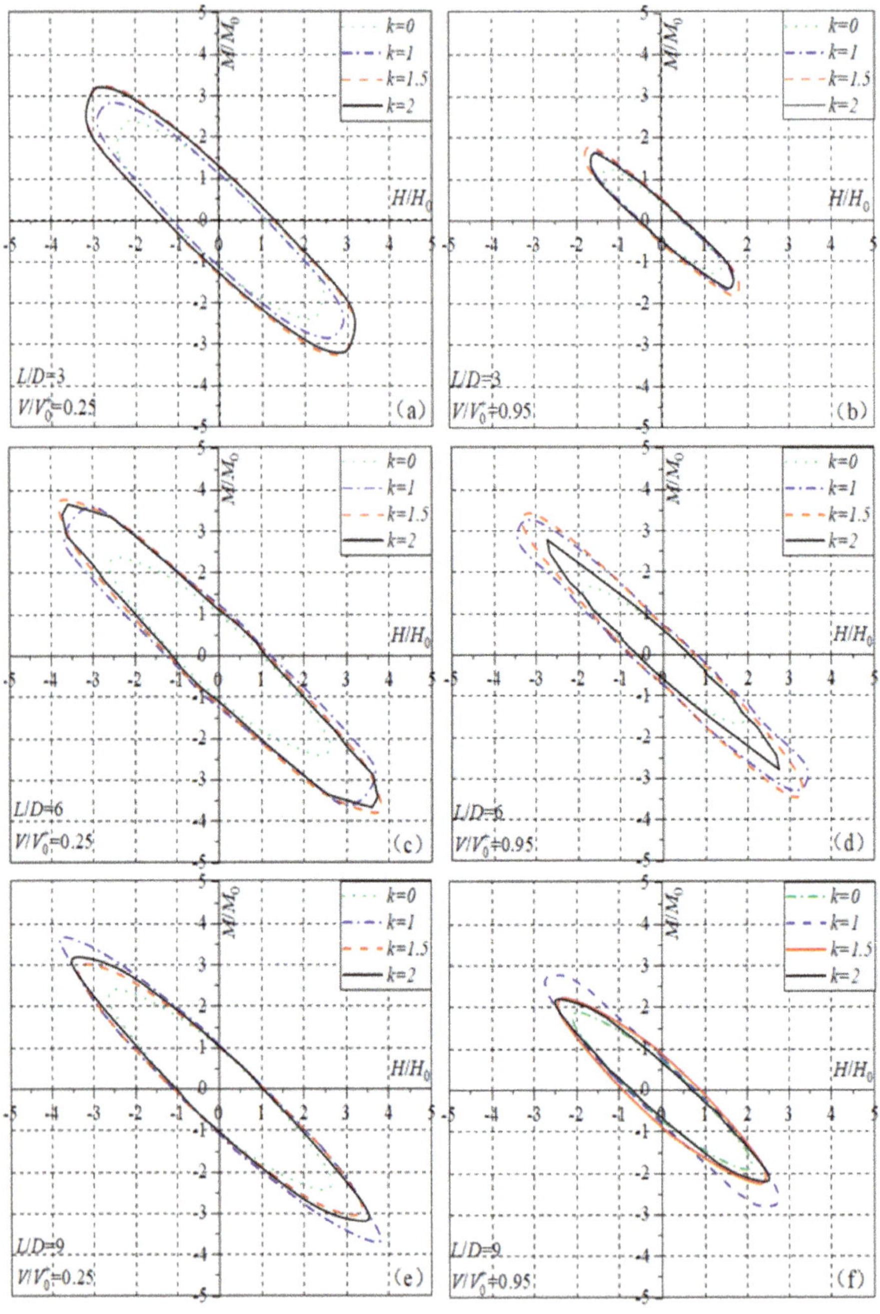

Fig. 9. Variations of the normalized failure loci with the non-homogeneous coefficient k in non-homogeneous clay.

6 Artificial Neural Network Model

The artificial neural network model coded in MATLAB was established to predict the capacities of the tripod pile foundation considering various lengths, spacings and diameters of the piles and the effects of combined loads and non-homogeneity of the soil. The parameters of length-to-diameter *L*/*D*, spacing-to-diameter *s*/*D*, loading ratio on *HM* plane a and non-homogeneous coefficient *k* are used as the covariates in the input layer, and the normalized ultimate bearing capacities corresponding to these parameters are generated as the outputs. In most cases, the simplest ANN structure can be used to obtain a good consistency between the predicted value and the actual value. This structure consists of an input layer, a hidden layer and an output neuron (Fig. 10a) A number of 1035 data sets from FELA simulations are fed to the ANN model in this section. The dataset is divided into three groups: 70% for training (Fig. 10b), 15% for verification (Fig. 10c), and 15% for testing (Fig. 10d). The ANN model is trained by the Levenberg-Marquart method, which is proved to be efficient and time-consuming. The prediction accuracy of the current ANN model is estimated by regression value (*R*). It can be seen from the figure that the final predicted value fits well with the results obtained from the simulation analysis, and the regression value *R* is around 0.95 (Fig. 10e).

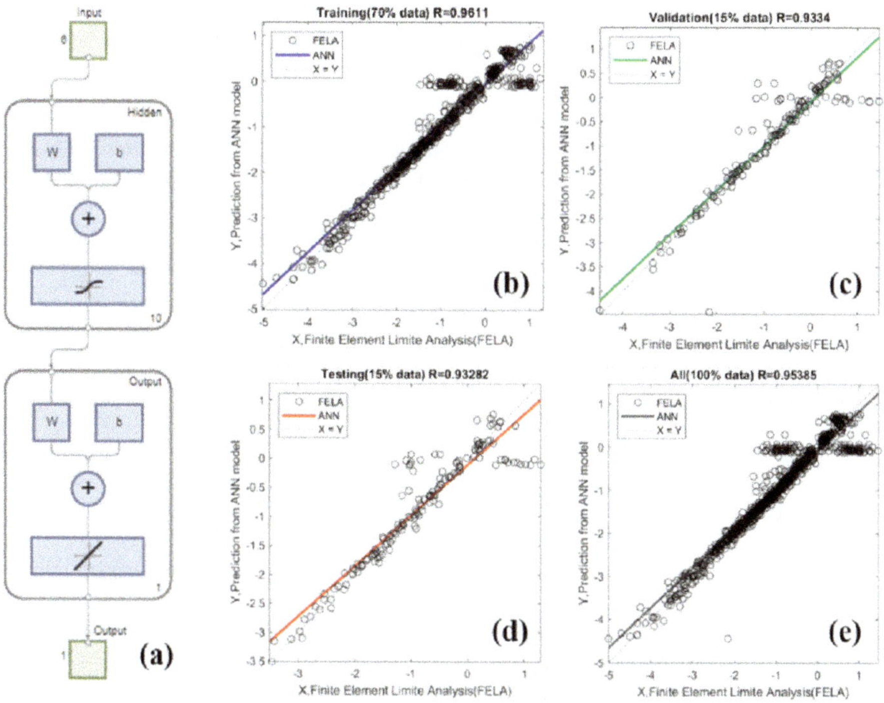

Fig. 10. The structure of the ANN model and its fitting results (a) Model structure, (b)Training datasets, (c)Validation datasets, (d)Testing datasets, (e)All datasets.

7 Conclusions

In this paper, the bearing capacities of the tripod pile foundation in non-homogeneous clay under the combined vertical, horizontal and moment loads are investigated via FELAs which implemented in the commercial software OptumG3. Three-dimensional failure envelopes constructed in *VHM* loading space are obtained considering various geometries of the tripod pile foundation and different non-homogeneous coefficients of the clay. An artificial neural network model is eventually established to predict the capacity of the tripod pile foundation. Some predominant laws are presented as follows:

1. The *HM* failure locus of the tripod pile foundation shows an inclined elliptical shape, and two typical collapse mechanisms are identified corresponding to the curve segments divided by the turning point.
2. The increase of the vertical load reduces the size of the failure locus on the HM plane, and the failure envelope in three-dimensional loading space is observed as a half of a super-ellipsoid.
3. The geometric parameters (viz. length-to-diameter ratio L/D and spacing-to-diameter ratio s/D) of the tripod pile foundation impact the size of the failure envelope as well as the pattern of the failure mechanism.
4. The increase of the non-homogeneous coefficient k expands the failure envelope of the tripod pile foundation for small vertical load levels, while for the large vertical load levels, the envelope shrinks with the grow up of the coefficient k.

Acknowledgement. This study is sponsored by National Natural Science Foundation of China (Grant No. 52378357 and 52008268), Doctoral Start-up Foundation of Liaoning Province (Grant No. 2022-BS-195), Chunhui Project Foundation of the Education Department of China (Grant No. HZKY20220416), Foundation of Liaoning Province Education Administration (Grant No. JYTMS20231608) and Shenyang Excellent Young and Middle-aged Science and Technology Innovation Talent Support Program (Grant No. RC230805). The authors would like to thank OptumCE for providing academic access to OptumG3.

References

1. Liu, R., Shi, J., Chen, G., Yang, X.: Lateral bearing capacity and calculation method of pile-bucket foundation in clay for offshore wind turbines. Ocean Eng. **283**, 115039 (2023)
2. Reese, L.C., Welch, R.C.: Lateral loading of deep foundations in stiff clay. J. Geotech. Eng. Div. **101**(7), 633–649 (1975)
3. Murff, J.D., Hamilton, J.M.: P-ultimate for undrained analysis of laterally loaded piles. J. Geotech. Eng. **119**(1), 91–107 (1993)
4. Randolph, M.F., Houlsby, G.T.: The limiting pressure on a circular pile loaded laterally in cohesive soil. Geotechnique **34**(4), 613–623 (1984)
5. Wang, L., et al.: Nonlinear analysis of laterally loaded single pile in layered soil based on conical strain wedge model. Ocean Eng. **280**, 114699 (2023)
6. Zhao, Z., Kouretzis, G., Sloan, S.W., Gao, Y.: Effect of geometric nonlinearity on the ultimate lateral resistance of piles in clay. Comput. Geotech. **105**, 110–115 (2019)
7. Zhou, S., Zhou, M., Tian, Y., Zhang, X.: Effects of strain rate and strain softening on the installation of helical pile in soft clay. Ocean Eng. **285**, 115370 (2023)

8. Zhao, Z., Zhang, H., Wu, F.: Development of lateral capacity-based envelopes of twin-pile group under combined V-H–M loading. IOP Conf. Ser. Earth Environ. Sci. **861**(3), 32061 (2021)
9. Matlock, H., Ingram, W.B., Kelley, A.E., Bogard, D.: Field tests of the lateral-load behavior of pile groups in soft clay. In: Offshore Technology Conference, pp. 163–174. OTC (1980)
10. Georgiadis, K., Sloan, S.W., Lyamin, A.V.: Ultimate lateral pressure of two side-by-side piles in clay. Géotechnique **63**(9), 733–745 (2013)
11. Georgiadis, K., Sloan, S.W., Lyamin, A.V.: Effect of loading direction on the ultimate lateral soil pressure of two piles in clay. Géotechnique **63**(13), 1170–1175 (2013)
12. Liu, B.: Study on bearing capacity behavior of pile foundation for offshore wind turbines using 3-D FEM. Master thesis of Dalian University of Technology, Dalian (2009). (in Chinese)
13. He, B., Jiang, J., Cheng, J., Zheng, J., Wang, D.: The capacities of tripod bucket foundation under uniaxial and combined loading. Ocean Eng. **220**, 108400 (2021)
14. Shen, K., Jiang, J., Wang, D., Zheng, J.: Capacities of tripod bucket foundation under uniaxial and combined loading considering adhesion factor. Mar. Georesour. Geotechnol. **40**(12), 1520–1528 (2022)
15. Kim, S.: Evaluation of combined horizontal-moment bearing capacities of tripod bucket foundations in undrained clay. Ocean Eng. **85**, 100–109 (2014)
16. Zhao, Z., Kouretzis, G., Sloan, S., Gao, Y.: Ultimate lateral resistance of tripod pile foundation in clay. Comput. Geotech. **92**, 220–228 (2017)
17. Gourvenec, S., Randolph, M.: Effect of strength non-homogeneity on the shape of failure envelopes for combined loading of strip and circular foundations on clay. Géotechnique **53**(6), 575–586 (2003)
18. Gottardi, G., Houlsby, G.T., Butterfield, R.: Plastic response of circular footings on sand under general planar loading. Géotechnique **49**(4), 453–469 (1999)
19. Vesic, A.S.: Design of Pile Foundations. National Research Council, Washington (1977)
20. Koumoto, T., Meyerhof, G.G., Sastry, V.: Analysis of bearing capacity of rigid piles under eccentric and inclined loads. Can. Geotech. J. **23**(2), 127–131 (1986)
21. Georgiadis, K., Georgiadis, M., Anagnostopoulos, C.: Lateral bearing capacity of rigid piles near clay slopes. Soils Found. **53**(1), 144–154 (2013)
22. Yu, J., Huang, M., Li, S., Leung, C.F.: Load-displacement and upper-bound solutions of a loaded laterally pile in clay based on a total-displacement-loading EMSD method. Comput. Geotech. **83**, 64–76 (2017)
23. Hansen, J.B.: The ultimate resistance of rigid piles against transversal forces. Bull. Danish Geotech Inst. **12**, 5–9 (1961)
24. Zhang, H.: Bearing capacity of tripod pile foundations under V-H-M combined loading mode. Master thesis of Shenyang Jiangzhu University, Shenyang (2023). (in Chinese)

Research on the Evaluation Indicator System of Urban Infrastructure Resilience

Jiaxi Liu(✉), Kun Zhao, and Chuanmeng Wei

School of Civil Engineering, Shenyang Jianzhu University, 25 Hunnan Middle Road, Shenyang, Liaoning, China
jxliu@sjzu.edu.cn, 469357429@qq.com

Abstract. Urban infrastructure is the guarantee for cities to maintain normal socio-economic activities. Resilience assessment of infrastructure helps to ensure the safe operation of infrastructure. In view of the actual needs and important problems of urban safety at this stage, this study starts from the concept of resilience development and urban safety resilience, analyzes the characterization of safety resilience, and proposes a characterization model of urban infrastructure safety resilience. Based on the three basic elements of risk source, disaster bearing body and disaster reduction body, the framework of urban infrastructure safety resilience is constructed. This study analyzes and summarizes the main influencing parameters of the functional changes of infrastructure in each stage of resistance, adaptation and recovery, and establishes the indicator system of urban infrastructure safety resilience evaluation, which provides a reference for the evaluation of urban infrastructure safety resilience and urban safety improvement countermeasures.

Keywords: Urban safety · Infrastructure · Resilience Assessment · Resilience Characterization

1 Introduction

Urban safety plays an important role in the smooth running of society. For a long time, China's urban security has been faced with major challenges such as natural disasters, accidents and security incidents. In order to effectively deal with complex urban security issues, urban construction concepts [1] such as urban livability, green city, sponge city, smart city, and urban resilience have emerged one after another. The concept of urban resilience has broken through the traditional rigid urban safety management thinking. In addition to focusing on the city's own resilience when disasters occur, it also emphasizes the ability to adapt, recover, absorb and learn through systematic integration of elements. However, most of the existing urban resilience studies focus on urban resilience, community resilience and climate change in response to sudden disasters. While there are few studies on the resilience of urban infrastructure security. As the main disaster bearing body of the city, infrastructure is of great significance to the safe operation of the city. Based on the above practical background, this paper summarizes the development of the concept of resilience and the research trend of urban resilience, analyzes the

I. Tuns et al. (Eds.): CIBv 2023, LNCE 510, pp. 442–452, 2024.
https://doi.org/10.1007/978-3-031-60765-3_35

elastic representation model of urban infrastructure, summarizes the current situation of urban infrastructure resilience evaluation, and constructs an urban infrastructure safety resilience evaluation index system based on the elements and characteristics of urban infrastructure safety resilience.

2 Urban Resilience

2.1 Resilience Concept Development

The term elasticity originates from the concept of elasticity in engineering. Its meaning is "return to the original state", which refers to the ability of the object to return to the original state after deformation under the action of external forces. Canadian scholar Holling [2] proposed the concept of "ecological resilience" on the basis of the idea of engineering resilience. He believes that resilience is the return of a system to its original equilibrium after external interference. Berkes [3] revised the concept of resilience based on ecology, and considered that there are multiple equilibrium states in the system resilience. Building upon the intricate social ecosystem, scholars in the field of disaster science have delved deeper into the internal composition and mechanisms of resilience systems, leading to the emergence of evolutionary resilience. Meanwhile, elasticity theory has undergone three transitions from engineering elasticity to ecological elasticity and finally to evolutionary elasticity, continuously enriching its connotations.

In recent years, with the development of related research in the field of urban safety, how to quickly restore urban functions and reduce the loss of life and property has become the focus of its attention. The concept of resilience has also been applied to urban safety research. Mileti [4] defined safety resilience as the ability of a region to sustain a certain level of daily life in the face of major natural disasters without significant help from the outside world. The United Nations International Strategy for Disaster Reduction (UNISDR) [5] defines safety resilience as the capacity of a system, community, or society to withstand or adapt to a disaster in order to achieve and sustain an acceptable level of functionality. According to the Resilience Alliance, urban resilience refers to a city's ability to assimilate and absorb external disturbances while preserving its fundamental characteristics, structures and key functions. Some domestic and foreign scholars have expressed the concept of urban safety resilience as follows (Table 1).

The city is composed of many elements. The urban system is a rich, open and complex system. If the subsystem mutates, it will lead to the disorder of the whole complex system of the city. The emergence of safe and resilient cities can effectively ensure the safe operation of urban systems and enable cities to cope with complex security issues.

When the city suffers a disaster, the city can withstand, adapt and quickly restore its original ability. When the impact is small, the city can absorb the impact; when the impact is moderate, the city can reduce the impact; when hit by big shocks, the city can withstand and recover quickly.

On the whole, when the impact of disasters on its subsystems is damaged, the safe and resilient city can meet the basic operation of the city through its own adjustment, and quickly restore balance, reduce the loss of people's lives and property, maintain social stability.

Table 1. Concept expression of urban safety resilience

author	rationale	Definition of urban safety resilience
Mileti [4]	ecological resilience	Safety resilience refers to the ability of the region to maintain a certain degree of daily life without great help from the outside world, suffering from major natural disasters
UNISDR [5]	ecological resilience	Safety Resilience is the ability of a system, community, or society exposed to a disaster to resist or change in order to achieve and maintain an acceptable level of functioning
Resilience alliance [6]	ecological resilience	Urban safety resilience is the ability of the city itself to digest and absorb external disturbances and maintain its main structure and key functions
Rockefeller Fund of the United States [7]	evolution resilience	Urban safety resilience is a system composed of individuals, communities, institutions, and industries. It still has the ability to survive, adapt, and develop regardless of sudden or slow disturbances
Fang Dongping [8]	evolution resilience	Urban safety resilience is the ability of urban systems and their subsystems to maintain or quickly restore their functions when disturbed, and to better respond to future uncertainties through adaptation

2.2 The Connotation of Urban Safety Resilience

Safety resilience can provide continuous guarantee for the system in a changing environment, so that the system function can maintain normal operation. Safety resilience curve reflects the process of system function decline and recovery under the influence of external disasters. Most of its representation models are based on the conceptual framework of safety resilience curve, as shown in the figure. The horizontal axis is the time axis, and the vertical axis is the system function. By taking preventive measures before a disaster, you can restore system functions to the pre-disaster level within a period of time after a disaster. The curve can reflect the safety toughness of the system to some extent. Starting from the concept of safety resilience curve, some researchers consider the resilience elements such as system state and recovery time, and use mathematical formulas to establish a safety resilience characterization model. For example, Bruneau [9] introduced the safety elasticity curve in the study on the safety elasticity of community response to earthquake disasters. With the help of mathematical formula (1), the safety resilience of the system is expressed, and the safety resilience R is defined by integrating the function Q(t) of the system on time t (see Fig. 1).

$$R = \int_{0}^{t_1} Q(t)dt \tag{1}$$

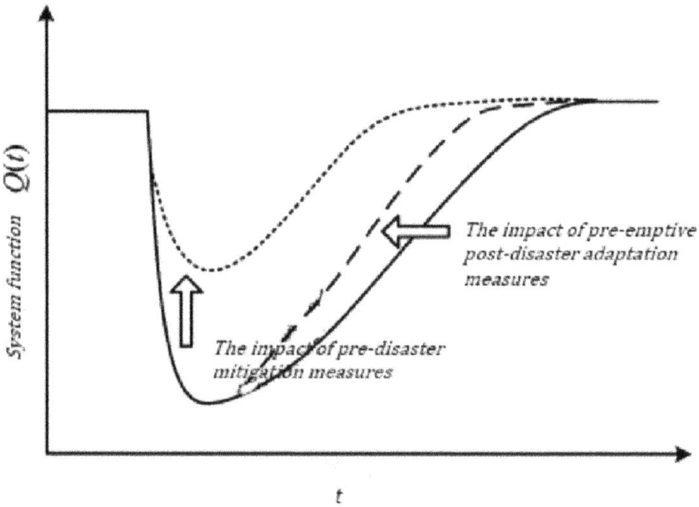

Fig. 1. Safety resilience curve.

These curves represent the safety resilience of the city to a certain extent, but there are still many limitations in this way of representing safety resilience. For example, it ignores the links within the system, overemphasizes the external conditions of the system function, and cannot well point out the direction of improving the safety resilience of the system. Therefore, it is necessary to construct a system that can reflect the essential characteristics of the system. At present, the theoretical model with consensus in the public safety science and technology community is the public safety triangle model proposed by Fan Weicheng [10] et al. It takes emergencies, disaster-bearing carriers, and emergency management as the main body, and links them with disaster factors. It reveals the basic elements of public safety science, and successfully introduces the model into the research of urban safety resilience.

3 Urban Infrastructure Security Resilience

3.1 Urban Infrastructure Resilience

Infrastructure projects are of great significance to urban safety. Based on the relevant literature, this paper divides the types into six categories: transportation, energy, communication, water resources, society and others [11] to build resilient infrastructure, it can effectively deal with urban uncertainty risks and improve urban comprehensive disaster prevention ability.

At present, there is no consensus on the concept of infrastructure security resilience. Infrastructure resilience is defined by Gay et al. [12] as the ability of an infrastructure system to reduce performance losses and recover to a certain level of performance under disturbed conditions. In the UK Cabinet's "Keeping the Country Running: Natural Disasters and Infrastructure" report, infrastructure resilience [13] is defined as "having a good network system to ensure it has the necessary resilience, reliability and redundancy; establish a good elastic organization so that it has the ability to recover from disasters".

Infrastructure can be defined as a infrastructure in disaster's capacity to absorb shock, adaptation and transformation, so as to quickly recover its basic functions. It is mainly manifested in two aspects. First, when it is impacted by external disasters, the infrastructure has certain adaptability and can resist the absorption of impact energy. Second, when it is damaged by the outside world, the infrastructure recovers to the operational level as soon as possible through its own adjustment ability to ensure the daily life of urban residents.

3.2 Characterization of Infrastructure Safety Resilience

In order to establish a sound urban infrastructure security resilience system, promote the effective integration of urban disaster prevention system and infrastructure security resilience needs. Based on the correlation between the three elements of urban risk source, urban infrastructure and urban emergency management in the process of disaster response and the triangular model of urban public safety, this paper constructs a characterization model of urban basic safety resilience based on [14] "risk source + disaster bearing body + disaster reduction body".

Based on this model, the characteristics of urban infrastructure as a disaster-bearing body and the characteristics of urban management safety resilience are analyzed. The safety resilience of infrastructure should have the following characteristics: (see Fig. 1).

(1) Redundancy: The urban infrastructure system and its coupled systems have a certain safety margin and can maintain normal functions when the system changes.
(2) Robustness: Refers to the characteristics [15] of the infrastructure system to maintain its own function under certain parameter disturbance.
(3) Adaptability: All elements of the urban infrastructure disaster bearing system have the ability to flexibly adjust and adapt to changes in the external environment.
(4) Diversity: Urban infrastructure projects have a high degree of richness, which can improve their resilience to unknown disasters.
(5) Rapid stability capacity: The city can respond quickly and provide stability under the impact of disasters.
(6) Learn resilience: After a disaster, cities can quickly adjust their resilience and learn from it to improve their ability to respond to disasters.
(7) Cooperativity: All government departments adjust relevant resources, strengthen the linkage guarantee, and ensure the maximization of residents' interests (Fig. 2).

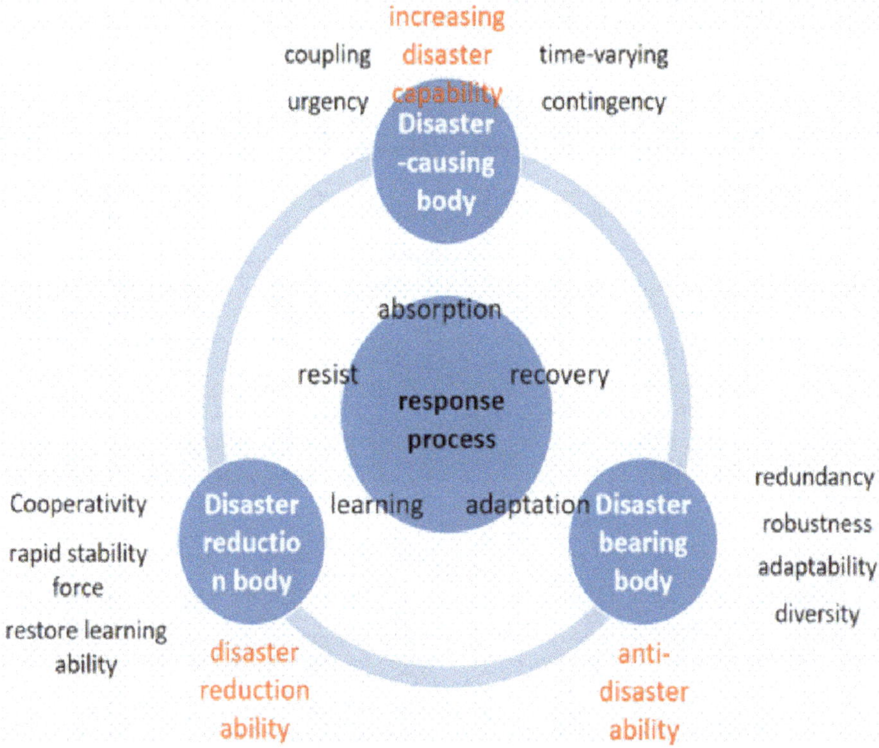

Fig. 2. Infrastructure security resilience characterization model.

4 Establishment of Infrastructure Safety Resilience Evaluation Index System

4.1 Evaluation Index Selection Analysis

At present, the resilience evaluation of urban infrastructure in China is still in the primary state, and there is not a general evaluation criterion. The trilateral model constructed by Fan Weicheng and other scholars is a "trilateral" model generally recognized by public security technicians in our country. On this basis, the disaster factor is regarded as the node of the connection boundary, and the disaster factor is taken as the node of the connection boundary, thus reflecting the essential characteristics of public security.

Based on the framework of urban infrastructure resilience characterization, this paper selects natural disasters such as typhoons, rainstorms, earthquakes, fires, natural disasters such as typhoons, earthquakes, and accidents such as public facilities and equipment accidents as the components of [16] disaster-causing body/risk source under the framework of urban infrastructure resilience characterization. Urban infrastructure and urban safety resilience management correspond to carrier and disaster reduction body respectively. The combination of the three forms the first evaluation index of a comprehensive evaluation index system.

4.2 Establishment of Evaluation Index System of Infrastructure Safety Resilience

See Table 2.

Table 2. Infrastructure security resilience evaluation index system

first-level	second-level	third-level	the nature of indicators
risk source	natural disaster	typhoon	quantification
		Rainstorm (flood)	quantification
		snow hazard	quantification
		earthquake	quantification
	multiple disaster accidents	Earthquake and landslide	quantification
		Floods and mudslides	quantification
	security incident	fire	quantification
		traffic accident	quantification
urban infrastructure	traffic facilities	road area per capita	quantification
		Safety and durability level of highway bridge	quantification
		Number of inter-city material delivery channels	quantification
	public facilities	heat facilities	quantification
		water facility	quantification
		power supply facilities	quantification
	communication facilities	mobile subscription	quantification
		Fixed broadband household penetration rate	quantification
	monitoring and early warning facilities	Urban public area monitoring coverage	quantification
		Percentage of early warning system coverage	quantification
		Intelligent monitoring management rate of municipal pipeline network	quantification
	emergency support facilities	Situation of shelters	quantification

(*continued*)

Table 2. (*continued*)

first-level	second-level	third-level	the nature of indicators
		emergency material reserves	quantification
		construction of medical disaster relief institutions	quantification
urban emergency management	emergency rescue capability	rescue team situation	characterization
		rescue equipment investment	characterization
		frequency of professional rescue team drills	quantification
	management system construction	emergency management system	characterization
		city government leadership	characterization
		supporting security investment	quantification

4.3 Analysis of Infrastructure Safety Resilience Evaluation Index System

1. Risk source

   The risk source is the disaster-causing body, which mainly refers to the risk factors that cause urban safety, including natural disasters, multi-disaster accidents, and safety accidents. The purpose of this part of the index is to judge the susceptibility and danger of various disaster risks in the target city through the statistical analysis of historical disasters and accident disaster information, and to set up corresponding indicators to express disaster risks for related multiple disasters. The actual urban disasters are very complicated, and there are complex relationships among different disasters in terms of time, scope and influence. Considering comprehensively, the coupled disasters based on trigger relationship such as earthquake and landslide, flood and debris flow are selected in this paper. Secondly, the safety disasters such as traffic accidents and fire that are easy to occur in the city are selected.

2. Urban infrastructure

   Urban infrastructure mainly refers to the elements that bear the role of risk sources, including transportation facilities, public facilities, communication facilities, monitoring and early warning facilities, and emergency protection facilities. This part of the indicators mainly reflects the safety and resilience characteristics and needs of urban infrastructure as a disaster-bearing body. In the transportation system, indicators related to material transportation, rescue and evacuation streamlines are set up to examine the adaptability of urban principles to disasters. The indicators of public facilities and emergency support facilities reflect the adaptability and resilience of urban disaster-bearing systems, and show the resilience characteristics of urban lifeline systems such as water supply, heating and power supply. In the monitoring and early warning indicators, the ability of information transmission and data acquisition is evaluated through the construction of early warning, monitoring and communication facilities. Secondly, set up emergency support facilities indicators around the shelter situation, emergency material reserves, medical construction and other aspects to ensure the basic daily life of the people after the disaster.

3. Urban emergency management

   Urban emergency management, that is, disaster reduction body, is the main body that can play the role of disaster prevention, disaster relief and disaster reduction in the time development of urban safety (before, during and after the disaster). It mainly includes two aspects: emergency rescue capability and management system construction. In view of the behavioral needs and streamline characteristics of post-disaster crowd evacuation and emergency rescue, in the emergency rescue part, the rationality and perfection of urban emergency rescue capability are investigated through indicators such as rescue team situation, rescue equipment input and professional rescue team drill frequency. The construction of the management system further evaluates the feasibility of the effective implementation of urban management measures through indicators such as urban government leadership, emergency management system, and support guarantee investment.

5 Conclusion

The construction of resilient infrastructure can greatly improve the ability of urban infrastructure to respond, recover and adapt to sudden urban security problems, which is of great significance for the sustainable development of cities. At present, the theoretical research and application exploration of urban infrastructure resilience in China are still immature, and there are research dilemmas in theory and method. However, the overall research progress is still in the stage of exploration and rise, and the research work needs to be systematically improved. It is an urgent problem to form a complete infrastructure resilience construction system and promote the safe development of China's cities. Through the research on the resilience assessment of urban infrastructure in this paper, the following conclusions are drawn:

(1) A characterization framework of urban infrastructure safety resilience based on "risk source + hazard bearing body + disaster reduction body" is established. Based on the review and analysis of domestic and foreign research on urban infrastructure resilience and theoretical research in the field of urban safety resilience, this paper proposes a characterization model of "risk source + disaster bearing body + disaster reduction body" for urban infrastructure safety resilience. The model reflects the safety resilience characteristics of urban infrastructure to effectively respond to disasters, maintain urban functions, and quickly recover from disasters, emphasizing its ability to cope with uncertainty. Based on this, a framework for characterizing urban infrastructure resilience is established.

(2) The resilience evaluation system of urban infrastructure is constructed. Based on the characterization framework of urban infrastructure safety resilience based on "risk source + disaster bearing body + disaster reduction body". Taking natural disasters, accident disasters and other urban prone disasters as risk sources, urban infrastructure as disaster bearing body, and urban management safety resilience as disaster reduction body, under the mutual support and demonstration of literature, taking the ability of urban infrastructure to resist, absorb, adapt and restore learning as the core selection index, the evaluation index system of urban infrastructure safety resilience is established, which includes 3 first-level indicators, 10 s-level indicators and 28 third-level indicators. A complete evaluation index of urban infrastructure safety resilience is obtained.
(3) The selection of evaluation indicators and the elements of resilience expression are analyzed. This paper expounds and analyzes the risk characteristics of risk sources, the functional evolution of urban infrastructure from disaster damage to recovery stage and then to steady state stage, the post-disaster government emergency rescue capability and emergency system management, and realizes the correlation expression of various elements of urban infrastructure safety resilience, which lays a foundation for the later research on urban infrastructure safety resilience assessment.

References

1. Chen, A., Shi, Y.: A review of the concept evolution and evaluation methods of resilient cities. Ecol. City Green Build. (01), 14–19 (2018)
2. Holling, C.S.: Resilience and Stability of Ecological Systems. Rev. Ecol. Syst. **4**, 1–23 (1973)
3. Linking social and ecological systems: management practices and social mechanisms for building resilience. Cambridge University Press, New York (2000)
4. Mileti, D.: Disasters by Design: A Reassessment of Natural Hazards in the United States. Joseph Henry Press, Washington, D.C. (1999)
5. Williams, P., Nolan, M., Panda, A.: Disaster resilience scorecard for cities: frequently-asked questions. UNISDR (2014)
6. Alliance, R.: Urban resilience research prospectus (2007)
7. Spaans, M., Waterhout, B.: Building up resilience in cities worldwide - Rotterdam as participant in the 100 resilient cities programme. Cities **61**, 109–116 (2017)
8. Fang, D., et al.: Urban resilience - reflections based on 'systems in three-dimensional space.' J. Civ. Eng. **50**(07), 1–7 (2017)
9. Bruneau, M., Chang, S.E., Eguchi, R.T., et al.: A framework to quantitatively assess and enhance seismic resilience of communities. Earthq. Spectra **19**(4), 733–752 (2003)
10. Huang, H., Li, R., Fan, W., Shan, C.: Analysis of the characteristics of safety resilient cities and its enlightenment to the safety development of Xiong' an new area. China Saf. Prod. Sci. Technol. **14**(07), 5–11 (2018)
11. Bi, W., Tang, Y., Mao, T., Sun, X., Li, Q.: Review of resilience management of urban infrastructure systems. Chin. J. Saf. Sci. **31**(06), 14–28 (2021)
12. Wu, S., Liu, S.: Research status and development trend of infrastructure resilience-analysis based on international literature review. Int. Urban Plann. **38**(02), 31–38 (2023)
13. Civil Contingencies Secretariat Cabinet Office: Keeping the country running: natural hazards and infrastructure. London, UK (2010)

14. Yue, Q., et al.: Theoretical framework of urban safety characterization "library-network-flow-spectrum-method" based on "risk source + hazard-affected body + disaster reduction body." Eng. Mech. **39**(11), 52–62 (2022)
15. Zhou, S., Huang, H., Li, R.: Urban infrastructure resilience assessment and sensitivity analysis. J. Wuhan Univ. Technol. (Inf. Manag. Eng. Edn.) **42**(03), 189–195 (2020)
16. Wang, Y., Fan, S., Fan, L.: Research on evaluation index of urban safety resilience under multiple disasters based on multi-factor and multi-level evaluation. Build. Sci. **37**(01), 82–88 (2021)

Experimental Study with Tests on Non-structural Concrete Elements with Recycled Concrete Aggregates

Ligia Georgiana Hanuseac(✉) and Marinela Barbuta

Gheorghe Asachi "Technical University of Iasi, Faculty of Civil Engineering and Building Services", 1, Prof. Dimitrie Mangeron Blvd, 700050 Iasi, Romania
ligyahanuseac@gmail.com

Abstract. This study presents a practical application in which the different samples made by ecological concrete are analyzed at mechanical characteristics tests of compression loads, bending loads and splitting loads for different samples made by eco-friendly concrete. In the concrete recipe for the experiment was used fly ash as cement replaced and it mixed different sort of recycled concrete aggregates (RCA) waste crushing and sorted they as replacement of sand and natural aggregate. The mechanical characteristics such as: compressive strength, tensile strength by bending and split tensile strength of concrete whit fly ash and RCA waste have been determined. The mechanical characteristics of concrete samples were influenced by the dosage and the mixture of RCA waste. Failure in compression loads, bending loads and splitting loads of concrete samples was gradual and ductile.

Keywords: CO_2 · RCA waste · eco-friendly concrete · mechanical characteristics · sustainability

When we talk about the revolution and evolution of eco-friendly concrete constructions, we refer to four very important aspects: economic factors, mechanical characteristics, sustainable use of natural resources, and reduction of CO_2 emissions. It is necessary for these four factors to work together in a more seamless manner [1–3]. Firstly, it is crucial to understand that capturing and storing CO_2 does not reduce CO_2 generation; it prevents it from being released into the atmosphere [4]. Therefore, it is imperative to focus on finding ways to achieve a significant reduction in CO_2 emissions in the cement industry and the development of eco-friendly concrete recipes [5, 6]. According to International Energy Agency (IEA) global energy-related CO_2 emissions grew by 0.9% in 2022, last year's growth was much slower than 2021's rebound of more than 6%. Emissions from energy combustion increased, while emissions from industrial processes decreased. Increased deployment clean energy technologies such as renewables, electric vehicles and heat the pumps helped prevent an additional 550 Mt in CO_2 emissions. In UE buildings sector emissionsof CO_2 fell markedly in 2022 [7].

Carbon capture and storage (CCS) face several barriers, both technical and economic [8]. Apart from its enormous cost and slow implementation, it also faces challenges due to competition with more feasible solutions such as applying best available techniques

I. Tuns et al. (Eds.): CIBv 2023, LNCE 510, pp. 453–470, 2024.
https://doi.org/10.1007/978-3-031-60765-3_36

(BAT), replacing clinker, or substituting fossil fuels [9]. Another noteworthy aspect is the replacement of a portion of cement in conventional concrete recipes with materials like fly ash, slag, or silica fume [10]. By reducing the amount of cement and replacing it with fly ash, for example we can also reduce the CO_2 emissions generated during cement production by the same percentage [11, 12]. These materials are by-products from other industries and can replace a portion of the cement used in the concrete recipe [13]. Another important aspect to consider is the significant amount of waste generated annually from demolitions worldwide [14]. Large quantities of waste are generated annually, which, if not managed correctly, ends up polluting the environment and affecting human health [15]. This waste has a strong negative impact on the environment, which is why researching the characteristics and performance of concrete containing recycled concrete aggregate (RCA) is highly relevant [16]. Similarly, optimizing the proportions of cement, sand, aggregates, and water can result in more energy-efficient concrete recipes with reduced CO_2 emissions, thereby reducing the environmental impact [17]. These concerns are included in ongoing research efforts [18].

In conclusion, sustainable development means using natural resources without compromising their regeneration for future generations [19]. Ultimately, eco-friendly concrete is a revolutionary material that has a significantly lower environmental pollution impact compared to conventional concrete [20].

1 The Study Experimental Research

It is crucial to ensure that all four aspects work together and that concrete recipes are optimized [21]. The study is a practical application of optimized eco-friendly concrete recipes that aim to reduce CO_2 emissions and replace natural aggregates with crushed recycled concrete waste from demolitions (RCA) [22]. After optimizing the recipe, we produced non-structural concrete elements and subjected them to mechanical testing to determine their mechanical characteristics [23].

1.1 Materials

In the initial research phase, the reference (control) sample used was the recipe for conventional concrete with a C20/25 class, with a maximum aggregate size of 16 mm, established according to NE012-1:2007, the Code of practice for the execution of concrete works, reinforced concrete and prestressed concrete [24].

Materials used in the preparation of eco-friendly concrete recipes:

Aggregates – Ballast aggregates from local origin, they are obtained from the processing of river ballast in order to produce sands and gravel, having applicability in the field of construction materials. Originating from natural rocks of various sizes, resulting from the natural crushing of rocks (gravel pit aggregates) they must meet the following criteria: have high compressive strength, good surface roughness, not be cracked, be clean, have no unfavorable reactions with cement according SREN: 12620 + A1, 2008. The humidity of the aggregates was between 2% and 6% and the medium density of aggregates in the loose state, sort 8–16 mm is 1455 kg/m^3. The sizes of aggregates:

- Sort 0–4 mm: quantity 803.16 kg/m^3;

- Sort 4–8 mm: quantity 384.12 kg/m^3;
- Sort 8–16 mm: quantity 558.72 kg/m^3;

Cement –Carpatcement CEM II/A-LL 42.5 R; according to SR-EN 197-1:2011, the notations in the cement type designation mean:

- Cem II: composite Portland cement.
- A: the percentage of Al_2O_3 in the clinker mass.
- LL: limestone with a total organic carbon content below 0.20% of the clinker mass, a $CaCO_3$ content greater than 75% of the calcium oxide mass, and a clay content lower than 1.2g/100g of total limestone. According to Table 1 in SR EN 197-1:2011, CEM II/A-LL contains 80–94% clinker and 6–20% limestone.
- 42.5R: Indicates achieving a compressive strength of over 20 MPa at two days, between 42.5 MPa and 62.5 MPa at 28 days, and an initial setting time of over 60 min.
- Cement: 360 kg/m^3.

Thermal Fly ash – also called by-product because it comes from the combustion process of fuels.

Water- In the recipes used clean water, the water PH was between 6 to 7.

Additive Glenium Sky 617 - is an additive based on Superplasticizer/powerful water reducer polymers;

RCA - Waste from recycled aggregates – scrapped old concrete used as a substitute for the natural aggregate: In this way, old crushed concrete was reused as a substitute for natural aggregates in the new concrete, minimizing the negative impact on the structure. Studies conducted on this category demonstrate that concrete mixtures with recycled aggregates have performance equivalent to that of reference concrete in most properties. The waste used is of local origin, resulting from the demolition of industrial halls. The reinforcement was separated from the concrete, which was then crushed in two stages. After the first crushing, coarse aggregates with dimensions over 100 mm were obtained, followed by another crushing and sorting process, resulting in aggregates with dimensions:

- Crushed aggregates 0–4 mm.
- Crushed aggregates 4–8 mm.
- Crushed aggregates 8–16 mm.

1.2 Processing of Aggregates Resulting from Waste

The waste of local origin, resulting from the demolition of some industrial halls, suffered 2 crushing processes. The aggregates were sorted (0–4 mm; 4–8 mm; 8–16 mm) and introduced in the recipe (Fig. 1).

Fig. 1. Recycled concrete aggregates (RCA) waste

Starting from the standard recipe BM1, in rest of recipes was replaced 10% of cement quantity with the fly ash. (origin CET Iasi). From this composition were prepared concrete with waste at which the same dosage of fly ash was maintained (10%) and the aggregate sort 0–4 mm, 4–8 mm, 8–16 mm was replaced with different dosages of recyclable waste.

1.3 Types of Test-Pieces

The test-pieces were prepared according to the standards (SR-EN:12390-1, 2002), (SR-EN:12390-1, 2002/AC:2006) [25] (Table 2).

Table 1. Table of recipes

Recipe	Cement kg/m^3	Water l/m^3	Aggregates sort 0–4 kg/m^3	Aggregates RCA sort 0–4	Aggregates sort 4–8 kg/m^3	Aggregates RCA sort 4–8	Aggregates sort 8–16 kg/m^3	Aggregates RCA sort 8–16	Fly ash kg/m^3	Additive l/m^3
Recipe conventional										
BM1	360	180	803.16	-	384.12	-	558.72	-	-	3.6
Recipe whit 10% fly ash and combined sorts										
ANT1	324	180	803.16	-	-	100%	558.72	-	36	3.6
ANT2	324	180	-	100%	-	100%	-	100%	36	3.6
ANT3	324	180	562.21	30%	-	100%	391.1	30%	36	3.6
ANT4	324	180	562.21	30%	-	100%	223.49	60%	36	3.6
ANT5	324	180	562.21	30%	-	100%	-	100%	36	3.6
ANT6	324	180	321.26	60%	-	100%	-	100%	36	3.6
ANT7	324	180	321.26	60%	-	100%	391.1	30%	36	3.6
ANT8	324	180	321.26	60%	-	100%	223.49	60%	36	3.6
ANT9	324	180	-	100%	-	100%	391.1	30%	36	3.6
ANT10	324	180	-	100%	-	100%	223.49	60%	36	3.6

Table 2. Samples

Sample	Dimension **[mm]**	Number of test pieces
Cube	150 × 150 × 150	9 cubes for each recipe
Cube	100 × 100 × 100	6 cubes for each recipe
Prism	100 × 100 × 500	6 prisms for each recipe

For each recipe were made the following attempts:

- determination of density: 3 cubes with a side of 150 mm (the cubes from the determination of fc are used).
- determination of compressive strength on cubes, f_c; 3 cubes with a side of 150 mm that tested at 7 days, 14 days and 28 days.
- determination of tensile strength by bending, f_{ti}; 2 prisms of 100x100x500 mm, tested at 7 days, 14 days and 28 days.
- determination of tensile strength by splitting, f_{td}; it is performed on the pieces of prism tested for f_{ti};

2 Experimental Tests Performed on Concretes with Recycled Concrete Aggregates

2.1 Determination of Density

The density was determinate by formula:

$$\rho_a = (m_2 - m_1)/V \times 1000\,[\mathrm{Kg/m^3}] \tag{1}$$

where:

- ρ_a – apparent density, in kg/m^3;
- m_1 – mass of empty vessel, in kg;
- m_2 – the mass of the container filled with concrete, in kg;
- V – the volume of the vessel, in dm^3 (checked by filling with water);

The test pieces used to determine the density were cubes with a side of 150 mm. The samples were tried 7 days, 14 days and 28 days after pouring the concrete [24, 26] (Table 3).

Table 3. The density

Recipe	Medium density [kg/m^3]
BM1	2353.086
ANT1	2346.408
ANT2	2184.095
ANT3	2252.766
ANT4	2264.359
ANT5	2296.465
ANT6	2223.514
ANT7	2215.665
ANT8	2222.800
ANT9	2188.506
ANT10	2202.050

After determining the density, it resulted that this concrete is a heavy one with densities between 2184 kg/m3 and 2353 kg/m3. It is noticed that in the first part of the graph the densities are close to those of the control sample. The **ANT1** recipe recorded the highest density, being the recipe in which the 4–8 sort was replaced in a proportion of 100%. The recipe with the lowest density, **ANT2**, has replaced all sorts with waste resulting from concrete crushed in a proportion of 100%. Another recipe that has a high average density is the recipe **ANT5** with medium density 2296 kg/m^3, being the recipe in which the 4–8 and 8–16 sort was replaced in a proportion of 100% and 0–4 sort was replaced in a proportion of 30% (Fig. 2).

2.2 Determination of Compressive Strength

The compressive strength was determinate by formula:

$$f_{ck} = F_C/A_C\ [N/mm^2] \tag{2}$$

where:

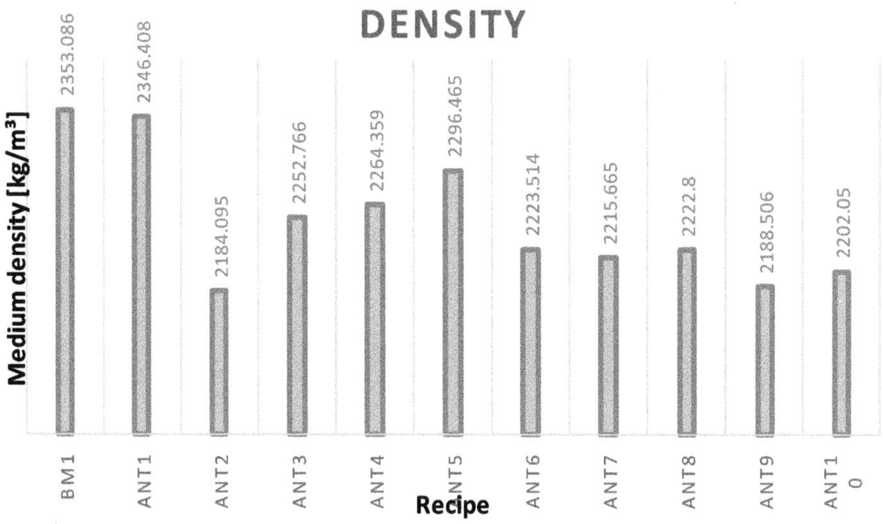

Fig. 2. Chart of density

- F_c - is the maximum force;
- A_c - is the area of the surface perpendicular to the direction of loading.

The cubes were tested perpendicular to the direction of pouring concrete at compression, with the help of a 300 kN hydraulic press, which achieves an evenly distributed load on the surface of the test tube [27, 28].

The press increases the unitary compressive efforts by approximately 5 $N/mm^2/sec$ (Fig. 3) (Table 4).

Fig. 3. Determination of compressive strength - mode of disposal

Table 4. Compressive strength-test result at 28 days

Recipe	Compressive strength f_{ck} [N/mm^2]
BM1	35.000
ANT1	25.086
ANT2	24.404
ANT3	21.289
ANT4	25.694
ANT5	24.537
ANT6	26.077
ANT7	25.837
ANT8	25.253
ANT9	24.877
ANT10	23.924

It was found that concrete with demolition waste have a very good behavior at the actions of the compressive force. The highest value of the compressive strength is given by the recipe **ANT6** ($f_{ck} = 26.077$ N/mm^2). The lowest resistance was recorded by the recipe **ANT3** ($f_{ck} = 21.289$ N/mm^2). Being concrete suitable for making belts and foundations.

We can observe from the graph (Fig. 4.) that the difference in compressive strengths between the control sample and the eco-concrete samples is approximately 9 N/mm^2. The ten eco-concrete mixtures exhibit similar compressive strengths ranging from 23 N/mm^2 to 26 N/mm^2. The main factor influencing strength is the quantity of fly ash used in the concrete recipes. It is important to note that the compressive strength of concrete is influenced by several factors, and the quantity of fly ash in the concrete mix is one of them. Fly ash, which is a residue obtained from the combustion of coal in power plants, can be used as a partial replacement for cement in concrete. It can contribute to improving the properties of concrete, including compressive strength. However, it should be mentioned that the compressive strength of concrete is also influenced by other elements, such as type and quality of aggregates and crushed procedures. The highest value 26.077 N/mm^2 of the compressive strength is given by the recipe **ANT6** being the recipe in which the 4–8 and 8–16 sort was replaced in a proportion of 100% and 0–4 sort was replaced in a proportion of 60%. The lowest value 21.289 N/mm^2 of the compressive strength is given by the recipe **ANT3** being the recipe in which the 0–4 and 8–16 sort was replaced in a proportion of 30% and 4–8 sort was replaced in a proportion of 100%. Based on the analysis, we observe the following aspect: the compressive strength is influenced by the sort of the recycled concrete aggregate (RCA) particles and the higher the replacement quantity. The concrete class C20/25 refers to a minimum characteristic compressive strength value of 20 N/mm2 and the classification of concrete according to SR EN [35, 36, 37] and according to this criterion, all the concrete recipes designed and studied belong to class C20/25.

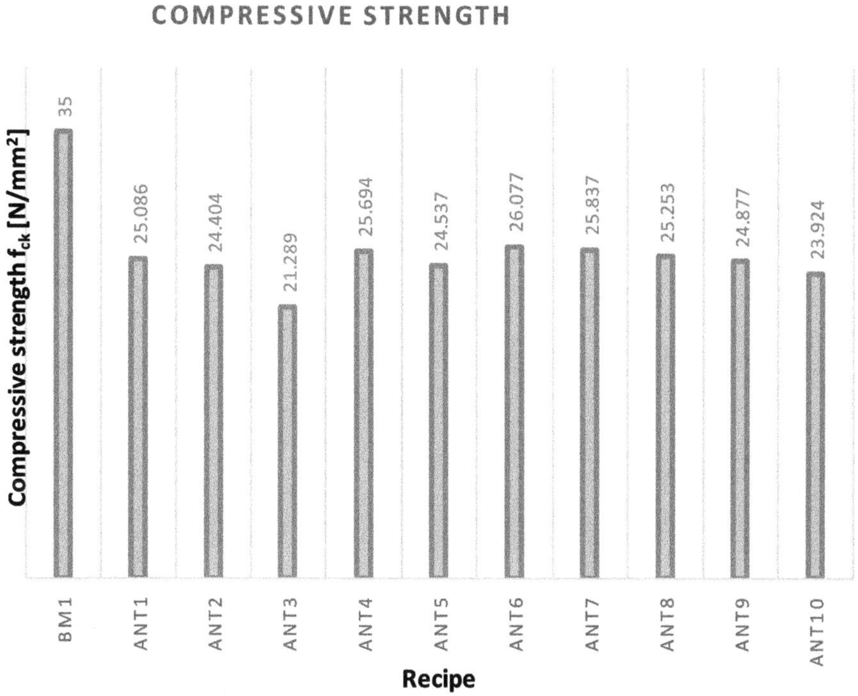

Fig. 4. Compressive strength-test result at 28 days

2.3 Determination of Tensile Strength by Bending

The tensile strength by bending was determinate by formula:

$$f_{ti} = 0.857x[(F_{ti}l)/(bh^2)]\ [N/mm^2] \tag{3}$$

where:

- F_{ti} – applied force, in N;
- l – the distance between the supports, in mm;
- b – the average width of the cross-section, in mm;
- h – the average height of the cross-section, in mm.

It is made on simple leaning prisms subjected to bending by applying a force concentrated in the middle of the opening [39]. The rupture occurs in the section of the moment of maximum bending, through a crack that appears in the large area (Fig. 5) (Table 5).

It was found that concrete with demolition waste has a very good behavior at the actions of the bending tensile strength. The highest value of the bending tensile strength it is giving by the recipe **ANT1** respectively **ANT6** ($f_{ti} = 1.66$ N/mm2). The lowest bending tensile strength was recorded by the **ANT2** and **ANT7** recipes ($f_{ti} = 1.30$ N/mm2).

Fig. 5. Determination of tensile strength by bending- mode of disposal

We can observe from the graph (Fig. 6.) that the difference in Tensile strengths by Bending between the control sample and the eco-concrete samples is approximately 0.55 N/mm^2. The ten eco-concrete mixtures exhibit similar compressive strengths ranging from 1.3 N/mm^2 to 1.66 N/mm^2.

Table 5. Tensile strength by bending- test result at 28 days

Recipe	Tensile strength by bending f_{ti} [N/mm^2]
BM1	2.21
ANT1	1.66
ANT2	1.30
ANT3	1.47
ANT4	1.59
ANT5	1.46
ANT6	1.66
ANT7	1.30
ANT8	1.48
ANT9	1.45
ANT10	1.53

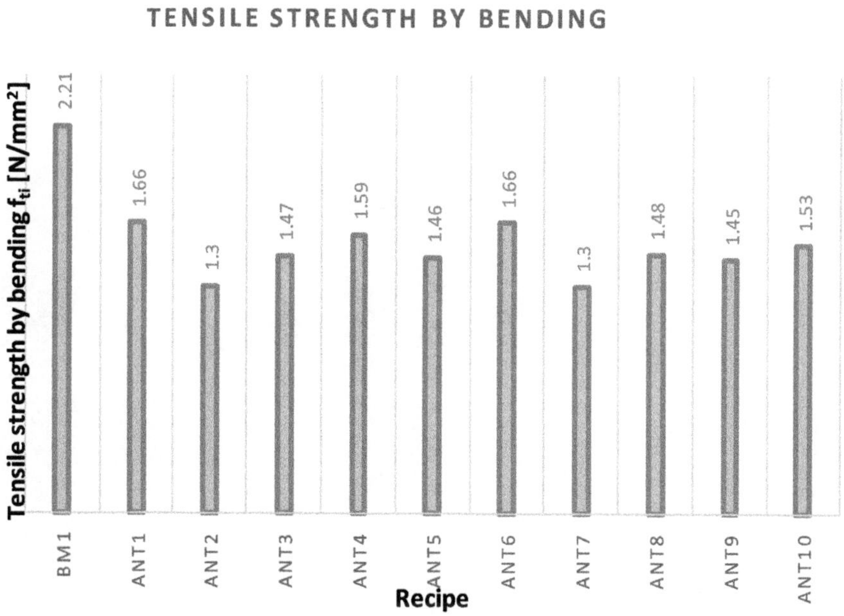

Fig. 6. Tensile strength by bending -test result at 28 days

2.4 Determination of Tensile Strength by Splitting

The tensile strength by splitting was determinate formula

$$f_{td} = 2P/\pi bh\ [N/mm^2] \tag{4}$$

where:

- P – the breaking force, in N;
- b–the mean width of the breaking cross-section, in mm;
- h– mean cross-sectional height, in mm.

A hydraulic press was used for the test, recording the value of the breaking force [29]. The force is applied perpendicular to the direction of casting of the composite material, continuously and evenly until breakage (Fig. 7) (Table 6).

Fig. 7. Determination of tensile strength by splitting-mode of disposal

Table 6. Tensile strength by splitting-test result at 28 days

Recipe	Tensile strength by splitting f_{td} [N/mm^2]
BM1	1.22
ANT1	0.90
ANT2	1.24
ANT3	1.06
ANT4	1.31
ANT5	1.36
ANT6	1.00
ANT7	1.13
ANT8	0.95
ANT9	1.01
ANT10	0.97

It was found that concrete with RCA has a very good behavior at the actions of the splitting test. The highest value of tensile strength by splitting is presented by the recipe **ANT5** (f_{td} = 1.36 N/mm2). The lowest resistance was recorded by the concrete sample with **ANT1** recipe (f_{td} = 0.9 N/mm2).

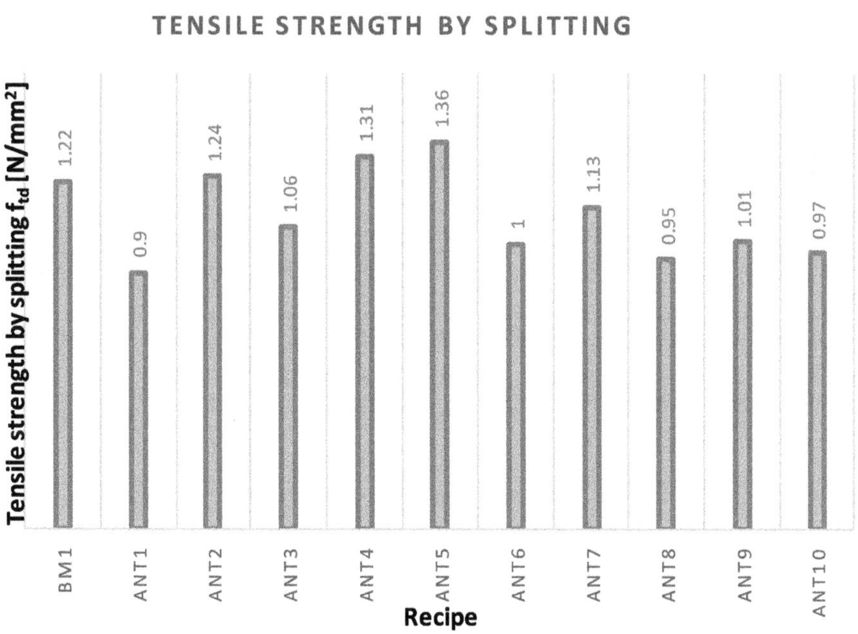

Fig. 8. Tensile strength by splitting -test result at 28 days

We can observe from the graph (Fig. 8.) that the difference in t ensile strengths by splitting between the control sample and the eco-concrete samples are very small, and in a few of cases the eco-concrete has better results, ex: **ATN2, ATN4** and **ATN5**.

3 Discussions and Results

After analyzing the results obtained from the experimental study, it was found that all the recipes recorded remarkable values. If we consider the compressive strength, and tensile strength by splitting the eco-concrete recipes maintained the concrete class C20/25. However, only the tensile strength by bending (f_{ti}) recorded values that changed the concrete class of these recipes (Table 7).

Table 7. Conclusions

Recipe	Medium density [kg/m^3]	Compressive strength f_{ck} [N/mm^2]	Tensile strength by bending f_{ti} [N/mm^2]	Tensile strength by splitting f_{td} [N/mm^2]
BM1	2353.086	35.000	2.21	1.22
ANT1	2346.408	25.086	1.66	0.90
ANT2	2184.095	24.404	1.30	1.24
ANT3	2252.766	21.289	1.47	1.06
ANT4	2264.359	25.694	1.59	1.31
ANT5	2296.465	24.537	1.46	1.36
ANT6	2223.514	26.077	1.66	1.00
ANT7	2215.665	25.837	1.30	1.13
ANT8	2222.800	25.253	1.48	0.95
ANT9	2188.506	24.877	1.45	1.01
ANT10	2202.050	23.924	1.53	0.97

The recipes, **ANT5** and **ANT6** obtained the best values after the attempts. These recipes have replaced the aggregate of sorts 4–8 and 8–16 in a proportion of 100% and sort 0–4 in a proportion of 30% and 60% respectively.

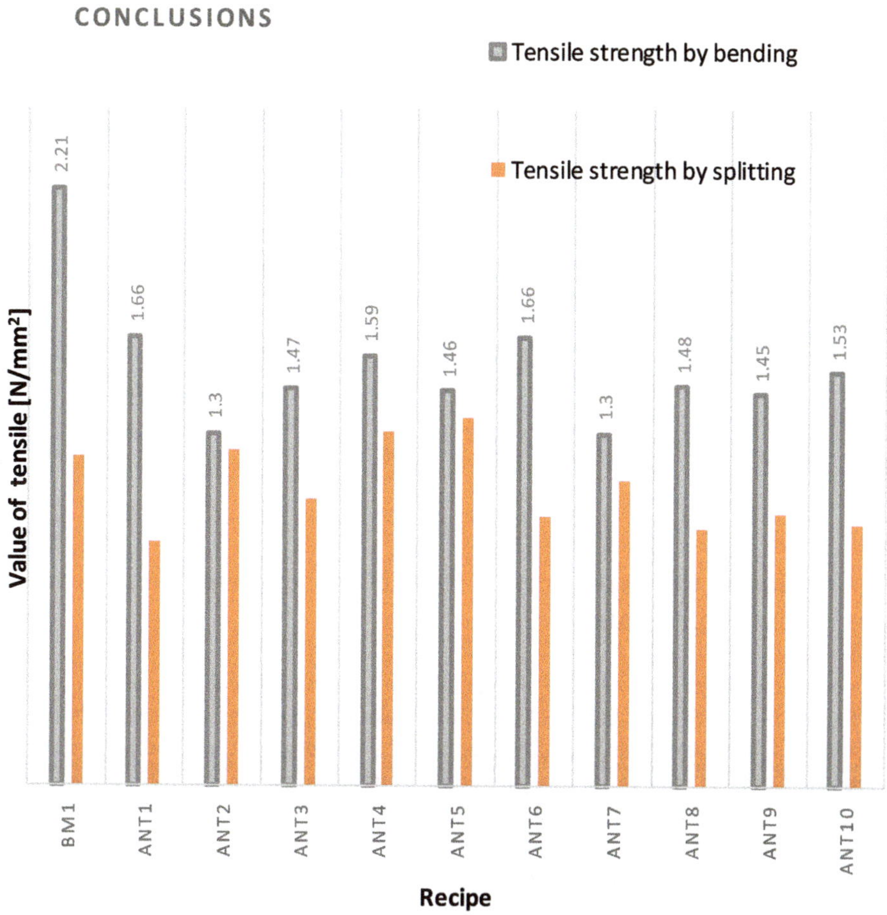

Fig. 9. Conclusions **tensile strength by bending** and **tensile strength by splitting**

We can observe from the graph (Fig. 9.) the recipes, **ANT5** obtained the best values after the attempts, the values of the **tensile strength by bending** and **tensile strength by splitting** are verry close that the value of **BM1** recipe, in this recipe have replaced the aggregate of sorts 4–8 and 8–16 in a proportion of 100% and sort 0–4 in a proportion of 30%. This draws our attention to the use of the recipe for other practical trials and uses.

A percentage calculation of the differences in values obtained gives us a much better perspective and conclusion on the analysis of the results. We can observe from the table (Table 8.) the mechanical residue values are close to those of the BM1 control recipe. The recipes **ATN5** snd **ATN6** recorded the best values.

Table 8. Conclusion in procent

Recipe	Medium density [kg/m^3]	P% values decreased	f_{ck} [N/mm^2]	P% values decreased	f_{ti} [N/mm^2]	P% values decreased	f_{td} [N/mm^2]	P% values decreased
BM1	2353.08		35.000		2.21		1.22	
ANT1	2346.40	0.28	25.086	28.31	1.66	24.88	0.90	26.23
ANT2	2184.09	7.18	24.404	30.28	1.30	41.17	1.24	-1.64
ANT3	2252.76	4.26	21.289	39.17	1.47	33.48	1.06	13.11
ANT4	2264.35	3.77	25.694	26.57	1.59	28.05	1.31	-7.38
ANT5	2296.46	2.41	24.537	29.88	1.46	33.94	1.36	-11.47
ANT6	2223.51	5.51	26.077	25.48	1.66	24.88	1.00	18.03
ANT7	2215.66	5.84	25.837	26.17	1.30	41.17	1.13	7.37
ANT8	2222.80	5.54	25.253	27.86	1.48	33.03	0.95	22.13
ANT9	2188.50	6.99	24.877	28.91	1.45	34.39	1.01	17.21
ANT10	2202.05	6.42	23.924	31.66	1.53	30.77	0.97	20.49

4 Conclusion the Experimental Results on Concretes with Recycled Concrete Aggregates

In conclusion, all recipes obtained very good results in mechanical tests, being suitable for the realization of concrete blocks, used to make masonry, can be used for building foundations, lightweight structural elements, slabs and floors: It is suitable for casting slabs and floors in residential or commercial constructions, where sufficient strength is needed for normal traffic and moderate loads, precast elements: can be used for manufacturing larger precast elements such as precast beams, columns, or walls, similar with another studies with eco-concrete, wastes and sorts of RCA [30–32].

The contribution that these recipes bring to the construction industry is significant, even though we have studied the mechanical resistances of these concretes, they contribute significantly to environmental protection by reducing CO_2 emissions and using RCA demolition waste.

Other conclusions: the sorts of RCA refers to the size of the aggregate particles. When RCA particles are coarser, they provide better interlocking and filling of voids in the concrete mixture, resulting in improved strength. RCA can also enhance the bond between the aggregate and the cement paste, leading to higher compressive strength. Moreover, increasing the quantity of RCA in the concrete mix can contribute to a higher compressive strength. This is because RCA, when used as a replacement for natural aggregates, can have a higher strength compared to traditional aggregates. By incorporating a larger amount of RCA, the overall strength of the concrete can be increased. It is important to note that while the coarseness and quantity of RCA can positively impact compressive strength, other factors such as the quality of RCA, water-cement ratio, curing conditions, and mix design also play significant roles.

References

1. Proske, T., Hainer, S., Rezvani, M., Graubner, C.-A.: Eco-friendly concretes with reduced water and cement content – Mix design principles and application in practice. Constr. Build. Mater. **67**, 413–421 (2014)
2. Garg, C., Jain, A.: Green concrete: Efficient and eco-friendly construction materials. Int. J. Res. Eng. Technol **2**(2), 259–264 (2014)
3. Yu, R., Spiesz, P., Brouwers, H.J.H.: Static properties and impact resistance of a green ultra-high performance hybrid fibre reinforced concrete (UHPHFRC): experiments and modeling. Constr. Build. Mater. **68**, 158–171 (2014)
4. Kurda, R., de Brito, J., Silvestre, J.D.: CONCRETop method: Optimization of concrete with various incorporation ratios of fly ash and recycled aggregates in terms of quality performance and life-cycle cost and environmental impacts. J. Cleaner Prod. **226**, 642–657 (2019)
5. Suhendro, B.: Toward green concrete for better sustainable environment. In: Procedia Engineereing, 2nd International Conference on Sustainable Civil Engineering Structures and Construction Materials, vol. 95, pp. 305–320 (2014)
6. Simion, I.M., Ghinea, C., Maxineasa, S.G., Taranu, N., Bonoli, A., Gavrilescu, M.: Ecological footprint applied in the assessment of construction and demolition waste integrated management. Environ. Eng. Manag. J. **12**(4), 779–788 (2013)
7. IEA Homepage. https://iea.blob.core.windows.net/assets/CO2Emissionsin2022.pdf
8. Berra, M., Mangialardi, T., Paolini, A.E.: Reuse of woody biomass fly ash in cement-based materials. Constr. Build. Mater. **76**, 286–296 (2015)
9. Bărbuţă, M., Harja, M.: Properties of fiber reinforced polymer concrete, Buletinul Institutului Politehnic din Iaşi, Tomul LIV (LVIII), Fasc. 3, Sectia Construcţii. Arhitectura (2008)
10. Mittermayr, F., et al.: Sulfate resistance of cement-reduced eco-friendly concretes. Cement Concr. Compos. **55**, 364–373 (2015)
11. Liikanen, M., Grönman, K., Deviatkin, I., Havukainen, J., Horttanainen, M.: Construction and demolition waste as a raw material for wood polymer composites assessment of environmental impacts. J. Cleaner Prod. **225**, 716–727 (2019)
12. Blaisi, N.I.: Construction and demolition waste management in Saudi Arabia: current practice and roadmap for sustainable management. J. Cleaner Prod. **221**, 167–175 (2019)
13. XinpengWang, R., et al.: Recycled coarse aggregates from pelletized unused concrete for a more sustainable concrete production. J. Cleaner Prod. **219**, 424–432 (2019)
14. Badur, S., Chaudhary, R.: Utilization of hazardous wastes and by-products as a green concrete material through S/S process: a review. Rev. Adv. Mater. Sci. **17**, 42–61 (2008)
15. Thandavamoorthy, T.S.: Wood waste as coarse aggregate in the production of concrete. Eur. J. Environ. Civ. Eng. **20**(2), 125–141 (2016).https://doi.org/10.1080/19648189.2015.1016631
16. Soares, D., de Brito, J., Ferreira, J., Pacheco, J.: Use of coarse recycled aggregates from precast concrete rejects: mechanical and durability performance. Constr. Build. Mater. **71**, 263–272 (2014)
17. Durga Sindhu Sree1, K., Chandra Sekhar, T.: Impact of partial utilization of the recycled aggregates on the characteristics of the concrete. Int. J. Adv. Res. Trends Eng. Technol. (IJARTET) 5(10) (2018)
18. Okafor, F.O.: Waste concrete as a source of aggregate for new concrete. Niger. J. Technol. **29**(2), 5–11 (2010)
19. Shakeeb, S., Farhanuddin, S., Sameer, S., Khan, M.: Experimental investigation on concrete by replacing natural coarse aggregate partially with recycled coarse aggregate (RCA) & addition of natural coir fibres. Int. Res. J. Eng. Technol. (IRJET) (2017)
20. Tripathi, C., Chandraul, K., Singh, M. K..: Experimental analysis on split tensile strength of concretewith twd and RCA. Int. Res. J. Eng. Technol. (IRJET) (2019)

21. Aminur Rahman, M., Imteaz, M.A., Arulrajah, A., Piratheepan, J., Disfani, M.M.: Recycled construction and demolition materials in permeable pavement systems: geotechnical and hydraulic characteristics. J. Cleaner Prod. **90**, 183–194 (2015)
22. Cakir, O.: Experimental analysis of proprieties of recycled coarse aggregate (RCA) concrete with mineral additives. Constr. Build. Mater. **68**, 17–25 (2014)
23. Sai Krishna Mohan Chowdary, V.: Experimental study on strength behaviour of self compacting concrete using recycled aggregates. Int. J. Innovative Res. Adv. Eng. (2015)
24. SR EN 1992–1–1 – Design of concrete structures. General rules and rules for buildings
25. SR EN 12390 – 1 – Testing hardened concrete. Shape, dimensions and other requirements for specimens and moulds
26. SR EN 12390 – 2– Testing hardened concrete. Making and curing specimens for strength tests
27. SR EN 12390 – 3 – Testing hardened concrete. Compressive strength of test specimens
28. SR EN 12390 – 5 – Testing hardened concrete. Flexural strength of test specimens
29. SR EN 12390 – 6 – Testing hardened concrete. Tensile splitting of test specimens
30. Ligia Georgiana Hanuseac (Ghalayini), Dumitrescu, L., Barbuta, M., Baran, I., Bejan, G.: Eco-Mechanical Index of Lightweight Concrete Mixtures with Recycled Materials. In: 13th International Conference Interdisciplinarity in Engineering (INTER-ENG 2019), Procedia Manufacturing, pp 667–674 (2020). https://doi.org/10.1016/j.promfg.2020.03.095
31. Hanuseac, L., Barbuta, M., Bejan, L., Rosu, R., Timu, A.: Experimental Study on Hollow Blocks with Wastes. In: 14th International Conference Interdisciplinarity in Engineering, Proceedings, vol. 63, p. 79 (2020). https://doi.org/10.3390/proceedings2020063079
32. Timu, A., Barbuţă, M.: Green concrete with fly ash and plastic waste, Buletinul Institutului Politehnic, Iaşi, nr. 3 (2018)
33. Cătălina-Mihaela Helepciuc (Grădinaru) (2016) Betonul ecologic – A alternativă viabilă pentru protejarea mediului înconjurător, Creaţii universitare, 66–71

Characterization of Ceramic Waste Plaster Mortars by Advanced Analytical Methods: NMR and SEM-EDX

Anamaria Zaharie[1](✉), Monica Luminița Pleșa[1], Lucia Daniela Manea[1], Fechete Radu[1], Elena Jumate[1], Claudiu Aciu[1], and Dorina Simedru[2]

[1] Technical University of Cluj-Napoca, Memorandum Street, No. 28, Cluj-Napoca, Romania
annazaharie@yahoo.com

[2] INCDO-INOE2000, Subsidiary Research Institute for Analytical Instrumentation Cluj-Napoca, Donath Street, No. 67, Cluj-Napoca, Romania

Abstract. The practice of reusing ceramic waste is an environmental, economic and technical advantage. Brick and masonry waste contributes to the production of sustainable materials whose physical and mechanical performance can be improved. According to recent studies on plastering mortars with ceramic waste, by replacing up to 25% sand with brick waste, respectively 15% brick powder waste with cement, does not negatively influence the mechanical strengths. The present research work proposes to investigate by non-invasive SEM-EDX and NMR methods, plaster mortars with 35% ceramic waste for partial replacement of aggregates, respectively mortars with 30% ceramic waste partially replacing cement. The aim of these investigations is to examine the topographic surface, the elemental composition of the samples as well as the characteristics of pores and connections within the structure of the ceramic waste mortar samples. The results obtained identify an increase in the structural performance of the recipe with 35% brick waste, as well as relative pore water content. These investigations provide us with a com-prehensive view of the characteristics of mortars with technical performance within the limits of the standards in force and whose aim is to reduce CO_2 emissions and the consumption of resources as much as possible and to obtain a sustainable material.

Keywords: Ceramic waste · Plastering mortar · NMR · SEM-EDX

1 Introduction

The economic development of recent decades and especially the impact of this evolution on the environment and energy consumption, has made it necessary to change the approach to all processes/steps in the construction sector. Reducing the use of energy and natural resources and the reuse of waste are mandatory objectives of the new approach to construction. According to directives adopted at European level, in 2020 non-hazardous construction and demolition waste must be recycled for reuse and recovery to a minimum of 70% [1]. In 2021, 37% of the resulting carbon emissions are produced by the construction sector [2].

I. Tuns et al. (Eds.): CIBv 2023, LNCE 510, pp. 471–489, 2024.
https://doi.org/10.1007/978-3-031-60765-3_37

Analises using sources like Scopus presents key aspects of ceramic waste (CW) interest (see Fig. 1).

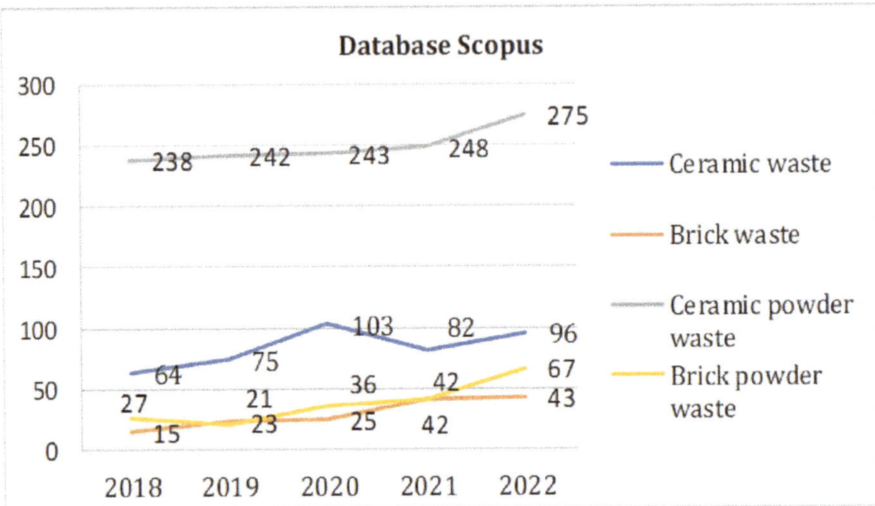

Fig. 1. Number of scientific publications indexed in the Scopus database [3].

These analyses provide insights into research trends and publication efforts. Applications of ceramic waste, such as in plaster mortars; contribute to environmental sustainability by reducing landfill impact. Bibliometric analysis aids in evaluating research impact, offering valuable information for strategic development and efficient policies in ceramic waste management. Figure 1 provides a comprehensive view of the potential of using CW, highlighting research gaps and sustainable material options in the field of CW. The grey graph also indicates the interest of engineering and materials science researchers in the use of brick powder waste.

According to technical studies, replacing up to 35% of the aggregate with brick waste (BW) does not negatively influence the mechanical strength of waste ceramic mortars [4]. The construction industry sector is responsible for 9% of carbon dioxide emissions [5]. Waste in the form of brick powder (BPW) has the potential to increase mechanical strength [6]. The replacement of cement with BPW has the potential to reduce the amount of carbon dioxide emissions.

The use of aggregates from construction demolition (such as masonry waste) has significant potential to replace natural aggregates, helping to conserve natural resources and reduce landfill space.

In order to use ceramic waste on a large scale in mortar or concrete as a substitute for aggregates, it is necessary to understand the behavior over time of these new types of construction materials. Currently there are various studies analyzing the physical-mechanical characteristics of mortars with masonry waste aggregates.

This paper aim is to investigate the characteristics of plaster mortars with ceramic waste by non-invasive methods NMR and SEM-EDX, at microscopic level in order to observed the influence of CW in the compositions of plastering mortars and the evolution of porous structure of studied mortars in order to identify recipes whose performance comply current standards.

2 Materials and Recipes

2.1 Materials

Classical materials have been studied in order to obtain new plaster mortar recipes. The binder used is Portland CEM I 52.5R Cement, according to SR EN 197-1:2011 [7], the aggregate is river sand with a grain size between 0–4.00 mm. The ceramic wastes used are brick wastes (BW) and brick masonry wastes (BMW) under different grain sizes. In order to replace the river sand, BW and BMW were used with a grain size of 0–4.00 mm and BPW was used to partially replace the Portland cement, see Fig. 2.

Fig. 2. Ceramic waste studied a) Brick waste (BW) 0–4,00 mm, b) Brick masonry waste (BMW) 0–4,00 mm, c) Brick powder waste (BPW).

2.2 Recipies

Five plaster mortar recipes were made, of which four with ceramic waste. 30% BPW was proposed for cement replacement and 35% BW and 35% BMW for partial sand replacement used in the CSIV standard plaster mortar recipe. A recipe was also made in which the sand was replaced with 20% BW and the cement with 15% BPW. Table 1 shows the proposed recipes for this experimental study.

Table 1. Recipes of plastering mortars

Recipe	Note	Materials				
		Cement [kg]	Sand [kg]	CW [kg]	Water [l]	Additive [l]
CSIV Plastering mortar	RS	385	1541	0	217.15	0
Plaster mortar with 35% BW as a substitute for aggregates	35BW	385	1001.35	533.35	258.57	0
Plaster mortar with 30% BPW as a cement substitute and additive	30BPW-A	269.50	1541	115.50	200	0.19
Plaster mortar with 20% BW as aggregate substitute and 15% BPW	35MIXT	288.75	1005.70	365.95	203.40	0
Plaster mortar with 35% BMW as an aggregate substitute	35BMW	385	1001.35	533.35	250.25	0

The proposed recipes were cast in the Construction Materials Laboratory of the Faculty of Civile Engineering from Cluj-Napoca according to SR EN 1015-2-2001 [8], in order to obtain normal consistency of fresh mortar according to SR EN 1015-3:2001 [9]. The fresh mortar was kept in metal prism of size 40 × 40 × 160 mm until testing. A humidity of (65 ± 5)% and a temperature of (20 ± 4) °C was ensured for the process of hardening the samples.

3 Methods

In order to observe the influence of ceramic waste on the composition of mortars, determinations were made in order to identify the respective mechanical characteristics, non-invasive determinations such as Nuclear Magnetic Resonance (NMR) and Scanning Electronic Microscopy coupled with Dispersion X-ray Spectroscopy (SEM-EDX).

The determination of the bending tensile strength, the compressive strength and the resistance to the support layer were carried out according to the standards in force [10, 11]. In order to highlight the evolution of the mechanical strengths of the mortars studied, the tests were carried out at different time intervals, namely 3, 7 and 28 days. The results were disseminated in other scientific articles published at various international conferences according to the bibliography [4, 6, 12].

Nuclear Magnetic Resonance (NMR) is a performant investigation increasingly used in the analysis of materials such as mortars [13, 14].

Nuclear magnetic resonance has proven to be one of the most widely used methods for characterizing porous materials, especially for characterizing pore size and distribution. It is based on the fact that water molecules enter the pores and can diffuse within them.

Thus, water molecules exhibit a random Brownian motion and, during this random motion they can undergo various spin-network relaxation processes (Longitudinal

Relaxation Time-T_1) and spin-spin relaxation processes (Transverse Relaxation Time-T_2). The T_2 relaxation time reflects the rate of decrease of the MRI signal, caused by diffusion and disordered phenomena in the studied system.

Generally, the relaxation time T_2 is less than the T_1 relaxation time, which means that the NMR signal fades faster during T_2 than during T_1. The T_2 time depends on the physical and chemical properties of the molecules or atoms that make up the studied system. The size and mobility of the molecules can influence the T_2 relaxation times - the higher the mobility or faster diffusion in the solution, the shorter the T_2 relaxation time [15].

The phenomenon of nuclear relaxation is influenced by the size of the pores, by the ratio of their surface area to their volume, and the spin-spin relaxation times, usually called T_2, can be easily measured by a fast technique called CPMG (Car-Purcell-Meiboom-Gill). This technique can achieve a large number of echoes (usually between 1000 and 3000) in a single measurement, known as "singleshot measurement", making the process fast and efficient.

The CPMG technique uses a series of radiofrequency pulses and a set of multiple echoes to produce MRI signals. This process can suppress signal line deformation effects caused by magnetic field inhomogeneities and increase the signal-to-noise ratio. This technique provides us with information on molecular structure, molecular mobility, and molecular interactions in solution [15].

The analysis of the samples by NMR method was carried out in the Physics Laboratory of the Faculty of Materials and Environmental Engineering using the equipment in Fig. 3.

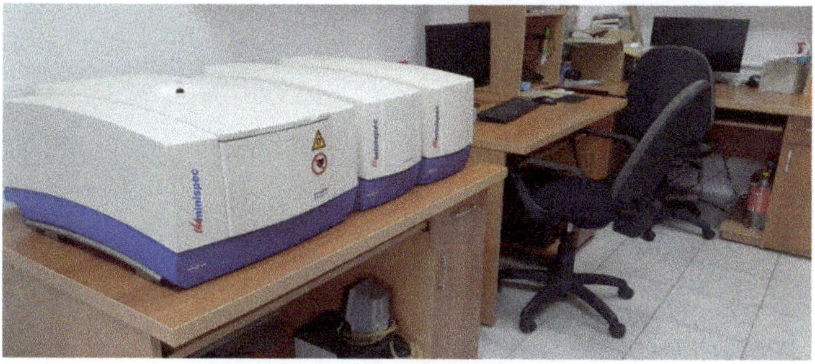

Fig. 3. NMR spectrometer used in mortar analysis.

Regarding NMR testing, a first set of investigations was carried out on plaster mortars with ceramic waste at the age of 3, 7 and 28 days after casting. A sample was taken from each mortar recipe and inserted into the 9 mm diameter cylindrical specimen and subjected to testing.

Following testing at 28 days, RS, 35BW, 30BPW-A, 35MIXT, 35BMW mortar samples were immersed in water and kept for 1 month under hydration. The purpose of this dive is to identify the dehydration process of the samples tested in the first 28 days, respectively the filling of the small, medium and large pores following the dive, see Fig. 4.

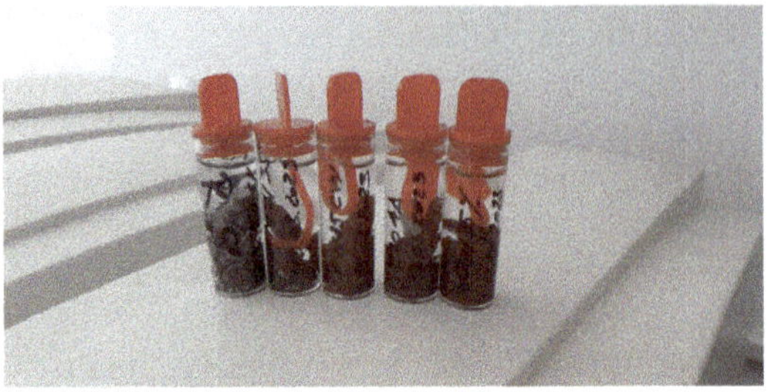

Fig. 4. Hydration of mortar samples with water and testing 1 month after immersion

SEM-EDX is an analytical technique to examine the topographic surface and the elemental composition of a sample [16, 17]. The analysis is performed by high-resolution images that are obtained by detecting the rays reflected by the analyzed sample or produced inside it [18].

The experimental study on SEM-EDX analysis was carried out within INCDO-INOE 2000, the Research Institute for Analytical Instrumentation Branch in Cluj-Napoca. The scanning electron microscope (VEGA3 SBU, Tescan) coupled with an energy dispersion X-ray spectroscope (Quantax eds, Bruker) was used for the study (Fig. 5). EDX data was processed using BrukerEsprit 2.0 software.

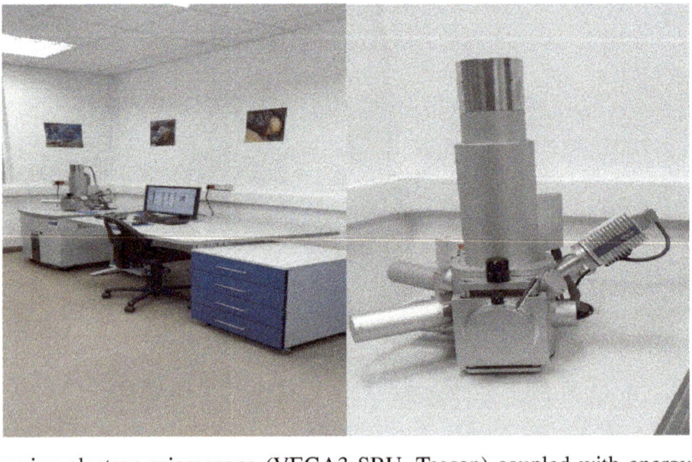

Fig. 5. Scanning electron microscope (VEGA3 SBU, Tescan) coupled with energy dispersion X-ray spectroscope (Quantax eds, Bruker)

The proposed recipes were studied 42 days after casting. Analyses were performed on pieces of sample of approximately 5 mm^2 which were bonded to a sample holder by means of a double-sided carbon strip, which is why carbon was excluded from the EDX analysis results (Fig. 6).

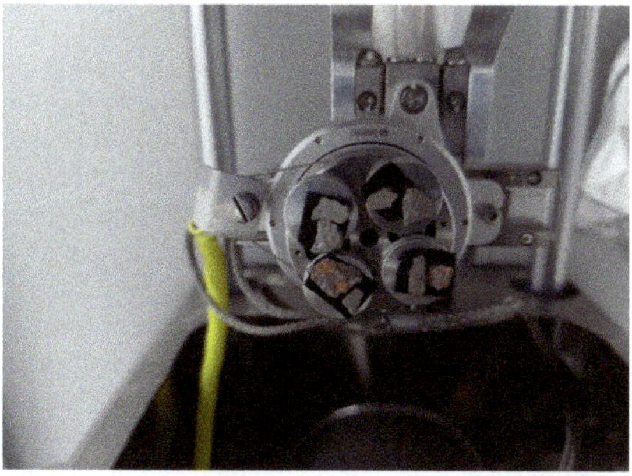

Fig. 6. Deposition of mortar samples on the sample holder of the scanning electron microscope for SEM-EDX analysis

4 Results and Interpretations

4.1 NMR Analysis

Figure 7 there are comparatively presented the distributions of T_2, measured for all five studied recipes compared 3 days after casting.

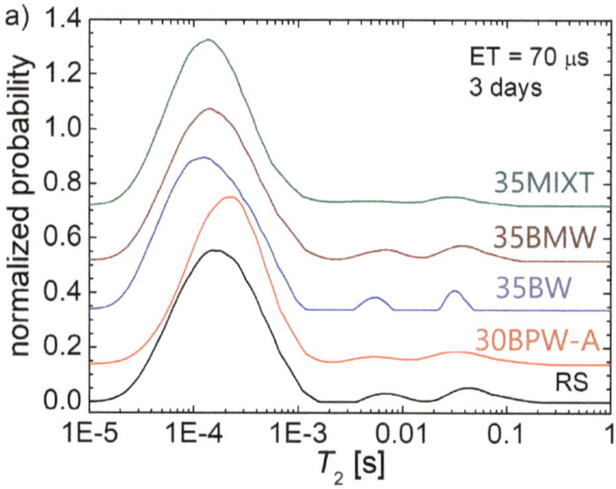

Fig. 7. Distribution of T_2 relaxation times of the five samples analyzed 3 days after casting.

The highest homogeneity of the distribution of T_2 is observed 3 days after casting, when the sample for which we have a replacement of the aggregate with 35% full brick waste (35BW) is noticed. Graphically represented with a blue color, the largest amount of water is found as bound water but at fairly low values of T_2 of ~(100–120) μs and whose pore structure is well formed even 3 days after preparation. The small and medium pores are quite well separated, which indicates that this structure is quite homogeneous. Additionally, this structure of small and medium pores is observed in the RS recipe as the amplitude is not as large and the width of these two peaks corresponding to the water in the small and medium pores is greater, fact that indicates its heterogeneity.

From the point of view of stiffness, if this is due to bound water, it can be assumed that the least rigid mortar sample 3 days after preparation would be the one in which cement was replaced with brick powder waste and superplasticizing additive was added (30BPW-A), represented by the red curve of the graph. This feature may also be due to the fact that the additive was used in the recipe, which instead of leading to an additional stiffening of the sample, probably led to a slight increase in its own mobility. According to Fig. 7 It is found that samples 35BW and 35MIXT show quite high compactness.

The evolution over time of mortars 7 days after preparation, shown in Fig. 8 highlights a porous structure best observed for sample 35BMW (brown graphical curve).

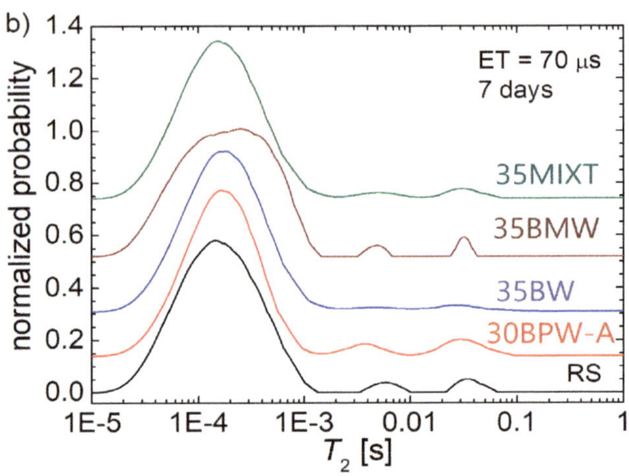

Fig. 8. Distribution of T_2 relaxation times of the five samples analyzed 7 days after casting.

The RS sample also shows a good porous structure followed by the 30BPW-A sample. Surprisingly, the sample for which the aggregate was replaced with 35% brick waste (35BW) no longer shows this porous structure. In addition, the sample for which the aggregates and cement were replaced with brick waste (35MIXT) shows a fairly large amount of extremely small pores or two types of bound water 7 days after preparation.

The final samples, 28 days after preparation, show the greatest varieties of rigidity and homogeneity, (See Fig. 9).

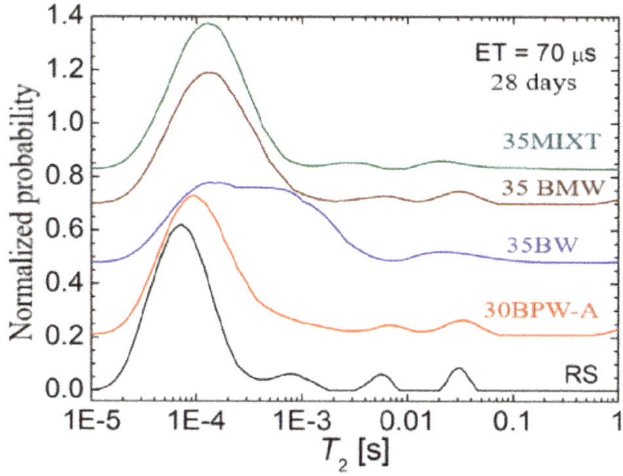

Fig. 9. Distribution of T_2 relaxation times of the five samples analyzed 28 days after casting.

Thus, the RS sample indicates a large amount of water as bound water, located on T_2 time below 100 μs, which indicates an increased compactness of this sample and which can be considered the most complex sample analysed at 28 days. The RS mortar structure is a porous structure in which we have small to very small pores, small pores and medium pores.

This pore structure can be observed as well in sample 30BPW-A, in which we have peaks corresponding to water from medium and small pores and the existence of extremely small but heterogeneously distributed pores can also be interpreted. The brick masonry waste mortar sample (35BMW) shows this structure of medium and small pores, relatively well represented for the 35MIXT sample but not so obvious.

In the case of the 35BW sample, it is observed that the small and medium pores have migrated and they are so heterogeneous that it is no longer possible to differentiate them (Fig. 8), in addition, a large amount of water located in the very small pores that are heterogeneous distributed. Thus, the less rigid sample in terms of bound water is 35BW and the more rigid RS.

Following 1 month hydration of RS, 35BW, 30BPW-A, 35MIXT and 35BMW plaster mortar recipes, it is observed that the T_2 relaxation time distributions change as a result of filling all types of pores with hydration or water products.

Thus, if the 28-day analysis of the recipes presented in Fig. 9 with the 56-day analysis presented in Fig. 10, it is noted that all the components of the mortar have undergone a hydration process but at the same time we have a process of dehydration of small, medium and large pores. Through the hydration process, these pores are filled although we expect that the hydration water of the recipe components, which is reflected by the appearance of a main drip at low values of T_2, will not undergo great changes.

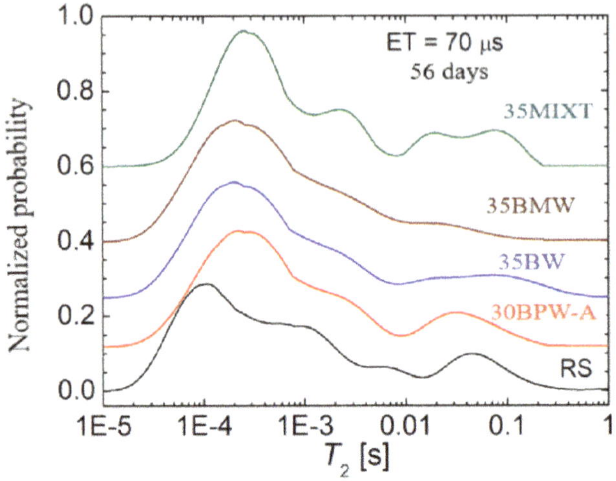

Fig. 10. Distribution of T_2 relaxation times of the five samples analysed 56 days after casting

Instead, the distribution of T_2 that reflects the amount of water in small, medium and large pores undergoes essential changes. Thus, if we look at the black colour curve that represents the distribution of T_2 measured for the RS recipe, it is observed that a slight displacement of the main peak occurs. Located at ~(70–80) μs according to Fig. 9 it is displaced following hydration at ~100 μs according to Fig. 10, although a very large amount of water fills the small pores as seen from the shoulder occurring at ~1 ms, whose value is close to that of the sample analysed at 28 days, the drop of the sample analysed at 56 days is wider according to the black colour curve in Fig. 10.

The amount of water in the middle pores does not increase significantly, instead a much larger amount of water is observed in medium to large pores as seen from a drip with a much larger integral area, located at ~(40–50) ms and expanding.

Therefore, we can conclude that we have components of T_2 with a fairly long relaxation time, ~(200–300) ms. This is not seen from Fig. 10, where the black curve for the RS recipe indicates that the large pores at 28 days were completely dehydrated, highlighted by hydration at two months.

By graphical representation of the red curve in Fig. 10, in the case of recipe 30BPW-A, it is observed that the small pores are hydrated. The displacement of the main drip at ~(200–300) μs versus 100 μs indicates a slight increase in the mobility of the rigid dynamic components in this recipe. The biggest change is given by the appearance of an extremely wide drip at values of ~(1–2) ms, which indicates the presence of small pores, pores that have not been highlighted much for the 28-day sample, instead the average pores are no longer as visible, probably the drip corresponding to such pores has become wide, so the amount of water in the average pores is approximately the same for both test periods.

In the case of the 30BPW-A recipe, a wide drip located at about 30 ms is observed, corresponding to medium to large pores, with a slight probability of even having large pores, thus, this drip is extended from ~10 ms to ~300 ms, which indicates a rather narrow heterogeneity of these types of pores. This distribution is quite similar to the one obtained in the case of the RS recipe.

In the case of the 35BW recipe, small pores are present as a shoulder, but compared to the distribution measured at 28 days, it seems that the relative value of the water content in these pores is no longer as large. Medium and medium to large pores are observed as two not very large but extremely wide peaks of amplitude, which indicates that the presence of solid brick waste probably leads to the largest increase in inhomogeneities at the microscopic level for this recipe.

From Fig. 9, the brown curve of the 35BMW recipe, indicates the appearance of the main drip at ~200 μs, which is almost the same localized as the other two recipes 30BPW-A and 35BW. Extremely wide peaks located at T_2 values on the order of milliseconds are observed. In this case, the pores may be associated with small to medium pores of the order of tens of milliseconds, or they may be associated with medium and medium to large pores. Their amplitude is seen to be quite small but instead they are extremely wide, which indicates that in the case of the 35BMW recipe the heterogeneity is the lowest. There is some inhomogeneity due to the use of this type of waste that leads to the formation of several types of pores from small, medium and large.

Apart from the RS recipe, the most complex structure analysed following hydration is the 35MIXT control recipe, the green colour curve in Fig. 10, for which we have a wide drip located at ~(200–300) ms and which represents the water related to the mortar components. A drip is also identified that may be associated with small pores located at ~(2–3) ms, best defined if we compare all the recipes shown in Fig. 10. Also, in Fig. 10 we can observe the average pores represented by a drip located at ~(15–20) ms, as well as the medium to large pores represented by a drip located at ~(80–100) ms.

4.2 SEM-EDX Analysis

SEM-EDX investigations of RS, 35BW, 30BPW-A, 35MIXT and 35BMW plaster mortars are aimed at obtaining information on the topology and elemental composition of the samples related to these mortars.

From Fig. 11a. it can be seen that the surface of the RS sample at small amplifications (~58x) is compact, slightly inhomogeneous. The surface of the sample is inhomogeneous, the appearance of pores of different sizes can be observed.

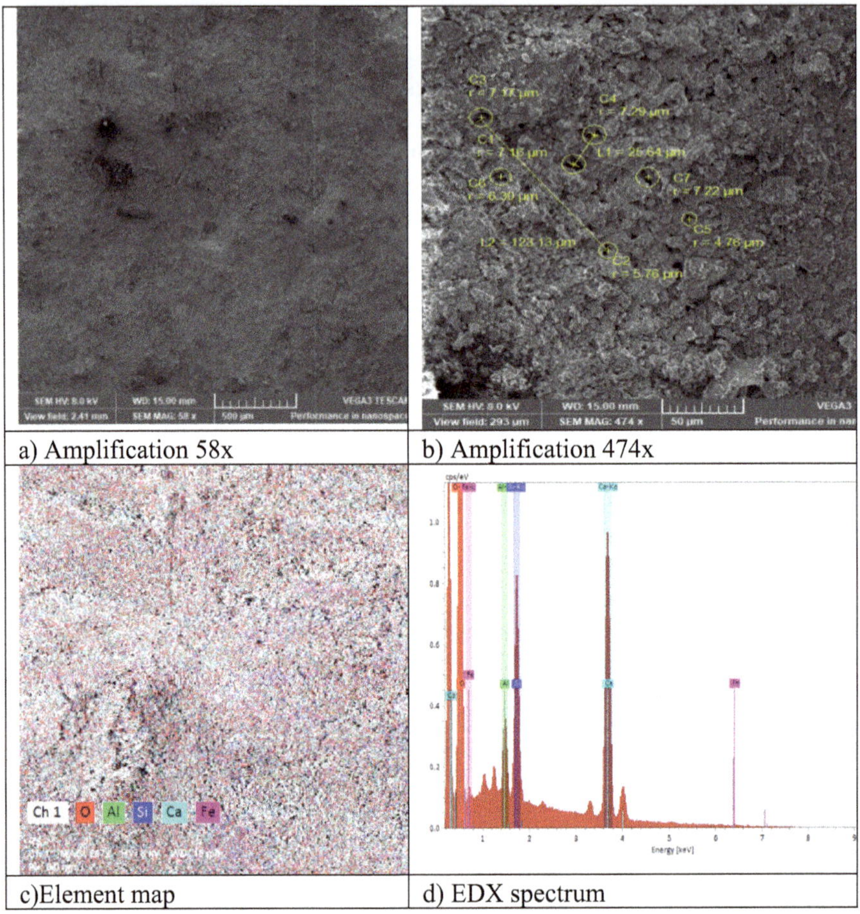

Fig. 11. SEM images obtained for the RS sample in the amplification range a) 58x- b) 474x, c) Element map, d) EDX spectrum.

According to Fig. 11b the pore radius varies between ~4.76 and ~7.29 μm and the distance between them varies between ~25.64 and ~123.13 μm. The majority elements present on the surface of the analysed sample are O, Ca and Si according to Fig. 11c and Fig. 11d.

Table 2. Elementary concentration of RS sample

Chemical Element	Al	Ca	Fe	O	Si
RS recipe	1.67	35.23	2.42	52.80	7.88

The Table 2 presents the chemical concentration of the sample. The Ca/Si ratio is 4.47. According to the literature, the value of the Ca/Si ratio for C-S-H in cement is

1.2 and 2.3, [17]. A higher Ca/Si ratio value like the one observed in the RS sample shows the presence in the sample of other Ca-based phases, which can be observed in the element distribution map in the areas where the Ca color is predominantly identified (See Fig. 11c).

The 35BW plaster mortar sample was first analyzed at small amplifications (33x).

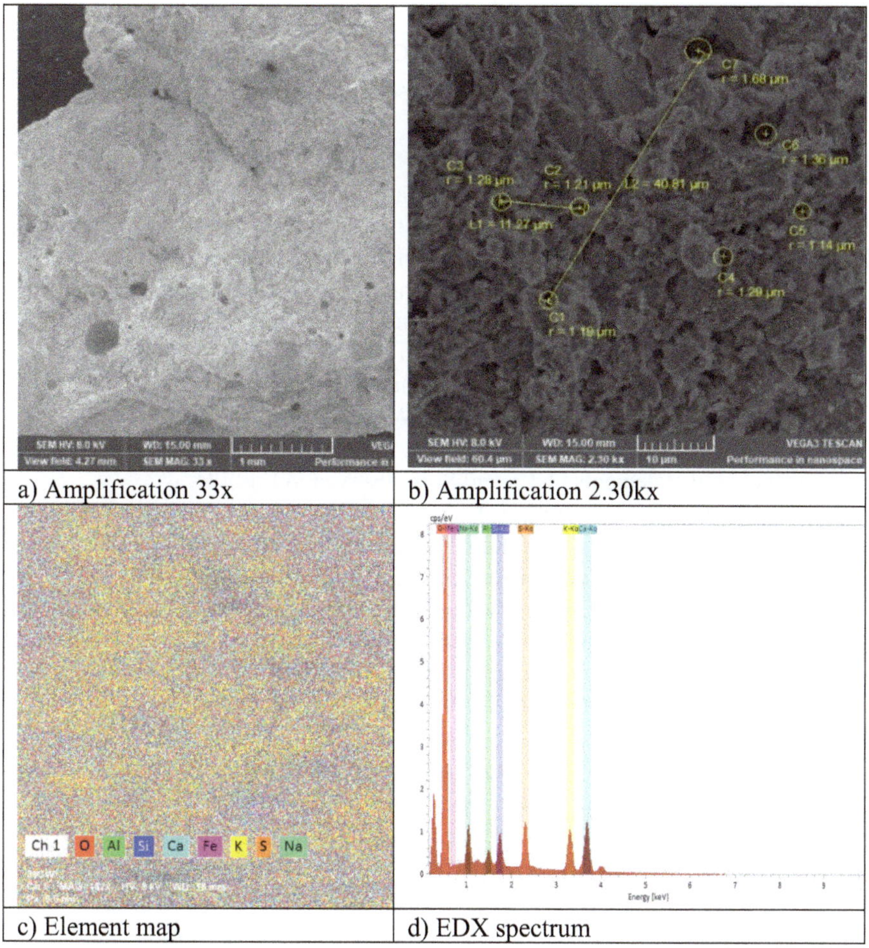

Fig. 12. SEM images obtained for the 35BW sample in the amplification range a) 33x- b) 2.30KX, c) Element map, d) EDX spectrum.

The image obtained in Fig. 12a indicates a slightly inhomogeneous compact surface with air gaps on the surface. The presence of pores on the analysed surface was confirmed at an amplification of 2.30kx, the pore radius being between ~1.14 and ~1.68 μm and the distances between them between ~40.81 and ~11.27 μm as shown in the Fig. 12b. Compared to the RS (CSIV standard recipe plaster mortar), the mortar with 35% solid brick waste (35BW) is a mortar with smaller radiuses and distances of the pores identified at the sample surface.

On the surface of the 35BW sample, a significant presence of chemical elements such as oxygen (O), calcium (Ca) and potassium (K) is observed, see Fig. 12c.

Table 3. Elementary concentration of 35BW sample

Chemical Element	Al	Ca	Fe	O	Si	K	S	Na
35BW recipe	0.98	22.48	1.41	50.07	3.95	12.20	5.84	3.06

The Ca/Si ratio of the sample is 5.56, this ratio indicates the predominance of Ca on the surface of the 35BW plaster mortar, which can be observed from the analysis of the elemental map and the EDX spectrum shown in Fig. 12d and Table 3.

SEM images obtained for the mortar sample with 30% brick powder waste and additive (30BPW-A) presented in Fig. 13a indicates an inhomogeneous surface.

The pores analysed at a 935x amplification of the surface of the 30BPW-A mortar, shown in Fig. 13b have radius between ~2.62 and ~3.20 μm and distance of ~7.18 μm. Relative to SEM analysis of RS mortar, 30BPW-A mortar has smaller pore sizes.

From the element map and from the EDX spectrum of the 30BPW-A sample, the predominance of the elements O, Ca and Si is observed in Fig. 13c.

The Ca/Si ratio of this sample is 2.6, a value above the limit of the C-S-H range provided, [19]. Compared to the RS from Table 2, the 30BPW-A mortar has a lower Ca concentration and a higher Si concentration presented in Table 4.

The topology of plaster mortar with brick powder waste and solid brick waste (35MIXT), is shown in Fig. 14a, with SEM analysis performed at 42x in which a compact and slightly homogeneous surface is identified. Regarding the characterization of the pores existing on the 35MIXT mortar surface, it was reanalysed at an amplification of 1.05kx (See Fig. 14b).

From Fig. 14b we identify the pore sizes present at the surface of the 35MIXT sample, whose radius is between ~2.38 and ~3.85 μm with a measured distance between them of ~91.07 μm.

The EDX analysis identified on the surface of the 35MIXT sample a preponderance of O, Ca and Si. The presence in the sample of other Ca-based phases, which can be observed in the distribution map of the elements in the areas where the colour assigned to Ca is predominantly identified in Fig. 14c, the Ca/Si ratio of this sample being 2.38 (See Table 5).

Regarding the reuse of brick masonry waste in order to obtain plaster mortars, the results are following the analysis of the images in Fig. 15a and Fig. 15b, achieved at an amplification of 24x and 1.67kx.

At small amplifications, a compact but slightly inhomogeneous topographical surface with air voids on the surface is identified.

In the 35BMW mortar sample (35% brick masonry waste as a replacement for aggregates) reanalysed in detail according to Fig. 15a and Fig. 15b identify the presence of

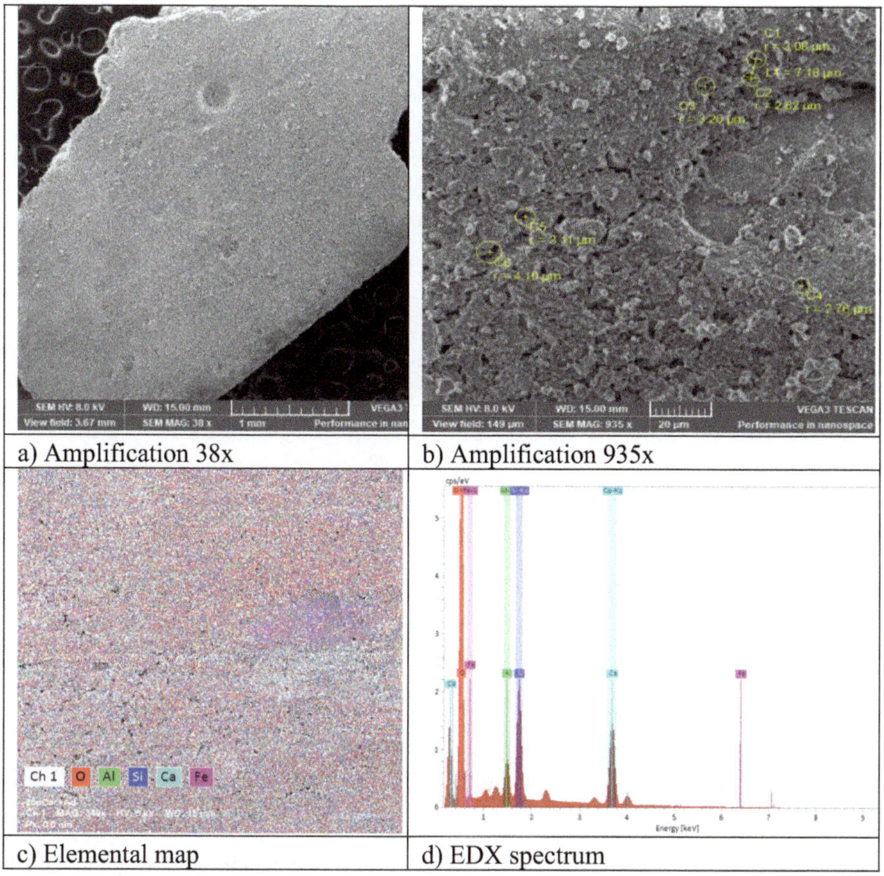

Fig. 13. SEM images obtained for the 30BPW-A sample in the amplification range a) 38x- b) 935x, c) Element map, d) EDX spectrum.

Table 4. Elementary concentration of 30BPW-A sample

Chemical Element	Al	Ca	Fe	O	Si
30BPW-A recipe	2.59	33.40	2.19	49.04	12.78

pores whose dimensions are recorded in Fig. 15b The pore radius ranges from ~1.52 μm to ~2.59 μm, whose measured distance is 46.90 μm.

The map of the elements and the EDX spectrum of the 35BMW mortar sample, identified in Fig. 12c indicates the predominance of type O, Ca and Si elements.

From Table 6 we cand identify the Ca/Si ratio is equal to 3.42, a value that shows the presence in the sample of other Ca-based phases.

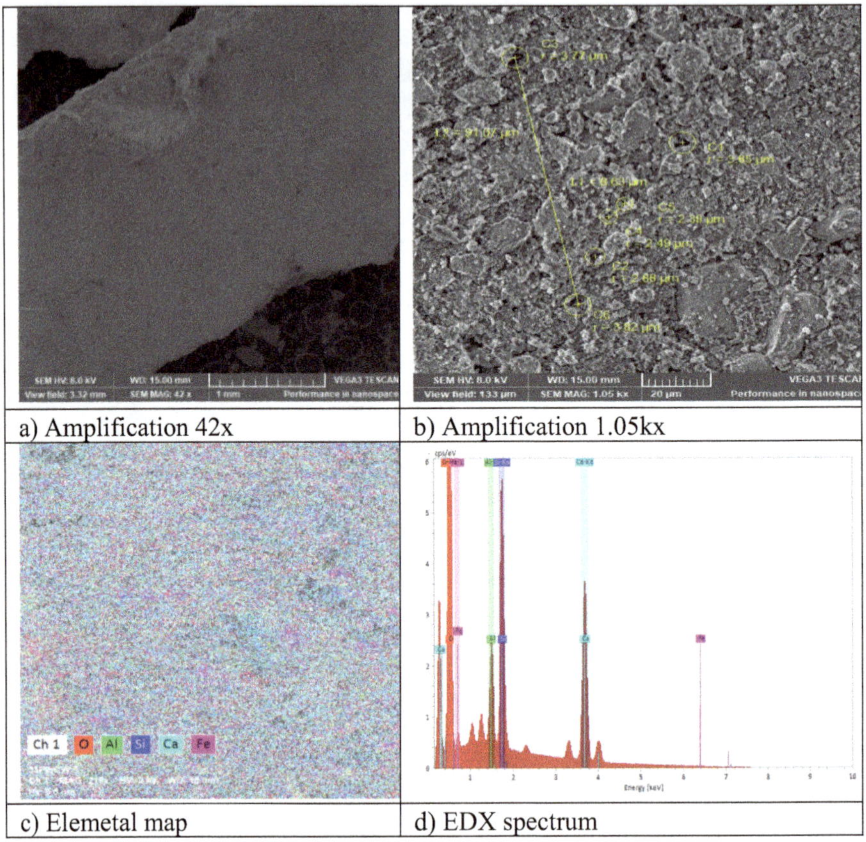

a) Amplification 42x
b) Amplification 1.05kx
c) Elemetal map
d) EDX spectrum

Fig. 14. SEM images obtained for the 35MIXT sample in the amplification range a) 42x- b) 1.05kx, c) Element map, d) EDX spectrum.

Table 5. Elementary concentration of 35MIXT sample

Chemical Element	Al	Ca	Fe	O	Si
35MIXT recipe	2.96	30.82	3.04	50.26	12.92

Comparing the 35BMW mortar sample with the traditional CSIV (RS) mortar established as the standard recipe, the 35BMW sample shows a smaller porous surface and therefore denser.

The literature divides the pores present in cement-based materials into: gel pores (<10 nm), small capillary pores (10–100 nm), large capillary pores (100–1000 nm), and air voids (>several μm), [17]. In the case of the RS sample, the pores observed belong to the air gap category, the values being close to the values of the large capillary pores.

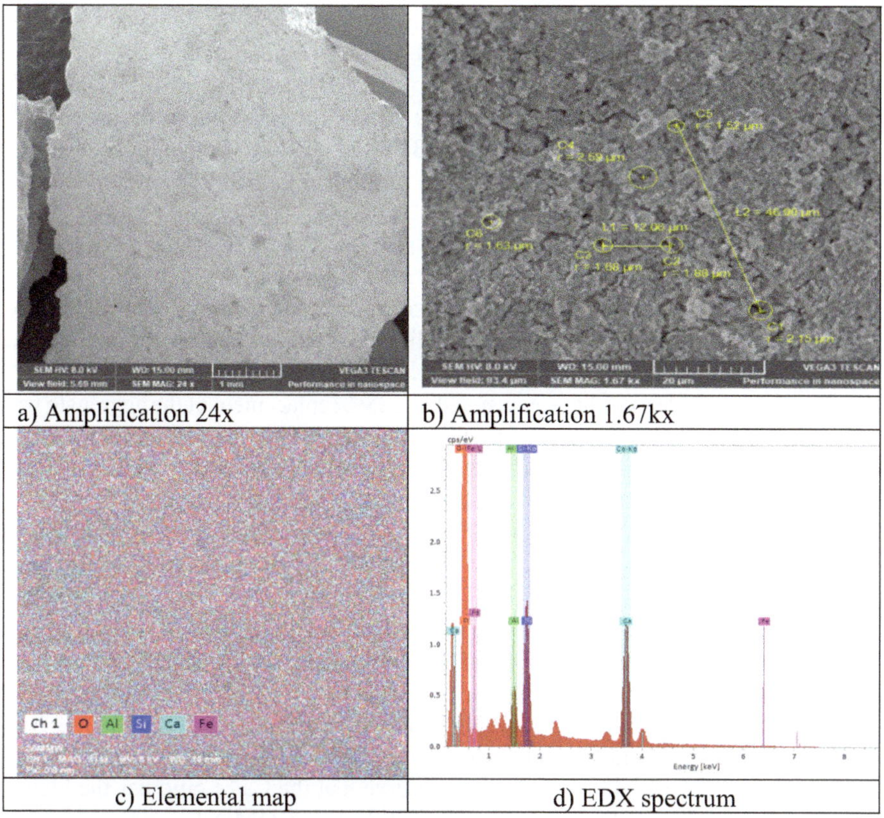

Fig. 15. SEM images obtained for the 35BMW sample in the amplification range a) 24x- b) 1.67kx, c) Element map, d) EDX spectrum.

Table 6. Elementary concentration of 35BMW sample

Chemical Element	Al	Ca	Fe	O	Si
35BMW recipe	2.05	33.79	2.44	51.85	9.88

The size of the air gaps and the distance between them is an indicator of the compactness of the studied materials, the smaller the size of the air gaps and the greater the distance between them, the more compact the sample.

5 Conclusions

The investigations proposed in this study are complex and current, making significant contributions to understanding and improving the technical performance of plaster mortars.

Regarding the porosity of the samples analysed by NMR at 3 days, the smallest porous structure is the 35MIXT mortar sample followed by the 30BPW-A.

The results obtained 7 days after casting of the T_2 distribution identify a complete disappearance of the porous structure for the 35BW recipe. Also, in the case of the recipe where cement was replaced with 30% BPW, an increase in inhomogeneity of the material is observed over time, probably due to the fact that BPW is more inert than cement.

The evolution of the 35BW mortar recipe at 3,7 and 28 days suggests that the pore definition starts to be configured only 28 days after casting.

After kept for one month at hydration of 35BW samples, a relatively lower value of water in small pores is identified and the medium, medium to large pores do not show an amplitude as large as in the 28-day analysis of the same sample.

The observations lead to the conclusion that a 35% replacement of the aggregate with 35% BW produces the occurrence of smaller pores, for which the main drip, located at the lowest values of T_2. In the case of the 35MIXT recipe is the most complex structure in terms of the existence of all types of pores: small, medium and large.

Comparing the results of NMR, the closest to the RS recipe as pore structure is the 30BPW-A recipe but the best performing is the 35BW recipe whose pore water content is lower.

From SEM-EDX analysis the 35BW is the optimal recipe in terms of all the technical performances obtained. In terms of pore size, recipe 35BW has the smallest radius value and the largest distance between pores, indicating the most compact mortar surface studied, description which is also resulting from NMR analysis.

The Ca/Si ratio is an indicator of the presence of C-S-H groups in cement. Analyzing the Ca/Si ratio for the studied mortars, we observe that the Ca/Si ratio has the highest value for 35BW, followed by RS, 35BMW, 30BPW-A and 35MIXT.

Therefore, in terms of all the analyses carried out on these mortar recipes, it was found that replacing 35% aggregate with waste brick leads to increased performance of plaster mortars.

The complex and non-invasive NMR and SEM-EDX investigations, used in the scientific program, completed and justified the values and the evolution of the physco-mechanical characteristics in time, proving that the work represents a fundamental research of plaster mortars with a strong applicative character.

Due to the potential of ceramic wastes in the production of environmentally friendly and ecological materials, it is necessary to promote the research and development in this field, as they can bring benefits both from a technical perspective and from an economic and ecological point of view. Therefore, the future research direction involves multi-criteria analysis of these recipes.

References

1. Parlamentului European şi a Consiliului: Directiva 2008/98/CE a Parlamentului European şi a Consiliului din 19 noiembrie 2008 privind deşeurile şi de abrogare a anumitor directive. Jurnalul Oficial al Uniunii Europene (2008). (in Romanian)
2. U. N. E. Programme: 2022 Global Status Report for Buildings and Construction. Global Alliance for Buildings and Construction, Kenya (2022)

3. Anamaria, B.Z.: Contributii privnd valorificarea deseurilor in scopul obtinerii unor materiale de constructii. UTPRESS, Cluj-Napoca (2023)
4. Zaharie, A., Ţinţişan, M.L., Siomin, A.C., Manea, D.L., Pleşa, M.L.: The use of ceramic waste in the construction materials industry based on the concept of sustainable development. In: Moldovan, L., Gligor, A. (eds.) The 15th International Conference Interdisciplinarity in Engineering. Inter-Eng 2021. Lecture Notes in Networks and Systems, vol. 386, pp. 182–191. Springer, Cham (2021). https://doi.org/10.1007/978-3-030-93817-8_18
5. Czajkowska, A.: The role of sustainable construction in sustainable development. In: MATEC Web of Conferences 174, vol. ECCE2018, p. 9 (2018)
6. Zaharie, A., Tintisan, M.L., Siomin, A.C., Manea, D.L., Pleşa, M.L.: Cement materials obtained by partial replacement of cement. In: IOP Conference Series: Materials Science and Engineering, vol. 1242 (2022)
7. A.d. S. d. Romania: SR EN 197-1:2011 (2001)
8. A.d. S. d. Romania: SR EN 1015-2:2001 (2001)
9. A.d. S. d. Romania: SR EN 1015-3:2001 (2001)
10. A. d. S. d. Romania: SR EN 1015-11:2020 (2020)
11. A. d. S. d. Romania: SR EN 1015-12:2016 (2016)
12. Zaharie, A., Ţinţişan, M.L., Siomin, A.C., Manea, D.L., Pleşa, M.L.: Reuse of mixed masonry wastes in plastering mortars. In: IOP Conference Series: Materials Science and Engineering, vol. 1283 (2023)
13. Cadar, D., Manea, D.L., Jumate, E., Popa, F., Moldovan, D., Fechete, R.: Structural and dynamic characterization of two-component waterproof mortars by 1H NMR, FT-IR, mechanical and SEM investigations. Constr. Build. Mater. **378**, 131182 (2023)
14. Ning, L., Zhao, Y., Bi, J., Wang, C., Shen, M., Li, Y.: Effect of aggregate size on water distribution and pore fractal characteristics during hydration of cement mortar based on low-field NMR technology. Constr. Build. Mater. **389**, 131670 (2023)
15. Ardelean, I.: Rezonanta Magnetica Nucleara pentru ingineri. UTPRESS, Cluj-Napoca (2013)
16. Chu, D.C., Kleib, J., Amar, M., Benzerzour, M., Abriak, N.E.: Determination of the degree of hydration of Portland cement using three different approaches: Scanning electron microscopy (SEM-BSE) and Thermogravimetric analysis (TGA). Case Stud. Constr. Mater. **15**, e00754 (2021)
17. Song, Y., Zhou, J.W., Bian, Z.N., Dai, G.Z.: Pore structure characterization of hardened cement paste by multiple methods. Adv. Mater. Sci. Eng. **2019**, 18 (2019)
18. Békési-Gardánfalvi, M., Hofmann, T., Fehér, S.: Microscopie electronică de baleiaj cuplată cu spectroscopie. Revista Restauratorilor Maghiari din Transilvania **17**, 13–21 (2017)
19. Richardson, I.G., Cabrera, J.G.: The nature of C-S-H in model slag-cements. Cement Concr. Compos. **22**, 259–266 (2000)

Author Index

I. Tuns et al. (Eds.): CIBv 2023, LNCE 510, pp. 491–492, 2024.
https://doi.org/10.1007/978-3-031-60765-3

www.ingramcontent.com/pod-product-compliance
Ingram Content Group UK Ltd.
Pitfield, Milton Keynes, MK11 3LW, UK
UKHW022217051026
15624UKWH00001B/198
9783031607646